The Compleat Conductor

The Compleat Conductor

GUNTHER SCHULLER

New York   Oxford
OXFORD UNIVERSITY PRESS
1997

Oxford University Press

Oxford   New York
Athens   Auckland   Bangkok   Bogotá   Bombay
Buenos Aires   Calcutta   Cape Town   Dar es Salaam
Delhi   Florence   Hong Kong   Istanbul   Karachi
Kuala Lumpur   Madras   Madrid   Melbourne
Mexico City   Nairobi   Paris   Singapore
Taipei   Tokyo   Toronto

and associated companies in
Berlin   Ibadan

Copyright © 1997 by Gunther Schuller

Published by Oxford University Press, Inc.
198 Madison Avenue, New York, New York 10016

Oxford is a registered trademark of Oxford University Press

Library of Congress Cataloging-in-Publication Data

Schuller, Gunther.
The compleat conductor / Gunther Schuller.
p.   cm.
Includes bibliographical references and index.
ISBN 0-19-506377-5
1. Conducting.   2. Orchestral music—Interpretation
(Phrasing, dynamics, etc.)   I. Title
MT85.S46   1994
781.45—dc20   93-36065

1 3 5 7 9 8 6 4 2

Printed in the United States of America
on acid-free paper

Contents

Preface

The idea for this book first came to me some thirty years ago when, having begun my own conducting career, I realized, with total astonishment, that what I had been asked to play as a horn player under various famous maestri, and what I had heard in hundreds of performances as a listener and eager-to-learn young composer, very often did not correspond at all to what composers had written in their scores. A nobody at the time in the world of music and working with some of the most famous and reputedly best conductors of the time, I was mightily puzzled by the discrepancies between their 'interpretations' and what I was seeing in the scores as I began to study them from a conductor's point of view. I knew, of course, about traditions and that there were good ones and bad ones (the latter especially in opera). But these deviations from the text were so prevalent and countermanded the most fundamental elements of composition, particularly tempo and dynamics, that because of my awe of these conductors I was hard put to reconcile their 'interpretations' with what I was beginning to appreciate as the true essence of those works, understood through or gleaned from their composers' wonderfully comprehensive notation(s).

Little by little I began to take the courage of my own findings (which were only those things that I found clearly indicated in the scores of the masters), and to insist in my rehearsals and concerts on 'realizing,' that is, bringing to life, that which the composers in their great wisdom had written. I began to experience the inspiration that will come from a thorough, minutely detailed study of a score and an unquestioned respect for it. Since much of what I was conducting in those early days was contemporary music—the new music of that time—and since there were usually no established traditions or recordings to fall back upon, I was learning to respect rigorously the content of a score—by whomever—and the score became a kind of sacred document for me. In all the intervening years I have seen no reason to change my views on this matter, whether in standard or contemporary repertory.

I was encouraged greatly along this path by working (as a horn player, not by the way as a conducting student) with great musicians such as Rudolf Kolisch and Edward Steuermann (in chamber music) as well as Toscanini, Reiner, and Monteux, who were more exacting in respect to basic notational elements and more consistent, i.e. self-disciplined, than most of their contemporaries with whom I worked and/or observed. Not that Toscanini, Reiner, and Monteux were in any way automatons who rigidly, slavishly, unimaginatively rendered the music. Quite the contrary, not only were they appropriately flexible in matters of tempo—sometimes two of them even perhaps too much—but more important for me, I could see clearly that their 'interpretations' came *out of*, i.e., *from*, a thorough study and understanding of the score. Their egos rarely interfered with the music, and their re-creative imagination was inspired by the *full content* of the score.

More troubling was when conductors, such as Szell or Walter, preached one thing and practiced another. It took a while to sort that out—eventually in favor of the composer. When I also understood that Walter and Szell, to name just two famous maestri, also performed no contemporary music—*truly* contemporary, that is, and perhaps complex and difficult—I began to realize that I was dealing not only with a certain personal/musical phenomenon, but with a deep-rooted professional reality, in which new music of challenge was frowned upon, pushed aside, and therefore the lessons learnable therefrom never learned.

I took up conducting rather late—in my early thirties—initially conducting my own works, but soon, encouraged by the musicians I worked with in New York, generally the cream of the crop, I branched out not only into other contemporary music (Babbitt, Carter, Sessions, Varèse, Schönberg, Webern, Křenek, Stravinsky, etc., etc.) but also the classical and Romantic masterpieces I loved so dearly and which I already 'knew,' although somewhat superficially, from having played much of that literature (but as I have indicated, unbeknownst to me) often in erroneous and misguided interpretations.

I began to create some small consternation by doing Beethoven symphonies in *his* metronome tempos (or very close thereto) and even more by insisting on real *p*'s and *pp*'s, requiring the whole range of distinctions between the typical eight or so dynamic levels, rather than the commonly used 'loud-soft-and-in-between' variety. As a composer, subtle dynamic differentiations had real meaning for me. They were not just some haphazard abstraction that could be modified or ignored, but one of the composer's most basic tools with which to color, to decorate the music, to delineate form and structure, to clarify lines and layers of music, to create variety of expression—in short, to create real music.

Rather quickly I began to see, to my surprise, that in most of the conducting world 'nobody gives a damn about the composer'—living or dead! In fact, those very words were initially going to be the title of this study. But as I began to work on the book I realized that such a title was too flippant, too provocative, and unsuitable for what I soon also realized had to be a scholarly and exhaustively documented study to carry any weight.

I formulated the book in three major parts, the first two modest in size, the third rather extensive. (Indeed the third part became more and more extensive as I worked at it, more extensive than I had originally planned.) On the face of it, the listings and citings and comments in Part III would, I then believed— and still believe—be considered, were they to appear alone, the ravings of a grumpy frustrated curmudgeon: to which the instant reaction would be: 'All that *can't* really be true!' I knew from the outset that I would have to establish not so much my credentials (they are ultimately unimportant), but rather the criteria by which one could objectively, consistently, reasonably evaluate the work of conductors. Thus I decided to devote one chapter (Part I) to my own ideas and criteria about the art of conducting, as an *interpretive, re-creative* art—my 'philosophy,' as it were—and a second chapter (Part II) to a kind of history of conducting, as written and thought about by some of the greatest practitioners of this curious and difficult performance art.

The research for Part II led to many interesting surprises, especially in the re-reading and re-study of Wagner's writings on conducting, in my view so often misinterpreted and misused for various polemical and ideological purposes. I was also delighted when it became evident that most of those writings—from Mattheson to Walter—actually confirmed my own thinking, my own 'philosophy' of conducting.

The results of my findings (Part III) are quite depressing—indeed so depressing that, predictably, many readers will reject quickly and completely what I have written. In that event, I invite those individuals to peruse carefully the more than 300 recordings I have listened to in exhaustive detail. A single listening to a Beethoven symphony could take two hours or more while I kept voluminous notes on all aspects of the performance, positive and negative. If at the end of such an expenditure of time and effort and hard listening—not to mention the purchasing and borrowing of these several hundred recordings— readers can still argue with my findings, I will be truly amazed. Recordings and scores do not lie—the former are as fixed and unequivocal as the latter.

Fortunately, other readers will salute my work, happy that finally someone has gone to the trouble of sorting these matters out and having the courage to differentiate right from wrong, as well as the courage—and the challenge—to name names. (Without names, Part III could be seen as a fairly worthless exercise in some sort of indulgent self-gratification.)

A third group—perhaps younger would-be conductors—will (it is hoped) learn much from this book. As I say, the factual evidence stated in it is clear and immutable: there are the scores on one side and the documentary recordings on the other. If we can all return to truly believing in the art of conducting as a kind of mission of bringing the great masterworks of the past and the present to life in respectful, selfless, non-egotistical, yet imaginative and creatively re-creative ways, we will have well served our Muse—Mistress Music.

Since I am a composer of some reputation, many may wonder why I did not devote my critical energies to contemporary or 20th-century works. The reasons are several. One is that contemporary music is almost always performed much more correctly (at least technically), much more respectfully, than the famous classical and Romantic pieces of the standard literature. Readers—and composers—may be very surprised at such a statement. But it is *absolutely true*—for simple (although saddening and maddening) reasons. In new music—especially in premiere performances—conductors and other interpreters are more or less relegated to 'realizing' to the best of their abilities what a composer actually wrote, what a composer has specifically notated. There are usually no recordings, no previous traditions, no 'interpretations' by 'legendary maestri,' to fall back upon, to be influenced by. Most often the only thing that exists is the score, the text, whereas in older music typically a whole history of traditions, of received interpretations, tends to exist, sanctified by critics and other pundits, with the net result that hardly anyone bothers to look at the original score, certainly not in detail.

The second reason I did not dwell on the performance of new or more recent music is that, as much as I love, admire, and devote my energies to the many great works of the 20th-century, my larger concern is with the masterpieces of the past. For not only are they as great as the masterpieces of our time—indeed, most people would argue that they *are* greater—but also the famous older works of the past constitute the foundation, the legacy, of our musical literature, against which, both historically and interpretively, our present-day work as re-creative, performing musicians ought to be judged. If we can show so little respect for the masterworks of the past, as is typically the case, we can hardly claim to understand and represent those of the present.

Accordingly, I chose the eight works discussed herein, partly because they are world-famous beloved masterpieces, staples of the symphonic repertory, and partly because they each embody, in addition to general interpretive issues, specific conductorial/performance problems, in many cases unique to these works. Initially, I had planned to include Beethoven's Eighth Symphony, the Schubert "Unfinished," the Dvořák "New World" and G major symphonies, Weber's *Oberon* Overture, Mendelssohn's *Hebrides* Overture, Ravel's *Rhapsodie Espagnole*, and Debussy's *Afternoon of a Faun*. But soon it became clear that such inclusions would make this already sizable book too voluminous and involve an unacceptable amount of redundancy, given that most interpretive misdeeds fall into a few basic categories.

The format and structuring of the book caused concern, for I realized that built into the very concept of the book was a certain amount of redundancy. As the book grew and expanded, Oxford University Press and I became rather alarmed at the magnitude of the expansion. I found it more difficult than I had originally expected to deal with the (unavoidable) massive attention to detail, and to document irrefutably my findings regarding the interpretations contained in the hundreds of recordings by nearly as many conductors. This may or may not require an apology, but perhaps an explanation. Inevitably, the book is in fact about details—details of performance, of interpretation. A *pp* is not a *p*, anymore than a *f* is a *ff* or a *poco ritardando* is a *molto ritardando*—even though in the real world these subtle differentiations are largely disregarded. Because these *are* crucial distinctions, which have everything to do with the appropriate or inappropriate outcome of a performance, it was essential that a great part of my analyses of the recordings (in Part III) *had* to deal with such details.

On the question of specifying names and providing considerable detail as to *how* conductors 'interpreted' a particular passage, that became absolutely mandatory, not only because that was (is) the essential rationale of the book but also, if one is going to mention one name, in all fairness one must list any and all others to the extent they are relevant. It would be patently unfair to mention (isolate) one conductor, say, negatively, without mentioning the others to whom the same comment is applicable. Hence, the often longish lists of conductors' names. One alternative, incidentally, would have been what has sometimes been done in the past, namely, to mention just a few of one's favorites—or unfavorites—and deal only with them. In my view, however, such an approach is essentially unfair and meaningless, since it is, in eliminating *a priori* a host of names, far too subjective, indeed incapable of even remotely claiming objectivity. Another alternative—although impossible to seriously contemplate doing— would have been to list every positive and negative critical observation I made during the thousands of hours of painstaking listening to the over 400 recordings—in which case the book would have been about 2500 pages long! Indeed I could have written a small book on each recording.

Some readers may take exception to my penchant for occasionally resorting to musicians' colloquialisms in an essentially serious didactic book. Frankly, I don't mind mixing such common parlance now and then with an otherwise analytical, scholarly approach to the subject, not only to better make the point but also to retain the flavor of such typical 'musicians-talk' comments.

Finally, by way not of apology but explanation (and warning), if some of my writing seems 'extreme'—if I appear to some to be a 'radical'—it is because I hear and see things that most others don't seem to hear or care about. I believe it is high time that some forceful, realistic writing on the subject of conductors and conducting—and interpretation—take place.

I owe thanks to so many who have in one way or another helped me with this project. First of all, my loyal staff (in my publishing and record companies) who often had to pitch in with various research chores, above all, Pamela Miller and my secretary, Derek Geary, to whom fell the brunt of typing up the text and integrating into it the hundreds of music examples. These illustrations, incidentally, became necessary when early on I realized that, needless to say, not every reader will carry around with him (or have access to) a score of, say, the Beethoven Fifth. If my job was to show the difference between what Beethoven actually composed and what most performances in fact deliver, it was mandatory that I *show* what Beethoven wrote. That way direct comparisons can be made; but therefore, also, the need for the many music examples contained herein.

I should also like to thank Alice Abraham, head of the WGBH (Boston) record library, through whose kindness I was able to listen to hundreds of recordings for untold hours—LPs and CDs—rather than having to buy, beg, borrow, or steal them. I should like further to thank Richard Dyer for lending me certain important recordings and out-of-print reference books central to my research, and my student and friend, Yoichi Udagawa, for help in various organizational and research matters.

I wish also to tender my appreciation to Mr. Arbie Orenstein (for his help and advice on the Ravel segment), the librarians at the New England and Oberlin conservatories and the Harvard Music Department, where I was able to study certain valuable, now-very-rare 78 recordings that have never been reissued in any form and probably never will be; and likewise to David Brodbeck for identifying the sources of certain crucial quotations by Brahms, Lewis Lockwood, Robert Pascall, George Bozarth and Walter Frisch for various helpful kindnesses. Special thanks also to Dell Hollingsworth of the Harry Ransom Humanities Research Center at the University of Texas in Austin for making available to me Ravel's autograph score of *Daphnis et Chloé*. Similarly, I owe thanks to Ms. Kate Rivers at the Music Division of the Library of Congress for help in reproducing the numerous autograph facsimiles contained in the book.

My thanks also go to Dan Kastner and Dr. Stanley Hoffman of Scores International (Boston) and their staff for producing the hundreds of music examples scattered throughout the book; likewise to John Scholl, also in Boston, whose technical know-how in preparing the facsimile plates was of enormous help.

Deep gratitude also to Robert DiDomenica, composer, teacher, and flutist extraordinaire, who diligently read through the entire manuscript, and not only advanced a number of excellent suggestions but also caught a number of serious lacunae in my over-all discourse.

I wish also to thank Sheldon Meyer, my editor at Oxford University Press for over thirty years, and editor Leona Capeless for their belief in the importance of this book and their loyal support in midwifing it through all its various difficult birthpangs.

Last but not least, I wish to dedicate this book to the memory of my late wife, Marjorie, who was at once my severest critic and my strongest supporter, and with whom I spent not only half a century together in an ideal personal and musical relationship, but in consort with whom—as aspiring teen-age musicians—I had my first epiphanal revelations, listening to and sharing the great masterworks of the past (and the present) with each other.

G. S.

Newton Centre, Massachusetts
October 1996

NOTE: As a special accessory to *The Compleat Conductor*, there is available in all major record stores a CD of the Beethoven Fifth and Brahms First symphonies, recorded by the author and a remarkable, hand-picked New York orchestra (GM Recordings (2051), 167 Dudley Road, Newton Centre, MA 02159).

Part I

A Philosophy of Conducting

A struggle, more or less unconscious,
between the creator and the interpreter
is almost inevitable. The interest of a
performer is almost certain to be
centered in himself.
> —T.S. Eliot, *The Sacred Wood*

Conducting is surely the most demanding, musically all-embracing, and complex of the various disciplines that constitute the field of music performance. Yet, ironically, it is considered by most people—including, alas, most orchestral musicians—to be either an easy-to-acquire skill (musicians) or the result of some magical, unfathomable, inexplicable God-given gifts (audiences). It is actually neither, the skills required in conducting at the highest artistic levels being anything but easy to acquire—many conductors never achieve them at all—while what the public mostly perceives as the magic and majesty of the baton is, but only in the best hands, a result of many years of intensive study and hard work, as well as talent, of course.

Talent in and of itself is not enough. Talent is also a much misunderstood or misinterpreted commodity. Talent may be innate, inborn, even inherited; but talent, no matter how great, needs to be developed, nurtured, and honed. For the skills inherent in fine conducting comprise a whole network of specific abilities and attributes: physical/gestural, aural, analytical, intellectual—even psychological and philosophical. For the conductor must not only know all there is to know about a score, down to the most minuscule details, but must develop the gestural skills to transmit that information clearly to an orchestra and the psychological dexterity to relate effectively (especially in rehearsals) to an orchestra—itself a complex collection of talented individuals, personalities, and artistic egos.

When audiences overrate and musicians underrate conductors' abilities and accomplishments, it is because the former tend to confuse conducting with gestural histrionics, and the latter with mere time beating. There are, of course, skillful time beaters, even among world-famous conductors, and equally skillful podium exhibitionists. But these for the most part demean the art of conducting, making it much less than what it can and should be. To delineate and analyze what conducting *as an art*—not merely as a profession or a career or a business—is and should be, will be the burden of Part I of this book.

The talents and skills—innate and acquired—that ultimately comprise the art of conducting are awesome. This undoubtedly explains why so few conductors attain them. There also ought to be a sense of moral obligation, a sense of unalterable respect for the great literature comprising our Western musical heritage; a sense that the art of conducting must be seen as a sacred trust to translate into a meaningful expressive acoustical reality, with as much insight and fidelity as is humanly possible, those musical documents—the scores, the texts—left us by the great composers.

These triple demands—talent, hard work, and an aesthetic morality—are unfortunately in short supply today. Perhaps they always were to one degree or another, but it seems to me that in this age of hype, promotion, public relations, and careerism—in an era when commerce and profit motivation dominate almost the entire social arena—in such an environment conducting has turned more and more into a business, into a commercial enterprise, with the predictable and commensurate lowering of artistic standards: in short—apart from a few glorious exceptions—with a rather serious debasement of the art of conducting.[1]

Those will surely seem very strong words to many readers, especially those who sit at home with their '50-masterpieces' record collections, idolizing the 'rich and famous' among the conductors, the superstars—the Bernsteins, the Karajans—to whom they attribute virtually god-like qualities. Close critical examination of what is actually produced on most podiums or in the recording studios of the world reveals that much of the business of conducting is more and more driven by extra-musical, extra-artistic considerations and motivations. Not that lack of talent and artistic integrity, charlatanism and musical fraudulence are limited to our era or are somehow an invention of our times. Interpretational excesses and abuses of the great literature have long existed, probably since conducting became a distinct profession. But the financial and material stakes of success in this era of electronic communication advances and mass markets are so much higher today, so much more tempting and therefore so much more potentially corrupting than ever before in our cultural history. Artistic standards and artistic integrity have declined dramatically in recent decades, succumbing to careerist and commercial pressures at an alarming rate. The idea that a musical artist, a conductor, ought to serve the music—rather than the music serving the musician—is occasionally given lip service, but is rarely put into practice.

1. A remarkably knowledgeable and courageous, no-holds-barred exposé of the serious degradation and venality in the conducting business, the wheeling and dealing of the power-brokering managements that control most of the music business, is Norman Lebrecht's *Maestro Myths: Great Conductors in Pursuit of Power* (New York, 1991). It is sobering reading, to say the least, and is highly recommended to anyone concerned about the integrity of the art and profession of music.

Such service ought to extend to the music of our own time, and I beg, not just the first decade and a half of our century. The ideal 'compleat' conductor is also an ardent advocate for the best in new music, with a deep and unshakable commitment to performing the great music of his contemporaries. This implies taking risks, the kind of risks courageous conductors such as Koussevitsky, Stokowski, Mitropoulos, Reiner, Steinberg, and Dorati took in varying degrees and ways. I must say that I can have only a diminished respect for a conductor or any musician, no matter how good otherwise, who feels little or no commitment to the music of his own time.

It may perhaps surprise the lay reader that there is something like a philosophy of conducting and that the art of conducting can be defined. But then it is doubtful that even the majority of conductors (and would-be conductors) ever think seriously about such matters. Indeed, musicians drift into conducting nowadays without much thought of what in fact they are undertaking, with little or no awareness of what the conductor's art entails—or should entail. The desire to lead an orchestra derives much more from some ego-driven ambition which has little or nothing to do with serious music-making, or with a sense of humility and devotion in serving the art. The idea that it is an enormous privilege, but also an awesome responsibility, to conduct Beethoven's *Eroica* or Stravinsky's *Sacre du Printemps* or Brahms's Fourth Symphony, is rarely considered. More often than not it is seen merely as a stepping stone in a career, and thus to fame and fortune.

It is much more usual that a young musician one fine day wakes up to the vague notion that he—or lately she—would like to direct and control a musical performance, rather than merely play a part in it. Typically, this vague notion soon turns into an irresistible desire and then a driving obsession. In rarer instances, the decision to turn to conducting is prompted by chance, an accident of fate, an incidental encounter: most commonly when a resident conductor becomes incapacitated and a musician, who may never have conducted before, is quite suddenly prompted to take over at a rehearsal or—less frequently—at a concert (as in the case of Toscanini, for example).

To be sure, there are those who are, by their talent and personality, destined to become conductors: musicians who have inborn intellectual, expressive, and physical aptitudes to interpret composers' works, and to elicit such interpretations from a collection of orchestral musicians. But even that group of conductors rarely takes the time at the first hint of conducting impulses to analyze critically what the art, the craft of conducting, actually involves. Rarely is thought given to the immensity of such a decision and to the awesome challenges it entails. I'd like to think that if young aspiring conductors were to think more deeply about these matters, many would be deterred from pursuing this particular, most demanding of musical professions.

As it is, musicians tend to declare themselves conductors by merely announcing as much to the world and ironically—sadly—the world generally accepts the pronouncement without question, regardless of whether the particular individual has conducting talents, regardless of whether the individual has the technical, intellectual, and emotional capacities to translate a musical score into an appropriate interpretation via appropriate conductorial gestures. Thus it is that young musicians with immense, irrestrainable egos become 'conductors' in an instant, overnight as it were, while more modest egos with more modest ambitions (but just as much, or perhaps even more, talent) remain relatively unrecognized and anonymous.[2]

Certainly it takes a healthy ego to develop the courage to stand before an orchestra of seventy-five or eighty musicians, to impose his or her musical/interpretive will on that orchestra, to in a sense dominate those musicians, and to dare to 'interpret' the great masterpieces of the Western tradition. When such a 'healthy' ego—let's call it a 'modest' ego, contradictory as that may sound, or in Bruno Walter's phrase, a "selfless ego"—is infused with an equally healthy respect for and gratitude toward the musicians who labor and toil under his baton, then one is likely to get what I will simply call for the moment very fine, high-level music-making (to be defined more precisely in the ensuing discourse).

But rather than dwelling on the negative aspects of the conducting profession and how casually most musicians tend to drift into conducting careers, let us explore in detail what in fact the art of conducting at the highest levels entails, what it means—or should mean—when someone says, 'I am a conductor,' and what is required in order to earn the right to conduct the masterworks of the past and the present.

As suggested earlier, the 'compleat' conductor must possess a whole range of diverse talents and acquire a broad and deep knowledge of the literature that goes far beyond that required for any other type of performing artist—instrumentalist, pianist, singer, whatever. But all these talents must be encompassed in one all-embracing basic attitude: a deep humility before the art of music that contains in it a profound love for and unswerving commitment to serving that art; a humility that considers it a privilege, an honor, to bring to life the masterworks of our musical heritage, and to communicate through them to our fellow human beings. With such an unostentatious approach, the many other talents a conductor needs to possess will evolve in proper perspective.

Ranging from the somewhat philosophical to the specifically technical, the requisite talents and skills needed to be a fine, perhaps even great, conductor are: an unquenchable curiosity about the miracle of the creative process and about how works of art are created; a profound reverence and respect for the document—the (printed) score—that embodies and reflects that creation; the

---

2. It is commonplace today for any famous artist—singer, pianist, violinist—to plunge suddenly into a conducting and recording career, fostered enthusiastically, of course, by their managements. The fact that such artists, commendable in their original careers, may be little qualified to conduct seems to matter not at all in the modern marketplace of music, where fame established in one domain can apparently substitute for real ability in another. Perhaps the saddest manifestation of this trend was the instant 'career'—as conductor—of pianist Glenn Gould, whose recording of Wagner's *Siegfried Idyll* is probably the most inept, amateurish, wrong-headed rendition of a major classic ever put to vinyl.

intellectual capacity to analyze a score in all of its myriad internal details and relationships; a lively musical, aural imagination that can translate the abstract musical notations of a score into an inspired, vibrant performance; and on a more practical level, a keen, discerning ear and mind; a versatile, disciplined, expressive baton technique; an efficient rehearsal technique; a precise and thorough knowledge of the specific technical limitations and capacities of orchestral instruments (strings, woodwinds, brass, percussion, harp etc.) not only as functioning today but in different historical periods; and finally but not least, a basic respect for the role the musicians—artists in their own right—play in the creation of the sounds that are ultimately transmitted to the audience, artists without whose vital contribution (as many conductors in their self-glorification tend to forget) their own talents and efforts would not be expressible.

It is to be expected that some will question the sequence in which I have presented these conductorial requisites. Some will even question the very notion that humility might be a primary element in a conductor's make-up, a notion (allegedly) irreconcilable with the (alleged) need for a conductor to dominate his musicians, to make them do his musical bidding. The answer to the second question, however, is simple: persuasion that results from a commitment to the integrity of the art and a consummate knowledge of the music at hand will always bring out the best in musicians. Moreover, the artistic humility I refer to is not (and need not be) without a healthy sense of the conductor's own worthiness. It is simply a humility which recognizes that the conductor's first priority is to serve the music, to be a medium, a vehicle, through which the work of art is revealed and expressed.

This humility then translates into a fierce determination to know completely and profoundly the work in all of its aspects, to explore the letter *and* the spirit of the work, to plumb its expressive and emotional depths, in order to reveal its essence. Given human fallibility and variability, absolute perfection is probably not achievable. But it is certainly the goal that conductors must strive for—in order to have the right to interpret, to realize, the works of the great masters, whose genius is many, many times greater than their own.

For if conductors arrogate to themselves the notion that they are going to interpret the masterworks of the past and the present, then they had better realize that that is not only a staggering task, but one that imposes a profound responsibility to what Beethoven and Wagner so aptly called "die heilige Kunst" ('the sacred art'). And that responsibility—that moral and aesthetic obligation—in turn demands that conductors achieve their so-called interpretations through, i.e. from within, the work of art, in boundless respect and reverence for it; and that they not, in reverse order, willfully or inadvertently impose some self-indulgent, over-personalized 'interpretation' on that work of art.

Indeed, if I had my druthers, I would in this context abolish the term—and the idea of—'interpretation' altogether and, following Maurice Ravel's sage advice, substitute the word 'realization.' Ravel, one of history's most meticulous, most precise, most detail-loving notators of music, urged: "Il ne faut pas interpreter ma musique, il faut le réaliser." (One should not interpret my music, one should realize it.) On another occasion, in a similar vein, Ravel told Marguerite Long, the pianist for whom he wrote his G major Piano Concerto (she premiered it in 1932 under Ravel's direction): "Je ne demande pas que l'on m'interprète mais seulement qu'on me joue." (I do not ask that one interpret my music, but simply that one play it);[3] the implication being "as written, as notated," or as Toscanini put it so often: "come è scritto."

Indeed, as the term, and the idea of, 'interpretation' has evolved over the last two centuries, it has become a dangerous concept, inimical and antagonistic to the art of music, less concerned with the music, the compositions and their true intent, than with the interpreter's self. Interpretation is about as far away from pure 'realization' as it possibly can be, with basic respect for the work per se virtually nonexistent.

Interpretation has come to mean in most circles; 'Don't trust the work of art; don't let it speak for itself; we will decode it, explain it for you.' The modern-day interpreter, consciously or unconsciously (mostly through arrogance and/or ignorance), is altering the work, the text without, of course, ever admitting as much. He insists that he is only making it 'more intelligible,' giving you its 'true meaning.' But in fact he is selectively picking out, emphasizing a set of components, of features, from the work as a whole. In effect, the interpreter is one who translates and transforms.

And as Susan Sontag pointed out long ago, "The interpreter, [even] without actually erasing or re-writing the text, is altering it. But he can't admit doing this. He claims to be only making it intelligible, by disclosing its true meaning." . . . "the effusion of interpretations of art today poisons our sensibilities. To interpret is to impoverish, to deplete . . . in order to set up a shadow world of 'meanings.' "[4]

Indeed, most interpretation, as I see it and as it is practiced nowadays (and was vigorously practiced even in earlier times by the likes of Bülow and Mengelberg) is nothing more than a refusal to let the work of art stand on its own. Whether in hermeneutic interpretations by critics, historians, and writers or actual acoustic re-interpretations by conductors and performers, the work, the composition, is not permitted to be itself, to come to us in a pure 'realization.' As a result, hardly anyone—least of all audiences—can now *really know* the work itself, distinguish it from its myriad interpretations and translations.

The sad irony here is that anyone who has not experienced a true 'realization' (as opposed to an 'interpretation') can have no idea what an *incredible* experience and revelation that can be. It is folly to think that we as performers, as re-creators, can elevate the work of art. It is the work of art that can elevate us. And *that*, once encountered, is the ultimate experience, the ultimate artistic achievement. (See also Felix Weingartner's thoughts along the same lines, cited in Part II, p.102).

3. Marguerite Long, *Au Piano avec Maurice Ravel* (Paris, 1971), p. 21.
4. Susan Sontag, *Against Interpretation* (New York, 1966), pp. 6,7.

The state of conducting, alas, is today—with some notable and wonderful exceptions—far removed from such artistic/moral/ethical considerations. Musical
integrity, respect for the composer's work, idealism, and a sense of humility
toward the art of music are in very short supply. We might expect from conductors at least a simple basic respect for the dead. But when we speak of the
Mozarts, the Beethovens, the Brahmses, the Debussys of our musical heritage,
we ought to double our respect for those particular dead, and make it a matter
of honor and pride scrupulously to respect their creations, their scores. The art
of conducting ought to consist of faithfully retracing the manifold steps by which
the composer originally created the work, of re-tracing and re-living the creative,
visionary journey on which the composer embarked in the first instance.

That is an immense challenge, a task which takes many kinds of knowledge
and, in its highest form, extraordinary skills. The integrity of that re-creative
process is what is at stake here, and what this book will be about: and by integrity, I mean a kind of 'morality' of conducting as an art and as a coherent
philosophy, not as a mere profession or, worse, a business.

But even such stringent criteria, as demanding as they may sound, are as a
broad philosophy still too vague and general, and apt to hide a plethora of
common conductorial sins and aberrations. Every conductor, after all, thinks of
him/herself as embodying the highest moral artistic integrity and possessing all
the requisite skills to interpret the great masterworks of our literature. We must
therefore consider more precisely the specific core skills with which the conductor can effectively respond to—and achieve—the stated challenges.

A simple definition of the art of conducting could be that it involves eliciting
from the orchestra with the most appropriate minimum of conductorial (if you
will, choreographic) gestures a maximum of accurate acoustical results.[5] But in
order to know what those "most appropriate" gestures and "accurate acoustical
results" might be, one must have a precise and deep knowledge of the score and
the creative process that produced it. That is not as easy a task as it might seem
at first blush. To begin with, we are all—we conductors (with some notable
exceptions: Ozawa now, and Reiner, Monteux in the past, come to mind)—to a
lesser or greater degree limited by the physical disposition of our bodies, our
own physical structures. We are in a profound and virtually inescapable sense
prisoners of our bodies. Almost all of us have some more or less serious limitations as to what we can do with our hands, our arms, our shoulders, our head,
our eyes—in short our body equipment. Almost all of us are to one extent or
another variously inept in one area or another. A perfect conducting machine,
like an Ozawa or, in quite different ways, a Carlos Kleiber or a Reiner or a
Leinsdorf or a Bernstein, is an extraordinary rarity. Most of us are either too tall,
or too short; our arms are too long or too short, or too stiff or too loose, or too

5. One of the most remarkable conductors of the recent past, Fritz Reiner, who very much practised
what he preached, once put it similarly in an interview: ". . . the best conducting technique is that
which achieves the maximum musical result with the minimum of effort" (*Etude*, October 1951).
One wishes that Leonard Bernstein, Reiner's pupil, but later one of the world's most histrionic and
exhibitionistic conductors, would have taken his teacher's advice to heart.

something. Most of us are not free enough in our arms and torso to control fully the minutiae of movements which so crucially affect the musical/acoustic results emanating from an orchestra; and most of us are too habituated to certain physical movements to be free at the precise moment to alter or control them.

Our physical attributes profoundly affect our conducting abilities, positively or negatively as the case may be. That is not to say that one cannot learn, particularly with experience, to become gesturally more controlled, more relevant, more relaxed, more 'appropriate.' Much can be achieved in this realm with good training, and there are many tricks and methods by which one can learn effectively to re-train and discipline one's body, one's physical equipment, as it were. But I believe—and know from many years of experience and of observing several hundred conductors (many of them world-famous) with whom I worked as an orchestral musician—that for most conductors there are ultimately some physical limitations or idiosyncrasies which, no matter how one tries to overcome them, cannot be entirely outgrown. What we are thus left with is the goal of developing our physical, manual, gestural skills—one of the essentials of our conducting craft—to their highest possible potential, so that we may accurately reflect and transmit to the orchestra (and thence to the audience) that which the music requires us to express.

But that physical expression is but the exterior manifestation of what we know and feel about the music (the score). All the physical, choreographic skills in the world will amount to nothing if they represent an insufficient (intellectual) knowledge of the score and an inadequate (emotional) feeling for the music— in other words a knowledge of *what* to represent, of *what* to 'realize.' A beautiful baton technique can achieve little if the mind that activates that baton doesn't know what there is to know in the work and what, in fact, its notation expresses. The clean baton technique of a conductor who, for example, does not hear well harmonically or whose mind and ear cannot keep a steady tempo may still be a beautiful thing to watch, but from a strictly musical point of view it is a useless skill. It is equally true that a first-rate mind and ear can achieve very little if the technique needed to express what is in that mind and ear is deficient.

Pursuing this thought further, we must therefore understand precisely what it is we have to know in a composer's work in order to translate it into an accurate living representation and expression of the music. The answer in the broadest and deepest sense is: we must know *everything* it is possible to know. By that I mean what is perhaps, in absolute reality, an impossibility, but certainly a magnificent goal to strive for, and to continue to strive for as one matures. We must know (or at least try very hard to know) essentially why every note and every verbal annotation in that score is there, what their meanings and their functions are in the over-all work. In that sense the art of conducting ought to be the art of collaboration—between conductor and composer—even dead composers.

Perhaps now the reader can begin to appreciate the magnitude of the task and the complex demands the art of conducting makes. To know how and why every note in, say, Beethoven's Fifth Symphony is there; to know how and why those several thousand musical choices and decisions which produced that extraordinary masterpiece were made: that is our task, and I would say, our aesthetic/moral obligation. This in turn means a complete functional harmonic, pitch, and intervallic analysis of the work; an analysis of its thematic/motivic content and inner relationships; an understanding of the work's internal tempo relationships (within movements and from movement to movement); its tempo stresses and strains (most likely induced by its harmonic rhythm and expressive needs); its phrase and period structuring, in the small as well as the large sense (again intimately tied to the underlying harmonic rhythms); its structuring in terms of primary, secondary, and tertiary materials; its homophonic and (where appropriate) polyphonic structuring; its instrumentation (including a historical understanding of the then-prevailing instrumental capacities and limitations); an understanding of Beethoven's use of dynamics (both as means of structural delineation and expressive, decorative profiling); and finally, beyond the score itself (to the extent that available documentation allows), the background to the creation of the work, and any artistic, cultural (perhaps even social) influences on its creation.

Conductors often delude themselves into thinking that, our conventional musical notation being limited in some respects, there is much that one cannot know about a work because its notation simply cannot reveal or prescribe everything. While it is true that our musical notation has its limitations, I would still argue that there is much more to be gleaned from our notation than we generally assume. It is true that the ultimate, most subtle nuances and personal refinements of interpretation are in fact not, in an absolute sense, notatable. (And this book will not be about such subtleties and refinements of interpretation.) Indeed, tempo, tempo modifications, dynamic and timbral indications cannot be absolute or objectively precise; they remain relative and thus prone to subjective evaluation. But it is just as true that they are more than adequate to achieve an ideal realization of a work and that a sensitive musician with sound musical instincts, probing the essence and style of a given work—especially in post-Haydn/Mozart repertory—can extract insights from the notation of the score that will provide him with very precise ideas as to how to conduct the work. Indeed, the problem in conducting and interpretation is not that our notation is 'inadequate,' but that 50 percent of it is ignored by most conductors. In short, there is much more reliable evidence in scores than we generally suspect, especially in scores by late 19th- and early 20th-century composers, most of whom have taken more than the usual pains to meticulously express their intentions in their notation. This 'evidence' then is tantamount to very specific instructions— instructions which in my view we dare not disregard or reject, which we must respect, or at least try to honor.

I am not so foolish as to argue that scores, often the only relevant document left to us by the no longer living composer, are *absolutely* reliable. Composers *do* make mistakes, often by omission or in the haste of creation. Publishers and editors also make mistakes and contribute errors other than those made by the composer. And some composers are extremely precise and detailed in their notation, while others often assume a prior knowledge of their style and notational habits. But all that notwithstanding, we ought as conductors and performers to honor the basic premise that the score is a precious, unique, sacred document, which in essence should be relied on for all the information it can yield.

Generally the music world makes a mystique of conducting, as if it were based on some mysterious, divine gift, bestowed upon only a few 'chosen' musicians each generation. The fact is that the highest levels of conducting are achieved by dint of hard work, intensive study, including close scrutiny of the score, and an absolute commitment to expressing with the utmost fidelity the information the score contains. An interpretation that does not start with the score, that fails to evolve out of the score in *all* its notational, prescriptive details (not just those that the interpreter deems convenient to consider), in short, an interpretation that starts with the *interpreter* rather than the work (the score) is, I believe, fundamentally invalid. The premise, too often affirmed today, alas,— even by (or perhaps especially by) famous conductors—is to start at the other end of the process: to arrive at an 'interpretation' before the score is fully assessed, or biased by extra-notational influences, such as a famous (but not necessarily representative) recording or someone else's prior interpretation, or some handed-down tradition, or—worse yet—personal whim and fancy. Before we start 'interpreting' and imposing ourselves on the score, before we start intruding upon the music, we ought to adhere to the discipline of thoroughly studying every note, every dynamic marking, every phrase, every instrumentational detail of that score. Our 'interpretation'—or 'realization'—must ultimately be derived directly and primarily from the source, arise out of the score, accumulate, as it were, *from and through* the score.

As a working method in the process of revealing the score to the orchestra and thence to the listener, the specifics of how all the elements of music (the composer's tools) are used—harmony, melody (or theme or motive), rhythm, dynamics, timbre (orchestration), form and structure—must be separately and then collectively explored and understood. In general, we call this analyzing the score. But 'analysis' can have different meanings for different constituents: musicologists, composers, conductors, for example. I will therefore be very precise and speak of analysis as particularly applicable to conducting. In the ideal and fullest sense this analysis and understanding will comprise all the vertical (harmonic) and horizontal (melodic or thematic) relationships: how these intersect and influence each other until every note, every rhythm, every orchestrational detail is seen (and heard), until the entire criss-crossing network of myriad, kaleidoscopic musical interfacings is understood and felt. Thus, the harmonic rhythm of a work can illuminate its phrase structure, or the timbral or sonoric profiling of the work can delineate its formal and textural aspects, or the dynamic refinements can underscore and reveal the orchestral colors with which a composer is 'painting' his music. There is no true masterpiece in which these elements—these composers' intellectual or intuitive choices and decisions—do not symbiotically interrelate and ultimately correlate into a vast and complex musical network.

When we say that the *Eroica*, the *St. Matthew Passion*, Brahms's Fourth Symphony are perfect masterpieces, what we are really saying is that in those works (and others of that calibre) the composer has made thousands of minute final decisions and choices, selected from a veritable infinity of options, and which we in retrospect upon hearing the work hear as the 'best possible choices,' as 'inevitable'—and thus 'perfect.' Indeed, that is one simple elementary way of describing the composing process: i.e., a composer, having just written the 5th or 572nd or 1003rd note, now has to write the 6th or 573rd or 1004th note; and out of all the possible options in respect to note, pitch, and rhythmic choices, orchestrational decisions, dynamic considerations, etc., the composer now selects that one note he considers to be 'the best' or the 'most logical,' the most consistent with what has come before and what may follow. And when that choice, that decision, is made by a Beethoven, a Mozart, a Brahms, a Tchaikovsky, a Ravel, a Stravinsky, a Schönberg, a Webern, a Berg, it is more often than not at such a level of intuition, intelligence, imagination, vision, originality— and daring—that we feel in retrospect it *was* the only 'right' choice, the 'best' choice and seemingly 'inevitable.' (The fact that the composer might ten years later, as he develops and matures, make an even 'better' decision—or, as sometimes happens, revise and 'improve' a previous work—does not alter the fact that at the initial moment of inspiration and creation, that composer's choice was in fact his 'best choice.')

It is a conductor's job to understand the process by which a thousand and one such 'inevitable' choices are made by the composer and, as I say, to retrace those steps of creation, to re-create in his conducting that decisional process, not in some merely mechanical rendering but in a manner that is emotionally, expressively inspired by that process. Let me quickly add here, lest I be misunderstood, that I am not hereby arguing for an interpretation that slavishly follows the letter but ignores the spirit of the work. Nor am I saying that there is somehow, even if 'one does everything right,' such a thing as a (let alone *the*) 'definitive interpretation.'

On the first point, a mechanically, technically accurate performance may be clinically interesting, but unless its accuracy also translates into an emotional, expressive experience—for the listener, the musicians (including the conductor)—it will be an incomplete realization, one that will not—indeed cannot— adequately represent the work. On the second point, the very idea of a 'definitive' rendition is a complete fiction, one which certain critics evidently like to accord their favorite interpreters and which, I suppose, certain conductors feel they are able to achieve. Nonsense! There can be no such thing as a definitive interpretation, and for many reasons. To begin with, it is impossible for anyone to know *all* there is to know about a work, that is, to have unequivocally total, objective knowledge of a work and what was felt and heard in its creator's mind and ear. This in itself ought to preclude anyone's claiming that a given performance represents the definitive interpretation. All we can actually get in musical judgments and understandings is an opinion; and the best we can hope for is that that opinion be a richly informed one. Furthermore, the words 'definitive' and 'interpretation' are self-contradictory, since the word 'interpretation' by definition means a particular rendition out of several or many alternatives. But beyond that, even a single conductor's interpretation of a given work will not be, and cannot be, totally consistent. It will be subject to a host of variables, starting with his own constantly changing emotional and physical feelings from day to day, but extending to such matters as the different style and sound characteristics of different orchestras (not to mention the highly variable emotional and physical feelings of the musicians in those orchestras), different acoustics in different halls, the effects of weather and atmosphere on human beings as well as instruments and acoustics, and last but not least the variables in the receptiveness of different audiences on different days under different conditions (including, of course, those very same critics who feel the need to declare an interpretation 'definitive'); and so on, virtually without end.

The most that we ever say about a performance is that *in our opinion*—already a huge qualification—a certain performance seemed 'ideal' or 'good,' *and* for such and such reasons. It is hoped that those reasons will be adducable from the score, from the work itself, and not from some exterior motivation.[6]

So we shall not be speaking here about 'definitive' performances, but only, where appropriate, about 'ideal' or 'good' ones, and—of course—of many 'not so good' ones. For the moment, however, the point is that, while several different renditions of a given work may each be valid, representative, good, ideal—*if* they are based on a close reading of the score—all the variables of conditions and temperaments mentioned above ought not to allow us to assume therefore that *any* arbitrary, personal interpretation can also be valid and thus be sanctioned. The excuse that our musical notation is limited or incapable of 'telling us all' will simply not do, because, as already mentioned, closer inspection of our notational system and how composers have used it through the centuries will reveal that there is always much more that *is* objective and clearly stated (and therefore ought to be binding to the interpreter) than that which is left open or unstated.

I will deal in considerable detail with questions of tempo, tempo modification, and metronomization later in this chapter—complex subjects, to be sure, not to be settled in some simplistic 'yea' or 'nay' argumentations. As much as I may plead—along with Beethoven and Berlioz and many other composer-conductors—for a basic respect for metronome markings, with all the attendant qualifications, I must make it very clear that I do not believe that an exacting adherence to metronomic indications will *by itself* guarantee a good, a great, or a 'correct,' performance. (Mr. Norrington, Mr. Gardiner and a host of others, please note!) Tempo and tempo modifications are but one of many aspects of

6. The claims for 'definitive performances' are constantly increasing, especially in the realm of recordings, and most especially in connection with the 'period instrument' movement. Every conductor is well-advised to read Richard Taruskin's brilliant exposé of the pretentiousness and ludicrous claims made by Hogwood, Norrington, the Hanover Band, and some other present-day 'authenticists' ("The New Antiquity," *Opus*, (October 1987), pp. 31–43, 63).

'interpretation' which, *in conjunction* with many other considerations, (such as dynamics and color), can ultimately produce an 'ideal' performance.

Mere 'correctness,' in fact, accomplishes very little. The truth is that I have in my lifetime heard many performances with which, in terms of a certain kind of correctness and factual evidence, I had intellectually to disagree—performances by, say, Furtwängler, Mitropoulos, Walter, to name a few very famous ones—which nonetheless were in various ways transcendent, even sublime, aesthetic experiences and in some profound ways revelatory performances.

In the end, my preference is ultimately for a transcendant rendition which also involves the utmost respect for the composer and his score. For let us never doubt that respect for and full explicit knowledge of the score are compatible with a 'great' interpretation/realization. It is only lesser minds and talents that would have us believe otherwise.

It is in this realm of artistic integrity, transcendant perception, and deep respect for the composer's creation that the conductor's art in its highest aspirations and attainments will distinguish itself. Therein will lie the true 'interpretation.' And such conductors are the real poets, the really *creative* interpreters, the visionaries of the realm.

As stated, this book will not deal with the most sublime and subtle refinements of the art of conducting, and for one very good reason: verbal description and explication cannot effectively deal with such subtleties of expression. They can only be savored in the reality of a performance. For it is the mystery and power of music—in this regard, unique among the arts—that only in purely musical terms can those highest forms of expression be made manifest. The art of music and of musical interpretation at that very highest level is beyond words, even, I hazard, beyond those of the greatest poets. Moreover, having already suggested that it is those subtle variables of interpretation—as long as they do not go against the letter and spirit of the score—which may give legitimacy to *different* performances of the *same* piece. It is at that very highest level of performance that a wealth of interpretive choices and decisions become available at least to the really sensitive intelligent and imaginative re-creator. It is in this realm that there is not one *pp*, but many subtly different *pp*'s; not one *f* but many different kinds of *f*'s; not one slur but many kinds of legatos, etc. etc. The more basic point, however, is that it is a *pp*, not a *p* or a *mf!*

The same is true of all other dynamic distinctions or articulations. Take *sf*, for example: there are many—I am tempted to say—dozens of different *sf*'s. As one tiny example I offer the m.128 *sf* in the first movement of Beethoven's Seventh Symphony (and its parallel, m.340). This *sf* can be performed in many different ways, with different feelings and emphases, all of them within the realm—the species—of *sf*. For example, one can give this *sf* a very hard-hitting effect with a strong, incisive attack in the strings (if it were in the winds, with a strong, incisively tongued attack). Or, one can give this *sf* a deeply expressive, weighty feeling, infinitesimally delayed. Or, it can be a warm, rich singing *sf*, as one can see Carlos Kleiber elicit from the Concertgebouw Orchestra in a film produced by German Unitel television. (This type of *sf* is especially effective—by way of contrast—after the sharply iterated chords four bars earlier.) Or, one can extract a more pointed, lightly stinging *sf* from the orchestra with a right-hand stab.

Another example of how *sf*'s can have varied meanings (and impact differently) in differing contexts is to be found in Beethoven's *Eroica* in the first movement. In mm.25–34 Beethoven writes twelve *sf*'s, which in all performances and recordings are pounced on *ff*, with a vengeance. Such an interpretation is wrong—or, at least, not *necessarily* right. An alternate interpretation, just as reasonable and logical, is to remember that *sf*'s are contextually related, depending on the prevailing dynamic level in which they are situated. Thus a *sf* in *p* is not the same as a *sf* in *f*. Here, in the *Eroica*, those *sf*'s are all contained within a basic *p* level; thus these are mild *sf*'s, hardly more than slightly heavy accents in *p* (the *p*'s being clearly indicated in mm.23 and 27). As further confirmation that this is at least as viable a realization of these particular *sforzandi*, is the indication *cresc.* in m.35, a crescendo which leads to a full-orchestra *ff* two bars later in m. 37. If, however, one is already delivering heavy, pounding, all-out *sf*'s in mm.28–34, how can one make a crescendo in the next two bars to a higher dynamic level? One can't.

These are but two small examples of one commonly used expressive indication, but by extension almost every notational device we have can be expressed in subtly different but still appropriate, legitimate ways, that are, *within* the parameters of its intended meaning.

It follows from such examples and considerations that it is not in this higher (highest) realm of performance and interpretation that the problems of *misinterpretation* and willful disregard of the composer's intentions lie. Such arguments as to whether one should (or should not) subtly linger on a given note to bring out its special place in a melodic line or its harmonic function; whether or not it is permissible in a Trio of a classical symphony's Menuet or Scherzo to slightly relax the tempo; whether or not to subtly emphasize rhythmically a particular harmonic cadence or ingenious harmonic modulation or sudden shift; whether or not it is permissible (or advisable) to make a subtle ritardando in the last measure of a development section just before the recapitulation—these are all (a) viable interpretive options either way; (b) a matter of individual taste (good taste, of course); and (c) impossible in any case to referee in some absolute unequivocal way. In most such matters we do not have hard indisputable evidence (such as an early 19th-century recording or definitive written documentation, precise descriptions of performance practices etc.). What we *do* have are many theories, too often those of self-appointed pundits, who claim to know the answers and who manage to promote their ideas as verified truths.

No, this book will be about much more elementary considerations, matters which in fact our notation in most cases *can* (and does) clearly reveal, can state unequivocally—and, alas, matters which many famous maestri ignore(d) (out of ignorance), reject(ed) (out of arrogance), or misunderstand (stood) (out of inadequate study).

Let us return for a moment to those earlier mentioned prerequisites of learning a score, of what constitutes fine, intelligent, artistic conducting: harmonic, thematic, structural analysis; tempo and tempo relationships; full understanding of the work's instrumentation and its dynamic functions. In one sense these skills are beyond questions of baton technique, that is, those sets of knowledge are not acquired *through* one's technique, although, obviously, once acquired they need to be *expressed* through one's baton technique, one's manual skills. Those gestural representations must be intimately geared to the expressive needs of the compositions, and *only* to those needs. It follows therefore that how a conductor moves his arms, how he or she expresses the music's content physically—choreographically—can only be determined once the piece has been thoroughly studied and understood. This idea parallels—it is another way of expressing—the earlier-mentioned thought that a conductor's interpretation must develop out of and follow a study of the score, not be arrived at *before* or *extraneous* to such study.

Once that intimate knowledge of the score is achieved, it is time for the physical aspects of conducting to come into play. But that alone will not be sufficient unto the task of realizing the essence of a score. The conductor also needs, as I mentioned earlier, a keen ear. For, unless the gestures through which a conductor interprets that score are also balanced and tempered by a keenly listening ear—listening not only for wrong notes and mistakes, but for all the previously analyzed aspects expressed in that score—then those gestures and that interpretation may fall wide of the intended mark.

I like to think of *that* listening ear as the 'third ear,' an ear which 'sits' well outside the conductor's body and listens not only to the totality of what the orchestra is producing but also to the effect the *conductor's* conducting is having on that orchestra and on the music. It is therefore a highly critical, a highly discriminating ear; it is a regulatory ear. But it must also be a *self-regulatory* ear. It must be as much directed at one's self (the conductor) as at the orchestra. Thus the 'third ear' is an ear which, critically assesses whether how and what someone is conducting corresponds in fact to what is intended by the composer in his score.

One often hears that a certain conductor "has a terrific ear" (or—more often, from orchestra musicians—"a lousy ear"). And there certainly are conductors with 'better' ears: better trained, physiologically better, more innately gifted. But what is rarely realized—or discussed or taught in conducting classes—is that all the 'excellent ears' in the world are irrelevant if those ears do not know *what it is they should be hearing.* In point of fact, one's ears are useless equipment if one's mind, the musical intelligence, does not inform the ears *what to hear,* what to be listening for. But beyond that, the statement that a certain conductor "has a terrific ear" is meaningless because it is ambiguous, unless the statement also defines what *kind* of ear is terrific. For there are in my view at least seven different 'ears'—seven different aural capacities—which a conductor should command. The reality is that most conductors have at best one 'ear,' and many seemingly none.

The seven kinds of ear—the seven hearings, all directed by the mind—which the compleat conductor has, are for (1) harmony; (2) pitch and intonation; (3) dynamics; (4) timbre; (5) rhythm and articulation; (6) balance and orchestrational aspects; and (7) line and continuity. I cannot think of a major conductor, working today, who possesses all seven, with the possible exceptions of Carlos Kleiber, Haitink, Skrowaczewski, and Gardiner.

It is often said that Boulez has "a terrific ear." But Boulez's terrific ear extends only to pitches and to some extent to intonation. There are many other things which Boulez's ear does not hear. While Monteux's ear was much better, in fact, phenomenal in respect to pitch and intonation,—in his quiet, secure, unostentatious way—he also heard almost everything else that needs to be heard. Stokowski had an extraordinary ear for sonority, for the sheer sensuality of sound, and virtually none for pitch and intonation. His mind was not interested in the latter. Some conductors, like Szell, Haitink, Skrowaczewski, for all their differences in approach, have in common a keen ear for sonoric and orchestrational balance. Other conductors (Furtwängler was one) have shown a fantastic ear for line and continuity.

Most conductors do not have innately the ability to hear in the full sense I have just described. Some conductors learn by way of years of experience to become more accurate, more discriminating listeners. Since, as I suggested, the ear can only hear what the mind, the brain, directs it to hear, it follows that the ear must be wide open and the mind clear-thinking and uncluttered.

It is one of the most difficult challenges, for conducting students, for example, to be aurally/mentally free enough to hear precisely, critically, the results of their conducting. Consider the fullness and complexity of the conductor's task. On the one hand, he has to activate the orchestra by the movement of his hands and arms in certain specific ways to produce the desired sounds, while simultaneously listening critically to the orchestra to assess whether it is playing 'correctly,' whether the sounds emanating from the orchestra correspond to what he has in mind, which in turn will, it is hoped, correspond to what *the composer* had in mind. Then in addition—again simultaneously—a 'third ear' (and eye) should be critically watching the conductor, to see whether what *the conductor* is doing corresponds in fact to those "most appropriate" gestures, which will produce in the orchestra the ideal desired result. One of the most common faults among young learning conductors, for example, is to conduct with huge emotion-laden beats, when the dynamic the composer has written is, say, *p*. No orchestra in the world will play a true *p* when the conductor is belaboring it with three-foot-long beats or huge flailing motions. What is even worse is when the conductor then criticizes the orchestra for playing too loud!

Since orchestra musicians are quick to criticize conductors for not hearing a wrong note—inadvertently played wrong or mistranscribed in the part (such things happen often enough)—I want now to defend the conductor, not every conductor, of course, but the conductor in general. For, while conducting, it is not always possible, even for the best ears (and minds), to hear everything. When you are really deeply immersed in the feeling, the over-all expression, of the music—a given measure or passage—you may not, in your intense feeling of the expression, hear a tiny rhythmic, intonational or background note problem. In concentrating actively on one matter, it is entirely possible for even the 'best ears' to miss some other matter, because one is aurally distracted by the chosen primary concern. For example, it is reasonable and logical that, if a conductor concentrates on one particular instrument (or note, or ensemble balance), he may not then hear some partly hidden, light mistake in another instrument or in another aspect of the work.

A very common conductorial problem is giving an upbeat in one tempo and the succeedng downbeat and further beats in another tempo. This drives orchestras crazy; and the conductor in question will have totally lost the respect of the musicians after two or three such inept moves, particularly if he fails to realize that the resultant rhythmic shakiness is *his* fault, not the orchestra's.

Orchestra musicians, by the way, have a conductor analyzed usually within the first five or ten minutes of a first rehearsal, at least in respect to basic abilities. Musicians also know that, even with very famous and popular conductors, many times *they* save the conductor from serious embarrassment by *not* playing what the maestro conducts. The point is—and all musicians know this, while audiences mostly don't—that a conductor's baton makes no sound, and a conductor's mistakes therefore will go unnoticed by the audience (and even most critics), but not by the musicians. But if the musicians were to actually *play* the conductor's mistakes, everyone would hear them. In several orchestras with which I played during my twenty-year career as a hornist, it was a standing joke: "If only we had the nerve to play what some conductors conduct, their careers would be over in a flash."

Orchestra musicians are, of course, not always paragons of righteousness and complete devotion or commitment to the music—or, for that matter, to the conductor. While an orchestra, when the chips are down—at a concert, as opposed (sometimes) to a rehearsal—will generally give its best, concentrate, and try to remember all that has been rehearsed, in other respects orchestra musicians, as a lot—probably most—seem to have little intellectual interest in the music itself. I don't know why it is, for example, that it is a rare musician who reads the program notes provided by the (sometimes excellent) program annotator for that week's concert.

I also find it very curious that most musicians, as I have mentioned elsewhere, are hardly ever interested in looking at or studying a score, except perhaps once in a while to correct a wrong note in their part. They generally seem uninterested in the background of the composer and the composition, especially when it comes to new (newer) music. I am no longer amazed or surprised as I used to be in my younger years at how many orchestra musicians, when asked, will hardly even know the name of the composer they just played (unless it is one of ten or fifteen top names). In some curious way the music on their stand is a kind of anonymous abstraction, with no personal relationship to them, and of only moderate intellectual interest. It is something to be rendered—to be consumed, as it were—and then promptly forgotten. On to next week!

Certain critics—and many conductors, soloists, and chamber musicians—have of late attacked the notion of scrupulous faithfulness to the score, and have in turn airily defended the taking of individual interpretive liberties with the composer's text.[7] This has caused a fair amount of confusion among performers, teachers, and students and has tended further to usurp interpretive standards in the conducting profession. Mostly the discussions have been polarized into two extreme and opposing viewpoints: if you follow and respect the score, you are considered an "academic," "a cerebral intellectual," an "unfeeling conducting mechanic"; if you are free and indulgent in your interpretation, you are likely to garner praise for your "profound interpretive skills" and "musical insights." Specifically, the typical argument goes: strict readings of the score inevitably equate with 'stiff,' 'inexpressive,' 'pedantic' performances, while the individualized, highly emotional, less scrupulous reading of the score is seen as an inherently desirable ideal in which the conductor (or artist) can freely express his or her view of the work. Such black-and-white either-or formulations of the argument fall far short of recognizing the true complexity of the matter; but unfortunately that is how the subject is generally treated and argued, almost always on polemical, ideological rather than objectively artistic and substantive grounds. The truth is infinitely more subtle and complex.

In any case, can we really claim that some conductor's version of what a composer wrote is automatically, inherently better than the original? And *who* is to say it is better? Can one logically assume and argue that someone's interpretation and 'translation' of a text is preferable to what was originally created and painstakingly written down by the composer? Can we rightfully claim that we are 'improving' the music? An automatic 'yes' to this question would appear rather ironic and paradoxical, especially in the case of those composers to whose acknowledged greatness we all constantly pay lip service.

Obviously, no one in his right mind—least of all this writer—would wish to argue for mechanistic, inexpressive, inept performances on the one hand or indulgently permissive ones on the other. To polarize the argument thus is to miss the real point altogether. The best approach, as usual, lies somewhere between these extremes. For ultimately, as suggested earlier, there is no such thing as the absolute 'definitive,' 'correct,' interpretation; indeed any piece of music, any phrase, any musical idea, can at the highest levels of performance be 'correctly' rendered in several ways. However, the differences between such several justifiable ways are apt to be fairly subtle and, more important, must initially be based on an objective, intelligent, enlightened, all-embracing reading of the score.

The difficulty in this discussion lies in the fact that no human being, no artist, no conductor can ever be totally objective in artistic/interpretive matters, or—to put it another way—can ever avoid being subjective to some extent. Clearly, the argument generally mounted by the opponents of textual fidelity—to wit, that

7. The most recent collection of such views is contained in Jeanine Wagar's *Conversations with Conductors* (Boston, 1991), in which a number of famous maestri condone and defend the purposeful disregard of tempo (especially metronome) indications and favor liberal orchestrational re-touching of scores.

someone is too 'objective' in his performance, too cold, too intellectual, too inexpressive, too reliant on the score — is itself false and specious, because even that alleged 'objectivity' is bound to incorporate a greater or lesser degree of subjectivity. No decision we make as performers can ever be totally objective, bound and influenced as we are by both the limits and the qualities of our talents, by our backgrounds, our training, and our cumulative experience. A certain amount of subjectivity and the predilections of our personality will always come into our performing and color it, characterize it, in subtly distinctive ways. And that is good and not to be deplored.

We are, after all, what we are; and conductors are what they are. No conductor is purposely bad or purposely good. Every conductor is trying to evolve out of his talents the highest and most personal expression. Unfortunately, this often fails because (a) there is among conductors' views of themselves a sizable gap between perception and reality, that is, between their perception of themselves and the reality as seen by others; and (b) conductors now increasingly try 'to be different' in order to carve out for themselves some special career niche. In today's highly competitive musical marketplace, monopolized and controlled by 'charisma'-obsessed managers and agents as well as highly developed marketing and promotion techniques, to be eccentrically 'different' is virtually to assure popularity, fame, and the concomitant financial rewards.

This alarming trend can best be seen and heard in recordings (as Part III of this book will amply show), in that conductors, battling it out in the fiercely competitive recording market, have now learned that they will stand out, will be reviewed and discussed more readily, and will thus attract more attention the more they can interpret a work *differently* from the several dozen recordings of it that are already in the market place. This has become more than a trend in recent years: it has become an obsession and a specific skill, eagerly supported by managers and, of course, most record companies. At that point the composer's score becomes, alas, a total irrelevance, an annoying burden. In this perverse view of things, the music becomes fair game to be exploited for whatever career gains it can provide. Beyond the immediate negative effects of specific personal mis-, under-, or over-interpretations by conductors, there is an unfortunate cumulative effect as well: the varied distinctive qualities and characteristics of the great symphonic masterpieces are submerged in one generalized, (ironically) depersonalized, generic, amorphous, androgynous performance style. Instead of the personality of the *composer*—and the true personal and special essence of the work in question—we get the personality of the *conductor*. When several hundred conductors impose their interpretive whims and fancies on the works of, say, Beethoven, Brahms, Tchaikovsky, Schumann, or Dvořák, all those pieces begin to sound alike; they are covered by a blanket of subverting interpretations, which make it impossible to hear the true, dramatic, often startling differences between and among those composers. Beethoven sounds like Brahms; Brahms sounds like Beethoven, and both of them sound like Tchaikovsky, and so on. The extraordinary discipline, economy, terseness of construction—in a sense, even the simplicity and directness—of Beethoven are lost and made to sound like the more luxuriant, effusive, romantic emotionality of Brahms or Tchaikovsky. He who has not heard that intrinsic difference between Beethoven and Brahms—their occasional similarities and close idiomatic relationship notwithstanding—can probably not imagine the considerable gulf between the two. And that distinctiveness, the work, after all, of two totally different personalities originating in two totally different periods, can be brought to life only through the most scrupulous, admiring and respectful realization—not interpretation—of the text, the score. Then the true essence of each composer's musical language will be able to send its unique and distinctive message, and the listener will then know that the composers, especially the very great ones—the giants of our tradition—knew best what they wanted and how to put it in a clear and effective notation.

The excessive personalizing of interpretation (with utter disregard for the score) has been allowed to fester under the mistaken notion that the conductor/performer is more important than the composer, that the composer and his works are there to serve the careers of conductors, when in fact it should be the other way around. Little recognition is given to the simple fact that, if it weren't for composers and their creations, conductors (and performers) would have nothing to conduct and to play. The immense success—meaning rounds of applause and standing ovations—many conductors garner conducting a Brahms or Beethoven symphony would be clearly impossible if Brahms or Beethoven hadn't composed those symphonies in the first place.

Fortunately there are great conductors who approach each score and each performance with an innate, unswerving respect and reverence for the score, aided and tempered by musical intelligence and a never-ending quest for deeper knowledge, thereby upholding the highest standards of the art of music. But then the *business* of music, unable to tolerate that kind of artistic and professional integrity, dismisses a Haitink, for example, as "lacking in charisma," as "unexciting"—in my estimation not only a totally inaccurate view but an irrelevant point. For 'exciting' in such critics' and audiences' minds (audiences being primarily swayed by the critics in such matters) usually means 'different,' 'flashy,' 'sensational,' 'eccentric,' 'exhibitionistic'—and well publicized.

This sad trend has reached such proportions in the last decade or two that even relatively serious and intelligent conductors tend to doubt a score more than they trust it. One hears more and more from conductors that Bartók's metronome markings "are all wrong"; Shostakovich's tempo markings are irreconcilable with his metronome indications, and thus "are quite unreliable"; "Schumann was a poor orchestrator and his dynamics are mostly all wrong" (this notion had, of course, already started in Mahler's and even Wagner's times); "composers make mistakes in their score"—the implication is that *all* composers do and do so most of the time— and therefore their scores "are not to be trusted"; "musical notation is inexact" and cannot be precise, and therefore we must "reinterpret" and "improve" the scores; and on and on.

For Strauss's and Brahms's music such notions have had a disastrous effect on the performance of their works. Both rarely used metronome markings and their music is subject, therefore, to a wide range of tempo interpretations (misinterpretations). In this same category fall the never-ending arguments about and widespread disregard of Beethoven's metronome markings.

But it isn't just in regard to tempo and metronome indications that composers' scores are being challenged or ignored. Dynamics—that other precious element through which composers refine and clarify their musical message—are roundly ignored, rejected, mistrusted, and subjected to personal revision. Even worse is the rampant disregard of phrasing and articulation, especially phrasings (and therefore bowings) in string sections, a situation sometimes merely tolerated by conductors, but more often than not induced by them.[8] Brahms's symphonies are a particular victim of this trend, in which twice as many bows, producing, of course, a bigger, louder, more 'exciting' sound, are somehow automatically considered better than adhering to the composer's original dynamics and conception.[9]

In this and many other less obvious but equally dangerous ways, many conductors have cumulatively and collectively spread the notion—with little resistance, by the way, from orchestral musicians—that the composer's score is to be treated with considerable suspicion, that it is quite all right to ignore the salient details of a score, and that conductors usually know better what a composer intended than the composer himself. This arrogance, rampant as it is now, is quite indefensible and brings a degradation to the art of performing that must be arrested before we lose all sense of musical/artistic integrity.

Add to this already chaotic situation (1) the recent record-industry-promoted / hyped obsession with so-called authentic instruments and allegedly "historically informed" performances; (2) the fantastically enhanced and powerful promotional marketing tools employed today by most musical institutions and managements (tools which were generally unavailable as recently as fifty years ago);[10] and finally (3) the gradual, year-by-year, imperceptible corruption of our ears

---

8. Concertmasters, responsible for the strings' bowings in most orchestras, are often not in a position to question or resist a conductor's interpretive wishes and quickly accede to his demand to 'play louder' make 'a bigger sound' or a 'fatter tone,' by using twice as many bows as the composer may have indicated. Also, unfortunately, many concertmasters decide their bowings only linearly, i.e. merely on the basis of the string parts, without looking at the score and considering what else is going on vertically, contrapuntally, and contextually.

Incidentally, concertmasters (and other section soloists) need to be constantly reminded nowadays that not every passage marked "solo' is to be played loudly. (Solo in Italian means 'alone,' not 'loud.') This is especially necessary in passages marked p.—and there are thousands of those—in which case perhaps the conductor ought to intercede, not to have the concertmaster play more loudly, but rather—God forbid!—to ask that the accompaniment be played more softly. What a novel idea!

9. I am quite aware of the fact that phrasings and bowings in many composers' works do not always coincide, and intelligent judicious bowing choices are therefore necessary. But surely it is not a defensible (or the only) solution to disregard a priori the composer's phrasings/bowings and automatically 'upgrade' them to louder, more excitable decibel levels.

10. But see Joseph Horowitz, Understanding Toscanini (New York, 1987), for a bold and brilliant analysis of how a major musician (Toscanini) was marketed and promoted in the 1930s and '40s to a culture-god and cult figure.

by ever louder, artificially produced performing levels, made possible by the microphone and modern electronic technology, [11] and it becomes clear that we have in music, but especially in the field of conducting, wide-spread philosophical and aesthetic chaos. What *sells* is what counts, and if the public wants the music louder or faster, well why not give it to them—and to hell with what the composer wrote. Nobody in the end gives a damn!

I should like to make clear once again that I am not here defending, let alone advocating, some kind of pedantic, lifeless, rigid, mechanistic, technically 'accurate' performing and conducting. I have heard too many such performances in my lifetime, and I abhor them. They serve no useful purpose. Conducting/performing without feeling, without expression, without imagination, without illumination of the score, is a completely pointless musical activity. The ideal conductor is one who combines feeling and intellect in a symbiotic unity: when he thinks he feels, and when he feels he thinks. Indeed, to some extent the conductor's personality *must* express itself in the performance, not only because that is good and an important part of music-making, but because it is—as I have suggested earlier—inevitable, unavoidable. It is only a question of *how much* and with what effect that personality will impose itself upon the work and intervene in its re-creations. It is, as in most things in life, a matter of degree. The secret of great artistry and true integrity of interpretation lies in the ability to bring to life the score for the listener (and the orchestra) through the fullest knowledge of that score, so that the conductor's personality expresses itself *within* the parameters of the score. It illuminates the score to the fullest; it does not alter it or distort it. And the conductor's personality is not substituted for that of the composer.

That is clearly a much more difficult challenge than merely indulging one's musical whims and predilections. To know the score fully and to probe its inner essence in the thorough manner suggested here is a staggering task. It is a much easier (lazier?) approach to say: 'Oh, I think I'll do it this way,' and when asked why, to respond: 'Well, I feel it that way.' To work from *within* the score towards a realization is a formidable task. It takes tremendous discipline and conscience to evolve an interpretation that is faithful to the specifics of the score, faithful especially to the dynamics and to the tempos. It takes considerable discipline to *not* make ritards too early—or too much; to *not* make crescendos (or diminuendos) too early—or too much; to hold to the articulations and phrasings the composer has written; to respect fully the meaning of the verbal annotations the composer has incorporated in the score; and to know the sounds, the sonorities,

---

11. Most recordings of string quartets today, for example, are electronically "enhanced" (as the industry euphemistically puts it) and amplified so as to sound as big and as loud as a full symphony orchestra in the throes of the most climactic moments of Strauss's *Alpine Symphony* or Mahler's Eighth. With such abuses—along with the ear-splitting dynamic levels of rock music, the ever louder television commercials, the ever noisier film soundtracks (replete with ever more car crashes and explosions, building detonations etc.), and other similar modern acoustic plagues—we are well on the way to ruining our aural sensibilities altogether. Indeed, to some extent we probably already have.

the instruments that the composer heard in his time and for which he wrote his music.[12]

Most great composers, particularly as one moves into the mid-19th century (let alone early 20th century) repertory, were meticulous, precise notators of their music, at least to the rather considerable extent our notation allows. In the scores of the great masters—and even the minor masters—there is much more that is precise, accurate, clear, and objective, and therefore to be trusted, than there is vague and undetermined.

Consider, for example, Beethoven's wide range of dynamics, from *ppp* through *pp, p, mp, mf* to *f, più f, ff* , and finally *fff.* That sounds pretty precise and discriminating to me. And how well Beethoven knew human nature, human frailty; his scores are peppered with *sempre pp's* in the many extended passages during which the music is to remain quiet and soft throughout. Beethoven knew well that most musicians (and conductors) would be unable to resist the temptation to increase the dynamic level after a few measures. And he was so right! Beethoven's *sempre pp* markings are generally ignored, or simply considered dispensable and irrelevant.

Beethoven often structured his forms in enormous sound plateaus with a single dynamic.[13] These take tremendous discipline to control, and almost inhuman energy in the case of the *ff* plateaus, or fastidious dynamic control in the case of the *pp's.* Mozart and Beethoven composed much of their music in block structures, where entire phrases or periods are set in one dynamic, say *p,* and then suddenly followed by a contrasting section of *f.* It takes a disciplined orchestra and conductor to not anticipate the incoming *f* with a crescendo.[14] Since Mozart (in his later works) and Beethoven also wrote long crescendo passages, leading from an initial *p,* say, to a resolving and climaxing *f,* it behooves one to observe carefully the difference between these two structural ap-

12. To accomplish a truthful rendition it is not always necessary to resort to "period instruments," as some of the authentic instrument propagandists would like to make us believe. (More on that subject later.)

13. See, for example, the long *pp* plateau in the development section of the *Pastorale's* first movement; or the almost minute-long, relentless *ff* in the development section of the Eighth Symphony's first movement, eventually topped by a *più f* and a climactic *fff.*

14. I have, for example, almost never heard the four measures before the first *f* in the *Marriage of Figaro* Overture played without a preceding crescendo, especially in the horns and oboes. If once heard without such a crescendo, that is, with a *subito f* in m.12, the effect is dramatic and unforgettable. Similarly, I have seldom heard correctly the three dynamic steps that occur twice in the second movement of Schubert's *Unfinished Symphony* at the end of the two woodwind solos (clarinet and oboe), in that wonderful sequence of two-bar phrases in *p, pp, ppp* successively—in changing instrumentation as well—followed by the dramatic full orchestral *f.* By way of example, in a quite recent recording by a world-famous conductor and an almost as famous much-praised orchestra, these particular two passages were played not only without Schubert's *p-pp-ppp* dynamics, but with such a bold, arrogant, pushy, and unvaried *mf* that the immediately ensuing full-orchestra ff, which is supposed to represent a maximum mood, dynamic and structural contrast, became completely meaningless and destructive, and would not even have registered a changing effect at all, had not an overly loud and vicious timpani hit exaggeratedly italicized the moment. This is great interpretation?

proaches. They are not willy-nilly interchangeable. Since these great masters—
and many others in the classical period—used dynamics not only as mood and
character defining devices, but as a means of delineating form and structure, we
do serious damage to their work when we ignore these dynamic distinctions.[15]

In short, as a conductor and performer I would much rather do with dynamics, i.e. a strict and respectful attention to dynamics, what most conductors want
to do by distorting tempos. The flexibility and mobility that, I suppose, they
wish to bring to a performance—that is usually their reason, when asked—is
better achieved through observing the composer's dynamics, while playing
around with the tempo in any willful, arbitrary or (as so often) exaggerated
degree only does damage to the music, to its form, its flow, its continuity, its
coherence.

But dynamics are not only a composer's major clue as to structure and intended mood; they also allow the diverse instrumental colors of the orchestra to
shine forth. In the endless noncommittal *mf* or the (only slightly better) stereo-
typical 'loud-medium-soft' dynamics on which many orchestras today seem to
rely entirely—tolerated or actually encouraged by their conductors—instrumental colors and the whole timbral range with which great composers invariably
infuse their works are simply not realizable. Playing with undifferentiated loud
or medium dynamics—one perceptive critic has called it the "modern
industrial-strength sound"—creates a kind of unremitting timbral gray which,
when it becomes habituated in an orchestra, destroys one of the most important
elements that makes orchestral music fascinating and worth listening to. Indeed,
it is that element that constitutes the very *raison d'être* of writing for an orchestra: its multi-colored timbral palette. Moreover, when sonoric/timbral refinement or variety is suppressed, feeling and emotional content are also quelled.
As one superb and famous jazz musician, Milt Jackson, once put it, "If you
don't get the right sound, you can forget about the feeling." How profoundly
true—and how simply put!

Also, when conductors continually ignore or abuse the dynamic markings
given by composers, conscientious orchestra musicians—there are always at least
a few in any orchestra—feel defeated and eventually give up trying to play the
true dynamics. They join the pack, as it were, and the result is dynamic/timbral
anonymity and anarchy.

It is saddening to observe how many American orchestras (with all their technical virtuosity and rhythmic vitality) suffer from the disease of dynamic paraly-
sis. In most cases it is not entirely the orchestra musicians' fault, however. It is
an abuse either directly engendered by the music director or tolerated by him. It
is also, as mentioned earlier, part of a larger problem that in recent decades
has seriously corrupted our aural sensibilities and criteria, a problem produced

---

15. One of the most brilliant uses of dynamic/structural delineation occurs in the final phase of the
development section of the last movement of Schubert's Octet (m. 223). Here the *ppp* written by
Schubert is not some accidental or arbitrary marking, but in point of fact signals quite unequivocally
both the dynamic low-point of the entire *allegro* section of the movement and the long development
section's climb back to the recapitulation.

by modern electronic technology, particularly the microphone, the amplifier, and the loudspeaker. The very names of these devices tell the story: a *micro* sound is *amplified* and sent over a *loudspeaking* apparatus. While the technology exists in recordings, radio, television, film, videos to also transmit the merest whisper of a sound, the most delicate pianissimos and refined dynamic nuances, the commerce and marketing of music use the technology almost exclusively to intensify listening levels at the highest end of the decibel scale. I am not sure what perverse desire in the human animal compels it to celebrate sheer loudness and almost unbearable noise levels. Is it some crude notion that louder (and bigger) is better? Is loudness a narcotic to which we have now become so addicted that we need ever increasing doses of it? Is it that we are no longer content to let our ears be the sole recipients of musical communication, but that the rest of our body needs literally to *feel*, to experience physically, the acoustic vibrations? Or is it that we are simply reacting competitively and in self-defense against the ever-rising noise levels in the material world around us?

I don't really know the answer. Maybe it is a combination of all such manifestations. I only know that, in the musical performing realm and the orchestral realm in particular, musicians are generally enthusiastic about playing loudly—from *mf* on up—and most reluctant (and often even unable) to play softly. Distinctions among *p*, *pp*, and *ppp* have in recent decades become a remote, esoteric rarity. In my own relatively short life-time I have seen the discipline of observing dynamics gradually deteriorate to the point where today aural sensitivity to dynamic differentiation and coloration is virtually nonexistent. It has to be rehabilitated and nurtured almost everywhere at every rehearsal, and conductors who insist on dynamic nuancing often have a difficult task ahead of them. (In extreme cases, where dynamic abuses have held sway for a long time, a guest conductor's attempt to elicit precise and refined dynamic shadings, may even be deeply resented—at least by some of the musicians.)

A by-product of the tendency of orchestras to play at ever higher, upward-spiraling dynamic levels is the bad habit among orchestral string players of over-bowing. It seems to be expected of string players to use the full length of the bow (usually at full bow pressure) regardless of the prescribed dynamic level or note duration. This is, of course, not an entirely new problem.[16] I recall from my childhood my father, a member of the New York Philharmonic's string section for forty-two years (1923–65), fuming about some of his colleagues' undisciplined bowing, never 'saving the bow,' allowing unsightly crescendos on relatively short up-bows, allowing premature diminuendos on down-bows, using the whole bow indiscriminately from frog to tip in even short note durations, and so on. And this was in the days of Toscanini (the ultimate orchestral disciplinarian) and Barbirolli, when orchestral discipline, both personal and musical, was generally at a higher level than it is today.

Today, in most American orchestras—less so in European ones—good, intelli-

---

16. Richard Wagner complained as early as 1869, in his *Über das Dirigieren*, about the careless bowing habits of string players in the German orchestras of his time.

gent bow distribution is a virtually lost art. Probably it is very little considered or seriously taught any more in our conservatories, music schools, and string studio teaching. In any case, bow distribution—speed of the bow, placement of the bow, weight or lightness of the bow—seems to be an unknown subject, buried under careless, bad habits. This is amazing since anyone should be able to hear that, if on an up-bow, playing, say, a whole-note at a moderate tempo, one uses the whole bow with full normal weight, moving from the tip to the frog, the result will inevitably be a rather considerable crescendo. If the note in question is to remain at one dynamic level, then there is a problem—but one of which most string players nowadays seem to be unaware. Conversely, the opposite happens on a down-bow, in which case—with all other conditions (above) being equal—the result automatically will be a diminuendo, a loss of tone. Furthermore, the relation between weight and speed of bow movement is hardly understood and certainly rarely applied; i.e., if one wants to use a full bow—without a crescendo—one can lighten the weight of the bow (as in the French or Belgian school of violin playing). Conversely, if one wants to use a heavy bow weight—again without a crescendo—then one simply has to use less bow and less bow speed.

Unfortunately, string playing in these respects has deteriorated so much since my father's days in the New York Philharmonic, let alone since Wagner's day in the nineteenth century. Paradoxically, while left-hand technique has flourished, that is, mobility, accuracy and speed of left-hand movement on the fingerboard has improved dramatically over the last three to five decades, bow distribution—intelligent control of bow speed and bow length—has reached a level of negligence and disregard that is very disturbing and musically destructive. The present inattention to bow distribution has reached epidemic proportions not only in this country, but in the general cultural Americanization of much of the world around us it is becoming a serious problem even in European and Asiatic musical circles. It seems as if there were some kind of law that one must, regardless of the musical context—long note, short note, loud note, soft note—use the entire length of the bow. Furthermore, string players seem almost completely oblivious of the fact—the acoustic, technical reality—that a too-swift down-bow will automatically make a diminuendo, while a too-swift up-bow will do the opposite, produce a crescendo. That can, of course, be controlled by intelligent bow distribution, bow speed, bow weight and, above all, by careful, thoughtful listening. But again, it seems that far too many string players no longer listen to the musical/dynamic result their bowing produces.

Take a tiny musical example, one of several thousand similar ones that constantly occur in music, such as [musical notation]. If, for instance, the decision had been made to bow the phrase as follows: [musical notation] then, if the player is not careful, that is, conscious of how the bow distribution can negatively (or positively) affect the result and unthinkingly uses full-length bows (from frog to tip and vice versa), the dynamic result will be [musical notation]. If the original dynamic and pre-ferred bowing had been [musical notation], then, with careless bow-ing habits, the result will be [musical notation]. Such bad habits are rampant now in American orchestras, a problem to which, alas, many conductors and music directors are either paying no attention or to which they are actually contributing by encouraging loud full-bow playing. If not attended to or if allowed to deteriorate further, it will result in the final degradation of quality and taste in music performance.

Inevitably, some readers will question: 'What difference does it make, if conductors start crescendos two or three bars earlier than written? Doesn't that make the performances more exciting, more expressive, more human?' or 'What difference does it make if conductors prepare each climactic moment or obvious return of thematic material with a big ritardando? Isn't that a natural impulse? Why is that wrong?' 'What difference does it make if conductors conduct pieces too slow—or for that matter, too fast? As long as the orchestra is playing all the notes, doesn't the music still come through?' And here is the trickiest retort of all: 'You say, Mr. Schuller, that that performance was all wrong: it didn't respect the score, it didn't reflect the intentions of the composer. Well, I don't know that score—I don't even read music—but I must tell you that I thought it was a *terrific* performance; I thought it was *exciting* and *that conductor*—well, he's just tops in my book.' How many times I have heard such questions and statements, particularly the last one!

The answer to all those (very typical) questions is: Yes, it makes a *big* difference. And all those deviations from the score do *not* necessarily make the performance 'more natural,' 'more human.' They may create that illusion—or delusion; they may fool the unknowing, unwary listener into thinking that it was 'exciting,' 'moving,' 'authentic,' when in reality the excitement was superficial and the work was grossly misrepresented. To paraphrase a famous saying about a surgeon and his patient—the conductor may have 'had a great personal success but the work he performed on died.'

No, the only acceptable answer to those questions is: The composer and his score *have* to be respected, especially when that composer is a Beethoven, a Tchaikovsky, a Wagner, a Brahms, a Stravinsky, or any of the other fifty to a hundred composers whose masterpieces make up the bulk of our repertory. To answer the question even more provocatively—and to answer it with another question: Are conductors X and Y, both world famous and popular, really better musicians than Brahms or Beethoven? Are they such fantastic musicians that they have the right to disregard or override most of the basic information contained in a composer's score? The answer is an unequivocal no; they aren't; and they don't have that right!

Let us take those not entirely hypothetical questions and examine them one by one. To start crescendos two or three bars early—or to make crescendos where there aren't any—makes a big difference in the quality of the performance, not only from the point of view of correctness, of fidelity to the score, but from the point of view of expression, of emotional response, of musical excitement. Any comparison between recordings, for example, that distort and deviate from the score, say, of the Brahms First Symphony (Bernstein), and those that re-create pretty faithfully what Brahms actually wrote (Toscanini or Weingartner),[17] will reveal that the more faithful performance is in fact the more exciting, the more rewarding, and the one that will stay in your memory longer.

And how could it be otherwise? Brahms certainly knew what he was doing. The evidence of his scores is that they are most meticulously notated; virtually every detail that can be captured in notation is accounted for. And as we all know, this was a composer who took nearly twenty years to complete his first symphony, going through endless revisions and rethinkings before presenting to the world his "Tenth Symphony," as Bülow called it, referring to it as a worthy successor to Beethoven's nine symphonies. Such a composer is obviously thoughtful, conscientious, mindful of making the work as perfect and as clear to the performer as possible. No one in his right mind will want to argue that Brahms was a careless composer, given to sloppy notation, casual in his musical orthography. And then there is Brahms's sheer musical, creative talent, his fertile musical imagination, his infallible musical instincts. Can we really casually ignore or dare to question that, after much thought and careful consideration of all the alternatives, Brahms put a crescendo in measure five of a phrase and *not* in measure two; that he marked a certain passage or a given instrumental part *pp* rather than *p* (a decision over which he may have labored a half hour or even several weeks or months), a *pp* which we then willy-nilly ignore and play as *mf*? No, Brahms certainly knew what he wanted and knew how to notate it. He knew it better than any of the highly touted baton-wielders of today. Taking his scores, say, of the four symphonies on faith results in performances that are infinitely more exciting than the approximations and deviations that presently fill our record catalogues.

But let us examine the question of the 'early crescendo' in even more detail. What is wrong with a premature crescendo is that almost always it peaks too early—it almost has to, doesn't it?—making the arrival point of the crescendo, its 'resolution,' so to speak, which is the whole purpose of the crescendo, not a high point but an anti-climax (or shall we call it a premature climax?) Moreover, early, exaggerated crescendos—especially if they are habitual and occur at every possible instance—distort the form and continuity of the music (most

17. For a detailed discussion of these (and many other) conductors' Brahms First Symphony recordings, see Part III pp. 279–377.

likely unbeknownst to the casual uninformed listener); and if done to excess, as is usually the case with such self-indulgent 'interpretations,' they tend to become repetitiously tiresome, and the constant exaggerated and overdrawn effects eventually cancel each other out.

There is another aspect to crescendos (and diminuendos) that is fascinating and worth serious consideration. If one studies the way composers have used crescendos in the last two hundred years,[18] and also studies the way the best performers and conductors have rendered crescendos, one will find that they usually reflect a geometric rather than an arithmetic curve. That is to say, a crescendo—the same principle can be applied to accelerandos—is usually best carried out when the curve, the incline, of the crescendo increases the longer it lasts; to put it another way: very little at first and gradually increasingly more.[19] Graphically, one can represent the right way, that is, the crescendo most often called for as in Fig.1a, and the wrong way (in most cases) as in Fig. 1b. The 'geometrically' shaped crescendo is not only (in most instances) more elegant, more satisfying, but—lo and behold—more exciting (if excitement is what is wanted), because as the crescendo is initially held back and then gradually released to run its course, its ultimate resolution, when it finally arrives, is all the more exciting, dramatic, and rewarding.

But there is another purely practical, even mundane, side to this question. A premature crescendo—too early and too much—is wrong from a technical point of view, on the simple ground that if one has crescendoed too much too early, it leaves no room to crescendo further. One can't, after all, retract the crescendo, dip down dynamically, and start over again. If one has arrived too early at the top of a crescendo curve, one has no choice but to remain in that dynamic plateau and await the point where the crescendo really should have peaked. Conversely, if one husbands one's crescendo curve and feels part of the way through that one has perhaps fallen behind, it is always possible, and quite easy, to catch up.[20]

A good exercise for musicians (including conductors) would be to practice "Rossini crescendos" for fifteen minutes a day for a few weeks. The pacing of

---

18. Crescendos as a specific notational device only came into widespread use in the late 1700s, although they were surely employed in music, especially impromptu, long before that time.

19. This is particularly apt for the marking *crecs. poco a poco* (as distinct from merely *cresc.*), for here the composer really means to indicate a very gradual, well-paced crescendo over a longer stretch of time.

20. By analogy, the same problems and solutions exist in respect to diminuendos, accelerandos, and ritardandos. Most of the time diminuendos, when not simply ignored, are done too much too quickly; and the same with accelerandos and ritardandos. Nor do most performers and conductors pay much attention to the very careful annotations of certain composers (among them Brahms, Ravel, Schönberg) who distinguish between, say, *poco rit.*, *rit.*, and *molto rit.*; or between *poco stringendo* and *molto stringendo* and other analogous markings. Even less consideration is given to the question of whether a ritardando, for example, retards *into* a slower tempo or whether it slows *beyond* the new tempo. Admittedly, however, some composers are ambiguous or inexplicit about this particular tempo modification. Nonetheless, musical intelligence can usually deduce the right approach from the music's structural context.

Fig. 1a

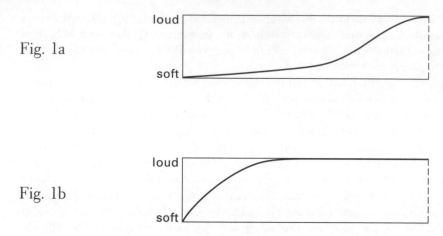

Fig. 1b

these long crescendos—often as long as sixteen (or more) bars—would perhaps cure musicians of making hasty crescendos and premature climaxes.

I do not know what physiological and psychological human impulses cause most musicians (and so many conductors whose work I have studied) to want to crescendo early and quickly, to rush forward into a crescendo without ever considering how far that crescendo is to go (in terms of duration) and how high it is to aim (in terms of dynamic level). Is it some deep-rooted human instinct, some uncontrollable emotional need, or is it just a lack of discipline, a form of carelessness? Perhaps it is a combination. But I *do* know that such impulses, such casual treatment of the matter, is in direct contradiction to how composers have traditionally used crescendos, and that such errant impulses can with care and attention be controlled.

Finally, there is the purely phenomenological aspect to this question. By what logic, by what reasoning can one assert that starting a crescendo several (or many) bars before the point indicated by the composer is automatically, inherently, better then adhering to the composer's indications? By what logic can we assume that to do the composer's bidding is somehow less good, is—as I have heard many conductors and critics argue (or imply)—somehow pedantic, or uninspired and dull? The answer is that there is no such logic: there is no rational argumentation that can establish that the taking of unintended, unwarranted liberties with a piece of music is *inherently* preferable to not doing so. I also know that, with a little self-control, training, thoughtfulness—and respect for the composer and the score—one can easily discipline oneself to render unto the composer his due, i.e., that which he intended and called for in his notation.

If someone is now going to argue, as surely some readers and critics will do, that this is a much too intellectual and cerebral approach, then I can only respond that all great art—all superior creativity, and by extension all superior *re*-creativity, i.e. in our present context, performing/conducting—is the result of an exquisite balance of the intellectual (that is, of the mind) and of the emotional/ instinctual (that is, of the heart, of the soul, of the spirit). I know of no great work of art—music, painting, poetry, cinema, choreography, whatever—that was not created out of a symbiotic interaction between the mind (intellect) and the heart (emotion, feeling, intuition). The problem for the performing arts is that this balance of mind and soul must also be faithfully rendered and re-created in the execution of the work, a balance which, alas, it is given to few interpreters to achieve.

As for the second of my quasi-hypothetical questions—to make a ritardando before every climactic moment or significant return of thematic material, especially in the earlier Viennese classical repertory—that is also ultimately a distortion of the form and over-all continuity, especially when overdone. There are exceptions, when a subtle relaxation of the tempo before a recapitulation—emphasis here on the word 'subtle'—is appropriate. In fact, anything, that is done subtly and not done at *every* possible opportunity is likely to be acceptable. Furthermore, such decisions—to ritard or not to ritard, for example—ought to be under constant reconsideration, ought to be handled with an open, flexible mind. What I find objectionable is when conductors commit their immense tempo distortions automatically, involuntarily, without thinking, without questioning. They institutionalize these distortions, and thereby in the end, after endless repetitions, turn them into meaningless gestures, into clichés, that are stripped of all conviction and significance.

It also makes a big difference if conductors take pieces at wrong tempos. All good music is conceived by its composer at some particular basic tempo. In fact, the tempo of a piece and the content of that music are intrinsically interconnected; one cannot exist without the other. A certain melody or theme is born of a certain tempo (and not any other), and conversely a certain tempo feeling, a pulse, will generate in a composer a specific melody or theme. The two are at the moment of inspiration creatively interlocked; they arise out of the same musical impulse, and they are therefore inseparable once created. Of course, minor or subtle tempo deviations are not only permissible and desirable but probably inevitable. A passage marked at, say, $\quarternote = 108$, is not particularly harmed by being done at $\quarternote = 104$ (or even 102) or $\quarternote = 112$. Most musicians, indeed, can neither control nor hear such minute tempo variations. What is *not* permissible, is indeed unforgivable, is when the tempo taken is twenty or thirty metronome points off the mark; when an *andante* is replaced by an *adagio*, an *allegretto* by a *lento*. It represents a serious distortion of the music, when a composer, like Brahms, writes *andante*, as he does at the beginning of the second movement of his First Symphony, and certain conductors, instead of adopting a tempo of $\quarternote = 56$–69, conduct at a tempo of $\quarternote = 30$—one conductor (Bernstein) even at $\quarternote = 24$. Such a slow pulse amounts to one beat every two seconds, hardly *andante* (which in Italian means walking, from the verb *andare*); it and is more like an *adagississimo*. At such a slow tempo, thirty or more metronome points away from the intended tempo, the music loses all its original intended character and meaning, and becomes in fact a kind of deranged, overblown aberration of the original. In any case, when an interpretation undermines and disrupts (distorts) the construct and the flow of the music, then it is simply wrong, bad and self-indulgent.

Beethoven's metronome markings have been a particular target of attack by conductors (and orchestra musicians) through the many years since his symphonies and string quartets entered the repertory. Various arguments have been (and continue to be) presented to invalidate Beethoven's metronomizations, ranging from: (1) his use of a faulty metronome, (2) his deafness, (3) his belated assigning of the metronomic timings many years after the fact,[21] (4) his determination of these tempo settings at the piano when already deaf for many years, all the way (5) to their alleged 'unperformability' and 'impracticability'. One can easily dispose of most of these objections by pointing out that they are speculative, fallacious and not based on documentable evidence. Indeed, some of the evidence, conveniently ignored by those who would prefer speculation to evidence, clearly supports the appropriateness of Beethoven's metronomizations. For example, the evidence is that Beethoven's metronome was *not* faulty. It still exists and its accuracy has been tested and confirmed.[22] Second, if Beethoven's metronome had really been faulty, how is it that it would have been only selectively faulty? For the objections raised about his 'too fast' tempos are only in respect to certain movements or sections of his symphonies and quartets. Could the metronome have been functioning properly for the last movement of the Fifth Symphony and not the other three? Or, since many conductors have considered the Fifth's last movement 'too slow,' could Beethoven's metronome have been wrong in both directions, too slow in one movement, too fast in another?

As for Beethoven's deafness as a reason for erroneous metronome markings, it is an argument even more ludicrous than the one about the faulty metronome. What makes anyone think that a composer who could create such masterpieces in deafness—Beethoven surely heard and conceived his music in his inner ear and obviously not at the piano, and didn't need to hear it at the piano—would need to hear them in acoustical reality to determine what tempo designations they should bear? Second, he could *see* the metronome's pendulum swings and oscillations. Third, even if Beethoven's metronome were malfunctioning, are we to believe that he did not know that there were sixty beats (pulses) to a minute (at metronome $\downarrow$ = 60) and that he was unable to calculate an 80 or a 166 or a 132 from that knowledge?

I will not be so foolish as to argue that Beethoven could not have made a mistake in his metronomizations, but I seriously doubt that he could have erred by twenty or thirty (or even forty) points, as many conductors would have us believe. All composers have probably erred in metronome settings at some time or other—some more than others (the case of Stravinsky is almost always cited as 'evidence' of composers' general vagrancy and inconsistency in setting tempos and themselves adhering to them).

One of the most outrageously exaggerated pronouncements along these lines

21. By the time Beethoven sent his publisher the new metronome markings, his first seven symphonies and ten string quartets had all been created.
22. See Peter Stadler, "Beethoven and the Metronome," in *Music and Letters* 48 (1967); also L. Talbot, "A Note on Beethoven's Metronome," *Journal of Sound and Vibration*, Vol. 17, no. 3 (1971).

was uttered a few years back by Christoph von Dohnanyi, as interviewed and quoted by Jeanine Wagar in her book *Conductors in Conversation*, to wit: "I'm very skeptical of using the 'correct' metronome markings in a stubborn way, especially since *there is not one contemporary musician who writes the right metronome markings!*' [italics mine]. The metronome was invented in Beethoven's time, 150 years ago, and *composers still can't use it correctly* [italics mine]. Stravinsky never used his own markings literally, and neither did Schöenberg."[23] (Incidentally, one might ask the maestro: "What about Berg and Webern?"—two of Dohnanyi's favorite composers. Does he ignore their metronome markings too?)

Composers *are* capable of errors in metronomization, working in the isolation and abstraction of their studio, but I dare say not to the degree (twenty to thirty points off) or in the frequency (almost all the time) as so many conductors (and Dohnanyi) contend. So it is possible—although certainly not proven—that Beethoven, assigning metronome markings to a work written a decade earlier, could have misjudged the tempo. But again, I would argue, hardly by the number of metronome points by which conductors are wont to deviate—selectively—from his tempo designations.

As for some of Beethoven's very fast metronome settings making the music unperformable, that, too, is nonsense. The judgment of alleged unperformability is the result of laziness, incompetence, and lack of musical imagination. A fair number of conductors (such as Toscanini, Leibowitz) and performers (the Kolisch Quartet) showed many years ago that Beethoven's metronome markings are all technically realizable and expressively felicitous. So have a number of conductors in more recent times, especially among the younger breed of 'period-instrument authenticists,' although not many of them have managed 'expressively felicitous' performances.

Many specific arguments, allegedly demonstrating Beethoven's metronomic incongruities and discrepancies, have been made by many conductors over the years. To examine them all would go beyond the scope and intentions of this book—nor do most of them merit serious rejoinders. Therefore, let a few examples stand for virtually all others. One of the favorite metronomic *bêtes noires* of many German conductors is the (to them) apparent 'irreconcilability' of the *Allegro vivace* in the first movement and the *Allegro ma non troppo* of the Finale of Beethoven's Fourth Symphony, the former marked $\circ$ = 80, the latter $\downarrow$ = 80. What irreconcilability? The beat in both instances is 160, to the halfnote in the *alla breve* opening movement, to the quarter-note in the Finale—all eminently playable, by the the way. Beethoven's *vivace* designation is quite appropriate in view of the *alla breve* meter, a reminder to the performer that a very lively tempo is in fact intended. The *ma non troppo* modification in the last movement is just as logical and justifiable, given the 2/4 meter with its constant chattering sixteenth-notes. (Note, by the way, that these sixteenths can actually—and rather easily—be played even faster; there are many passages in the orchestral literature that are noticeably faster.) Thus the eighth-notes in the

23. Page 55. Such broad generalizations are unworthy of an intelligent musician such as Dohnanyi.

first movement are the same speed as the sixteenths of the Finale. What is irreconcilable about that?

Another often cited example comes from Beethoven's Ninth Symphony, a work whose metronome markings evidently irritate and offend many conductors. The particularly vexatious passage is the famous cello and bass recitative (m.8) in the Ninth's Finale—deemed unplayable and allegedly discrepant with its vocal counterpart (m.216). What discrepancy? Beethoven's tempo marking at the beginning of the movement is *presto* ($\downarrow$. = 66). At the recitative entrance, the score says "selon le caractère d'un Recitative, mais in tempo" (in the character [manner] of a recitative, but in tempo). The implication of this annotation is to play the recitative passage in a 'freer' declamatory way, as in fact in a vocal-operatic recitative, and Beethoven's *mais in tempo* is merely an admonition to maintain both the relative rhythms of the passage and the tempo that is adopted for the passage. Here *in tempo* does not mean *in the tempo of the initial* 'presto' ($\downarrow$. = 66). The vocal version of the passage is in fact not notated differently. So again, no discrepancy.

One might also ask why either recitative should be played and sung at a mournful, lugubrious slow tempo of, say, $\downarrow$ = 90 ($\downarrow$. = 30), a tempo often taken here. The text, "O Freunde, nicht diese Töne! Sondern lasst angenehmere anstimmen und freudenvollere" (Oh friends, not these sounds! Rather let us sound more pleasant and joyful ones!) is not a funeral oration, but a spirited call, an invocation to celebrate the "joys of human brotherhood." So, a slightly more relaxed basic tempo, relative to the opening, is in order—rather than the ponderous distended tempos taken here for generations by countless 'interpreters,' nor, incidentally, the comically fast tempos taken recently by certain 'authenticist' conductors.

Such examples bring us directly to the main issue regarding tempo and Beethoven's so often misrepresented views on tempo and metronomization. It is clear from the evidence of his own statements in his letters and as reported by his various biographers[24] that (a) Beethoven was an enthusiastic supporter of Mälzel's metronomization; and (b) that the correctness of tempos in performance was a matter of primary concern to him, because he felt it was inextricably connected to the very essence of the work at the moment of creation and to the essential character of the work as realized in re-creative performance.

Beethoven was unequivocal in his statements on these matters. He considered tempo an intrinsic part of his musical conception, and felt that the older 18th-century simple tempo designations, what he called "tempi ordinari," were no longer adequate to represent his musical ideas fully. In a fascinating 1817 letter to Hofrat von Mosel, Beethoven writes that he had "thought for a long time of giving up these nonsensical terms allegro, andante, adagio, presto," adding fit-

---

24. For example, Schindler reported in his *The Life of Beethoven* (1840) that "when one of his pieces was performed, Beethoven's first question was always: 'How were the tempi?' Everything else seemed of secondary importance to him."

tingly, caustically, that these "four principal tempi do not possess nearly the truth or importance of the four winds." Beethoven goes on to make a distinction between the "body" and the "spirit" of the work. "Tempo is really more the body of the piece, while those terms that indicate the character of the piece refer actually to its spirit."[25] He adds, incidentally, that Mälzel's metronome gives composers the best opportunity to realize these ideas.

Beethoven was thus articulating a relatively new and, from the point of view of many more conservative late 18th- and early 19th-century composers, perhaps radical idea that tempo and character should be expressed independently of one another.[26] It is for this reason that Beethoven used so many modifiers of the four basic tempo designations, modifiers that became more and more elaborate in his later years attempting to be more and more exacting (remember his *Andante con moto assai vivace quasi allegretto ma non troppo*) [from the "Kyrie" in his Mass in C minor], and which—as I contend elsewhere in this book[27]—confirm the metronome markings as much as the metronome markings confirm the Italian tempo designations.[28]

My credo regarding tempo and tempo modifications *(tempo rubato)* comprises: (a) respectful adherence to tempo (including metronome) markings; (b) meeting the inherent challenges rather than discarding or rejecting them out of hand; and (c) a subtle tempo flexibility that meets the varied expressive demands of the music as it changes mood and character, especially in larger extended forms. The emphasis here is on the word 'subtle'; it is a matter of degree. In other words, tempo modifications must never destroy the continuity and form and thus the essential meaning of the work, or, as Bruno Walter once put it, "noticeable changes in tempo [speed] that are not demanded by the composer are, therefore, misrepresentations; whether they result from intellectual presumption or from sheer license, they deviate from the composer's intentions, and thus from the purposes of reproductive art."[29]

Which brings me to a related point. Conductors—and musicians in general— should learn that motion in music and tempo are not (necessarily) the same thing. Increased rhythmic activity is often used by composers to create the im-

25. *Beethoven's Sämtliche Briefe*, ed. Alfred Kalischer, Vol. 3, (London, 1909), p. 205.

26. See in Part II of this book how Wagner took off on this idea in his seminal *Über Dirigieren* (1869), Ludwig Spohr, one of the most celebrated composers and violinists of the early nineteenth century—and also one of the first baton-using conductors—expressed himself quite firmly on the matter of tempo and metronomization. "When one hears a wonderful musical work diminished in its effect by a wrong tempo, the wish arises that, finally, the practice of indicating tempos in the manner of Mälzel or Weber would become widely used. Then, of course, conductors would have to conscientiously follow such instructions and not, as now, simply follow their own feelings."

27. See discussion of the Italian tempo designations of the first two movements of Beethoven's Fifth Symphony in Part III. (pp. 159–60)

28. For more on the subject of Beethoven's metronomizations and the consistency with which they relate to specific character and category types in Beethoven's works, see Rudolf Kolisch's brilliant article "Tempo and Character in Beethoven's Music," *The Musical Quarterly*, Vol. XXIX, no. 2 (April 1943). Reprinted in a new translation in *Musical Quarterly* (Spring 1993).

29. Bruno Walter, *Von der Musik und vom Musizieren* (Frankfurt, 1957). *Of Music and Music-Making*, English translation (London, 1961), p. 32.

pression, the feeling, of increased momentum without actually speeding up the
basic tempo. Similarly, decreased rhythmic activity can be used to give a feeling of
slowing down the music. Performers should therefore not alter the tempo when the
composer has already composed the acceleration (or deceleration) into the music,
unless the composer himself has indicated a change of tempo. For instance, if a
composer has slowed down the rhythmic activity in the music, as for example in:

the conductor should not ritard the basic tempo (unless, again, the composer
has also called for that). In such an example, the slowing down—and the degree
of slowing down—has already been given by the composer, and no additional
deceleration of the tempo is wanted or required. This seems to be a hard lesson
for conductors to learn.

The last hypothetical but all too typical question I presented above is at once
the easiest to answer in discussion and the most difficult to resolve in practice.
The lay listener who doesn't really know what the composer wrote—who can't
read music, and who probably has never heard a true, correct performance of
the work—has a right to his opinion, of course, but his judgment is ultimately
invalid or, as *proof* of anything, irrelevant. There is really no way that the lay
listener can know that the performance the conductor just gave with such conviction and flair was in fact *wrong*, and to some extent or another a misrepresentation of the composer's work. There is little to be done about such a situation,
unless that listener by virtue of some unusual motivation or outside influence
(perhaps even a critic's review) would convince himself that only serious study
of the work would enable him to have a trustworthy opinion about the relevance
of that performance. But that is a utopian dream. Even the vast majority of
professional musicians never study an orchestral score. Why should we expect a
lay listener to do so?

The point needs to be stressed. The sad truth is that most people who have
anything to do with the performance of music, whether as listeners or as performers, have at worst absolutely no idea what it actually says in a score, say, of
a Beethoven or Brahms symphony, or at best have only the most superficial
acquaintance with the work and its score. This applies to the lay listener, the
music critic, managers, board members of musical organizations (who traditionally choose conductors and musical directors), the run-of-the-mill orchestral musician, and, alas, the vast majority of conductors. The problem, therefore, is that,
since no one really knows or, possibly, cares what the great masters have actually
written in their scores, the performers—but especially the 'dictators of the baton'—can do almost anything they want in their 'interpretations,' and no one
will be any the wiser. Thus a conductor and an orchestra can play a passage
that is marked *pp* at a healthy *mf*, and no one will know the difference or object.
Thus a conductor and his orchestra can make an accelerando (or a crescendo)
too early and too much—or make one where there shouldn't be any—and
hardly anyone will know the difference or object. Thus a conductor and his

orchestra can play a piece in a wrong tempo, and hardly anyone will be the wiser.

I fault especially the critics for this situation. In over fifty years of reading musical criticism in daily papers I have seldom read a review that mentions specific conductorial misdemeanors: a wrong dynamic in such and such a passage, a wrong tempo, an unwanted or exaggerated accelerando, an orchestrational distortion or deviation [30] — not offered, by the way, as a mere opinion ("the conductor took much broader than usual tempos") but as a fact, unequivocally represented in the score. (Writers for professional journals, magazines, or quarterlies tend to do much better in this respect.) Reviews of concerts consist usually of generalities, representing one performance ideology or another, but rarely are there specifics. No wonder conductors feel they can do more or less anything they want with a composer's music.

Musicians, as a lot, are not much better. Those relatively few who have actually studied a score carefully and know not only what's in it but how it should be performed, generally are not in a position to critique the conductor, except behind his back. The rest have never looked at a score, generally don't know or care what's in it, and are content to follow the bidding of a conductor, especially a famous one, through wrong or right tempos, disregard of dynamics, distortions of rhythms, whatever — except again to grumble about their lot and about conductors, but never specifically to challenge the conductor *on the basis of what a composer's score actually prescribes.* (In nearly twenty-five years of playing professionally in orchestras, I almost *never* heard an orchestra musician criticize a conductor on a specific conductorial misinterpretation *on the basis of information found in the score.*) Musicians' complaints rarely rise above the personal level, as for instance when a conductor's wrong tempo (too slow or too fast) makes it technically difficult to play a given passage; it is never a complaint based on the fact that the conductor's tempo was *intrinsically* wrong, in direct contradiction of the information contained in the score.

Various arguments have been presented over the years on behalf of the performer's right to 'interpret' the music as he or she best feels or understands it. In these claims all the arguments of the 'inadequacy of musical notation,' 'the impossibility of absolute objectivity in interpretation,' and 'the impossibility of ruling out the impact of the performer's individual predilections, capacities and limitations,' are trotted out as if they were somehow incontestable scientific facts. In truth, they are usually just opinions that are shaped into certain formulations to attain a certain polemical goal. Very often arguments on both sides — on behalf of performers' liberties or on behalf of faithfulness to the composer's score — are carried only so far as to serve that arguer's purpose. The debate rarely takes place on a level playing field. My own stance is midway between the two

---

30. The one exception that comes to mind is Scott Cantrell, nowadays critic and classical-music editor of the *Kansas City Star*, who often exposes specific deviations from dynamic and tempo markings in his reviews, as well as praising performers who are more scrupulous in these matters.

opposing arguments, hoping to give both the composer and the performer their equal due. If I occasionally lean in the direction of the composer, it is only because I feel that the composer's rights have been more trampled upon than the performer's. It seems to me that *that* is an almost unarguable point, since (a) there is no practical way of stopping performers from interpreting or misinterpreting composers' works; and (b) the vast majority of writing and discussion on this subject has been (and still is) primarily in defense of the performer's interpretive freedom, rarely in defense of the composer.

As I have tried to make clear earlier, polarized emotional arguments on behalf of one viewpoint or the other not only achieve very little but are not even particularly relevant, in view of the fact—this may seem startling to many readers—that ultimately there is no inherent incompatibility between the performer's and the composer's rights and purposes. Both can be served adequately when the interpretation derives *from* the score rather than *apart* from it. For fidelity to the score, the work, and intelligent, respectful interpretation are not incompatible. Within the confines of fidelity there is considerable interpretive freedom and room for multiple interpretations, but of course, not for interpretations that subvert the real meaning and intention of the composer. Such fidelity—inspired by the score—ultimately serves all constituents of music: the composer, the conductor (the performer), and the listener (the audience). To put it another way, intelligent, inspired textual fidelity best serves, intellectually and emotionally, the work itself, the performance, and the listener's aesthetic experience.

Let us examine more closely the most common historical arguments that have been presented on both sides of the issue over many years. One of the first arguments presented by performers (or critics defending performers) is that a composer's notation is limited, is inadequate, is incapable of precisely, objectively defining the composer's intentions. The assumption drawn therefrom is that this gives the performer a license to interpret personally and freely what the composer has wrought. To bolster this argument many writers have pointed to the inconsistencies found in composers' scores regarding, for example, tempo markings and dynamics. A variety of historical information is then trotted out to show—or at least to suggest—that even composers do not agree on the exact meanings of their notational practices. In regard to tempo markings, for example, it is pointed out (as the distinguished author-critic-historian Ernest Newman does in a fascinating article[31]) that Mozart used the tempo indication *andante* for both "La ci darem" in *Don Giovanni* and Papageno's "Der Vogelfänger bin ich ja" from *The Magic Flute*, both in a 2/4 meter, making the further point that both pieces cannot possibly be sung in the same tempo, that in fact the latter song's *andante* "calls for a pace something like 75 percent faster" than the former's.

There are so many things wrong with this argument, presented as some kind

---

31. Ernest Newman, "Interpretation," *The International Cyclopedia of Music and Musicians*, ed. Oscar Thompson, Tenth Edition, (New York, 1975); pp. 1076–79.

of 'evidence,' that one hardly knows where to begin to answer it.[32] For starters, I would be interested to know to whose performances of those two Mozart excerpts Newman was referring. I must say that in all the fifteen years I played both of these operas at the Metropolitan Opera with conductors such as Busch, Walter, Szell, Reiner, Böhm, and a host of outstanding Don Giovannis and Papagenos, I never once heard interpretations of these "arias" that were so divergent in tempo as Newman cites. It would mean that, if Don Giovanni were singing "La ci darem" at, say, ♩ = 60, then the 75 percent faster tempo for Papageno's "Der Vogelfänger" would had to have been ♩ = 105, which would be clearly impossible. Conversely, if the latter piece were sung at ♩ = 80 (the generally accepted tempo of this arietta), then "La ci darem," if 75 percent slower, would have to have been at ♩ = 20. Both impossibilities, so that on that score alone Newman's argument is fantastically off the mark.

The most divergent tempos in which those two Mozart pieces are generally sung/performed are approximately ♩ = 80 for "Der Vogelfänger" and ♩ = 60 ± for "La ci darem," giving the former a 33⅓ percent faster pacing than the latter, not 75 percent. And if the Don Giovanni duet were sung, say, at ♩ = 72 (and the other remained at ♩ = 80), then the tempo divergence would be more like 10 percent and quite acceptable. Even the 33⅓ percent is acceptable and reasonable, for let it be stated that the metronomic range within which a particular tempo marking (adagio, andante, allegro etc.) can vacillate—and has done so since the metronome came into use in the early 19th century—allows for about twenty points on the slower end of the tempo scale, considerably more on the higher end, as is shown below. (Fig. 2) The tempo range within, for example, an andante—over 30 points on the metronome—is indeed dependent, as almost everyone generally agrees, upon the mood, the texture, the density of the music and, in vocal music, upon the text used. Thus it is quite possible for any composer, especially in the 18th and 19th centuries, to have used the same Italian tempo term for several different moods and rhythmic textures. Which is to say that by itself the latitude within which certain composers may have used a given term does not invalidate the use of that term, nor does it necessarily give performers the license essentially to disregard it. (This is the point Kolisch makes in a more comprehensive way; see footnote 28.)

It is also interesting to recall that the old pre-electronic pendulum metronomes had tempo identifications which equated tempo terms (like allegro and adagio etc.) with a range of numbers similar to those in Fig. 2.

Newman in his postulation seems also to have completely ignored the fact that in Mozart's case the two different andantes are in some degree influenced by the internal rhythmic organization of the music: "La ci darem" contains sixteenth-notes, whereas "Der Vogelfänger" does not dip below eighth-notes (except in Papageno's pan-pipe calls). Such rhythmic considerations quite naturally

32. If I concentrate specifically on Newman's article at this point, and let it stand for many other such argumentations, I do so primarily because it is as succinct a summary of the various disputations on the subject as any I know of, and it is cogently argued, at least, and presented in a major widely read music encyclopedia and reference book.

Fig. 2

largo, largamente	beat = 40–60
adagio, lento, grave	beat = 48–66
andante, andantino	beat = 56–90
allegretto, allegro, allegramente	beat = 88–144
presto, prestissimo	beat = 132–180

will affect a tempo, particularly in a vocal piece. Furthermore, it seems clear to the point of obviousness that in the *Don Giovanni* piece Mozart used *andante* in the sense of 'moving,' 'flowing,' in a moderate tempo, to reflect the urgency of the mood: Don Giovanni's passionate pleading with Zerlina, and Zerlina's confused reaction, as she is torn between resisting the Don and succumbing to his advances. In the *Magic Flute* excerpt Mozart used *andante* to slow down the motion of the music, which in its rhythmic notation at first glance looks like lively *allegro* material. In citing the Mozart example Newman seems not to have realized that one of the points he is making about tempos, namely, that they tend to be contextually influenced (which, of course, is true) is not negated by his example, but actually neatly confirmed by it.[33] In short, Newman's argument using the two Mozart *andantes* is inaccurate, misguided, and ultimately meaningless.[34]

What Newman also neglects to mention is that, in contradistinction to his point about Mozart's andante, Mozart was astonishingly precise and inventive in modifying his generic tempo indications. For example, he used by my reckoning at least seven different *allegros* (see Fig. 3 below), a whole continuum of *allegros*, as it were, which as an interpreter, I must say, I have found not only extremely helpful but without exception in their discrete meanings unequivocally clear, especially in a textual context, as in Mozart's operas.

Newman also invokes Carl Maria von Weber's well-known listing of metronome numbers for each section of his opera *Euryanthe*, pointing out what he (Newman) regards as the list's many metronomic inconsistencies and discrepan-

---

33. It may be that Newman got confused in his mathematics or stated them clumsily. It is possible that he meant to say that "La ci darem," if sung at a tempo of ♩ = 60, would be paced at 75 percent the speed of "Der Vogelfänger," if that was sung at a tempo of ♩ = 80, rather than "75% faster".

34. His statement that "there is no general agreement as to whether *andantino* means rather faster or rather slower than *andante*" is equally confused. *Andantino* being the diminutive of *andante*, clearly indicates a slightly faster *andante* or, as David Fallows in *Grove 6* puts it, "a slightly more light hearted *andante*". It is the case, however, that *andantino* in the 18th and the early 19th century was regarded generally as a tempo slower than *andante* (see, for example, Neil Zaslaw's "Mozart's Tempo Conventions," in *IMSCR* (1972), p. 770. It is also true that even Beethoven was confused by the ambiguity of the term in his time, when it was in fact still used in conflicting ways, as Beethoven put it sometimes "close to *allegro* and on another [occasion] almost like *adagio*" (from a letter to George Thomson, one of Beethoven's publishers, dated February 19, 1813).

It was soon after Beethoven's era—and long before Newman's writing in the 1970s—that the term *andantino* acquired its present unambiguous meaning as stated above.

Fig. 3

allegro moderato
allegro maestoso
allegro spiritoso (allegro con spirito)
allegro molto (molto allegro)
allegro di molto
allegro vivace
allegro assai
(All terms differentiated from *allegretto* and *presto*.)

cies. Weber, for example, marks one *largo* ♩ = 84, another ♩ = 50; *andante con moto* is alternately 72, 66, and 80; *allegro* is 100 and 160 at one extreme, 60 at the other, and so on. To understand these apparent inconsistencies we must put this information in context. It is well known that Weber (unlike Beethoven) was ambivalent about metronomization and provided the markings for *Euryanthe* most reluctantly and only under duress. How seriously and thoroughly he thought about his metronome suggestions in each instance is thus perhaps open to question. Second, if one looks at the specific contexts in which Weber's variable metronome markings occur—contexts of mood, dramatic characterization and continuity, rhythmic/metric settings, etc.—one can see that they are not as vagrant and as inconsistent as they appear to be in bald numerical abstraction, as Newman presents them.

Other writers, historians, and critics have used various other historical accounts relating to metronomization to point to its inadequacy and risks: (1) such as Brahms's lifelong adamant rejection of metronome marks; (2) such as Wagner's resistance to the metronomization of his scores; (3) such as Beethoven's irritated disbelief in discovering that some lost metronome markings had been replaced by him with others quite different from the original ones, and so on. What Newman and other historian colleagues disregard or suppress in their disputations—this is what I meant earlier by curtailing the argument at the most propitiously self-serving moment—is the fact that composers, starting in Beethoven's time and ever since then, have become increasingly precise and consistent in their notational habits, so that the further one proceeds along the chronological/historical route to the present, the more one encounters an increased amount of detailed and reliable notational information. In effect, composers have learned to defend themselves against the vagaries and indulgences of performers; or at least they have tried to do so, notwithstanding the occasional inconsistencies, contradictions, and anomalies one may find occasionally with certain composers (e.g. Schönberg's sometimes overwrought fast metronome marks, particularly in his earlier years; Stravinsky's well-known tempo contradictions, not only in his revisions of his own works but in his own performances of them). Such occasional anomalies do not automatically invalidate all metronome markings. The fact is that composers, even great creative geniuses, are

fallible, and occasionally do make mistakes under various time and energy constraints, and are not necessarily entirely consistent in certain matters over the
span of an entire lifetime. Also, there is no doubt that some composers, then
and now, are more careless in notational matters than others. But all these facts,
even when added together, are not sufficient justification to argue that *therefore*
metronome marks and tempo indications are inherently unreliable and irrelevant, to be regarded either with great suspicion or to be conveniently ignored.
Just because a few composers have now and then displayed some human fail-
ings, it can hardly justify or rationalize the peculiar notion, for example, that
certain (of Beethoven's) metronome markings are rejectable, while others are
seemingly acceptable. (Examples of this run like a constant thread through the
analysis of recordings in Part III.)

There is no question that metronomization is not an absolutely reliable process; on the other hand, it can hardly be proven to be wholly useless and fallible. Metronome markings are, in fact, taking the entire literature in account
(not just the alleged 'problem days' of the early 19th century), more often helpful and reliable than not—by far. As I have already mentioned, it is not that
they are to be rigidly mechanically followed—no performer/conductor can guarantee that anyway—but that they offer valuable clues as to the relative tempo to
be taken and in most cases direct confirmation of the verbal tempo indications
given by composers. That composers in general for almost a century regard
tempo not as something rigidly or mechanically assumed and maintained, but
as something subject to variable interpretation is shown by the fact that most
composers use the term "ca."—"♩ = ca. 120" or "♩ = approx. 120"—in their
metronomizations, seriously invalidating Dohnanyi's earlier-quoted accusation.

Similar negative and mostly fallacious arguments have often been presented
over the years in regard to dynamics. The same Ernest Newman, for example,
argued that composers' varied and "inconsistent" use of dynamics makes strict
adherence to them by performers questionable, or at least very difficult and
problematic. Admittedly, dynamics are not 'precise' in the strictest sense of the
word, and in the prevailing system of notation cannot, in fact, provide an absolute, unequivocally explicit representation of dynamic levels. Nor can human
instrumentalists and their instruments reproduce with total accuracy precise dynamic levels, even if we had the notational means to prescribe them accurately.
But to proceed from that admission via a number of carefully selected examples
by certain famous composers, who used dynamics in idiosyncratic or inconsistent ways, to conclude that composers' dynamics, like tempos, are largely unreliable, and thus ought to be regarded freely or with suspicion—that they create
"insoluble problems" for the performer—is preposterous. Newman cites Verdi's
use of *pppp* and even *ppppppp* (in *Aida*), implying, without quite daring to say
as much, that this rather negates the usefulness and reliability of Verdi's entire
dynamic practice. He rightly points out that Verdi felt compelled to use such
extreme dynamic markings to shock his generally rather careless and, in matters
of notation, apathetic Italian orchestra players to play a normal *p*, suggesting
that a mere *p* marking would probably have resulted in "a hearty *f*." Unfortu-

nately Newman leaves the point dangling there, neglecting to mention that (a) the knowledge alone of the reasons for these extreme dynamics should already help in understanding how to implement them, that is, to take them with a little grain of salt, but at the same time not entirely ignore them; (b) that Verdi, who was after all not a fly-by-night mediocrity and was in fact a composer who was, especially in his later works such as *Aida* (and *Don Carlos, Othello,* and *Falstaff*) quite concerned with as comprehensive and precise a notation of his musical ideas as he could muster, still used *p* and *pp* in his scores, from which we can infer that *p* and *pp* also still meant something relatively specific to him, and that four *p*'s were intended to signify a comparatively softer dynamic level in the context of that work—even if we may have to decide that therefore his *p* is slightly louder than it is with other composers in different lands and at different epochs.[35]

My point is that the finest musicians (especially in chamber music or solo work) can differentiate more than the usual eight dynamics (from *ppp* to *fff*)— or the even much more common three dynamics: 'soft, loud, and in between'— and in fact do so instinctively all the time in the subtle dynamic nuancing that marks any truly fine player's performances.

Newman returns to a similar point somewhat later, suggesting "that a *p* or *pp* in a Wagner opera does not mean at all the same thing as *p* or *pp* in, say, [Mendelssohn's] *Midsummer Night's Dream* Overture." I am not entirely convinced that this is in fact true, for it might be nothing more than Newman's subjective perception of how Wagner and Mendelssohn should sound or indeed were sometimes performed during his lifetime. For all I know, an exquisite *pp* in Wagner's *Tristan* should be the same dynamic level *qua dynamic* as in Mendelssohn's Overture, while the sonic and acoustic amplitude might vary. But even if one grants Newman's point, it surely does not mean that musicians and conductors should therefore be allowed to render Wagner's *p*'s and *pp*'s as *mf*'s or *f*'s—nor for that matter to ignore dynamics in Mendelssohn's works, where the many subito *p*'s and *pp*'s, in his *Hebrides* Overture, for example, are also almost universally ignored or compromised—not to mention the *p* part in thousands of *fp*'s in Wagner's operas, especially *Der Ring* and *Parsifal*.

Finally, I return to the point that Newman and like-minded critics and his performer-colleagues almost never press their arguments to their ultimate con-

35. Similarly, at the other end of the dynamic scale, Verdi's scores are filled with five and six *f*'s. The implication there is that his simple *f* is softer than in most other composers' scores. Thus it may very well be that Verdi made a virtue of a painful necessity and thereby arrived at a much more differentiated range of dynamics.

The same problem is to be found in Tchaikovsky's Sixth Symphony (a brief discussion of which is found in Part III). It is also worth noting parenthetically that Milton Babbitt, and other composers such as Pierre Boulez, George Perle, Mario Davidovsky, and Robert DiDomenica have also frequently resorted to highly differentiated dynamics—as many as twelve—in certain works, dynamics which are eminently playable by conscientious performers, though perhaps not *absolutely* reliably and consistently. The irony here is that most musicians tend to complain bitterly about modern composers' use of such finely differentiated dynamic levels, but they do not complain about Tchaikovsky's use of virtually the same concept—they simply ignore *his* dynamic gradations.

clusions, for even they see the irrationality of that. For if one begins by questioning the reliability of the composer's notation in respect to tempos and dynamics and as a consequence suggests that one ought to permit musicians to take various liberties with the text, then where is one to stop compromising? At what point in the line of that argument should one put a halt to it? By what criteria is one to know how far away from the score one may or may nor depart? If we are entitled somehow to ignore *some* of a composer's tempo and dynamic markings, why not ignore more—or finally all—of them? Even the most ardent defenders of performers' privileges do not have the courage of their convictions to push the arguments that far. They prefer instead to raise a few smokescreen questions, loaded with dire implications and innuendo, and having gotten off those initial salvos, hastily retreat into the dark night of vestigial confusion, witnessing with pleasure from a distance the discomfort of composers, who, if they try to defend themselves and their notations, are summarily regarded as being overly defensive and paranoid.

Many conductors and their apologists (critics, sleeve-note writers, publicists) have defended conductors' liberties with regard to tempo and dynamic modifications, the retouching and revising of scores, by arguing that, in the main, such conductors are only trying to help the composer, to clarify his scores for the listener, to bring out 'important details.' Such arguments are as specious now as they were when Bülow first made them in the 1880s, especially those that try to justify the bringing out of 'important details.' There are several fallacies hidden in that benign-sounding 'rationale.' For what it usually means is that a conductor wants to bring out those details that *he* happens to think are important, which may not at all coincide with what the composer thought important. Second, what it unfortunately also too often means is that such 'details' are brought out in an exaggerated and overpersonalized manner which, again, reflects more the *conductor's* conception of the work than the *composer's*. Third, the italicizing of certain details may, under the best circumstances, illuminate some significant aspect of the work or of a given passage, but at the same time may do severe damage to another 'detail,' also of significance and intrinsic merit. Fourth, the bringing out of details almost always seems to mean—and listening to the hundreds of records in researching for this book has certainly confirmed this—'playing something louder,' or, if it is a matter of a tempo consideration, playing something 'much slower' than indicated or 'much faster.' It seems that it rarely occurs to such conductors and their apologists, if it is a question of balance or dynamics, to occasionally make *the other parts play softer!*

The defense of bringing out details through dynamic exaggeration has of late taken on a curiously ironic twist. It is now argued, for example, that given the larger size of today's orchestras, especially the larger string sections, certain 'details' in, say, the winds need to be brought to the fore, details which in Beethoven's or Brahms's smaller orchestras could readily be heard. (The same argument, by the way, is used to justify the doubling of wind instruments in classical symphonies.) Has it never occurred to such conductors that there are at least two other solutions to the 'problem'? One is to make those oversized, overstuffed string sections play softer and use less bow (and fewer bowings) and less vibrato; the other is to reduce the size of the string sections to what they were in the 19th century. (On that point Roger Norrington is absolutely right.[36])

In truth, 'the bringing out of details'—a phrase that on the face of it sounds so worthy, and harmless—can hide a multitude of sins. The problem is that unless it is done very subtly and judiciously, tempered by an innate respect for the score and its composer, it is likely to be anything but 'harmless.' If it is not done with a concern for the over-all coherence and integrity of the entire work, that is, considered and balanced against many other aspects of the work; if it is instead merely a personal, subjective, isolated fancy that happens to have attracted the attention of the conductor, then it is likely to do more harm than good and even if the 'bringing out of a detail' is by itself found to be helpful and correct, it can be the case—and so often is—that it damages or obscures some other equally 'important detail.'

Since discussions on these matters are almost always couched in ideological and polemical generalities, essentially meaningless to the lay reader—a senseless bickering among opposing camps—I would like to avoid that pitfall, and, at the risk of anticipating some of the comments and analyses in Part III of the book, mention a few specific examples of the dangers of too subjectively, too thoughtlessly, 'bringing out certain details.'

Willem Mengelberg has often been laudably described as a conductor who preferred isolating and emphasizing details in a score, as opposed to and even at the expense of preserving the consistency of the larger aspects and grand form of the work. We shall see how that harmless-sounding phrase—'isolating and emphasizing details'—can be a euphemism for distortion and willful arbitrary misinterpretation.

In the first movement of Brahms's Fourth Symphony,[37] Mengelberg felt that he needed to 'bring out' the first violins' line in mm.15–18 (as if that line ever needed further 'bringing out'). He did so by (1) having the violins play considerably louder; and (2) by stretching the tempo enormously in those four bars. The damage done by this emphasizing of a certain 'detail' is that the oboe entering in m.17 with what is in effect a continuation of the violins' line—an oboe passage which is under the best of circumstances difficult to hear—has now become virtually inaudible. Matters were not helped by the fact that Jaap Stotein, the Concertgebouw's oboist of the time, had a rather small tone, whereas the Concertgebouw violins were famous for their full rich sound. By bringing out one 'detail,' Mengelberg completely obscured another even more 'important' one. Moreover, his exaggerated slowing of the tempo, which Mengelberg probably felt would help delineate the formal outlines of the symphony's exposition, actually destroyed the very clarity and congruity of form he was trying to elucidate. In the meantime there are a hundred other 'details'—of balance, of dy-

---

36. See particularly Norrington's commentary on this and related performance practice matters in the sleeve-note for his Brahms First recording (EMI Classics).
37. Hear the recent re-issue of Mengelberg's 1938 Concertgebouw recording.

namics, of articulation, of form—throughout the work that needed Mengelberg's attention, but that he completely ignored or misinterpreted or was unaware of.

In the same symphony's second movement recapitulation (beginning m.64), after taking an already intolerably slow tempo in the main theme (in the violas), Mengelberg makes an enormous ritard in m.71, followed by a horn and bassoon quartet playing *mf* to *f*—Brahms's marking is *pp* (!)—which is then intended to be followed by a full contrasting *f* in alternating wind and string choirs. Mengelberg's various dynamic and tempo distortions are here presumably intended to 'bring out' significant 'details' of form, but, apart from showing a rather astonishing contempt for the score, they achieve quite the opposite results: the formal outline, so beautifully reflected in Brahms's score, is completely convoluted and subverted.

A third example of misguidedly 'making a point,' a very common one favored by many, many conductors—and let these three illustrations stand as well as for hundreds if not thousands of others—can be found in the last movement of Brahms's First Symphony. When the beautiful chorale in trombones, horns, and bassoons in the introduction (mm.47–51) returns in the *allegro* coda, most conductors have, evidently for generations, slowed down to a tempo close to that of the chorale's first appearance. This is done presumably to point out and emphasize that relationship for the listener, to produce a formal, expressive link between the two occurrences of the chorale. It may have achieved that goal with listeners—one can't even be sure, at that—but in the meantime it has completely subverted *Brahms's* intentions, *his* goal. In that coda Brahms indicates no ritardation of the tempo, intending for the 'chorale,' this time in a totally different musical, expressive (and tempo) context, to be a brilliant song of triumph, of exultation. Thus the (perhaps even well-intentioned) 'bringing out' of a formal, thematic detail can be seen to be in the end a total distortion and perversion of Brahms's intentions, so clearly expressed in the score.

Are such 'interpretations' the result of ignorance, malevolence, thoughtlessness? I don't think so. It is a matter of ego, and not the "selfless ego" Bruno Walter often spoke of. Earlier I have written of the counterproductive consequences of a conductor's ego imposing itself on the orchestra, on his fellow musicians. But there is another conductor's ego, even more harmful: that which assumes it can impose its own ideas, its own fancies and whims, on the score and the work of the composer. Conductors who perpetrate these impositions really think they know better than the composer what the composer 'had in mind,' and how the composer 'should have notated the work.' In that sense it is thoughtless, also insensitive to the rights of the composer. And who suffers from all this? The audience, duped by the errant conductor/interpreter, and the orchestra; although, sad to say, many orchestra musicians don't know or no longer care. They have come to realize that complying with the conductor's whims and distortions is an unavoidable part of their job.

When we speak about exactitude of interpretation and fidelity to the score, it is well if interpreters (conductors, performers, players, musicians, singers, coaches) understand that all composing, but especially at the highest inspirational levels, involves not only the creative faculties of the composer but his critical and self-critical faculties as well. Composing consists as much of re-jecting—critiquing certain ideas, certain possibilities—as it does of thinking of them, creating them, in the first place. A composer is always, at every step in the creation of a piece, making choices out of an infinite number of possibilities. This process involves acceptance as well as rejection of ideas. The greatest com-posers have the ability to know instinctively which musical ideas coming to them are suitable or acceptable to the piece (or more precisely to that particular moment in that piece) and which are not suitable or acceptable, and therefore to be rejected. This applies not only to the larger concepts and designs of a piece, but even to its minutest details. Every notational specification is sifted through the composer's mind, ear, and musical instincts, ultimately leading to a final choice, whether it be in regard to a note, a rhythm, a dynamic, the place-ment of a crescendo or an accelerando, or whatever.

What this means for us performers and conductors is that we must consider reliable (and in some sense perhaps even definitive) what a composer, after much exacting selfscrutiny of his ideas, both large and small, often accepting some and rejecting others and coming to a 'best' decision, has written unless we have overwhelming documentary evidence to the contrary. To put it another way, if the composer has pondered these thousands of questions and decisions and resolved them in a specific way, then we conductors and performers ought to take them on faith, and discipline ourselves to re-create those ideas, those conceptions and those feelings, as faithfully as possible. For myself I cannot express adequately enough in words what a thrill it is—a profound pleasure and honor—to do exactly what Brahms (or Beethoven or Tchaikovsky or Schumann) wants and has notated.

I am aware of the fact that lately in America in many circles certain terms, like 'discipline' and 'intellectual,' are considered to be dirty words. To exercise artistic, musical, personal discipline is considered by some to be 'square', 'dull,' 'uninspired,' 'elitist,' and even somehow 'un-American.' Such attitudes are unfor-tunate, because they are contradictory to the very concept of art and artistic creation, which are unachievable without a high degree of discipline. For let us not lose sight of the fact that composing at the highest level is also in part an act of discipline. And this aspect of discipline in creativity must be reflected and manifested in our interpretations and made audible for our audiences in our performances. In that respect the act of re-creating, of conducting, is perhaps even more fraught with difficulties than the act of creating. For, as I have sug-gested earlier, if all truly great creativity—all truly great works of art—embody in perfect balance the highest manifestations of both emotion (feeling) and in-tellect, of individual musical instincts and technique (or craft), then, by any reasonable standards and criteria, re-creations (performances) of those works ought also fully to reflect those twin impulses. That, of course, is easier said than done, for while we may eventually by virtue of painstaking study thoroughly comprehend a musical composition, it is still another matter to reveal and bring to life the feelings that lie behind the bare notes, behind the technical structure, and to connect the intellectual substance to the emotional essence.

It is on that premise that my philosophy of the art of conducting is ultimately based. But perhaps Weingartner said it best[38] in speaking of the homogeneity of conception in a great musical work and of the relationship between feeling and intellect in both the creation and the re-creation of music: "If this feeling is not strong enough, then the intellect takes its place, assumes a predominant role, and leads to an excessively analytic approach. In the opposite instance, feeling takes over in an unhealthy way and leads to unclearness, false sentimentality and emotional nonsense [*Stimmungsduselei*]. If neither feeling nor intellect is strong enough, then there results, according to the prevailing fashion, either mere metronomic time-beating or a mindless mania for [contrived] nuances [*Nuancierungswut*]. Neither, however, has anything to do with art, whose loftiest expression is the attainment of that delicate, more intuitive than calculated, balance between feeling and intellect, which alone can lend a performance its vitality and authenticity."

Very early in this discourse I alluded briefly to the conductor's ego and its place in the conductor's arsenal of 'talents.' As one of the most controversial and, often in the past, most sensational and most publicized aspect of the conductor's image, it is a subject to which we need to return—complex as it is. It is clear that a certain degree of *conviction*, based, one would hope, on comprehensive knowledge and talent, is a necessary part of a conductor's equipment, so to speak. It is necessary in order to impose a particular point of view, a particular 'interpretation,' upon an orchestra, in itself made up of a collection of distinct individuals and artistic egos. I use the word 'conviction' deliberately, because I would like to distinguish between conviction and ego. In fact I would like to make a further distinction between the human ego and the human egotist. A conductor's convictions and a healthy ego—as I referred to it early on—can be and should be conveyed by persuasion, not by domination. The ability to persuade musicians in turn should derive from a respect for the conductor based on his talent, his knowledge, and his behavior towards them, especially in rehearsals.

Such a condition is obviously a far cry from the situation which pertained half a century ago, when conductors' temper tantrums, their power to hire and fire virtually at will, their generally dictatorial attitudes dominated the field. I played as a hornist in those years with most of those tyrants—Toscanini, Stokowski, Reiner, Szell, Leinsdorf, Rodzinski, Dorati, Barzin, Morel[39]—and can testify first hand to the feelings of fear and insecurity (professional and financial) with which we musicians lived almost every day. I also played with many fine, even great, conductors—like Monteux, Mitropoulos, Goossens, Perlea, Busch, Rudolf, Kempe, Beecham—whose behavior and attitude toward musicians can only be described as benign, gentle, and courteous, who did not have to shout at and terrorize us to get the most wonderful musical results. But what is interesting is that among the conductors of both types there is no clear correlation between

38. Weingartner, *Über das Dirigieren* 1905, p.16; English translation *On Conducting*, translated by Ernest Newman, 1906, p.17.
39. If they were not absolute tyrants, they were (like Leinsdorf, Dorati, Barzin) at least extremely short-tempered, although in their later years they all mellowed somewhat.

their personalities or behavior and the quality of their talent: in both groups there were greater and lesser conductors, some who had inflated, domineering egos and others whom I would describe as having (in Bruno Walter's phrase) "selfless egos."

A conductor's attitude—whether benign or autocratic—is, of course, counterbalanced by an orchestra's collective attitude, which may likewise run the gamut from docility to hostility and belligerence. Many orchestra musicians regard *all* conductors as their 'natural enemy,' and in many famous orchestras the musicians' egos may be as highly developed and aggressive as the conductor's. It is a fact that virtually every conductor, even if famous or generally respected or popular, encounters at one time or another an orchestra with which he comes to grief, in which the working relationship with the orchestra, for often inexplicable reasons, simply turns sour. It is one of the great mysteries of the conducting profession—as well as one of its realities—that a conductor may be deeply loved by one orchestra and despised by another.

One of the most annoying mythologies in the realm of conducting is the notion of 'specialists' in one field or another. Thus we have 'Mozart specialists,' 'Stravinsky specialists,' 'Bruckner specialists,' 'Janácek specialists,' 'French repertory specialists,' 'Russian repertory specialists,' 'Baroque specialists,' etc., etc.. What this often means, alas, is that the 'specialist' in question is (a) more or less limited in his repertory to that specialty; and (b) allows himself major interpretational liberties in his 'specialty' by virtue of his assumption of the mantle of 'authority.' And because of this assumed authority and its attendant *renommé*, other musicians and critics accept uncritically—and even applaud—whatever the 'specialist' maestro produces.

I am not, of course, arguing against conductors having particular passions, particular predilections, particular stylistic or historical interests. All conductors are likely to have these; indeed it is to be fervently hoped that they have some such passions and predilections. But I would distinguish this from the sort of specialization mentioned above. The fact is that specialization in conducting, like bad tradition, is usually based more on lacks and limitations—deficiencies (technical, aural, intellectual)—than on any presumed special insights into the subject to be specialized in.

The truth is that if conductors would really learn and fully respect what is in the scores of the great composers—all composers (not just a selected few), including by the way a wide range of contemporary composers—they could be 'specialists' in the entire available repertory.[40]

---

40. I suppose, given the limited notational possibilities of the more remote repertory of, say, Baroque, Renaissance, and Medieval music, one might excuse conductors from becoming 'expert' in those traditions as well. It is a moot point in any case, since the early music repertory hardly ever figures in symphonic concerts, and is now—especially lately—left to the not always tender care of the 'early music specialists.' But in reality, with diligent study, appropriate research and reading, and healthy musical intuitions, even that more 'remote' literature could (can) become a part of a conductor's active repertory.

There is not so much vital information normally to be found outside a composer's score that would vitiate the primary knowledge that the score itself already contains. To put it another way, what a conductor needs to know to reproduce a work faithfully is, especially in the repertory from the early 19th century to the present, generally already contained in that score. Additional useful information may sometimes be gleaned from supportive sources—letters to and by the composer, contemporary accounts of performances, and such—but they rarely are important enough to supplant the information already contained in the composer's notations in his scores. Nor are such alternative source materials always reliable. The reliance on information *beyond* the score is too often an 'impressive' camouflage with which 'specialist' conductors manage to hide their limitations and deficiencies.

In this connection, it is necessary to assail the even more widespread notion that 'great conductors' generally do not—and in many minds, should not—need to deal with contemporary music or the more complex music of our time. This is the greatest myth of all surrounding the conducting profession. It is little understood and appreciated that a thorough knowledge of and versatility in contemporary music on the part of a conductor will actually help him/her gain important insights into—and a respect for—the older masters' works (unless that conductor is a 'contemporary music specialist' with no affinity for older styles, which is, alas, also often the case; *that* kind of specialization is, in my view, as much to be decried as the other kinds).

The major lesson to be learned from the conducting of contemporary scores (or, to qualify it a little, the best, the masterpieces, of contemporary music) is a greatly increased awareness of the meticulous care and extraordinary notational refinements which 20th-century composers generally lavish on their scores. Through that awareness, any conductor is bound to become more scrupulous in his respect for an earlier composer's notation. Ever since the important early works of Stravinsky, Schönberg, Berg, Webern, Bartók, Prokofiev, Szymanowski, Ravel (to name a few),[41] scores by 20th-century composers have become increasingly precise, evermore aware of every interpretational eventuality—and therefore intent as best as possible on precluding any unwanted interpretive choices. In this respect, in fact, 20th-century composers have been quite successful in perfecting their notational skills, to the point that interpretation in the willful, deviant, undisciplined way to which the 19th-century repertory is so often subjected, is virtually precluded. There is just too much detailed, precise information contained in a contemporary score to be simply ignored. Indeed, most conductors, it would appear, seeing such scores, become discouraged in contemplating the necessary technical control and intellectual discipline the conducting of such works requires, and quickly—and forever—disassociate themselves from contemporary music, returning to the 'relative safety' and interpretational liberties of the 19th-century Romantic repertory.

41. It should be added here, however, that Brahms too was a most meticulous and precise notator of his music, especially in his four symphonies, leaving very little to doubt or speculation. And yet how shabbily Brahms has been treated in matters of tempo and dynamics by most conductors and interpreters! (See Part III.)

In a long life of performing with and observing famous (and not so famous) conductors occasionally venturing into contemporary music—by that I mean truly
contemporary music, representing our own time, say, the late 20th century (and
not some trendy, anachronistic, neo-Romantic stylistic pleasantry)—I have been
startled to discover that even relatively fine conductors are unable to perform the
most elementary functions in conducting a contemporary work, such as holding
and controlling a tempo, accurately rendering a metric modulation, obtaining
rhythmic and dynamic accuracy, to cite but a few examples.

It is impossible to conduct a work by, say, Elliott Carter, Milton Babbitt, George
Perle, Jacob Druckman, Donald Martino, Pierre Boulez, Hans Werner Henze, Ol-
iver Knussen, and dozens of other composers without having, for example, the
technical/intellectual/emotional control to keep an absolutely steady tempo, no
matter what the metric or rhythmic/durational or contrapuntal complexities of the
music might be. To conduct Carter's *Double Concerto for Harpsichord and Piano*
or his *Penthode* for five instrumental quartets, for example, or Perle's *Short Symphony* requires the ability to maintain thorough tempo control at (or very close to)
the stated metronome indications: here there is no saying, 'Oh, well, he couldn't
have meant that metronome marking; so let's just do it a little slower'—which is
what most conductors do with Beethoven's metronome or Brahms's tempo indications. Moreover, since in these Carter and Perle works all other tempo variations
are determined and controlled through metric modulations, every one of such
tempo changes must also be rendered precisely, lest the whole continuity of the
piece be subverted and annulled. Let me emphasize that this is, in the case of the
Carter and Perle works mentioned—and, needless to say, hundreds of other similarly structured works by many other fine composers—not merely a matter of 'getting pretty close' to the tempo or 'more or less' managing those metric modulations. Since the whole work is structured in terms of these myriad tempo
relationships—nay, is *composed* through them, and represents the very content and
essence of the work—any deviation from the basic called-for tempos or from the
metric modulations that constantly modify them destroys the very essence and
structure of the work, rendering the performance not only wrong but pointless.

Similarly, in single tempo works like those of Babbitt's—*Composition for
Twelve Instruments* or *Relata I* come to mind—where all rhythmic variation and
complexity are already composed into the work and constitute its very essence
(at least one crucial element of its essence), where every rhythmic detail is
precisely calibrated and controlled, and thus constitutes the rhythmic continuity
and flow and line of the work, the conductor cannot deviate from or abandon
the basic tempo. This happens to require considerable conductorial tempo control, but is not by any means impossible to achieve.[42]

---

42. It should not be thought that such examples as cited here are a peculiarity of the 20th century,
a result (as some would have us believe) of the 'mathematicization' and 'excessive intellectualization'
of modern music. Precise tempo relationships and interrelationships between movements, sections,
set pieces of operas are as much a part of Mozart's *Marriage of Figaro* and Debussy's *Pelléas et
Melisande* or any number of classical symphonies as any contemporary 20th-century work. And as
for 'metric modulations,' they go all the way back to the *Ars Nova* of the 14th and 15th centuries,
not to mention Beethoven's symphonies, string quartets, and piano sonatas.

In a quite different way, the mature works of Webern are based on the strictest, most precise, tempos and tempo relationships, which include even such sophisticated concepts as, for example, notating and defining *rubatos*, or executing ritardandos during pauses and measures of rest, thus even determining the durations of silences in the music. Again, these subtleties of tempo control are so built into the very structure and essence of the music that any tampering with them—or failure to execute them—leads automatically to a serious distortion of the music.

What has been said here regarding a conductor's obligations in respect to tempo and rhythm in modern works could, by analogy, be claimed as well in the realm of dynamics, where again many of our greatest composers—from Stravinsky and Schönberg to the present—notate dynamics with a previously unknown subtlety of differentiation and precision, dynamics often functioning *structurally*, not merely decoratively or expressively, and which must thus be respected absolutely in a way that perhaps in the performance of 19th-century music, if ignored, is not quite as *structurally* damaging.

The healthy disciplining of one's conductorial craft through the performance of new music is not only an intrinsically worthy pursuit, but can be, as suggested, an eye and ear opener for the conductor in terms of the 18th- and 19th-century repertory as well. It is sad to realize that most conductors avoid the more challenging modern repertory like the plague, and, of course, are even further dissuaded from touching it by their managers and handlers. And once again, a double standard is applied as between conductors and musicians. Conductors are permitted, even encouraged, to avoid the contemporary repertory, are quickly forgiven if, when they do it at all, they have done it badly. (Indeed, who generally, except for the musicians and the composer, even knows that a contemporary work was inadequately rendered by the conductor?) On the other hand, musicians are *not* permitted to avoid contemporary music, for when it is programmed, the musicians simply *have* to perform it; they have no choice in the matter. Moreover, they are expected to play their parts perfectly. And what they play—or by chance fail to play—can be clearly heard, while, of course, what the conductor does can be heard, and thus be assessed, only by the most sophisticated, most knowledgeable of observers; to the rest it will go by unnoticed.

To pursue this train of thought further, it is rarely brought out that there is a kind of injustice in a situation which allows conductors virtually any kind of liberty of interpretation, while orchestral musicians are expected to perform with absolute precision and accuracy, allowing for no deviations from the text allowed—except for those imposed on them by the conductor. The irony here is that musicians are expected to perform 'perfectly' even within the relatively (or totally) distorted interpretations in which so many conductors indulge. More than that, musicians are not only expected to be technically precise and accurate in their performing, but play with great expression, warmth, interpretive insight, particularly, of course, in solo passages, whilst being locked into a rendition— too fast, too slow, too loud, too soft, too something—which does not correspond to the score to begin with. It is amazing to me that this double standard—one for conductors (and singers, by the way), another for orchestral musicians—is an accepted norm, is maintained throughout the musical world, tacitly justified, and rarely questioned—sad to say, even by musicians themselves.

I can testify to the virulence and widespread acceptance of this double standard in orchestral performance most personally. For, in my earlier career of over twenty years as a horn player in a number of major American orchestras,[43] in most instances as principal horn, I was expected to perform flawlessly, both technically and expressively, often enough within conductors' interpretations that were severely at odds with the information in the respective scores. Any number of musicians, then and now, can testify to the same experience. More than now, we musicians of the 1930s through the 1960s were in no position to protest these wayward interpretations in which we were so often imprisoned, because one could get fired by the conductor during a rehearsal, at the end of a concert, not at the end of a season with recourse to appeals, defense by orchestra committees, arbitration, and so on. It was simply understood—and is still largely accepted to this day—that a musician was (is) to perform more or less flawlessly in respect to rhythm, tempo, attack (and release) of notes, dynamics, ensemble blending as ordained by the conductor, whether his interpretation corresponded to the information in the score or not. In addition, as already mentioned, we were (are) expected to play with great feeling, with interpretive flexibility—not beyond the limits, set by the conductor, of course—and to contribute somehow meaningfully to *his* interpretation. And how we sweated and worried, tortured ourselves, to achieve these often artistically dubious results. I now marvel at the skill and chameleon-like adaptability with which the best musicians—then and now—walk this precarious musical tightrope.

If a rendition deviating from the text is allowable for conductors, why is it not also, permissible for orchestral musicians? Why can't a musician play in wrong tempos, insert *rubatos*, ignore dynamics, make crescendos too early, arbitrarily accelerate the tempo during crescendos, when conductors seem to assert such privileges unquestioningly, automatically? Not that musicians are entirely free of such musical misconduct. Most are similarly inclined to take unwanted liberties with the music when left to their own devices (as in chamber music). But nonetheless a different, much tougher standard pertains for them when they are in an orchestral situation, where they are forced to adhere precisely to the conductor's interpretations and whims, no matter how aberrant.

I believe that conductors who have come out of fine orchestras, who have had to perform at a high level in the kind of disciplined ensemble I have described, are often those whose interpretations are least willful, least arbitrary and self-indulgent. Conductors who have been first-rate orchestral instrumentalists—cellist (Toscanini), violist (Monteux), oboist (Mackerras), violinist (Munch)—

43. The Cincinnati Symphony Orchestra, the New York Philharmonic, the Metropolitan Opera Orchestra, as well as numerous other permanent or free-lance orchestras and recordings (with conductors such as Reiner, Stokowski, Walter, Leinsdorf, etc.).

tend to be those who treat composers' scores with an innate love and respect and understanding. On the other hand, in most cases conductors who never played an orchestral instrument professionally or who are (were) pianists—thus rarely, performing in ensemble, generally limited to performing soloistically, independent of any outside control—seldom possess those skills and feelings. That is why I do not necessarily subscribe to the widespread notion, propagated by many piano-playing conductors, that the piano is the preferred—or even mandatory—instrument for conductors to learn. It is obvious, of course, that skills at the keyboard will facilitate score reading at the piano, learning scores through piano transcriptions and piano reductions. The piano is also, obviously, a useful instrument to master in the long-standing European tradition of training conductors in the opera houses, starting as *Chorrepetitors* (coaches), working up to substitute and assistant conductor positions, and eventually, in some cases, to the position of *Erster Kapellmeister* or *Generalmusikrecktor*. Yet all those obvious advantages of being a proficient pianist and sight-reading score reader—Solti and Szell come to mind as outstanding masters of these skills—are outweighed by the much harder to obtain experiences in ensemble disciplines that the piano almost precludes and basic orchestral instruments offer.

Be that as it may, the fact remains that the often low-level imprecision and willfulness of most conducting is in drastic contrast to what is expected of orchestral musicians: absolute precision and adherence to the score (the part) *and* the conductor.

Another much-discussed—even much-belabored—issue is the question of whether to conduct from memory or not. This should be a matter of secondary importance, but unfortunately it has been made into a major issue by some critics, writers, conductors (obsessed with the fetish of conducting from memory) and their publicists. In my view it is very much a matter of personal choice; so I do not see any particular virtue attached to conducting from memory. As someone once put it, a great performance from a score is better than a mediocre one done from memory. It is also axiomatic that a great performance is great whether it is conducted from memory or not. And there have been in my experience very, very few conductors who had or have, while conducting from memory, more than a superficial knowledge of the score. Virtually all 'from memory' conductors whom I have observed, or whose recordings I have studied, know only the most obvious surface of the music and a few inner details that happen to be of particular interest to them. That complete intimate knowledge that I uphold as the ideal—that *understanding*, not necessarily the *memorization*, of every minuscule detail of a score—I have seen in only a very few score-less conductors, and then only among those who have (had) a limited repertory or specialize(d) in a certain repertory which they conduct(ed) dozens or hundreds of times. But even so, they also sometimes overlook(ed) important compositional and performance details.

The vast majority of conductors when conducting from memory primarily conduct the obvious melodic or thematic lines (mostly those in the upper register), some of the dynamics (at least the best among them do), but are rarely or only intermittently aware of harmonic or tonal-functional aspects, structural features, interesting countermelodies or motives, timbral balances, intonation, unusual orchestrational details. In other words, I have not seen in the majority of memorizing conductors a close correspondence between memory and true knowledge. Nor am I particularly impressed by someone conducting the Beethoven Seventh Symphony for the ninety-fifth time and conducting it without a score. If that conductor doesn't know the score by then—at least superficially—he or she has no right to step onto a podium and stand in front of an orchestra and an audience.

Audiences, of course, tend to be impressed by conductors who conduct from memory, little realizing that for some people memorization comes very easily, especially when it is mere surface memorization. The fact is that some people have an innately good and reliable memory, while others simply do not. Some people are capable of quick memorization, but also forget very quickly. Audiences also do not realize how much of what is in the composer's score such conductors are in fact missing. Some of the best conductors—like Monteux, Reiner, Solti—knew/know their scores very well, but nonetheless always work(ed) with the score. Others claim they *have* to memorize and conduct without a score[44] because they find working with a score, including turning pages, a hindrance, even a distraction. Still others—far too many in my view—conduct without a score out of vanity and to impress the audience (and the critics and the musicians).

There is also the fact that conductors working without a score *do* make mistakes, some more than others, even Ozawa, whose memory is the most phenomenal (and inexplicable) in the recent history of conducting, even far beyond Toscanini's or Mitropoulos's. But Ozawa's memory is also not infallible, and his knowledge of many scores is often of the across-the-surface-of-the-music superficiality I referred to earlier.

One of the real problems with conducting from memory, especially in the case of conductors who feel they have to conduct everything—entire programs, entire seasons—from memory, is that they tend to be so busy remembering what comes next in a piece of music that they do not fully hear what is being played at that moment by the orchestra, or to put it more precisely, *how* it is being played. The human brain can only deal with one mental activity at a time; some special minds perhaps with one-and-a-half or two. What this means in point of fact is that, if the brain is concentrating, as perforce it must do, in split-second timings on the next upcoming moments of music—let us call them point B—it cannot also concentrate with absolute clarity on where the music actually is at the moment—point A. The reason for this is that music being on the one hand an aural, auditory art and being only truly appreciable through the ear, but on the other hand the ear being connected to the brain and its

44. I find myself compelled to point out that 'conducting from memory' and 'conducting without a score' are not necessarily the same. The latter may mean—and, alas, often enough does—that the conductor in question, far from having memorized the score, especially in all its fullness, is simply conducting 'without a score.'

receptivity being governed by the brain (that is, the brain provides what it is to hear), if the brain is at point B, it can hardly direct the ear to hear what is going on at point A. In other words, this ineluctable fact prevents the conductors, intent on remembering what comes next in a piece of music, from paying attention to and conducting in detail that which is being played at that very moment. This is how and why many memorizing conductors fail to conduct, or they overlook, or simply are not aware of, important aspects of the music at that precise moment. This also explains why a conductor like Ozawa conducts a performance in a pre-programmed (i.e. memorized) way, including memorization of the planned conductorial gestures, and can therefore not make instantaneous adjustments or changes of gesture to correct or balance something, or influence the performance correctively at any specific moment. Indeed, Ozawa has developed and propounded a whole theory of conducting which is based on the notion that the conductor need not, *should not*, in a performance adjust dynamics or dynamic imbalances beyond what may have been rehearsed. It is a theory of conducting which avoids confrontation with the actual reality of a performance in progress. In this way of conducting, the performance and the conducting constantly pass each other by, never really becoming one, never the one influencing the other, each going its rehearsed pre-programmed way.[45] In such conducting it would seem very difficult or downright impossible to really 'get inside' the music, to achieve the kind of spontaneity that results from the conductor and orchestra creating the performance *together* in a continuous series of subtle give-and-take interactions. In such pre-programmed conducting, the conductor cannot suddenly adjust to bring out spontaneously a note that is being played too softly—say, the E third in a pp C major chord—or to subdue an overly loud trumpet in an ensemble, or to bring out a bass line that needs to be heard more prominently.

Conducting with a score does not, of course, in and of itself guarantee that a conductor will be able to make such instantaneous adjustments; the ear has to hear such imbalances before the mind can tell its conductor to make the corrective gestures. But it is my experience that most score-less conductors are much less likely to hear such momentary deviations, because all their concentration is absorbed by the act of remembering.

Obviously I feel there are some risks—at the very least, potential risks—and some serious disadvantages to conducting from memory. At worst, its practice is designed to impress gullible audiences, managers, agents, boards of trustees, and perhaps even some critics. At best, in the hands of a few highly exceptional musician conductors, it may be effective but still not without risks. My personal best suggestion is to know the score completely down to its minutest detail—in essence to know it 'from memory'—but nonetheless to have the score on the stand, as a potential refresher of the memory, as a support element that is there,

---

45. I recall at times experiencing the same kind of 'locked-in,' unyielding feeling as an orchestral musician playing under Toscanini in the 1940s and 50s, a feeling I never experienced with Reiner or Busch or Monteux, for example, who always used a score, although they knew the music completely and thoroughly.

should it suddenly be needed. The ideal is to be free from the page, but still to have it there as a comfort. To me a score by Brahms or Beethoven or Wagner or Ravel is a thing of beauty in itself. To see, even if only in a split-second glance, that inspired (and inspiring) document, that lovingly, painstakingly defined text, directly at hand—not just in the mind's eye—communicating its beauties to us, is an exquisite experience all by itself, and a marvelous emotional/intellectual confirmation of one's grasp of the score. It is like having a close, intimate, trusted friend at one's side. That, I believe, is the most honest, unpretentious, unostentatious approach to conducting.

I have not thus far discussed the manual-technical aspects of conducting, the so-called baton technique. And I do not intend to do so now,[46] except to examine two specific technical questions: (1) the respective function or functions of the two hands, and (2) the widespread habit of conducting continually with upward motions. Regarding the first question, there appear to be two basic but quite divergent theories. Stated in its simplest form, the one theory propounds the notion that the time-beating function is directed by the right hand, while the expression of the music and the control of dynamics lie in the left hand; the other theory holds that both hands may be involved in both functions, but that the right hand must in any case convey the entire character and meaning of the music as well as the beat and tempo. The latter theory suggests that a good conductor of that method could conduct with only the right hand, expressing by the size and character of the beat all that is contained in the music.

I very much recommend and support the second approach, not only because it seems to offer the more complete, the more flexible, the more expressive possibilities, but because most of the best conductors I either played for or have observed through the years were (or are) of the latter school. The beauty of this concept lies for me in the fact that the right hand expresses and embodies all that is essential to the music's correct characterization, leaving the left hand free to confirm, to highlight, to make more specific, to isolate some musical-compositional detail, to 'decorate' and refine, as it were, the basic conductorial gestures of the right hand. This can range all the way from both hands locked in identical, symmetrical (confirming) gestures through separate and diverse (highlighting, decorating) gestures to total inactivity of the left hand.[47]

The reader may have noted that I did not mention the control of dynamics as part of the left hand's duties. The reason is that that must be already con-

46. For detailed information on these matters, I refer the reader and the conducting student to any of the following instructional books, all of superior quality and filled with highly useful (although not necessarily always concurring) information: Max Rudolf, *The Grammar of Conducting* (New York, 1950, 1980); Frederick Prausnitz, *Score and Podium* (New York, 1983); Hermann Scherchen, *Lehrbuch des Dirigierens* (Leipzig, 1929), and in English translation, *Handbook of Conducting* (London, 1933,1989).

47. I am well aware of the fact that the eyes, the facial expression, indeed the whole demeanor of the body, are important, even crucial expressive elements of the art of conducting. However, they are beyond the scope of this discussion, being far too personal, too subtle, too diverse, to permit any coherent generalizations or suggestions.

tained in the functioning of the right hand. In fact, it is a crucial element of fine conducting, even of merely technically competent conducting. For, if the right hand is, for example, rigorously, abstractly beating time[48]—especially with, say, overly large beats—then all the attempts at nuancing of dynamics with the left hand will be of no avail. A very common failing, for example, among young or inexperienced conductors is to conduct a *p* passage with huge uncontrolled right-hand beats while irritatedly shushing the musicians with the left hand. The two hands can, in addition to confirming, discriminate and subtly differentiate, but they must not contradict each other. An orchestra will then mechanically follow the right hand or—more likely—ignore the conductor altogether.

The second question—conducting continually with upward vertical arm motions—is perhaps even more problematic in that it is a habit which is very easy to slip into and very difficult to get out of; and as mentioned, it is very widespread among conductors, especially among German or German-trained conductors. Technically, it consists of pulling the hands and arms precipitously upward on the beat—in its worst form on *each* beat (whether in 4/4 or 3/4 or 5/4 or any other pattern)—instead of delineating the beat with downward or sideward movements. Graphically, the incorrect and correct movements could be shown as follows: say, in 4/4, (see Fig. 4a, b, c) (a) repeated four times more or less in the center of the beating zone with four fast vigorous upward thrusts (□ indicates the baton stopping abruptly) instead of (b), or much better, (c). The incorrect movements, as in (4a), are wrong for two basic reasons: First, they inherently prevent the use of any lateral, horizontal hand movements (left to right, as in (4b) and (4c)); second, because they do so, they make it impossible to create any musical line. Each motion, especially if beat vigorously (as in a *f* passage), produces a punchy staccato gesture which emphasizes and isolates each beat. There is no orchestra in the world that can play a musical line, a phrase, with such a flailing, upward-thrusting beat. The orchestra is virtually forced to play a series of hard-hitting *beats*, rather than a sustained directional *line*—unless, of course, in self-defense it were not to look at the conductor (a recourse to which many musicians resort).

The disuse of lateral beat-pattern movements, i.e. the avoidance of using the entire arms'-length width of the conducting zone, is probably the most serious and widespread problem among conductors from a technical point of view. If it were merely 'technical,' it might be tolerable, but unfortunately it produces the most unmusical, inexpressive, mechanical results.

It is bad enough that meters such as 2/4, 2/2, 2/8—and even worse, 1/4—allow

---

48. If Toscanini's baton technique had a flaw, it was that in his desire to obtain absolute rhythmic control in his musicians' performances, his right hand became an inexorable dominating time-beating and tempo-controlling device, which sometimes could become—especially in fast tempos and loud passages—constricting and stifling of expression.

Of course, like many great conductors, Toscanini conducted more with his eyes than with his hands, his eyes often offsetting the occasional rigidity and unyielding sternness of his beat. For Toscanini's eyes were like burning coals, holding his musicians totally in their sway and, through their intense concentration, energizing the music in a way that I never felt quite so powerfully with anyone else, except perhaps with Mitropoulos.

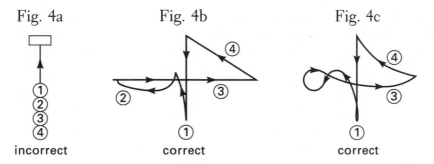

Fig. 4a — incorrect   Fig. 4b — correct   Fig. 4c — correct

for little or no lateral beat movement. The hands are relegated simply to moving vertically down and up. At least in any beat pattern of three or more, the hands can move laterally, expressing through the shape of the curved horizontal movements the mood, the character, the motion and direction of the music. In this context it is well to remind ourselves that the most important thing, from a technical point of view, is what a conductor does *between* the beats. Beating time is something that almost anybody can do—and unfortunately too many conductors are merely 'time-beaters'—but the real art of conducting resides in how you shape the music, give it its appropriate character and mood and essence by *how* you move from beat to beat, what you do *between* the beats.

And finally, a word about traditions is perhaps also in order. Traditions are, to be sure, a complex subject, for there are bad traditions (especially in the operatic world) and there are good traditions. To know which are good and which are bad, which to follow and which not to follow, in itself takes almost a lifetime of study. But knowledge of the traditions is a most important part of a conductor's training and study, even if only in the end to reject some (or most) of them. The study of performing traditions must be tempered by study of the score, although conversely the study of the score may also be informed by a study of the accumulated attendant traditions. There is no easy solution to this vexing problem, for in the end intelligence, common sense, a study of prevailing performance styles and practices as well as reliable source materials are the only solution, on the basis of which the conductor must then simply use his best judgment. In any case, to conduct a Brahms or a Beethoven symphony or a Wagner or Debussy work without an awareness of the respective received traditions associated with such works, is almost to forfeit one's right to conduct these works, even if—as I say—for good and proper reasons one decides ultimately to reject some of those traditions or some aspects of them.

Tradition is sometimes nothing more than bad habits or technical limitations ossified into permanence. But there are also good traditions, important traditions, which some superior musician or group of musicians evolved out of repeated experiences with a given piece of music or a given phrase. In years of studying various traditions—historical, regional, national—I have realized that those traditions that we consider valid and useful are usually ones that arose out of the demands of the score, the music itself, not out of the demands or limitations of a particular performer or group of performers. In any case, to learn to distinguish between bad and good traditions is another important obligation a conductor must assume, regardless of how arduous that task may be. My advice would be to initially consider all traditions with great suspicion, no matter how venerable or highly championed, that is to say, not to accept automatically any of them, but to study and research them as to their origin—when and by whom they were initiated—and appraise them, especially as to how, why, and to what extent they deviate from the composer's score. With that any reasonably intelligent conductor can assess which traditions are valid and which are groundless falsifications.

The reader will by now, it is hoped agree with me that conducting is a most demanding and challenging artistic discipline, particularly, as an interpretive re-creative art, and that it does not permit of the kind of casual flirtation and/or egocentric involvement which most conductors give to the task. It will also, I hope, have become clear that baton technique—bad or good—is not the most important factor in producing a performance, let alone a 'great' or recognizably 'authentic' performance. A good, clean, sensible technique does help—it is especially helpful to orchestral musicians and makes their life a little easier—but hundreds of conductors (over more than a century and a half) with poor or problematic techniques have proven that performancs are not thereby completely hindered, nor does that factor in and of itself predetermine the type and quality of interpretation/realization of a given work. The fact is that conductors with poor techniques have given great, profoundly moving interpretations, while conversely, conductors with excellent (clear, clean) techniques have given empty-headed (or wrong-headed), willfully wayward performances.

That is why I have hardly dwelt at all on conducting techniques in this book, concentrating much more on conducting as an interpretive art. In summary, much of what I have thus far written about and postulated as a basic requirement of good or great conducting amounts to score analysis, although I have avoided that particular term since it can, in the wrong hands and minds, be construed as 'academic' and 'overly intellectual.'

Analysis to me is simply the thorough study of the score, of its specific notation in all its elements: melodic/thematic, harmonic, rhythmic/metric, structural, textural, orchestrational, formal, etc. Analysis in that sense is an all-encompassing retracing of the steps of composition, yielding the fullest possible understanding of what went into the piece in the first instance and what therefore needs to be 'realized' in performing/re-creating it. Analysis can also frequently tell the performer *what not to do*. Analysis in that full sense will inform the conductor *not* to emphasize one thing only to obscure another; *not* to exaggerate the most obvious and commonplace; *not* to impair the expression of the spirit and essence of a piece which, be he reminded, should be the sum total of all the aforementioned interrelated elements and parts; *not* to overstress what is secondary or tertiary in structural importance, especially at the expense of something primary, and thereby destroying or subverting the *composer's* view of these balances and interrelationships.

I have emphasized questions of tempo, of dynamics, and of articulation, because these three realms are among the most basic requirements of a balanced, correct interpretation and because they are the three areas that are the most abused, ignored, and falsely rendered. Tempo questions relate not only to basic or opening tempos (of movements) given by the composer but to tempo modifications within movements or pieces *not* necessarily indicated by the composer. I can only reiterate by way of summary to all that has been said above that tempo considerations of all kinds, including *basic initiating tempos*, require full respect for the score, thorough study of the work and its (possibly) varying tempo modifications; for the element of tempo bears most profoundly and critically on the shape, continuity, and character of the work. The fact that a few, rare, sensitive, and experienced performers (conductors) can intuit a right tempo by 'feeling', as wonderful as that may be, is still insufficient as a basic approach, because it leaves a margin of uncertainty which can only be resolved and upheld to scrutiny by study and analysis.

No less critical and decisive are dynamics, for here the conductor's/performer's careless or willful intervention, altering the dynamics, can be severely damaging to the character and essence of the work—unbeknownst to the audience, mind you. The same can be said for articulation, phrasings, bowings etc.

My having emphasized these three performance elements should not lead the reader to believe that the other elements—structure, texture; orchestration (i.e. timbre, tonecolor, sonority), form, and continuity—are of little or no consequence in the re-creation of a work. Indeed, the ultimate achievement of a performance ought to be the reproduction of a work *in its totality*, that is, as I have postulated it, the retracing of *all* the compositional steps and bringing to acoustic life the myriad interrelated, both small and large, decisions a composer makes in creating a work. Even the smallest interpretive decision must be informed and illuminated by an intelligent and supportable, clearly articulated concept of the larger frame; the small events must have their correct place in the concept of the whole. And in terms of harmonic/melodic understanding, the conductor—and it is hoped through him—the players must have an awareness of the distinction between primary tonal elements in the harmonic frame of reference and the more secondary, perhaps decorative or elaborative, elements. Too many conductors concentrate only on, even exaggerate, these primary tonal elements, which are already all too obvious—having their own inherent prominence and needing no further 'bringing out,'—thereby ignoring less obvious tonal deviations (chromaticism, dissonance, suspension) that are often the most original and daring inspirations of a composer and, in my view, therefore require special intercedence by the conductor and performers.[49] As I have said earlier, it *is* that *strange* new note, or *unusual* voice leading, or *daringly original* orchestration by a Beethoven or a Brahms that may require our special attention, rather than conducting the already obvious.

When all of these interrelated elements are in their right place, meaning the

49. This may not be true in certain types of atonal contemporary works (by say Babbitt or Carter), where the intention and assumption of the composer are that all elements, all voices, are to be seen and heard as having equal prominence and relevance in the totality of the work.

composer's 'right' place, then—and only then—can there be a truly moving, illumined and illuminating performance. For there comes a moment—very rare though it may be—where pure feeling, based on a *complete* understanding of the text, takes over. The whole ensemble is sustained and transported by not even the conductor, but by the inherent content of the music as committed to notation by the composer, speaking directly to us. It is such a performance which, more than merely 'correct,' is *inspired* by the correct reading, and which can then reach heights of sublimity, and be a truly revelatory experience for all concerned: conductor, orchestra, and audience.

Having herein stated my own philosophy of the art of conducting, both as a prescription and a definition, it will be interesting to see in Part II what conductors, composer-conductors, and other writers have written on these matters over the several centuries since conducting developed as a distinct musical discipline.

Part II

A History of Conducting

Imperfection can spoil much more
than perfection can create.
> —Arnold Schönberg

A man paints with his brains and
not with his hands.
> —Michelangelo

Art hath an enemy called ignorance.
> —Ben Jonson

It is instructive and fascinating to study the various treatises, pamphlets, articles, books, and writings on conducting that have appeared through the years since Johann Mattheson published his *Der Volkommene Capellmeister* in 1739, a time when the art of conducting was still in its relative infancy. What is fascinating is how perspectives on conducting shifted over the years and with various authors—like Weber, Berlioz, Wagner, Strauss, Weingartner, Furtwängler, Scherchen, Walter, and most recently Max Rudolf, Frederick Prausnitz, and Kyrill Kondraschin—but also how certain fundamentals of the art of conducting have been perceived as constant and inviolate.

Mattheson (1681–1764) did not take the title of his book (The Complete Capellmeister, with emphasis on the word "complete") lightly. Indeed the title page announces that the book will "give thorough notice of all those things which he who would preside over an orchestra with honor and efficiency must know and must know completely." Mattheson characterized the conductor of his time—at least the ideal conductor—as a broadly educated artist, who was as knowledgeable in literature, poetry, painting, philosophy, and languages as in the various realms of music: harmony, counterpoint, orchestration, composition, and the art of singing. In what I think conveys remarkably modern insights into the aesthetic and psychological aspects of conducting—for example, the conductor's relationship with his musicians, deep commitment to the music of his time, and even the ethics and morality of the profession—Mattheson presented an all-embracing view of the art and profession of conducting that, to my knowledge, has rarely been expressed quite so comprehensively and so penetratingly. In its encyclopedic proportions (nearly one thousand pages in the English translation),[1] this remarkable tome deals with virtually all practical, theoretical and aesthetic precepts which, according to Mattheson, the 18th-century *Capellmeister* needed to understand and command.

Perhaps Mattheson's lofty view of conducting was expressed most challengingly and most succinctly in his dictum: "All sciences [and arts][2] are linked together chain-like in a circle. He who knows only his [one] craft, knows nothing, for he is but a pedant, even were he a general." ("Alle Wissenschaften hängen Ketten- oder Glieder-weise in einem Kreise aneinander. Wer nur allein sein Handwerk weiss, der weiss nichts, sondern ist ein Pedant, wäre er auch gleich ein Feldherr").[3] These same thoughts were echoed nearly a hundred years later by Robert Schumann when, in his "Über Dirigieren" he wrote: "What the composer created out of his inner self must be recognized by the conductor, who can only achieve understanding through vast knowledge. The spiritual greatness of music cannot be apprehended solely by learning the figured bass, or by studying and serving an apprenticeship, but by diligent study of every science connected with music."[4]

In contrast to Mattheson's immense tome, a much slimmer volume by Carl Ludwig Junker[5]—only some forty-eight pages in the original German—published in 1782, some forty years later than Mattheson's work, concentrates specifically on the conductor *as* conductor, separate from the composer. In four chapters, Junker deals brilliantly and succinctly with four aspects of conducting: (1) tuning an orchestra (evidently in the late 18th-century still a new practice, much in need of instruction and training); (2) the placement and disposition of the orchestra on stage; (3) tempo ("Von der Bewegung"), and (4) the politics of conducting (!) ("Von der Politik des Kapellmeisters")—the last a subject which Mattheson had also explored. Junker's most important chapter—and a subject of central concern to him (and us)—deals with questions of tempo, both the establishment of tempos and the flexibility within tempos. But, like Mattheson, Junker also deals with the philosophy, the morality of conducting, and the conductor's relationship with his players. He vigorously upholds the *art* of music and of conducting, deploring all commercialism: making music, playing, conducting merely "for money" ("Überhaupt soll die Kunst nicht nach Brod

1. A very cramped 500 pages in German *Steilschrift* in the original publication. *Der Volkommene Capellmeister* (Hamburg, 1739); in English translation, (Ann Arbor, 1981).
2. In 18th-century terminology the sciences (*Wissenschaften*) included the arts.
3. *Der Vollkommene Capellmeister* p.103. English translation p. 253.
4. Robert Schumann, "Über Dirigieren," *Neue Zeitschrift für Musik*, 1836.
5. Carl Ludwig Junker, *Einige der vornehmsten Pflichten eines Capellmeisters oder Musikdirektors* (Winterthur, 1782).

gehen") [6]—and in several amazing passages we find out that certain aspects of musical life and musicians' behavior haven't changed an iota.[7]

While Berlioz and Wagner also saw the art of conducting as a sacred trust, presupposing the broadest contacts with other arts, they both expressed their views in more specifically stylistic, interpretational and technical, even polemical and critically negative, terms. Berlioz's occasional diatribes against Habeneck[8] and Wagner's more than occasional fulminations against Mendelssohn, while to some extent possibly justified, rather narrowed the discussion from where Mattheson and Junker had left it a century earlier. Of course, Mattheson was not speaking of conductors in the narrower sense, but of the *Capellmeister* as composer, conductor, all-round *Musicus* and artist/scientist/philosopher. By the early 1800s many of the social, professional, aesthetic conditions of music-making, the status of composers and performers, had changed dramatically. Performers, composers and their music had been largely liberated from servitude to aristocratic (and ecclesiastical) establishments. A new professionalism and artistic autonomy were developing, with ever greater, more specific, more complex demands on performers and composers (especially the former). As a result, the composer-conductor as philosopher-scientist—which Mattheson and other earlier writers and theorists had proclaimed as the ideal—was giving way gradually to the more specialized, more narrowly focused artist/professional. Both Berlioz and Wagner emphasized the interpretational aspects of conducting from a later, particularly 19th-century 'Romantic' point of view, in which, oddly enough, literary and poetic inspiration no longer claimed as primary a role as in Mattheson's writing. The bulk of their discourse dealt with specific technical problems (often in relation to specific works), conductorial misdemeanors and bad habits of various kinds, and above all with questions of tempo and tempo liberties— justified and unjustified. The last subject, in fact, runs like a constant refrain through virtually all writing on conducting, from Mattheson, Junker, Berlioz, and Wagner through Weingartner, Strauss, Walter, Toscanini, Jochum, even Bernstein. It is the one subject that almost all writing on conducting, whether prescriptive, analytical, didactic or critical, focuses on—and a subject which to this day still provokes lively debate, especially in regard to Toscanini's legacy as well as the more recent effusions of the 'early music authenticists,' who have

6. "In no way may art be determined by the pursuit of bread," i.e. of money.

7. One of the more fascinating and priceless of these commentaries pertains to the careless, noisy tuning of an orchestra, evidently as much a bane in the late 18th century as in our own time. "Very annoying is the abuse," Junker writes, "so common with many orchestras, of not maintaining the necessary peace and quiet in tuning. Everybody argues, makes noises and runs around; and many, who ought to produce only a single, simple tone for purposes of tuning, indulge at that very moment in all sorts of leaps and cadenzas on their instrument" ("Ärgerlich ist noch der bey so mancher Kappelle gewöhnliche Misbrauch: Daß die bey dem Stimmen schlechterdings nötige Ruhe und Stille nicht zu finden ist, daß alles räsoniert, und lauft, und mancher der blos den simplen Ton der Zusammenstimmung angeben sollte, sich zu eben der Zeit, allerley Sprünge und Cadenzen auf seinem Instrument erlaubt"). *Plus ça change!*

8. François Antoine Habeneck, French conductor, the first to conduct Beethoven's symphonies in France.

now gone way beyond the late Baroque and early classic literature to invade the Romantic Beethoven, Schumann, and Brahms repertory.

It is illuminating—and ought to provide persuasive instruction to conductors of our time—that there is in almost all of the earlier writing a remarkable consensus on the subject of tempo flexibility, what Beethoven early on called "elastischer Takt" ("elastic time" or "elastic beat"). This consensus is in striking contrast to the mostly oversimplified and polarized discussions one encounters in many quarters nowadays. Moreover, it is often confused with another contentious subject, the question of *textual fidelity*. The proponents of faithfulness to the score—a philosophy Toscanini epitomized in three words: "come è scritto" ("as written")—tend to equate rigorous adherence to the score with relentless steadiness of tempo, when in fact the two approaches do not necessarily have anything to do with each other. Textual fidelity does not imply, let alone dictate, rhythmic rigidity—even though, of course, some misguided conductors, performers, and critics may make that assumption.

Let us examine the question more extensively, and try to establish once and for all what exactly is meant by 'flexible tempo,' 'textual fidelity,' what the originators of these concepts actually had in mind and what, therefore, we ought to use as our guidelines as interpreters of 19th-century orchestral literature. But first let us dispose of the simplistic and polarized arguments one has encountered so often, in which one side is accused of being pedantically rigid in tempo continuity, while the other side is counter-accused of being willfully free. Epithets are flung around to bolster each side's arguments: 'intellectual,' 'cold,' and 'unfeeling' for the one; 'indulgent,' 'permissive,' 'overly emotional' for the other. Parties on both sides of the debate set up straw men—most commonly Toscanini on one side, Furtwängler on the other—who become easy targets for polemical attacks.

Leaving aside the two major types of conductorial mediocrities, (1) those whose lack of talent prevents them from realizing a flexible 'tempo of feeling,' and (2) those who cannot maintain *any* semblance of a coherent tempo control (citing the worst examples never gets a debate anywhere), we find when we listen carefully and without prejudice to the best conductors representing each side, say, Toscanini and Furtwängler, that the former was never as inflexible as his detractors contend(ed) or as textually faithful as his admirers maintain(ed), while the latter was not always as freely indulgent as the anti-Furtwängler faction would have us believe, or as profoundly expressive as *his* adoring public claim(ed). Both conductors had too much respect for composers and their scores either to rigidify the flow of the music or sacrifice all structural coherence. What they actually did interpretively depended a great deal on the repertory being conducted. Both had remarkable strengths and weaknesses, and strong and weak areas of repertory.

Sometimes similar arguments of pro-free and pro-strict tempo are cast in terms of earlier conductor generations (usually meaning German 'Romantics' such as Muck, Nikisch, Mahler, Mengelberg, Pfitzner) vis-à-vis younger post-Toscanini generations (Reiner, Szell, Leinsdorf, Steinberg, Rodzinski). The assumption is usually that those German idealists, in the tradition of Wagner and Bülow, indulged in excessively subjective tempo liberties, as well as slow, sluggish, heavy basic tempos, while the younger mid-century conductors, much influenced by Toscanini and his phenomenal success, were the exponents of controlled, objective literalism, absolute precision, and streamlined bright tempos.

But these are, again, facile oversimplified generalizations that do not always stand up to closer scrutiny. The actual evidence of performances and recordings suggests that placing the above-mentioned conductors in two opposing camps turns out to be mostly inaccurate, misleading, and unfair. The evidence rather suggests that there was considerable variability and diversity of approach on *both* sides, and that a truer picture can only be gained in a case-by-case analysis, i.e. conductor-by-conductor, even work-by-work. The assertion that there were two monolithic, lock-step schools of conducting is simply not tenable. Just as Toscanini's performances reveal countless examples of subtle tempo modifications (largely informed by his Italianate, opera-influenced 'singing' conception), so also many of his best disciples were skillful practitioners of a judicious *tempo rubato*. On the other hand, listening to a broad sampling of performances and recordings by the 'German Romantic' school of conductors, one will find—surprising, I am sure, to many a reader—plentiful examples of strict tempo maintenance as well as astonishingly lively basic tempos.

No, the arguments over free tempo versus strict tempo, when reduced to the twin polarizations of an older, earlier (subjective) vis-à-vis a younger, newer (objective) generation are simply not tenable. Nor are the similar arguments when cast in Old World versus New World terms. For who conducts in slower, stretched-to-the-breaking point tempos nowadays than Giulini; or who was given to more *rubato* excesses than Bernstein; or who has so consistently twisted Furtwängler's *elastischer Takt* into trivializing mannerisms than Barenboim? And who was, on the other hand, more 'classically' consistent in matters of tempo and pulse than Monteux or Erich Kleiber or even Weingartner?

Perhaps it is time to examine what exactly the original mentors of these two (allegedly) opposing philosophies of tempo conception expressed. To a large extent it really all started with Beethoven, who was one of the first to use the terms "elastischer Takt" and "Gefühlstempo"(tempo of feeling). But even before Beethoven's time, we have indications that tempo fluctuations within movements were beginning to be considered a *sine qua non* of good performance. Even Mattheson, writing as early as 1739, dwells on the subject (although less copiously than later writers), presumably because in his time—the early 18th century—tempo modifications were not yet a consistent practice (except perhaps in vocal music). This is to some extent confirmed by the fact that orchestras were then presided over by either the leader of the first violins or the resident harpsichordist (or both). It was, in fact, the early 19th century's liberation from the previous classic and pre-classic forms and tempo conceptions that led to the decisive establishment and absolute need of baton conducting, by a conductor not sitting *in* the orchestra. Indeed, Baroque music, most importantly Johann Sebastian Bach's, was still so closely tied to popular dance forms set in steady, virtually danceable tempos that, again, there was no tremendous need for a separate conductor and time-keeper.

Nonetheless, Mattheson does allude briefly to the subject of tempo and tempo flexibility. After establishing the difference between the "measure" and the "beat" within a measure, on the one hand, and "movement," that is, tempo, on the other hand, he points out that the latter is commonly modified by adjectives like *affettuoso, con spirito,* adding somewhat cryptically, but nevertheless sagely, that "there would be more to understand through such descriptions than is written."[9] Moreover, movement (tempo)—what Junker some forty years later would call *Bewegung* (motion)—to quote Mattheson again, "can hardly be contained in precepts and prohibitions: because such depend primarily on the feeling and emotion of each composer, and secondarily on good execution or the sensitive expression of the performer." Mattheson explains that the mere indications of *allegro, lento, adagio, vivace* are too general to give precise tempo definitions. "Here each one must reach into his own soul and feel what is in his heart, since then . . our performing will to a degree acquire an extraordinary or uncommon movement, which neither the actual mensuration [meter] in and of itself nor its perceptible holding back or accelerating can impart, but which stems from an *imperceptible* impetus" (italics Mattheson's). "The effect is observable, but without knowing how it happened."

Mattheson continues: "I say observable because essentially the melody will be more or less altered in its subtle movement [tempo], appearing to be either faster or slower. But nothing appreciable will be taken away or added to the mensuration or the note values."[10]

Mattheson then praises and cites Jean Rousseau's[11] *Méthode claire, certain et facile pour apprendre à chanter la musique* (A clear, sure, and easy method to learn to sing music), published in 1678. Reiterating and confirming his notion of tempo flexibility, Mattheson quotes Rousseau: "Within the same mensuration, the tempo may turn out quite differently: for it is sometimes more lively, sometimes more languid, according to the various passions which one is to express. Thus it is not enough for the performance of a piece of music to give the beat and maintain it according to the prescribed tempo indications, but the conductor must also guess the meaning of the composer, that is, he must feel the various impulses which the piece is intended to express."[12]

Mattheson's own last word on the matter—in chapter 26, his final chapter—is: "Giving the beat is the main function of the conductor in performance. Such beat-giving must not only be done precisely, but as circumstances may require . . . the conductor can and should make little changes in the tempo, delay the pulse, yield, or in consideration of a certain specific feeling [*Gemüthsneigung*] and other reasons, accelerate the beat somewhat and drive it harder than previously."[13]

From both Mattheson's and Rousseau's statements it can be seen that Beetho-

9. Mattheson, *Der Vollkommene Capellmeister,* p. 171; English edition, p. 365.
10. Ibid., p. 183 (p. 367).
11. The Jean Rousseau referred to by Mattheson is not the more famous composer-encyclopedist-author-philosopher Jean-Jacques Rousseau, but an earlier Jean Rousseau, the foremost French gambist of the late 17th century, whose here mentioned *Méthode,* republished no less than six times (the last in 1707), was of considerable influence on the European continent.
12. Ibid., p.173 (p.368).
13. Ibid., pp. 481–82 (p. 866).

ven's "tempo of feeling" and "elastic beat"(*tempo rubato*) were not entirely unknown even as early as the second quarter of the 18th century, that is, in Bach's
and Handel's time. But by the end of the 18th century we begin to have indications that tempo conceptions were loosened even more to allow for a greater
degree of tempo modification. Such tendencies centered on the so-called *Affek-
tenlehre* (literally, the teaching of affections),[14] which by the late 18th century
developed into a musical aesthetic with a far-ranging influence on interpretive
practices. Eighteenth-century theorists, such as Johann Samuel Petri and
Friedrich Wilhelm Marpurg, began to speak of the expression of 'affections'
affecting the tempo *during* the course of a composition or movement, on the
assumption that, since music should closely reflect its emotional content, and
since no two emotions or passions can be alike, the tempo cannot be metronomically consistent throughout a work. Accordingly, the tempo must be modified
to reflect the various contrasting 'affections' or moods: sad, melancholy, angry,
energetic, joyous, and merry.

Both Johann Joachim Quantz and Leopold Mozart, both highly influential
composer/theorist/teachers in their time, also proclaimed the new tempo affections. The elder Mozart writes: "Jedes Tempo, langsam oder schnell, hat
seine Schattierungen"[15] ('Each tempo, slow or fast, has its gradations'). Quantz
went so far as to state that "an absolutely consistent tempo is nonsensical."[16]
Daniel Gottlob Türk, too, in his *Clavierschule* (1789), writes extensively about
tempo modifications.

Unfortunately we have little detailed evidence of how Haydn and Mozart
conducted or—more precisely—how, in their conducting[17] and interpretations

---

14. The German word *Affekt* combines into a single concept the terms "affecting" and "emotion".
15. Leopold Mozart, *Versuch einer gründlichen Violinschule* (Augsburg, 1756), p.30; English edition
translated by Editha Knocker, *A Treatise on the Fundamental Principles of Violin Playing* (London,
1948), p. 33.
16. Johann Joachim Quantz, *Versuch einer Anweisung die Flöte traversiere zu spielen* (Breslau,
1752), p.261. The official English translation reads "absurd and impossible." "Ich verlange nicht,
daß man ein ganzes Stück nach dem Pulsschlag abmessen sollte; denn dieses wäre ungereimt und
unmöglich."

Quantz, surprisingly, wrote relatively little about *tempo rubato*—he actually mentions the term
only once (p.146)—but much of what he teaches in his *Versuch einer Anweisung* carries such an
implication, in an attempt to arrive in performance at the true expression and 'affect' (*Leydenschaft*)
of the music. Perhaps, because of an increase in tempo and expressive liberties by performers in his
time, and, conversely, a worrisome tendency among younger musicians as well as professionals being
unable to *maintain* a good tempo, Quantz spends much more time and efforts on those concerns.
17. Haydn and Mozart 'conducted' mostly, if not entirely, from the keyboard or the *Konzertmeister*
position. As Adam Carse pointed out in his splendid study *The Orchestra in the 18th Century* (Cambridge, 1940), "conducting an orchestra, as we understand it now, was unknown in the 18th century.
Misunderstanding easily arises when the word 'conducting' in 18th century literature is interpreted
in its present-day sense, and is associated with the use of the baton." Mozart, Carse adds, "did not
hand over the baton to another conductor," as some 20th-century writers have suggested (see, for
example, Annette Kolb, *Mozart* [Chicago, 1956], p. 340), for "he had none to hand over" in that
he was leading from the harpsichord or the pianoforte.

Carse also cites a letter by Mozart, dated October 19,1782, in which he writes that when he felt
the orchestra in Vienna in his *Entführung* was getting a bit sleepy, he decided to resume his place
"at the Klavier and conduct it."

of their own works, they dealt with the question of tempo modification. But we do have clear indications from Mozart in his letters that he featured a pronounced *tempo rubato* in his piano playing, a *rubato* which, it turns out, is the same one we associate with Chopin.[18] In a letter to his father dated October 24, 1777, Mozart writes: "In *tempo rubato* in an *adagio*, the left hand should go on playing in strict time, where the left doesn't know anything about it" ("dass die linke Hand nichts darum weiss"), the implication being it was for the sake of expression.

It is clear from this and other evidence that the *Tempo des Gefühls* was not a 19th-century invention but a fairly well-established interpretative practice long before that.

Junker's aforementioned handbook on conducting, in a chapter entitled "Von der Bewegung" ("About Tempo"),[19] also makes it quite clear that earlier prac-

---

As for Haydn, we have Johann Nicolaus Forkel's word (in his *Musikalischer Almanach*) that Haydn at Esterhazy "spielt zugleich die erste Violine" ("at the same time he plays the first violin"), meaning that he led his symphonies with the violin; while at Salomon's concerts in London, we know that Haydn was engaged to "preside at the piano for his new symphonies."

18. Chopin characterized his concept of *tempo rubato* as follows: "The singing hand may deviate, [but] the accompaniment must keep time. . . The graces are part of the text, and therefore part of the time. . . Imagine a tree with its branches swayed by the wind; the stem represents the steady time, the moving leaves are the melodic inflections. That is what is meant by *tempo* and *tempo rubato*" (cited by Edward Dannreuther in his *Musical Ornamentation*, II, London, 1895, p.161).

19. Junker, *Einige der vornehmsten Pflichten eines Cappellmeisters oder Musikdirectors*, pp. 20–43. Junker was, like Mattheson, versatile and broadly trained, a prolific writer and composer. Although a minor figure in the latter realm, he was otherwise extraordinarily versed in a wide range of subjects, including theory, aesthetics, philosophy, the visual arts and iconography, *belles-lettres*, world populations, as well as being an avid chronicler of music and music-making in his time. Thayer in his biography of Beethoven includes an important account written by Carl Junker of the young Beethoven in a 1791 performance with the court orchestra of the Elector of Cologne. In a communication to Bossler's *Musikalische Correspondenz*, dated November 23, 1791, Junker wrote (excerpted here): "I also heard one of the greatest of pianists—the dear, good Bethofen. I heard him improvise in private; yes, I was even invited to propose a theme for him to vary. The greatness of this amiable, light-hearted man, as a virtuoso, may in my opinion be safely estimated from his almost inexhaustible wealth of ideas, the altogether characteristic style of expression in his playing, and the great execution that he displays. I know, therefore, not one thing that he lacks that conduces to the greatness of an artist. Bethofen, in addition to the execution, has [great] clearness and weight of idea, and [much] expression. In short, he is more for the heart—equally great, therefore, as an *adagio* or *allegro* player. His style of treating his instrument is so different from that usually adopted, that it impresses one with the idea that, by a path of his own discovery, he has attained that height of excellence whereon he now stands" (cited in Thayer's *Life of Beethoven* [Princeton, 1964], pp.104–5).

I first came upon Junker's splendid little volume in the Library of Congress in the late 1940s. Not as yet involved with conducting, much of Junker's book seemed to me at the time, although fascinating, of little practical relevance. More recently I have, however, come to cherish this rare and early documentation of conducting practices, and was particularly pleased to note that Richard Taruskin a few years ago quoted Junker extensively ("Resisting the Ninth," *19th Century Music*, Spring 1989), pp. 252–54), at the same time correcting Paul Henry Lang's earlier misquotes and/or out-of-context citations of Junker on the subject of steady tempos in an article on Beethoven symphonies. (An English translation of Junker's handbook is being prepared and will soon be published in this country.)

tices of regular, steady tempos, to the extent they existed, were being questioned and challenged. Indeed, Junker gives the subject of tempo flexibility a thorough treatment, with thoughts that are—and therein lie their importance—echoed decades and even centuries later by, among others, Beethoven, Wagner, Strauss, Weingartner, and Walter. Junker initiates the subject of tempo by the obvious suggestion that it is the composer who determines the tempo of a work or a movement: "whether it is lively or slow" ("in so ferne[die Bewegung] geschwind oder langsam ist").[20] "But even if the composer," so writes Junker, "can indicate various tempo gradations within these basic lively or slow tempos, how many modifications lie nonetheless between the two which he cannot indicate because he has as yet no notation (*Charaktere*) for them."[21] Junker continues: "For the conductor . . . an *allegro* cannot be tied to a single all-embracing fast tempo concept (*Begriff von Geschwindigkeit*), just as an *adagio* cannot be tied to a [single] slow one. The precise determination of tempo rests finally on good taste, rests on its own feeling of rightness" (literally truth [*Wahrheit*]), "which can only be fixed (*fixirt*) through previous study of the score."[22]

After urgently recommending the use of the score, a still very rare practice in the days when performances were more often than not led by the concertmaster or harpsichordist, reading respectively from a violin or continuo part, Junker characterizes one "who would conduct from a score without knowing it well, which he has not studied, which he has not read through thoroughly," is a "windbag (*Windbeutel*)."[23] At another point Junker puts the question this way: "Must every piece be performed through to the very end at the stated tempo, never even approaching greater speed or slowness? Or might this tempo, even in the middle of a piece, be slightly modified, might it be accelerated, might it be held back?"[24]

Junker is not satisfied to have these concepts remain as questions. He answers, it seems to me, with remarkable precision and succinctness. "To answer the first of the two questions positively without qualification would mean to deprive the art of music of one of its most powerful means of expression and emotion (*Rührung*); and would remove from her [the art of music] all possibility of different gradations and modifications of expressive movement (tempo, *Bewegung*). To answer the second question positively without qualification would have the river overflow its banks, would cause a thousand disorderlinesses (*Unordnungen*), and would deprive music of its truth. But as soon as the last sentence"—meaning Junker's second question—"is qualified and limited, it can be answered in the positive."[25]

20. Junker, *Einige der vornehmsten Pflichten* . . ., p. 20.
21. Ibid., p. 21.
22. Ibid., p. 21.
23. Ibid., p. 22.
24. Ibid., p. 36. "Muss jedes Stück, ganz bis zu Ende, in der nemlichen Bewegung, die sich niemals, weder einer grössern Geschwindigkeit noch Langsamkeit nähert, vorgetragen werden? Oder darf diese Bewegung, selbst in der Mitte des Tonstücks, etwas abgeändert, darf sie beschleunigt, darf sie zurück gehalten, werden?"
25. Ibid., pp.36–37.

Junker nails his point down by stating that "both composer and performer [conductor] must work hand in hand, and that modification of tempo as a second art remains absolutely necessary."[25] Finally, he adds: "Furthermore, as little as I hold with the use of an unqualified *tempo rubato* (because it is so often misused to produce an ear splitting racket [literally "clanking," *halsbrechendem Geklirr*], I nonetheless insist that in a good orchestra tasteful tempo variations are in order where appropriate, and where under certain circumstances they can have a good effect."[26]

Refining his point even further, Junker writes: "We admit that there is in these matters much opportunity for misuse. But we have made these suggestions of tempo flexibility only with the most *appropriate and scrupulous qualifications and limitations* (italics added). We say it again: only to him who is in all respects a reliable virtuoso can such *imperceptible* modifications of musical progression be entrusted; that such modifications are permissible *only as a requisite of the composition, not as a need of the performer*" (again italics added) (*als Bedürfnis des Stücks selbst, nicht als Bedürfnis des Spielers*). Junker adds very wisely that "just as slight expressive nuances can never be applicable to the same degree in an *allegro* as they might be in an *adagio,* so too the tempo rubato is less applicable and true in an *allegro* than an *adagio.* An *adagio,*" Junker reasons, "offers the performer more beauties in detail (*Schönheiten des Detail*), more nuances, than an *allegro.*"[27] To bolster his arguments Junker suggests that "there is no passion [feeling] whose movement would be so circumscribed as to be absolutely regular [uniform]; it constantly ranges through various modifications of tempo"[28]—echoes of Petri and Marpurg.

Thus, apart from being one of the earliest full discussions of the subject of *tempo rubato,* it is also one of the clearest and most detailed. In so far as Beethoven and Wagner (as well as later conductors) echoed Junker's thoughts, including his warnings against excesses and abuses of tempo modifications, one wonders whether these composer-conductors knew Junker's writings. Perhaps not; but in any case, the later authors confirm, both in conception and detail, what is one of the most important aesthetic-philosophical tenets of the art of conducting, one which is, alas, nowadays widely misunderstood and/or abused.

The publication date 1782 of Junker's pamphlet is the same year the twenty-six-year-old Mozart composed his *Die Entführung aus dem Serail,* as well as three of his five early piano concertos, and the same year Haydn composed his six Op. 33 string quartets and symphonies N° 76 through 78. One wonders to what extent Junker was influenced in his writings by observing the performing practices of Mozart or Haydn in Vienna, or Carl Philip Emmanuel Bach in Berlin, or indeed, Stamitz and company in Mannheim—just as he observed Beethoven as a pianist.

Interestingly—and seemingly contradictorily—Wagner, nearly ninety years

26. Ibid., p. 38.
27. Ibid., pp. 39–40.
28. Ibid., p. 37. "Es gibt keine Leydenschaft, deren Bewegung, sich selbst immer gleichartig, abgezirkelt seyn sollte; sie walzt sich durch verschiedene Modifikationen der Bewegung hindurch."

after the publication of Junker's pamphlet, wrote in his *On Conducting* about Mozart's "naive *allegro*" (or at other times "absolute *allegro*"), contrasting it with Beethoven's newer, expanded *allegro* concept, which Wagner called "sentimental." (Wagner explained that his terms "naive" and "sentimental" were adopted from a well-known essay by Gottfried Schiller about "naive and sentimental poetry," the terms equating in 20th-century English more with 'simple' and 'expressive' respectively).[29] Wagner was obviously under the impression that Mozart's (and presumably Haydn's) *allegro* movements, including his opera overtures—Wagner especially mentions the *Figaro* and *Don Giovanni* overtures—were performed in one steady, relentless tempo, and very fast. He cites Mozart's remark that these overtures "cannot be played fast enough," corroborating the point further with the well-known anecdote about Mozart, in rehearsal of the *Figaro* Overture, having finally driven his desperate musicians to achieve the full desired *presto*, then commended them encouragingly: "That was beautiful! But this evening still a little faster!"

We shall return momentarily to Wagner's pronounced views on *tempo rubato*, only to note now in passing that he clearly ascribed this concept to Beethoven and his innovative symphonic masterpieces, which he felt could no longer be played in one tempo per movement, in the manner of what he somewhat pejoratively called Mozart's "naive *allegro*" and what Beethoven had already called the "tempi ordinari" of the "barbarous period of music."

As for Beethoven himself, there is abundant evidence that he considered the *tempo rubato* a *sine qua non* of high-level music-making in his time. And many contemporaries of Beethoven have testified to the notion that for Beethoven metronome markings were valid "nur für die ersten Takte, weil Gefühle ihr eigenes Tempo haben" (only for the first measures, as feelings have their own tempo).

The music director of Vienna's Theater an der Wien, Ignaz von Seyfried, testified that Beethoven "was very particular about expression, the delicate nuances, the equable distribution of light and shade as well as an effective tempo rubato. . . ."[30] Beethoven's friend and biographer Anton Schindler in his *Life of Beethoven*[31] went so far as to annotate a 21–bar section of the second movement of Beethoven's Second Symphony with various *poco accelerandos*, *poco lentos*, *tempo I's*, and additional crescendos, reflecting what he claimed the master did when he conducted the work himself. If what Schindler recalled of Beethoven's performance is accurate—and we know that Schindler was not always a reliable witness—then Beethoven did indeed apply the 'tempo of feeling' to his own works. But before we allow ourselves to indulge in every arbitrary tempo whim, justified and vindicated by no less than a Beethoven, let us note the subtlety of Schindler's annotations and how frequently *poco* occurs in his transcription—a point to which I will have reason to return again and again.

Carl Maria von Weber was one of the most brilliant of the early conductors,

29. Wagner, *Über das Dirigieren* (Leipzig, 1869), p. 31.
30. Ignaz von Seyfried, *Beethoven Studien* (1832), cited in Sonneck, *Beethoven Impressions of Contemporaries* (New York, 1926), p. 41.
31. Anton Schindler, *Life of Beethoven* (Münster, 1840), p.164–65.

and also one of the first, along with Spohr and Spontini, to use the baton.[32] He was very much an advocate of flexible tempo and the flexible beat, not only clearly implied in his operas and chamber music works, but also lucidly revealed in a letter to the violin virtuoso-composer-conductor Heinrich Aloys Präger. As quoted by Felix Weingartner in his *Über das Dirigieren*, Weber wrote to Präger:[33] "The beat (the tempo) must not be a tyrannically impeding or driving hammer [*Mühlenhammer*], but rather [should be] to a piece of music what the pulse-beat is to the life of man. There is no slow tempo in which there aren't passages that demand a quicker motion, in order to prevent a feeling of dragging. Conversely, there is no *presto* that does not need at some points a quieter delivery, so as not to preclude, through rushing, the means of expressiveness."

"What I have here said should not, for Heaven's sake, give any singer"—to which Weingartner appends a footnote to say "the same, of course, goes for conductors"—"the right to the performance lunacy of willfully distorting individual measures and thereby produce in the listener the unbearable sensation akin to seeing a contortionist forcibly contort all his limbs. The forward movement of tempo as well as the holding back of tempo, both should never produce a feeling of jerking the tempo around, of moving abruptly and forcibly by fits and starts [*das Gefühl des Ruckenden, Stoßweisen oder Gewaltsamen erzeugen*]. In other words, [tempo modification] in a musical-poetic sense can only occur in terms of phrases and periods, informed by the particular expressions of passion and emotion [*Leidenschaftlichkeit*]."

"For all of this we have in music no means of notation. These lie solely in the feelings of man's heart [*Menschenbrust*]: and if they cannot be found there, then neither the metronome, which can only prevent the crudest of blunders, will help nor will such at best incomplete indications as I might be prepared to incorporate to enrich my material [my notation], were I not warned against this by many experiences as a result of which I am forced to consider these already as superfluous and useless, and fear them as being misrepresentative." For all that, Carl Maria von Weber did very much advocate the use of metronome markings. In speaking of the *Te Deum* of his composer-theoretician colleague Gottfried Weber[34] (no relation), he wrote "It is very much hoped that other composers will follow Herr Weber's lead in this matter."[35]

---

32. The use of a baton in conducting did not take hold as a consistent, common practice until the third decade of the 19th century. Before that, performances were led by musicians—more often than not the composer of the work being performed—either from the keyboard or the concertmaster position (sometimes both simultaneously), or with the director or *Kapellmeister* using a variety of methods, implements, devices, tools and instruments: violin bows; three- or four-foot-long batons, used to keep time by stamping them loudly on the floor; rolls of paper (silently or, when struck together, audibly); various-sized pieces of wood; diverse vocal sounds (grunts, hisses, etc.); rapping on the music stand; and, of course, foot stamping.

33. Felix Weingartner, *Über das Dirigieren* (Leipzig, 1895), p.80–89; (1905), p.43.

34. Gottfried Weber's *Te Deum* (1814) seems to have been the first published score to use metronome tempo markings; and in 1817 he published an article, "Über chronometrische Tempo Bezeichnung" (On chronometric tempo indications).

35. Carl Maria von Weber, *Writings on Music*, English edition, Martin Cooper, ed. John Warrack (Cambridge, 1981) p. 128.

Weber here is anticipating the same concerns expressed by Wagner a few decades later in his *Über das Dirigieren* (see below), in which he repeatedly warns against the excesses of arbitrary and unfounded, egotistically motivated distortions of tempo, thereby in time "totally mutilating our great classical music beyond recognition,"[36]—a prediction which, alas, came true in the Bülow and post-Bülow era.

While Berlioz mentioned matters of tempo only briefly in his *L'Art du chef d'orchestre*, Wagner, as we shall see, dwelt on the subject at considerable length. Berlioz was greatly interested in composers' adhering to metronome markings, and urgently recommended that conductors "consult the metronome indications and study them thoroughly." But then he added quite judiciously: "Naturally I do not mean to say by this that one should imitate the mathematical regularity of the metronome; all music executed in such a manner would have an icy frigidity and stiffness"; and even more wisely: "I even doubt whether it would be possible to maintain such empty uniformity for more than a few measures."[37] And finally, and perhaps most important: "The metronome is nevertheless an excellent aid in determining the initial tempo of a piece *and its main shadings*" (italics added).[38]

Wagner echoed the same thoughts, but put them even more forcefully. (Since Wagner's thoughts in *Über das Dirigieren* have been frequently misread, misunderstood, and even mistranslated, I intend to deal extensively—not selectively, as has often been done—with the fullness of his comments on questions of tempo and tempo modification.) After dealing with the general question of "the right tempo" at some length, declaring that this question "is the point at which it becomes clear whether a conductor is suited or unsuited to the task,"[39] he moves on to the subject of *tempo rubato*, declaring that tempo modification is "of immeasurable importance for the proper rendering of our classical music".[40] Wagner then cites a number of works—the first and last movements of Beethoven's *Eroica*, the *Egmont* Overture, Weber's *Freischütz* Overture, his own *Meist-*

36. Wagner, *Über das Dirigieren*, p. 54.

37. Hector Berlioz, *L'Art du chef d'orchestre* (Paris, 1844). English edition, *The Orchestral Conductor: Theory of his Art*, Mary Clarke, (London, 1856).

38. Ibid. Berlioz wrote about conducting in generally broader terms. How thoroughly he understood the new aesthetics of conducting—new as opposed to the older, merely time-beating functional approach—can be gleaned from his *Mémoires*, (1869) in which he listed the "qualities necessary to produce a good conductor" as "precision, flexibility, passion, sensitiveness and coolness combined, together with an indefinable subtle instinct" (Hector Berlioz, *Mémoires de Hector Berlioz*, Paris, (1869); English translation by Rachel and Eleanor Holmes, New York, (1932), p. 199). Berlioz was one of the first to emphasize in writing that the beat in conducting had to acquire, apart from its time-beating function, an expressive/aesthetic significance, combining the technical with the expressive.

39. Wagner, *Über das Dirigieren*, p. 28. Indeed, Wagner's book has become a polemicist's football, 'interpreted' to support whatever theories or practices of conducting a particular writer wants (wanted) to defend. This becomes all the more possible in the absence, obviously, of not only any recordings by Wagner but also surprisingly few accounts of his conducting and interpretations. It is only through a strict and correct reading of his text that Wagner's very detailed writings can be properly understood and appreciated.

40. Ibid., p.39

ersinger Overture (then only a few years old), as well as several of Beethoven's string quartets and sonatas—detailing how tempo modifications can be applied to these works, and are, in his view, absolutely essential to their correct interpretation. Several of these are especially interesting for our purposes here. For one thing, since Wagner's conducting and tempo conceptions have, as mentioned, often been misunderstood or misinterpreted in order usually to support or attack one theory of conducting or another, it will be useful to consider very carefully what precisely he *did* say. For another, the particular works I have chosen to single out here—Weber's *Freischütz* and Wagner's own *Meistersinger* overtures— will give us specific and detailed insights not only into Wagner's concept of *tempo rubato* but, in the case of the *Meistersinger* Overture, into the very construction of the work, and thus how he intended it to be performed in respect to tempo considerations.

Taking off from his earlier point, that is, of differentiating between Mozart's "naive *allegro*" and Beethoven's "sentimental *allegro*," and ascribing to these two types of inherently different expressive characteristics and, therefore, of performance needs, Wagner details, among other things, how in particular the *allegro* section of Weber's *Freischütz* Overture ought to be paced. Using one of his own performances (in Vienna in 1864) as a model for, as he claimed, the "correct interpretation" of the work, Wagner states that, "after thus restoring to the introductory *adagio* its frighteningly mysterious dignity, I was able to allow the wild *allegro* tempo to run its passionate course, without being bound in any way to consider the more delicate expression of the [later] softer second subject; for I had full confidence that I would be able at the appropriate moment to slacken [modify] the pace just enough to arrive imperceptibly at the right tempo for this theme."[41] Let us take special note in passing of the word "imperceptibly" (*unmerklich*). Wagner now digresses slightly to point out that "it should be quite evident, that the newer type of *allegro* movement consists of a combination of two essentially different parts. In contrast to the older naive or pure *allegro* structuring, it is this combining of the pure *allegro* with the thematic uniqueness of the songful *adagio*, in all its possible gradations, that so enriches our new modern *allegro*."[42] Wagner cites another Weber excerpt, the second subject in the *allegro*—the clarinet theme (Ex. 1)—of the *Oberon* Overture as a prototypical

Ex. 1

example of a passage that no longer fits into the earlier *allegro* type nor, by implication, *allegro* tempo. He elaborates further: "On the surface this songful theme notationally appears to fit neatly into the *allegro* scheme; but as soon as its true character is identified, it becomes apparent to what extent this scheme is capable of modification in order to accommodate *the composer's desire to have*

41. Ibid., p. 43. Translations of excerpts from Wagner's *Über das Dirigieren* are by this author.
42. Ibid., p.43.

both main characteristics equally represented"[43] (the emphasisis is Wagner's). It should be noted that, here again, Wagner expresses himself very carefully and precisely. His phrase "Welcher Modifikationen dieses Schemas eben fähig gedacht sein mußte" is so cautiously worded as to be almost convoluted; similarly, the phrase "beide Hauptcharactere gleichmässig verwendbar dünken zu können." Wagner is trying to make it clear that, given all his qualifiers, the tempo modification he suggests are to be slight, modest, so as not to impair the balance between the two "main characteristics." The implication here is clearly that exaggerated alterations of tempo not only are not needed but clearly not wanted.

Returning to his *Freischütz* performance in Vienna, Wagner reports that "after the utmost excitation of the [initial *allegro*] tempo, I used the *adagio*-derived long-sustained song of the clarinet (Ex. 2) imperceptibly to hold back the tempo here, where all figurative movement is dissolved into sustained (or trembling) [*zitternden*] sounds, so that, despite the renewed motion of the connecting figure

Ex. 2

(Ex.3) which so beautifully prepares the *cantilena* in E♭, the arrived-at tempo

Ex. 3

was only the slightest nuance removed from the basically never-relinquished main tempo. . . . The success of this rendition was immediately so apparent to the excellent musicians [in the orchestra] that for the, once again, imperceptible reanimation of the tempo with its pulsating figure [Ex. 4], it took only the slightest indication of the pace to find the orchestra enthusiastically ready for the return of the energetic drive of the main tempo with its succeeding *fortissimo*."[44]

Ex. 4

Notice again the use twice of the word "imperceptible," as well as the terms "slightest nuances" and "basically never-relinquished main tempo," and the entire sense of the passage to indicate that both the main *allegro* tempo and its imperceptibly more relaxed second subject counterpart are nothing more than slight variants of essentially the same tempo. It is as if Wagner imagines the conductor to sit on the fulcrum of a tempo seesaw, and with the slightest tilt to

43. Ibid., pp. 43–44.
44. Ibid., pp. 44–45.

one side bring forth the full energetic *allegro,* and with an equally slight tilt to the other side relax the tempo enough to accommodate the songful *cantilena* of the second subject, without losing the fundamental essence and feeling of the basic *allegro.* This is, of course, what Beethoven meant by "tempo of feeling." It is also what others, including Wagner, have called reflecting the spirit—not merely the letter—of the music.

In some ways, Wagner's performance suggestions for his own *Meistersinger* Overture are even more interesting and instructive than his *Freischütz* exegesis. For here we see not only a splendid example of how Wagner's concept of tempo modification was meant to function in performance, but also how Wagner, especially in the later operas, built this concept right into the compositional, creative process; in other words, tempo rubato is composed into the very fabric and structure of the music. Nowhere is this more apparent and crucial than in Wagner's last opera, *Parsifal,* in which, in his own performance annotations in reference to tempo changes and modifications, he consistently uses the words "nicht auffallend" (hardly noticeable) and "unmerklich" (unnoticeable).

Wagner begins by explaining that the basic tempo of the *Meistersinger* Overture is marked "Sehr mässig bewegt," translated into the older (Italian) nomenclature: *allegro moderato.* He continues: "No tempo is more in need of modification, especially in longer time spans and when the thematic material is treated in a strongly episodic manner. It is a tempo often chosen for the expression of diverse motives in manifold combinations, because its broad structure in a regular 4/4 meter easily supports such expression through the merest suggestion of tempo modification.[45] In addition, this moderately moving 4/4 meter is certainly the most multidimensional. When beat in strongly animated quarter-notes, it can express a real lively *allegro* such as my here-used main tempo, which is presented in its liveliest form in the transitional eight bars [Ex.5], leading

Ex.5

from the basic march theme to the E major section. But one can also think of it as a half period, combined out of two 2/4 bars, thus allowing for the introduction of a lively *scherzando* at the entry of the shortened theme [Ex.6]. It can

Ex.6

45. Wagner's mid-19th-century German is quite dense and grammatically convoluted, and thus not easily translated. In my translations here I have retained as much of Wagner's prolixity and complexity of thought as possible—difficult though it may be to read—on the grounds that inherently complex thoughts cannot be reduced to elementary simplicities, and also to preserve in its purest form and greatest precision Wagner's actual thought.

even be interpreted as an *alla breve* (2/2 measure), where it can express the older, real *gemächlich tempo andante* (mainly associated with church music), which is conducted with two moderately slow beats. It is in this sense that I used the latter tempo, starting in the eighth measure after the return of C major, to combine the main march theme, carried by the basses, with the broadly intoned or sung second subject, now stated in rhythmic diminution in the violins and cellos [Ex. 7]."

Ex. 7

"I introduced this second subject initially in a reduced form in a simple 4/4 meter [Ex.8]. When played now with the utmost gentleness [*Zartheit*], this

Ex. 8

theme takes on a haste that is almost passionate (something like an intimately whispered declaration of love). To retain that main characteristic of gentleness, the tempo must be held back by a little [*um Etwas*], since the theme's passionate haste is already decidedly enough expressed in its moving figurations, thus permitting it to be pushed to the farthest variant within the main tempo in the direction of the *maestoso* of the 4/4 march theme; and in order to carry this out imperceptibly (that is, without distorting the main character of the underlying basic tempo), this change of character is initiated with a measure marked *poco rallentando*. With the increasingly restless feeling of this theme [Ex.9], it was easy to lead the tempo back to its original livelier direction, in which finally

Ex. 9

it could lend itself to function as the above-mentioned *andante alla breve*, whereby I only had to take up anew a variant of the main tempo already previously developed in the exposition of the piece. I had caused the initial statement of the majestic march theme to evolve into a broader coda of *cantabile* character, which can only be correctly interpreted when set in the aforementioned *andante alla breve*. Since the theme preceding this full-bodied *cantabile* [Ex. 10] is the fanfare [Ex. 11], to be performed in powerful quarter-notes, this

Ex. 10

Ex. 11

tempo conversion must obviously occur at the end of the quarter-note movement which coincides with the more sustained line of the *cantabile* on the dominant chord. Now, since this broader tempo in half-notes undergoes during its considerable duration a lively intensification [*Steigerung*] as well as a tonal modulation, I felt I could leave the movement of the tempo all the more to the discretion of the conductor, without drawing particular attention to it, in as much as the interpretation of such passages, if left to the natural instincts of the performing musicians, leads all by itself to an enlivenment of the tempo [*ganz von selbst zur Befeuerung des Tempos hinführt*]. Relying on this, I felt, as an experienced conductor, no need to indicate any place other than where the tempo is to return to the original pure 4/4 meter, which should be clear to anyone with truly musical feelings, because of the addition of the harmonic progression's quarter-note movement. In the conclusion of the Overture this broader 4/4 time clearly returns with a recapitulation of the above-mentioned, powerfully sustained, march-like fanfare, to which are added the decorative figurations in double-time, thereby bringing the tempo full circle back whence it began."

It should be abundantly obvious to "anyone with truly musical feelings"—I might add and of musical *understanding*—that Wagner is here dealing with subtle variations and nuances of tempo (his constantly reiterated word is "imperceptible") which should never distort or go beyond the music's basic, inherent tempo. In effect Wagner rules out, and forcefully opposes, any excessive alteration of the tempo, a lesson which unfortunately many conductors, including Wagner's own pupil and disciple Bülow, as well as many famous conductors past and present, seem(ed) unable to comprehend.

I have dwelt on Wagner's central ideas in regard to tempo and tempo inflections because he has so often been blamed for the excesses and tempo distortions of other conductors, including almost the entire wing of German-tradition conductors. This became a really contentious issue in the Toscanini era, when that conductor's widely acclaimed objectivity and textual fidelity were constantly being pitted against Furtwängler's approach, perceived as much more subjective, fluid, almost improvisational. (As already mentioned, these oversimplified descriptions of the two famous conducting rivals are largely invalid, since, for example, Toscanini's 'textual fidelity' did not absolutely preclude, as many of his better recordings attest, his adopting a more fluid pacing, while, conversely, Furtwängler was quite capable of highly rhythmic, precise, clear-headed interpretations, as, for example, his superbly played recording of Tchaikowsky's Sixth Symphony, especially its Scherzo.)

For a wide range of reasons, including many extra-musical, personal and political ones, the two conductors were constantly being contrasted in America—Toscanini favorably, Furtwängler negatively—all of this brilliantly documented in Joseph Horowitz's excellent *Understanding Toscanini*.[46] As promulgated over a period of many decades by those who saw in Toscanini "the greatest conductor of all time," the general impression was fostered that the German conducting tradition, as represented primarily by Furtwängler and Nikisch—a tradition which Toscanini was avowedly determined to eradicate—went back to and was first articulated by Wagner.[47] (There was, incidentally, a particular irony in this since Toscanini was one of the finest and most respectful interpreters of Wagner.)

The lesson to be drawn from Wagner's writing on tempo rubato in *Über das Dirigieren* is that a composer's score must be inherently respected in all its details; that such fidelity to the score ought not necessarily to result in stiff, accurate-to-the-letter, rigidly metronomic renditions, but can instead, as dictated by the spirit and feeling of the music, incorporate the concept of a flexible tempo, of subtle inflections and nuances; and that such tempo variations ought never to go beyond the bounds of the basic tempo, ought never to lead to distortions and exaggerations of tempo. Rather than suggesting arbitrary interpretational license, Wagner's numerous and constant reminders of restrained tempo behavior are embodied and italicized in his frequently reiterated terms and phrases "imperceptible" ("unmerklich"), "a little" ("etwas"), "to hold back only as far" ("nur so weit zurückhalten"), "the least indication" ("leiseste Andeutung"), "without drawing much attention to it" ("ohne besonders hierauf aufmerksam zu machen"), and so on.

Perhaps the problem is that, as is so often the case in the history of humankind, a prophet's or leader's disciples are his worst enemies, the worst corrupters of his thoughts. Certainly this seems to have been the case with Bülow, Wagner's one-time favorite protégé, who by the end of his career with the renowned Meiningen Orchestra, became famous for indulging in every conducting excess imaginable, including, of course, the wildly emotional manipulation of tempos. Taking inordinate license with Wagner's (and Beethoven's) concept of tempo inflections, Bülow developed a conducting style which justified its excesses under the heading of "Lisztian license," podium charisma, and authoritarianism. Whereas Wagner's subjective approach, as originally postulated, derived from what he saw (and heard) as a kind of "deep structure" in all great music, that is, the underlying ebb and flow of long term harmonic rhythms,

46. Joseph Horwitz, *Understanding Toscanini* (New York, 1987).

47. It can now be clearly seen in retrospect that this pejorative assessment of Wagner's influence was inaccurate and unjust since, for one thing, Wagner drew his interpretational theories from none other than Beethoven, not only in the master's own conducting and his comments on the subject, but his compositions as well, and since, for another thing, the concept of tempo rubato and a subjective, spiritual approach to interpretation goes even further back to Mozart and Haydn and even earlier, as Junker's and the other aforementioned theorists' writings so clearly attest.

Bülow, by all accounts, took every short-term harmonic move as an excuse to stretch, or contract, tempos to the virtual breaking point. While Wagner's tempo inflections were governed (and restrained) by phrase structures and character contrasts of themes, as well as vertical (that is to say, harmonic) tensions, Bülow took such structural signposts as justification for enormous liberties, "to the point of caricature," as Felix Weingartner once put it. I suspect—and indeed there is much evidence to this effect—that there was a certain arrogance and conde-scension towards the public in Bülow's late work, as well as (according to Weing-artner) "a mania for notoriety." Bülow's exaggerations were in part motivated by a desire to teach what he considered the unsophisticated public something about music, about form and structure. An example cited by Weingartner in his 1895 *Über das Dirigieren* is probably typical of Bülow's late work. "Where a modification of the tempo was necessary to get expressive phrasing, it happened that in order to make this modification quite clear to his audience, he *exagger-ated* it [Weingartner's italics]; indeed, he fell into a quite new tempo that was a negation of the main one. The *Egmont* Overture was a case in point. Wagner tells us [in his *Über das Dirigieren*]," so Weingartner writes *a propos* of this passage (Ex. 12)—"which, as he [Wagner] says, 'combines in a drastically con-

Ex. 12

densed way a fearful severity with serene self-confidence' ["dieses aus schreck-lichem Ernste und wohligem Selbstgefühl so drastisch enggeschürzte Motiv"], and which, as a rule 'in the headlong rush of *allegro* was washed away like a withered leaf' ["wie ein welkes Blatt hinweggespült"]—that he induced Bülow to play it in the true sense of the composer, modifying 'with only a hint' ("nur andeutungsweise") [again Wagner's caution of moderation] the hitherto 'passion-ate' tempo, so that the orchestra might have the opportunity to differentiate this thematic combination, with its rapid fluctuation between great energy and thoughtful self-content.' All who have heard this overture under Bülow must agree with me that at the place in question he by no means made 'only a hint' of a modification, but leaped at once from the *allegro* into an *andante grave*, thereby destroying the uniform tempo that should be preserved in the *allegro* of that overture," adding "as in general in every piece of music that has a uniform tempo mark at the beginning."[48]

Other writers and conductors have taken Bülow to task for his exaggerated

48. Weingartner, (1895) p. 18; (1905), p. 13. We also have the word of Brahms in commenting on Bülow's conducting, thay was "always calculated for effect. Immediately a new musical phrase be-gins, he makes a small pause, and likes to also change the tempo a little." Brahms goes on to say "I have deliberately denied myself this in my symphonies. If I had wanted it, I would have written it

tempo fluctuations and other effect excesses, although he also had his defenders, such as the eminent musicologist/historian and editor Heinrich Reimann. But perhaps no one took more deadly aim at Bülow than the American author and "serious music" popularizer, David Ewen. One of the major Toscanini apologists in the 1930s and '40s, and therefore someone by predilection inclined to deprecate the whole German Romantic school of conductors, Ewen nonetheless came as close as anyone to characterizing Bülow's conducting style. Ewen castigated Bülow for being "a pernicious influence" who took "amazing liberties" with tempo and phrasing, who was given to "exaggeration and over-statement," who "tampered with the scores"[49] and influenced many of his German successors like Mahler, Richter, Levi, and Nikisch to tamper further with the early Romantic repertory (especially Beethoven and Schumann symphonies).

One who tried to stem the tide against excessive subjectivism in interpretation, especially that of conductors, was Felix Weingartner. In his *Über das Dirigieren*—interestingly the same title as Wagner's book—he presents as balanced a picture of Bülow's impact on the evolution of conducting as seems possible, praising him for his landmark contributions in eradicating the early 19th-century manner of mere elegant time-beating, at the same time pointing out his later harmful influence. Weingartner allowed that the guilt for these misdemeanors lay "both with [Bülow] himself and a number of his followers; and to expose these and attack them is as much a duty of sincerity as to acknowledge the gains with frank delight. It cannot be denied that, even while he was leader of the Meiningen Orchestra, there was often a *pedagogic element* in Bülow's renderings. It was clear that he wished to deal a blow on the one hand to philistine, metronomic time-beating, on the other hand to a certain elegant off-handedness and superficiality. But in his effort to be excessively clear, [Bülow] went too far. [Forgetting] that works of art and art performances exist for the sake of themselves and their own beauty," Bülow's "tendency" to "tendentiousness" made him prone "to make details excessively prominent." In so doing, Weingartner felt, Bülow had lost touch with the notion that each detail in a composition "has its full *raison d'être*, but only in so far as it is subordinated to a homogeneous conception of the essential nature of the whole work—a continuous conception that dominates all detail."[50]

Weingartner illustrates his concerns about exaggerated interpretations and tempo distortions most cogently in a passage regarding Mendelssohn's *Hebrides* Overture.[51] Weingartner describes a performance in which literally not one

in." We know also from many contemporary reports, including those of many of Brahms's own pianist students, that he was much freer in respect to rubato in his piano playing and that there was to be less tempo nuancing in orchestral performance than in performances on the piano.

49. David Ewen, *Dictators of the Baton* (Chicago, 1943), pp.23, 27.

50. Ibid., (1895), p.17, 23; (1905), p.12; 16.

51. Ibid. (1895), p.46; (1905) p.30. How fascinating it would be—especially in view of Wagner's uncomplimentary views of Mendelssohn's conducting—to know how he (Mendelssohn) conducted his own *Hebrides* Overture, a work that is more tone poem than straight classical overture, indeed more in the mould of Weber overtures, which Wagner so admired and whose appropriate perfor-

measure was played in the same tempo as another. "The second and fourth bars, repetitions of the first (Ex. 13) and third bars (Ex. 14), were rendered in a noticeably different tempo; and so it went in a similar vein to the end of the

Ex. 13   Ex. 14

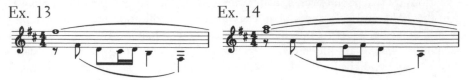

piece. The humanly utmost in unnaturalness (*Unnatur*) was achieved, so that this beautiful work was distorted and its true character obscured (literally: *verwischt*, wiped away)."

Weingartner goes on to describe the opposite performance approach, the one Wagner opposed so vehemently. "Of course it would be just as wrong to run through ["abzuspielen"] the piece in a metronomic quarter to quarter. But the modifications of tempo, a few of which Mendelssohn himself prescribed, must occur in such a way that the uniformity and coherence of the work . . . are not dissevered ["zerstückelt"]. At times the sea around Fingal's Cave is calm; at other times a stronger wind causes higher waves and the white foam of the surf breaks more violently against the shore, *but the picture of the landscape remains the same* [italics are Weingartner's]; and a real terrifying ocean storm, which would give the whole scene a totally different character, never occurs in Mendelssohn's overture. A tone of gentle, noble melancholy, which lends the Hebrides Islands their peculiar charm, is preserved throughout in the music. Does it not warrant the sharpest condemnation when that which a master has genuinely experienced ["empfunden"]) and expressed in consummately beautiful tones, is distorted by a conductor with all sorts of intrusions [*Zutaten*]?"

" 'And why all this?' I have asked myself on many occasions. Why this mania

---

mance conception he took such pains to describe in his writings. Alas, there seems to be no account of Mendelssohn's conducting of his *Fingal's Cave*; and so we are left with the tantalizing question of whether, as Wagner implies, his rendition would in fact have been in the "elegant" but emotionally uninvolved, mere "time-beating" manner.

One also wonders just *how* Mendelssohn performed Bach's *St. Matthew Passion* in that work's celebrated revival in Berlin in 1829—when he was only twenty years old. Again, no specific, detailed account of Mendelssohn's approach to the work seems to exist, possibly for the very good reason that Bach's great masterpiece had not been performed for nearly one hundred years, so no one in 1829 could have properly assessed its performance in terms of accuracy and stylistic authenticity. On the other hand, we do know that the work created such a sensation that it had to be repeated twice, and that these performances started the revival of popular interest in Bach's music. We also know from many accounts, including his own letters (for example, to his sister Fanny and his teacher Karl Friedrich Zelter), that his conducting was informed by a basic fidelity to the score and historical, stylistic authenticity. Mendelssohn by all accounts—and Wagner's diatribes against Mendelssohn and his school peculiarly confirm this—became a symbol of objective music-making, exposing classical clarity and unity in performance, regularity and fluency (meaning liveliness) of tempo. Although as a composer he portrayed the Romantic themes of his era, as a conductor he was a classical traditionalist, intent on preserving the ideals of classical forms and their interpretative purity—perhaps a Toscanini of his day.

of some conductors to make out of musical works something other than what they really are? Why this fear of maintaining a uniform tempo for a certain amount of time? Why this rage to superimpose nuances and expressions of which the composer never dreamt?"

"The reason for these strange spectacles [*Erscheinungen*] lies in a kind of personal vanity and egotism that is not satisfied to perform a work in the true sense and conception of its creator, but instead wants to show the public what it 'can make' out of the work. The conductor's mania to please and to be adulated [*Gefallsucht*] is thus put above the requisites and spirit [*Geist*] of the composer." [52]

As already mentioned, along with Weingartner who, both in his writings and conducting, tried to stem the tide of excessive interpretive license and conductorial self-indulgence, and more or less contemporaneous to him, there was Arturo Toscanini. Although not generally given to expressing himself in writing on conducting—neither in Italian nor in English—Toscanini did occasionally unburden himself, mostly in rehearsals or interviews and mostly critically/negatively, on interpretational matters. On tempo and tempo modification he once said: "The correct tempi; that's the important thing: the right tempi! The tempo must change, weaving in and out, *but always close and always returning*" (italics added) "Yes, in music just to have the correct tempo, with all that goes with it, means nothing. Niente!"

On another occasion, he unleashed this blistering cannonade against conductors: "Who do they think they are, those musical assassins, changing, distorting? They think they are greater than God!" [53]

We shall return again to Weingartner's writings, among the most insightful, intelligently balanced, and self-effacing in the entire history of conducting. But for now I would like to stay with our theme of the moment: tempo—or more precisely—tempo modification and textual fidelity. Richard Strauss's writings on conducting, contained in *Recollections and Reflections*,[54] are not as comprehensive as those of some of the other famous composer-conductors. They also tend towards the anecdotal and glib, even occasionally the cynical. Strauss did not offer very much advice on questions of tempo and tempo modification, but he

---

52. *Gefallsucht*, although generally construed to mean 'a desire to please,' actually can also signify 'the need to be admired.' Ibid. (1895), pp.47–49; (1905), pp.31–32.
Although initially written in 1895, this is as apt a description of Leopold Stokowski and Leonard Bernstein at their least and worst as can be found. Bernstein, one of the most overrated and adulated conductors of recent times, rarely practised what he preached—a sad fact given his enormous basic natural talent, musical and conductorial/gestural. In his *Joy of Music*, he wrote, for example, "Perhaps the chief requirement of all is that [the conductor] be humble before the composer; that he never interpose himself between the music and the audience; that all his efforts, however strenuous or glamorous, be made in the service of the composer's meaning—the music itself, which, after all, is the whole reason for the conductor's existence" (*The Joy of Music*, New York, 1954, p. 156). It is as perfect and beautiful a statement about the art and philosophy of music as can be found. It is all the more saddening and perplexing that Bernstein rarely followed his own credo.
53. As cited in Samuel Antek's *This Was Toscanini* (New York, 1963).
54. Richard Strauss, *Recollections and Reflections* (Zurich, 1949).

did warn against some of the (in his mind) more notorious bad habits and vul-garisms perpetuated by certain conductors. He in particular deplored the all too common "slackening of pace just before a great fortissimo" which, "for all its popularity," he called "quite unbearable" and "amateurish."[55] Similarly, Strauss decried slowing down the last bar in the development in Beethoven's Third *Leonore* Overture before the entry of the trumpet on the stage. "On the contrary this whole passage is to be played *accelerando*; after all Pizzaro, as he rushes at Leonore, knows nothing of the B-flat of the trumpet."[56] Well, I don't know about the accelerando, but it certainly does not make sense to make a big ritard, as almost all conductors insist on doing.

One of Strauss's wisest admonitions concerns "subjects which the composer himself has already drawn out, which should not be drawn out further."[57] Lastly he termed "dreadful" the ritardandi in the bars leading to the second subjects in Weber's overtures, thereby possibly disagreeing with Wagner. We cannot be sure because Strauss mentions no names, and it is possible, indeed likely, that the ritardandi he refers to were of the excessive kind, as practiced by Bülow and some of his disciples, whom Strauss heard in his younger years.

Bruno Walter's writings on conducting are amongst the finest and most com-prehensive in the entire literature on the subject.[58] Of a more philosophical turn of mind than the very pragmatic-minded Strauss, Walter brings together the most cogently technical advice with profound philosophical, aesthetical, moral principles involved not only in conducting but in all re-creative, reproductive, interpretive music-making. His chapter entitled "Of Tempo"[59] examines the subject brilliantly and exhaustively, recapitulating and expanding upon a num-ber of the more salient points about tempo rubato made by Beethoven, Wagner, and Weingartner.

Referring his readers first to Wagner's essay *Über das Dirigieren* and his cen-sure of undifferentiated tempos, Walter then points out how Wagner's teachings "were perverted and exaggerated" by a considerable number of conductors who "fell into the opposite error": an exaggerated, arbitrary, meaningless modification of tempo. The deficiencies of the erstwhile time-beaters had been replaced for the most part by virtuosos "who, not content with the autonomous life of a piece of music," thought they must "enhance it by an over-differentiation of tempo and delivery; too little was followed by too much."[60] According to Walter "the *right* tempo" is one that "permits the musical meaning and the emotional sig-nificance of a phrase to show to best effect," adding most significantly, "and that allows for technical exactness."[61] Walter in this sentence was the first to tie

55. Ibid., p.54
56. Ibid., p.60
57. Ibid., p.61
58. Bruno Walter, *Von der Musik und vom Musizieren* (Frankfurt, 1957). English edition, *Of Music and Music-Making* (London, 1961).
59. Ibid., English edition, pp. 29–45.
60. Ibid., English edition, p.30.
61. Ibid., English edition, p.30.

together, as a working principle of what constitutes "the right tempo," the notion of emotional significance/expression *and* technical exactitude, by which he means, I am sure, not only a tempo that permits a passage to be played securely, accurately, technically correctly, but one that represents a high degree of textual fidelity. Probing still further, he states that, since "in the course of a piece of music its content, mood and technical requirements change incessantly, the tempo has to be adapted" to these changes "to remain always right. The right tempo for a piece is relative."[62]

Walter, sensing that this is still too general, too imprecise a formulation, one that still allows for some of the excesses of tempo modification that he is trying to warn against, adds most tellingly: "Our problem [is] that of *tempo*, not *tempi*. For the well-constructed piece of music in organic form is defined by *one* main tempo which, though it may change in the course of the composition, maintains a continuity that accords with the symphonic continuity of the composition. From this the conclusion can be drawn that we must remain at the same speed until a change in the design of the music forces us to modify it." The concept of a right tempo "demands a flexible continuity for tempo," what Walter then calls "apparent continuity."[63]

Even more specifically—and I think most important—Walter stresses that the fact that composers indicate "noticeable changes in speed" with directions such as *ritardando, accelerando, più mosso* etc., should prove that "noticeable changes in speed, *other than those marked by the composer, will offend against his intentions*" (italics added). Driving the point home further, Walter says, "This means that all other modifications of tempo, as they correspond to the ebb and flow of the music, must be of the unnoticeable kind."[64]

"I should like to point out," Walter continues, that the conductor "who permits himself unwarranted changes of tempo" is implying "a re-evaluation of the meaning of those passages that were intended by the composer to have an even tempo. For *ritardando* and *accelerando* are not merely indications of motion; they also have the emotional significance of hesitating and urging." The sudden unwarranted change of tempo "gives the impression of a formal paragraph, of a division in the course of the music. It is scarcely necessary for me to stress that we reproductive artists must not indulge in such arbitrary, disruptive acts, affecting the soul as well as the form of a work.[65] Noticeable changes in speed that are not demanded by the composer are, therefore, misrepresentations; whether they result from intellectual presumption or from sheer license, they deviate

62. Ibid., English edition, pp. 30–31.
63. Ibid., English edition, p.31.
64. Ibid., English edition, p.31.
65. Ibid., English edition, p. 32. In this regard Walter was wrong, for it *was* and *is* quite necessary to stress that conductors ought not to indulge in "such arbitrary, disruptive" liberties.It is also sad to report that Walter himself often in his performances, especially of Brahms, did not follow his own advice. Although he came as close as anyone to giving a precise formulation of the twin notions of tempo and tempo modification, he evidently either felt he was not obliged to follow his own rules or he was not aware of breaking them. Having worked with Walter often in the 1950s and '60s, I can report that it was most often the latter.

from the composer's intentions, and thus from the purpose of reproductive art."[66]

Just as I have done earlier in this book, Walter presents the usual counter-argument to his strictures. " 'What pedantry!' I can hear many a musician exclaim: 'Should not a performance have spontaneity and the flair of improvisation? If we stay strictly in tempo, except where the composer himself interrupts the continuity, the restraint imposed on us will rob our interpretation of all immediateness. Why, should I not obey my heart, if it prompts me to hold back here, to press on there, to ritard one phrase and accelerate the next? My performance will sound spontaneous since I am playing the music in the way I feel it, and not as the composer compels me to play it.' "[67]

Walter's answer to that is "that the tempo directions of the composer are an integral part of the notation of the work; the changes in tempo demanded by him are part and parcel of the composition."[68] To which I would add that the *absence* of indicated changes of tempo by a composer are also part and parcel of the composition. Walter says as much a few paragraphs later: "We should be aware of changing the speed in the absence of the composer's direction just because our personal taste or the leanings of our heart would have it so. The introduction of tempo changes where they are not prescribed by the composer is in the nature of an encroachment on the composition as such."[69]

"And where would be the limit to such encroachments? Once we deprive a phrase of the sense given it by the author's directions—or give it a sense not asked for by him—why should we not go further and boldly obey the bidding of our 'heart', when in some passages it wishes to change the notes and rhythms prescribed by the author?"[70]

Walter's sternest reproach of those who would tamper with the composer's score comes a few sentences later when he states that "we must feel free" in the rendering of a work, "but free within the laws whose binding force we have recognized when we chose to be musical interpreters. If we should chafe under the immanent laws of a work of music as under a compulsion, we are not made to be its interpreter."[71]

In a profound and beautiful summation of the subject, Walter concludes: "For the criterion of our talent as reproductive musicians lies exactly in our capacity for assimilating the intentions of another so completely that not only are the demands of the work no burden to us, but that we feel them to be our own demands. Only thus shall we feel free within the limits of the laws imposed on the work by the author, and only thus will our music-making sound spontaneous, since we now are free to follow the bent of our own heart which has learnt to beat in unison with that of the composer."[72]

66. Ibid., English edition, p.32.
67. Ibid., English edition, p.32.
68. Ibid., English edition, p.32.
69. Ibid., English edition, p.33.
70. Ibid., English edition, pp.33–34.
71. Ibid., English edition, p.34.
72. Ibid., English edition, p.34.

In a somewhat confessional mood, Walter admits that in his younger years he also committed "the sins of willful interferences with, or capricious misinterpretations of, the composer's clear intentions." He ascribes these to a certain youthful artistic immaturity, noting that "it was my nature to fall in love with every beautiful detail of a composition and try to reproduce them with all the intensity of expression of which I was capable, and thus neglect the synthesis and unity of conception which are the main point of an authentic interpretation. My enthusiasm for details was stronger than my capacity for subsuming them under a higher order."[73] Eventually realizing that what was missing in his performances "was a regard for the work in its entirety, without which its greatness, seriousness, and unity will not reveal itself," and that thereby "I had done damage to the musical form," Walter says he eventually learned that there was "a method of interpretation higher, nobler, more in accord with the greatness of the work, than is indulgence in the sway of one's feelings."[74] Oh, that certain famous conductors of today would have learned that lesson!

Walter correctly relates tempo question to continuity: "The concept of tempo is invalid without the correlative one of continuity. The clearest proof of this is in those very changes of tempo that are indicated by the composer. What could his *ritenuto* mean if the tempo itself were irregular? Only the fact that continuity, that is, regular flow, is an essential attribute of the concept of tempo, gives proper significance to every *ritenuto* and *accelerando*, every *meno* and *più mosso*."[75]

Walter closes the subject of tempo with an excellent summary paragraph. "At all events, the concept of the *right tempo* stands and falls by the recognition of the principle of [what he calls] *apparent* continuity. If we deliberately deviate from [a tempo], obeying or disregarding the directions of the author as our fancy bids us, then anarchy will destructively descend on a domain of lofty order. And in the resulting distortion there will hardly be found a trace of the work, which was created according to a profoundly meaningful design, and should only be re-created in the same spirit".[76]

In another chapter, entitled "Of Correctness," Walter returns to the subject of tempo in a more general, oblique, but nonetheless extremely insightful way. Speaking of "correctness" as the "indispensable condition and prerequisite for any musical interpretation that does justice to the spirit and soul of the work," he then defines this "correctness" as including the "rightness . . . of time" and "compliance . . . with tempo indications," significantly adding that "it is only from such a basis that meaningful music-making can evolve."[77]

On the broader subject of textual fidelity, Walter did not feel that a certain non-excessive degree of interpretational freedom was incompatible with faithful adherence to the score. He rightly believed that "a work is capable of different interpretations and that, moreover, our own repeated performances of it need

73. Ibid., English edition, p.38.
74. Ibid., English edition, p.36.
75. Ibid., English edition, p.43.
76. Ibid., English edition, p.45.
77. Ibid., English edition, p.84.

not, when allowance is made for spontaneity, entirely agree with each other. Faithfulness to the spirit knows of no rigidity; the spirit of the work of art is flexible, *elastic* [italic Walter's], hovering."[78]

The last voice to be heard here on the subject of tempo and tempo fluctuations is that of the late Eugen Jochum, a disciple of Furtwängler and a fine conductor in his own right, with a special affinity for Beethoven and Bruckner. Echoing Wagner's theories on tempo variations, Jochum wrote, "The layman generally does not realize how great the variations can and must often be in a steady tempo within one movement in order to bring out the vitality of the musical flow. Yet the listener must always have the impression that the tempo does remain steady."[79] As Wagner had already explained in his *Über das Dirigieren*, Jochum cited the first movement of Beethoven's *Eroica* as "a particularly interesting example, which with its mighty proportions cannot be played through in the same unbending tempo, and can stand—even demands—extensive modifications. But these must have the right balance and not go to extremes: this is precisely the art of the interpreter, which," Jochum adds with an underlying touch of hopefulness, "presupposes a highly refined awareness of tempo."[80]

In summary, on the subject of tempo fluctuations—a subject so sorely in need of objective consideration—we have heard the thoughts of some of the greatest conductors and composer-conductors, all pointing to the same basic central notion: "Do it, but don't do it to excess." In our day when we have either conductors who play the Romantic literature with metronomic rigidity (even as they ignore the actual metronome markings) or conductors who take such "amazing liberties" that the music loses all meaning, all sense of balance and proportion, and becomes a willful ego display, the great conductors' views cited here ought to offer some sobering advice. Young conductors especially, whether still studying or coming up in the professional ranks, ought to heed the advice here given and not be misled by whatever some famous 'maestro,' whose records may be selling in the tens of thousands, did or did not do.

Another subject I have dwelt upon in Part I, which it will be useful to explore further through the writings of some of the great conductors and composer-conductors of the past, is the matter of following composer's indications regarding crescendos and diminuendos, not only where they are placed but of what duration and how far they should extend in dynamic range. Again, the advice of three major historical conductor figures is instructive, especially as their views on the matter are virtually unanimous.

We all know that crescendo and diminuendo were first prominently developed under Johann Stamitz's direction in Mannheim. It was a new manner of featuring dynamics—soon to be dubbed a "mannerism" by none other than Leopold

---

78. Ibid., English edition, p.125
79. Eugen Jochum, "About the Phenomenology of Conducting" (Hamburg, late 1930s), in *The Conductor's Art*, ed. Carl Bamberger (New York, 1965, 1989), p. 261.
80. Ibid.

Mozart[81]—in which all too soon dramatic increases and decreases of amplitude often became an effect for its own sake.

It was Beethoven who rescued the crescendo/diminuendo manner from mere *Effekthascherei* (sensational effects) and turned it through his innovative genius into a powerful, moving means of expression. This along with his introduction of daring rhythmic innovations (especially his dramatic use of syncopations) and a radically new declamatory style, as well as expanding the classical orchestra's dynamic and sonoric range, all required a whole new kind of conductorial leadership. Orchestral players became involved in a more diversely demanding individual role, the collectivity of which produced an intellectual and emotional complexity—for the player as well as the listener—which far transcended the ordinary standards of his time.

Whether in his own conducting Beethoven was able to control all aspects of this radical kind of new music-making is debatable. There are many accounts of his conducting—perhaps not all absolutely reliable—that describe Beethoven's conducting as "violent" in its gestures, and "frequently misguiding."[82] Anton Schindler, Beethoven's friend and biographer, suggested that as a conductor "the Master was neither good nor bad. His impetuosity did not permit him to arrive at the requisite tranquility and self-command." He would lose "himself in gesticulations which caused a wavering in the orchestra."[83] Ludwig Spohr, who saw and heard Beethoven conducting on several occasions, recalled in his *Autobiography* that Beethoven was "in his manner of conducting very awkward and helpless, and his movements lacked all grace",[84] but suggested that his inefficiency as a conductor was primarily due to his deafness. Speaking of Beethoven's conducting of the Seventh Symphony premiere, Spohr described the performance as "quite masterly, in spite of the uncertain and frequently laughable direction of Beethoven."[85] On the other hand, Seyfried recalls that Beethoven was "very meticulous with regard to expression, the more delicate shadings, and equalized distribution of light and shade, and an effective tempo rubato".[86]

Beethoven was also quite unequivocal and painstaking in his dynamic indications. Having noted musicians' tendency to crescendo too much too early and, worse, to automatically associate ascending passages with an increase in dynamic, and descending ones with a decrease, Beethoven sprinkled his scores with cautionary, reminding markings of *sempre pp* and *sempre f*. What is remark-

81. Leopold Mozart made frequent references to dynamics and musicians' use of them, not only in letters to his son but in his 1756 *Versuch einer gründlichen Violinschule* (English edition, *Violin Playing*, 1948). In the latter, for example, he writes (p.218): "It follows that the prescribed *piano* and *forte* must be observed most exactly, and that one must not go on playing in one tone like a hurdy gurdy. One must know how to change from *p* to *f* . . .each at the right time; for this means, in the well-known phraseology of the painters: Light and Shade."
82. Ignaz von Seyfried, *Beethoven Studien* (1832), cited in Sonneck, *Beethoven: Impressions of Contemporaries* (New York, 1926), p.40.
83. Anton Schindler, *Ludwig van Beethoven* (Münster, 1840), p. 44.
84. Louis Spohr, *Autobiography* (London, 1865), p. 188.
85. Ibid., p.187.
86. Seyfried, *Beethoven Studien*, p. 42.

able, even uncanny, about these reminders is that they are almost always placed in exactly the right spot, namely that moment when most musicians will in fact wish to anticipate a crescendo or diminuendo. A careful study of Beethoven's scores will reveal that in matters of dynamics he was a most meticulous and demanding notator, undoubtedly in large part to counteract the bad habits of musicians and future conductors, as well as the after-effects of the later excesses of the Mannheim school.

Wagner had much to say on the matter of dynamics, especially the proper realization of crescendos and diminuendos, no doubt inspired by Beethoven's works. Again Wagner[87] picks some excellent (and problematic) examples to make his points. One of these concerns a certain chromatic "ascending passage"in the first movement of Beethoven's Ninth Symphony (Ex. 15), con-

Ex. 15

trasting a certain "masterful" performance in 1839 by the orchestra of the Paris Conservatoire (conducted by Habeneck)[88] with his own many vain attempts with various orchestras—he mentions in particular Dresden and London—to achieve this passage correctly. "Never was I able," writes Wagner, in a remarkable instance of modesty, "to make the bow—and string—changes in this sequential ascending passage totally unnoticeable; nor was I able to avoid involuntary accentuations, because the ordinary musician has a tendency to become louder in ascending figures and conversely become softer in descending ones." That Wagner was equally worried about musicians' converse tendency, to make accelerandos with crescendoing and/or ascending figures, we know from many instances in his scores, most notably his "Nicht Eilen!" in the Prelude to *Tristan und Isolde* as the music nears its orgiastic climax—an admonition, alas, ignored since 1857 by hundreds of conductors who either think they know better than Wagner or who haven't the technical control to restrain the orchestra.[89] "By the

87. Wagner, *Über das Dirigieren*, p.13.

88. On the matter of the excellence of the Conservatoire orchestra under Habeneck, Mendelssohn and Wagner decidedly agree. Mendelssohn, in a long detailed letter about musical life in Paris (dated February 15, 1832), called the performances of the Conservatoire orchestra "the most accomplished performances to be heard anywhere." It is "the best I have ever heard." Mendelssohn also offers the statement, amazing to 20th-century minds, that they "rehearsed for two years (*sic*) before venturing a performance until there could be no question of a wrong note any longer" (see Mendelssohn, A *Life in Letters*, ed. Rudolf Evers (New York, 1990), p.176).

89. One of the more flagrant ignorings in recent years of this "Nicht Eilen!" in Wagner's *Tristan* Prelude occurred during an arrogantly indulgent, but unfortunately widely praised, performance in New York by Sergiu Celibidache and the (wonderfully playing) Curtis Institute Orchestra.

fourth bar of the quoted [Beethoven] passage," Wagner recalls, "we invariably got into a crescendo, so that the sustained G♭ in the fifth bar was involuntarily— yea, necessarily—played with a hefty accent, enough to detract greatly from the unique tonal significance of this note. What expression this passage receives when performed in this ordinary manner, contrary to the master's clear enough, expressively designated intentions, is difficult to make clear to an unrefined ("grobfühligen," literally, crude-feeling) ear and listener. Undoubtedly it will express a certain sense of dissatisfaction, restlessness, and frustration, but of precisely what kind, we can only learn when this passage is played the way Beethoven intended and the way I have heard it realized only by the Parisian musicians in 1839."[90] Notice the word "realized" (*verwirklicht*), Wagner hereby rejecting the "ordinary manner" and interpretation (or misinterpretation) and accepting only a "realization" of Beethoven's clearly indicated intentions. Note also Beethoven's *two "sempre pp's"*—in a five-bar span(!)—which, alas, then as now seem to be there to be ignored.

Wagner drives his point home, correctly connecting dynamic control and sensitivity with the true expression and spirit of the music, by implication even with tone color and timbre. He writes: "I recall how clearly and directly the impression of dynamic monotony (please forgive this apparently senseless expression for a difficult-to-describe phenomenon!) spoke to me in the unusually, even eccentrically, varied interval progression in this ascending figure, flowing into the infinitely, delicately sung, prolonged G♭ which, in turn, is answered by the similarly sung G♮, and almost magically initiated me into the incomparable mysteries of the spirit."[91]

Wagner then asks the only partly rhetorical question: How did the Paris musicians "achieve the solution to this most difficult problem," and provides the perhaps obvious answer: "through the most conscientious diligence, which is only given to such musicians as are not satisfied to constantly compliment each other, imagine that they know and understand everything, but rather who stand before the initially incomprehensible with a degree of humility and concern, and who try to solve what is difficult in that realm in which they excel: technique."[92] Wagner ends this particular discussion with a reference to our old friend 'tempo,' specifically the Paris performance's "right tempo." "Old Habeneck certainly did not have in this regard any special, abstract-aesthetic intuitions—he was not a man of genius ("er war ohne alle 'Genialität' "); but he found the right tempo, in that he, through persistent effort, led the orchestra to comprehend the true *melos* of the symphony."[93]

By "Melos" Wagner meant a singing quality, a linear expression, a continuity of line, which he felt was the true essence of all great music, and without which music was meaningless, soulless, abstract and stiff—as he once also put it: "something between grammar, arithmetic and gymnastic."[94] Wagner felt that it

90. Wagner, *Über das Dirigieren*, p. 14.
91. Ibid., p.14.
92. Ibid., p.15.
93. Ibid., p.15.
94. Ibid., p.15.

was the specific character of a melody (or theme), the melodic content of a piece, which to a large extent, along with the all-important harmonic considerations, generated a movement's tempo ebb and flow. The singing quality of a performance—he said of the Paris Conservatoire orchestra "that wonderful orchestra sang that symphony"[95]—was all important to him. Interestingly—returning for a moment to our earlier discussion of tempo elasticity, textual fidelity and the related Toscanini-Furtwängler feud of fifty years ago—Toscanini and Furtwängler excelled in making their orchestras sing: Toscanini in a more simply lyric Italianate manner, where the song, the melody itself, was italicized and valued; Furtwängler in a more deeply expressive German way, with enormous stretch in his melodic lines, where the melody was always rooted to the underlying infrastructure and its harmonic tensions.

Wagner related the control of crescendos and diminuendos very much to the delicate control of tempo gradations. That is to say, he found a correlation between the needs of both tempo and dynamics and their virtually infinite variety of shadings, which arise out of the variable meanings and feelings of the music, *but* which must not overstep the bounds of their basic order, and must in any case be held in check to reflect the intentions of the composer. That is *not* to say that crescendos should be accompanied by accelerandos or vice versa. Wagner makes these distinctions abundantly clear time and time again, both in his writings and his scores.

One specific and very telling comment in this regard is his reference to the relationship between the older forms and their inherent dynamic structure. He writes that until the Mannheim Orchestra discovered the crescendo and diminuendo, "the instrumentation of the old masters" (meaning primarily Mozart and Haydn) "reveals the fact that nothing which would effect a truly emotional/ expressive interpretation was allowed to be interpolated between the *forte* and *piano* sections of the [old naive] *Allegros*."[96] By implication Wagner is saying that the new "sentimental" Beethovenian *allegro*, as in the first movement of the *Eroica*, also required a dynamic flexibility and sensitivity analogous to the structural and (thus) tempo fluctuations. It is also true, of course, that composers, at least in the 19th century, had more notational means at their disposal to indicate dynamics than they had to indicate subtle tempo fluctuations. Moreover, composers could—and mostly did (see any number of Brahms symphony movements)—call for enormous dynamics fluctuations but at the same time very modest (if any) tempo fluctuations.

That Wagner was not unaware of the subtle but crucial relationship between dynamics and tone color, fundamental to all fully expressive music making— and by expressive I don't mean only loud and exciting—is shown in a passage of special interest to me as an ex-horn player, when Wagner speaks of his performance in Vienna of the *Freischütz* Overture. About the opening horn quartet he writes: "The horn players, under the sensitive artistic leadership of [the prin-

95. Ibid., p.15.
96. Ibid., p.33.

cipal horn] R. Lewi totally changed their approach [*Ansatz*, literally, attack] in the introduction from their previously accustomed rendition as a pompous, swaggering show piece to a soft woodsy fantasy [*Waldphantasie*], blending with the indicated pianissimo of the string accompaniment and, in a totally different [new] way, distilling a magical perfume [*zauberischen Duft*] over their songful melody." A moment later Wagner speaks of the horns' "delicately inflected" phrase rather than the "usual *sforzando*" (Ex. 16) and still later the cellos', as intended, "softest sigh" (Ex. 17),

Ex. 16   Ex. 17

rather than "the now so customary heavily accented attack [*Anstoss*]."[97]

As we can see, the disregard of dynamics, premature or exaggerated crescendos and diminuendos, accelerandos and ritardos, is not just a modern-day manifestation; it goes back through the centuries, eliciting frustrated warnings and implorings from Wagner all the way back to Gluck (among many others). Gustav Mahler, too, was not silent on these matters. As cited by one of his biographers,[98] speaking of orchestra musicians' bad habits, he said: "There are frightful habits, or rather inadequacies, which I have encountered in every orchestra; they . . . sin against the holy laws of dynamics and of the hidden inner rhythm of a work. When they see a crescendo, they immediately play *f* and speed up; at a diminuendo they become *p* and retard the tempo. One looks in vain for gradations, for the *mf*, *f*, *ff*, or the *p*, *pp*, *ppp*. And the *sf*'s, *fp*'s, shortening or extending of notes, are even less in evidence."

One wonders, however, given what one knows about Mahler's conducting and interpretations as rather willful and 'spontaneous,' especially in the earlier half of his career, whether Mahler was always as disciplined in respect to dynamics and tempos as he would have us believe from some of his statements on the matter.

The same customary carelessness, often destroying the real essence and meaning of the music, is as wide-spread today as it seems to have been in Wagner's and Berlioz's day. Some things, I guess, never change. The abuse, and misuse, of dynamics is perhaps the most common evil in orchestral playing today, (especially in the United States), being either tolerated or generated by our conductors. This is particularly ironic, since the technical abilities of modern players are so high that no claim could ever be made that subtle dynamic control is beyond their capacities. And to excuse this dynamic laziness by saying "it's more fun to play loud" or "it makes a bigger effect" or "it's more exciting" or—more philosophically resigned—"it's just human nature," is insufficient reason, and just plain laziness, carelessness.

97. Ibid., p.42.
98. Natalie Bauer-Lechner, *Erinnerungen an Gustav Mahler* (Leipzig, 1923), p.78.

It obviously *is* "human nature," since it has been a performance problem since time immemorial. For example, Gluck, the creator of operas that called for greater dramatic as well as musical coherence, made demands on performers that were wholly new at the time (18th century) and foreshadowed much of modern performance. Gluck, evidently a tough-minded taskmaster, insisted on the proper dynamics, on real *pianos, pianissimos, fortes,* and *fortissimos,* denouncing the "lazy" omnipresent *mezzofortissimo.*

Weingartner's writings remind us that the abuse of dynamics, both inadvertent and conscious, was still very common in the late 19th and early 20th century. In 1905 Weingartner, in the second edition of his book, was compelled to remind conductors (and musicians) "to observe most precisely whether an accent comes in a *forte* or in a *piano* passage, which will determine quite different grades of strength and expression for it.[99] It is also of the utmost importance whether a succession of accents occurs in a passage proceeding in uniform loudness or during a crescendo or diminuendo; in the latter case the accents also must, of course, have their own gradual increase or decrease."[100]

Strauss echoed these thoughts more succinctly when he said, flatly: "One should differentiate carefully between *sfz*'s in Mozart and in Beethoven," implying that the former should never be played roughly, observing more their function in part as "architectonic pillars"; while Beethoven's *sforzatos* "represent explosions of wildest despair and of defiant energy." He is speaking of *sfz*'s, not *fp*'s. Strauss points out that Mozart used *ff* only on rare occasions, that his *f*'s should never be treated roughly and should always retain a certain "beauty of sound."[101] Strauss here is reminding us that there are indeed many kinds of *f,* as different between Debussy and Tchaikovsky, or Wagner and Mahler, as between Mozart and Beethoven.

Echoing Weingartner's admonition about *sforzandos* or *sforzando pianos* in the context of the prevailing dynamic, Strauss singles out one of my own *bêtes noires:* how many times have I witnessed the astonished faces of musicians in rehearsals after correcting the dynamics in the transitional passage (Ex.18) to the E major cadence in the first *Allegro* of Beethoven's Third *Leonore* Overture!

Ex.18

99. Weingartner, *Über das Dirigieren* (1905), p.30. I would be a millionaire if I had a dollar for every time I have, even in my modest career as a conductor, had to point out in rehearsals that an accent or a *sf* is not always necessarily the loudest accent producible, that in fact it may be a *sf* in *p.*

100. Ibid. (1905), p.30. Two important and famous passages to which this admonition applies in particular—an admonition mostly ignored by orchestras and conductors—are in the first movement of Beethoven's Fifth Symphony, mm.38–43 (see also p. 129 in Part III), and Beethoven's Eighth Symphony, again in the first movement, mm.60–65 (and its recapitulation, mm.257–62).

101. *Recollections and Reflections,* p. 57

As Strauss puts it, the *sfp*'s here "should be kept *piano* for four bars, until *forte* is really marked in the score,"[102] that is, moderate *sfp*'s, hardly more than > accents (which Beethoven was not using because they were not yet in common notational use), so as not to disturb the basic dynamic of *p*.

What so many of these falsifying interpretive problems come down to is the intrusion upon the music of the conductor's/performer's ego. I have already addressed this issue, but it will be important and useful to hear what some famous conductors and/or writers have had to say on that subject. Mattheson expressed his concerns on the subject again in *Der vollkommene Capellmeister*. After speaking quite excitedly about bad conductors/directors—"unpracticed," "inefficient . . . swine" and "the lewd nature" of bad directing, he argues—more calmly—that "a director . . . must not be lazy with unconstrained words of praise, but must copiously employ them. But if he . . . must admonish and contradict someone, then he should do so quite seriously, yet as gently and politely as possible." (Toscanini might have benefited from reading this advice.) In the last chapter Mattheson lists "among the fundamentals of a musical director" that he "not completely reject the praiseworthy work of other people and *be only enamored of his own working*" [italic emphasis mine]![103] He then quotes the Latin proverb "Voto non vivitur uno" (It is not done with one).

On the subject of the conductor's ego vis-à-vis the composer and his works, Mattheson's final words (they literally constitute the final paragraph of his monumental tome) are "a sharp power of discernment is required to succeed in divining the sense and meaning of another's [meaning composer's] thoughts. For, anyone who has never learned how the composer might prefer to have it himself will scarcely be able to perform it well, but will often deprive the thing of its true force and charm so that the composer, if he should hear it himself, would hardly know his own work."[104]

Junker, while not dwelling in particular on the subject of the conductor's ego and its possible beneficent or negative impact, did make it clear that he considered a good conductor on a par with the composer in matters interpretational. While acknowledging that the composer, through his own individual notation and "the various types of colorations" available to him, can "better and more completely express these tempo modifications than the conductor," he adds that "it is equally true that the two of them, composer and performer," creator and re-creator, "must work hand in hand, and that varying a tempo, as an auxiliary art, remains indispensable."[105]

Beethoven, Berlioz, and Wagner, as creative, innovative composer-conductors, never questioned their inherent right—at least not in writing—to prescribe what they thought were the basic conductorial rules of behavior in interpreting

102. Ibid., p.60.
103. Mattheson, *Der vollkommene Capellmeister*, pp.480, 484; English edition, pp.864, 871.
104. Ibid., p. 484; English edition, p.871.
105. Junker, *Einige der vornehmsten Pflichten*, p.37.

their own and (by extension in the case of Wagner) others' work.[106] But by the time their progeny had had its way with an ever wider and still growing repertory, entailing ever broader, deeper, and more complex demands, the whole issue of mediocre conductors and/or intrusive egos had become a major problem which some felt honor-bound to comment upon.

Weingartner's comments on this subject are among the most telling in that he tackled it from several important angles. He eases into the subject a bit obliquely, generously giving his conductor targets a benefit of doubt. "If many of the errors"—elsewhere he referred to them as "perversions of style"—committed by conductors "could be supposed to be 'proofs of ardor' and of good intentions, it was in the end regrettable that by the behavior, artistic and personal, of some 'new-modish Bülows' so much attention was directed to the person of the conductor that the audience even came to regard the composers as creatures of their interpreters, and in conjunction with the name of a conductor people spoke of 'his' Beethoven, 'his' Brahms, or 'his' Wagner."[107]

How contemporary this all sounds! Back to Weingartner: "The saddest part of this business was that the chief arena chosen for all these deviations and experiments was our glorious classical music, especially the holiest of all, that of Beethoven, since Bülow had acquired the reputation of a master-conductor of Beethoven, and his followers wanted to outbid him even there".[108]

Elsewhere, in a slightly more philosophical vein, Weingartner speaks the great truth, so little remembered today: "To have given a fine performance of a fine work should be his [the conductor's] greatest triumph, and the legitimate successes of the composer his own." [109] Amen to that!

106. Wagner's ego was certainly boundless, but it is clear that in *Über das Dirigieren* and his other writings he was more concerned with the preservation of the art of music (whose most recent head he considered to be Beethoven) and the highest possible performance practice, than with asserting his own ego or his own conductorial skills. He was very deprecating about the conducting of Mendelssohn, whom he characterized as a talented, elegant but rather slick and entertaining conductor ("glatt und unterhaltend"). He saw in him—as a conductor—someone essentially superficial and empty, and blamed the miserable state of German music (as he saw it) on Mendelssohn's conducting and that of his school: Ferdinand Hiller (1811–85), Felix Otto Dessoff (1835–92), Wilhelm Kalliwoda (1827–93), Ernst Methfessel (1811–86), Carl Reinecke (1824–1910), Julius Rietz (1812–1877), and Franz Lachner (1803–1901), among others. (The only conductors Wagner felt he could praise were Liszt and his student Bülow.) But in all of Wagner's diatribes against Mendelssohn and his school of "time-beaters," there shines through a deep concern to protect the integrity of German music in the lineage of Beethoven and, beyond that, to advance the art of music and its performance practices—needless to say in the direction of a style of interpretation and conducting appropriate to his own musical vision, in particular his music dramas. What he particularly resented about Mendelssohn and his followers was their laying claim (unjustifiably, he thought) to Beethoven's legacy, corrupting it and, to add insult to injury, idealizing their 'objective,' 'antiseptic' aesthetic as Beethovenian classicism. Wagner saw that as a particularly loathsome subversion of Beethoven's true legacy, that is, to reduce Beethoven's music and aesthetic to the naive, neatly packaged forms and expressions of Mozart's time seemed to Wagner all the more outrageous in that it was widely accepted by a gullible public.

107. Weingartner, *Über das Dirigieren* (1905), p.30; English edition, p.29.
108. Ibid. (1895), p.55; (1905), p.36; English edition, p.35
109. Ibid. (1895), p.78; (1905), p.43; (1913), p.58; English edition, p.41.

Bruno Walter and John Barbirolli echoed these thoughts in a slightly different manner. Walter, for example, in his benign and elderly-wise writing on the whole relationship of conductor to his musicians, to the music he performs, and to his audience, points out that music has one "curious propensity," in that in its "acoustic representation music becomes a transmitter of personality: it transmits the ego of the performer more directly to the listener than can any other medium of direct communication from one human being to another."[110] (Dancers and actors might take issue with that singling out of music performance, but the point is otherwise well taken.) Walter goes on to say: "This explains the unequaled personal success of executant musicians of strong individuality, and their breathtaking though transitory impact, which successes of certain interpreters are in no wise a yardstick of true musical culture, as a misled public opinion often believes they are. Not on successes but on achievements depends the standard of the public cultivation of the arts. In proportion as the conductor attempts, and is capable of, satisfying the true purpose and aim of re-creation, he has proved himself the chosen apostle of creative genius and the *faithful servant of his art* [italics added]."[111]

Even more profound are Walter's further admonitions: "The contrast between the two extremes of the re-creative character is shown by the egotistical tendency in the one case, and the selfless manner, in the other, in which the ego is affirmed. The egotist strives, consciously or instinctively, to conquer, to dominate, to triumph, [to achieve] unlimited artistic aggrandizement, [to attain] his ends with ruthless energy. Under the egotist's direction a certain *sameness* (italics Walter's) will descend on all works, one that will detract from the wealth and variety of their creative content, but one which at the same time is capable of giving a strong, nay, overwhelming, impression of personality.

"Selflessness, on the other hand, with an equal investment of personal dynamism wishes to convince, help, advise, and teach. Such an ego does not prey upon others, but seeks to give of itself to . . . the composer and . . . to the players, and thus wield the influence of an educator. The selfless ego strives to extend its power over others, the self-centered ego strives to incorporate others into itself. Between these two extremes in the realm of reproductive art—let us call them the conqueror and the guardian"—brilliant, that!—"there is, of course, every possible kind of gradation and mixture, and the resulting differentiation between the various types of musician serves to enrich our musical life."[112]

Beautifully said. To which one can only add Sir John Barbirolli's wonderful admonition to young conductors: "Make your watchwords integrity and sincerity to yourself and loyalty to the man whose music you are seeking to interpret. Never think 'What can I make of this piece?,' but try to discover what the com-

---

110. Walter, *Von der Musik und vom Musizieren*, English edition, p.122. Walter's brilliant analysis of the ultimate ineffectiveness of tyrannical conducting practices—clearly he had Toscanini and some of his German-Hungarian colleagues in mind—should be read by every aspiring would-be conductor whose inclination it is to lord it over musicians.
111. Ibid., English edition, p.123.
112. Ibid., English edition, p.124

poser meant to say. The goal for all true musicians should be: service to that
great art which it is our privilege to practice." [113]

But let Felix Weingartner, that most eloquent and humble defender of the art
of conducting—even though he didn't always practice what he preached—sum
up this chapter. In his *Über das Dirigieren* he offers a brilliant—and, I believe,
up to that time (1905) unprecedented—aesthetic/professional credo that deals
directly with the relationship of the conductor to the composer and the composition. He writes: [114] "It is *impossible* for a conductor to improve the value of a
work; he can only from time to time lower it. For the *best* that he can do is to
perform the work at a level equivalent to the value, its quality. If the performer's
work is congenial to the composition, then that performer has fulfilled his task
to the highest possible extent. To do 'more' than that is not possible [*Ein 'Mehr'
gibt es nicht*], for there is no conductor in the world who can turn a bad composition into a good one through his interpretation. What is bad *remains* bad, no
matter how well it is played." (I might add this goes for a bad interpretation as
well.) "Indeed, an especially good performance will make the *weakness* of the
work stand out even more than a mediocre one. The sentence 'The work owed
its success to its excellent performance' contains a half-truth, for the performer
has the right to expect full recognition for his contribution to the work, but a
still higher recognition is due the composer, since it is he who has given the
performer in the first place the *possibility* to have a success with his work."

113. Sir John Barbirolli, "The Art of Conducting," *The Penguin Music Magazine* (London, 1947),
p.19.
114. Weingartner (1905), p.17; (1913), (pp.37–38).

Part III

Schwerere Verstöße sind
kaum zu denken.
    —Heinrich Schenker[1]

God is in the details.
    —Mies van der Rohe

Never trust the teller;
trust the tale.
    —D. H. Lawrence

We have now come to that part of the book in which the ideas and concepts discussed and proposed earlier by way of my own philosophy of the art of conducting (Part I), as well as the writings of many legendary historical conductor-composer figures (Part II), will be put to the test of actual practice, as exemplified in hundreds of recorded performances of eight selected major works of the repertory. Conversely, put to the test as well will be the efforts of a multitude of conductors, past and present, dispassionately, objectively, measuring the quality of their performances, their 'interpretations,' against the actual information and content of the scores, as left to us by the great composers. As mentioned earlier, this discussion will not in the main deal with those most refined subtleties of interpretation that mark the greatest performances, those rarefied, sublime, transcendent moments in which the performance reaches spiritual heights which words can no longer express—only the music itself can do so—and which, indeed, even our most sophisticated musical notations can neither capture nor elicit. Such subtleties of interpretation reside in that final highest realm of reproductive performance where inexplicable, indefinable—and unteachable—instincts and intuitions take over, giving a personal, inimitable touch to a performance (or a moment in a performance) which goes beyond the musical notation, beyond the text, and captures that essence, as Mahler once put it,

---

1. "More serious offences are hardly imaginable." Heinrich Schenker, *Beethoven: Fifth Symphony* (Vienna, 1925).

which lies behind the notes.[2] It is in this ultimate realm—that final highest 'percentage' of an interpretation—in which performances *can* legitimately differ, although only in minute and subtle variables. This is precisely so because, as just stated, performance at that realm goes beyond the limitations and finiteness of musical notation to where musical instincts and intuitions, to where taste and intelligence, must perforce take over, where the notation can no longer prove something 'right' or 'wrong.' This is, of course, that final realm where the 'truth'—the essence—of a piece of music lies, and where the greatest musicians can put their personal, unique imprint on a performance—that realm that gives ultimate meaning to the notion of 'interpretation.' The great caution here, of course, is—or should be—that this notation-surpassing interpretation must in the first instance be derived from *within* that same notation, must be built upon all the inherently valuable information which *is* in the score. If it surpasses the score, it must not have bypassed it.

To put it another way, those aspects of a performance that transcend the limitations of the score, that explore regions beyond the scope of musical notation, should represent the final stages of an interpretation which *in all other respects and at all other levels* is wrung from a faithful, rigorous, intelligent, disciplined reading of the text. If an interpretation—no matter how compelling, how exciting, no matter how sublime at certain moments—is achieved from outside the score's basic information, to the extent that it ignores the score, it is to that extent invalid. Ignoring the fundamentals, the hard facts, of the text— tempos, dynamics, phrasings, articulations, form and structure, the balance of primary and secondary voices, etc.—will largely (perhaps completely) invalidate whatever momentary individual felicities the performance may offer.

So we are left with those aspects of performance—and musical notation— which are fundamental, which are precise and unequivocal, which are measurable, and which in turn permit us to evaluate a performance in relation to at least those fundamentals. But that 'at least' is a great deal; it is in fact, in the repertory discussed and analyzed here, most of what one needs to know and do as a conductor. It is not all that mysterious, as some would have us believe; and thus, measuring the evidence of a performance against the evidence of the score is also not a particularly mysterious process. A recording does not lie; but neither does a score by Beethoven or Brahms. Two solid, objectively assessable entities—a recorded performance and a musical score—stand in reciprocal dependence upon each other, unadorned, unenhanced by the visual, choreographic aspects of conducting and the fantasies of hermeneutic interpretations, and thus, in this naked condition, are objectively comparable and quantifiable.

Staying with these criteria and principles, I will—to borrow a quote from

2. What Mahler is actually reported to have said is "What is best in music is not to be found in the notes," quoted by Bruno Walter in his *Erinnerungen an Gustav Mahler* (Leipzig, 1923). (I am certain that Mahler did not mean thereby to imply that what is found "in the notes" should be ignored.) Weingartner put it similarly in his *Über das Dirigieren* (Leipzig, 1895), urging conductors to "*see what is behind the notes.*"

Furtwängler's 1951 article "Beethoven und Wir" (Beethoven and Us)[3]—"do my best to avoid the expression of (mere) opinions, impressions, perceptions, to deal instead, as much as this is possible in the field of music, entirely with the facts contained in the score, as I am able to glean them from the score."

The conclusion of these analyses are, unfortunately, rather depressing. So many performances—so many conductors fail to show even the most rudimentary respect for the information contained in the scores. It is as if the majority of conductors learn the music they are conducting outside the score, one could say in disregard, in defiance, of it. Even those conductors who are more knowledgeable and respectful of what the score contains tend at one point or another to lapse into some arbitrary, idiosyncratic interpretational foible: a bad tradition, or an easy way out of a difficult performance problem, or a performance cliché or stereotype contradictory to the score but left unquestioned—or simply a plain misreading of the score. Happily, a few conductors—Carlos Kleiber, Bernard Haitink, Stanislav Skrowaczewski, Otmar Suitner, James Levine, occasionally Christoph von Dohnanyi, Claudio Abbado, and John Eliot Gardiner (among the present mature generation); Toscanini, Reiner, Dorati, Albert Coates, Weingartner, and Erich Kleiber of an earlier period—do survive this admittedly stringent test quite well. To me they are the true keepers of the flame of musical, artistic integrity, without sacrificing one iota of the drama, excitement, and emotion that all great music contains and seeks to communicate.

3. In: Wilhelm Furtwängler, *Ton und Wort* (Wiesbaden, 1955), p.223.

Beethoven: Fifth Symphony

There is no work in the entire literature which is more popular and more often recorded—some 160 recordings to date—than the Fifth Symphony of Beethoven. It is also, alas, one of the least understood and most consistently misinterpreted.[1] It has become so popular and, therefore, such a mandatory component of any conductor's repertory, that it is virtually taken for granted, both by conductors (and orchestras) and audiences, too often heard and performed perfunctorily if not mechanically, and worse, in self-indulgent distorted versions that bear little or no relation to the reality of the score. The Beethoven Fifth has become a musical commodity, a musical platitude, which hardly anyone hears with fresh ears, which hardly anyone performs with a sense of discovery, and which hardly anyone treats with the respect this monumental masterpiece deserves—although everyone, of course, claims to do so.

It is thus doubly ironic that the Beethoven Fifth is one of the most difficult works to conduct, one of the most problematic to perform and fulfill all of its creative aspirations. The difficulties begin in multiple fashion in the very first measure, and by the sixth measure the conductor and orchestra have encountered problems that, certainly unprecedented in 1808, present formidable technical and interpretive performance issues to this day.

Any conductor who lunges unthinkingly into the first few measures—probably the most universally familiar four-note motive ever created in all of music —is already in trouble. For, in order to know how properly to declaim that famous opening, one needs to understand its agogic placement by Beethoven in the over-all structuring of the work and, therefore, its correct rhythmic and gestural feeling. There are many ways to play this *incorrectly*. The first question that arises—or should arise (so often in the

1. Its closest contender, on both counts, is undoubtedly Gershwin's *Rhapsody in Blue,* for certainly no famous work has been more mishandled, bowdlerized, dismembered and misinterpreted.

numerous recordings I sampled it never does)—is whether the first three notes represent some sort of upbeat (anacrusis) gesture or a more downbeat-weighted figure. The answer to that question is not entirely obvious nor is the question all that easy to resolve, primarily because Beethoven chose to notate the first movement of his symphony, marked *Allegro con brio*, in a fast 2/4 meter (at ♩ = 108). This means that each pulse beat of the music is contained in a single measure, in turn requiring the movement to be conducted in 'one.' Conducting the movement in 'two' has to be ruled out despite the 2/4 time signature, because (a) it is virtually *physically* impossible, at the very least incredibly exhausting, to conduct a six-minute movement with a fast beat of ♩ = 216; and (b)—more significantly—the music is simply not composed in two beats per bar.

The opening fanfare motive of three short notes followed by one longer note (twice)—a four-bar motive on which virtually the whole movement is rigorously built—is gesturally expressed in single beats: one for the three-note groups, another for the single longer notes. Beethoven should properly have given the music a 1/2 time signature (most 20th-century composers would have done so), but 1/2 was a virtually unknown time designation in the 19th century. Beethoven would have made it a lot easier for us to understand the structuring of the piece, had he written it in 4/4 time, that is, instead of the main theme (mm.6–10) being written as in Ex. 1, writing it (in metric reduction) as in Ex. 2.

Ex. 1

Ex. 2

In the Ex. 2 version, given the generally accepted concept of an hierarchical division in a 4/4 measure of the four beats into 'strong' and 'weak' beats (the first being the strongest, the third the next strongest, the second and fourth significantly weaker), it would have been immediately clear where the main pulses of the music lay. It would also have made it clear that the five measures 6–10 (corresponding to the five quarter-note beats represented in Ex. 2) comprise a four-bar phrase (mm.6–9), i.e. a four-bar theme, in which m.6 (the sixteenth-note gesture ♪ ♩ ♩ ♩ in Ex. 2) is in an anacrusis (upbeat) position. As Beethoven finally notated the theme (in single one-beat measures), the four-bar structuring does not become immediately apparent; after all, the measures all look alike on the face of it. Someone might, for example, read the phrase structuring as in Ex. 3a (♩ representing strong, (♩) less strong, and ⌣ weak beats) rather than as in Ex. 3b. Indeed, thousands of conductors have in fact (mis)read the music that way (as in Ex. 3a) or as in Ex.3c, ever since it was written 190 years ago and performed many *thousands* of times since then.[2]

Example 3b represents the one and only correct phrasing of the theme. Ignorance of this fact or failure on the part of the conductor and the musicians of the orchestra to recognize this can only lead to a wrong and inaccurate performance, particularly since virtually the entire movement is rigorously based on this four-note motive and the ensuing four-bar periodizations.

Now, why is Ex.3b the only 'correct' phrasing of the theme? To answer that question and to confirm the accuracy of this statement, one has to analyze almost the entire structure of the first movement, or at least a number of key structural points. It will become clear in such an analysis that the entire movement, with a handful of notable exceptions, is composed (structured) in four-bar entities. This, by the way, should not come as a surprise to anyone since the vast majority of all classical and Romantic music is composed in four-, eight-, or sixteen-bar phrases, in short in binary or quaternary structuring—three-bar (ternary) structurings being quite rare. What *was* surprising—and new—was that Beethoven chose in this case to notate his four-bar phrases in a single-measure-per-beat meter.

Beethoven had, of course, experimented with such ideas before in, among other instances, the Scherzos of his first four symphonies, all in 3/4 time, yet in two-, four-, or eight-bar one-beat-per-bar periodizations. But those were Scherzos, set in very bright tempos: one would have expected them to be in 'one.' On the other hand, writing a Sonata-Allegro first movement of a symphony in 'one' was a radical break-through. It was an early example of Beethoven's penchant—

---

2. Examples of misreadings abound in both the analytical literature on the symphony and its performance history, including even those of very famous musicologists and conductors (such as Hugo Riemann, Sir George Grove, Theodor Müller-Reuter, Hans von Bülow, Felix Weingartner), as brilliantly recounted, for example, by Heinrich Schenker in his *Beethoven: Fifth Symphony* (Vienna, 1925), pp. 22–35. The most recent misreadings (misinterpretations) are those of Norman del Mar in his *Conducting Beethoven*, (Oxford, 1992).

by the end of his life almost an obsession—for notating his music in metric/
rhythmic extremes: fast tempos in large rhythmic units (whole notes to eighth-
notes), slow tempos in small units (eighth or sixteenths to sixty-fourths, reaching
even 128th-notes). In both directions—fast and slow—Beethoven's tempos were,
in addition, the fastest and slowest, respectively, that had ever been attempted.

This is not the place to engage in a complete structural analysis of the Fifth Symphony's first movement.[3] Suffice it to say—the reader may wish to trace the quaternary structuring in detail from beginning to end (conductors and instrumentalists
are certainly urged to do so)—that the entire movement is constructed out of four-
bar phrases with the following exceptions: (1) mm.224–28 and mm.386–90, two
five-bar phrases (more of that later); (2) the fermata measures 5, 24, 252, 482; and
a number of six-bar phrases which, however, are merely two-bar extensions or
stretchings of four-bar phrases, still inherently binary.

It is essential to know and accept this fact, at least in principle—on some
minor or complex phrase details intelligent minds may differ, as we shall see—
for without this knowledge it is literally impossible to perform the movement
correctly, nay, to even start it correctly, for in turn it is impossible to know how
the famous opening bars are structured agogically. The opening 'motto' phrase,
after all, appears to be a five-bar phrase. But is it?

The answer to this and all similar questions is to be found in the text itself,
in the score; and Beethoven supplies us with many clues. If we accept the fact,
based on the previous supposition, that m.6 (see Ex. 1) represents an anacrusis
measure, an upbeat gesture, then it might be logical to assume the same for
m.1. But if m.1 is an upbeat measure—a fourth beat in a four-beat structure—
and m.2 is a downbeat measure, then why are there *three* more bars before we
get to the next upbeat measure, m.6? The four-bar structuring would seem to
be destroyed right at the outset! The answer is that m.5, the second fermata
measure, is a sustaining extension of m.4 (which represents the third beat in a

---

3. Such an analysis would go beyond the scope and purpose of this study. Only those features of a
score that are absolutely vital to a conductor's understanding of the work will be dealt with here.
While making a complete detailed bar-by-bar harmonic analysis of a work—emphasis on the word
'complete'—is a most desirable undertaking, I cannot in good conscience claim that it is 'absolutely
vital' in order for a conductor to effectively discharge his duties as an interpreter. One cannot, after
all, conduct an F minor chord—all other things (dynamics, instrumentation, rhythms, metric placement etc.) being equal—differently from a G minor chord. Major harmonic stations in tonal music
are, of course, important to know and to hear; so is the *feeling* of the harmonies, not just their
theoretical/intellectual substance; and to that extent my analyses will deal with such matters when
applicable. Still, it is the musicians who produce the actual pitches (harmonies), not the conductor.
The conductor does, on the other hand, crucially affect and determine the rendition of a work in
respect to tempo, rhythm, over-all continuity, and dynamics. To put it another way, only a conductor
can set the tempo, start a piece off on its particular rhythmic and tempo course, the harmonies
simply coming along, as it were, supplied automatically by the musicians.

A more detailed harmonic/structural analysis of the Beethoven Fifth can be found in Schenker's
Beethoven: Fifth Symphony. Schenker was undoubtedly the first—at least in writing—to analyze
correctly the structure of the first movement, both in its long-term harmonic and melodic progression and its metric/rhythmic structuring. He was also the first to make a thorough study of Beethoven's autograph score and the first to provide a critical comparison with the first published edition.

four-beat phrase). But why the extra measure (m.5)? The answer is that Beethoven wanted the second fermata to be longer than the first one, and this was the only way available to him to indicate this in the musical notation of his time.[4]

Again, if Beethoven had chosen to cast the work in 4/4 and written the opening as in Ex. 4, its phrasing and structuring as well as the relative durations of the fermatas would have been instantly clear. Beethoven's original manuscript score (see Plate I) reveals that he initially cast the opening phrase in four bars, but a year or so later—probably in 1809, and long before the first printed edition of the symphony—added an extra measure (as well as at two similar places, the present m.252 and m.482).

Ex. 4

That this conception of the structure of Beethoven's opening motto (and therefore the periodization of the entire movement) was not always understood or agreed upon is evident from some of the earlier writing on the subject. Even Weingartner, in his *Suggestions for the Performance of Beethoven's Symphonies*,[5] fails to see the logic and simplicity of the extra fermata bar being an extension of the previous measure. Instead, he invents an elaborate and complicated theory of explication, which also attempts to explain and rationalize the two aforementioned later five-bar phrases, but at the same time unfortunately completely and mistakenly reorganizes the structuring and periodization of the entire movement.[6] Weingartner also does not seem to have known of Beethoven's original four-bar disposition of the opening motto, writing his exegesis before the time when the facsimile reproduction of the Fifth Symphony autograph became widely available. (However, he could have studied the original manuscript in the Prussian State Library in Berlin, which acquired the autograph in 1878, having been bequeathed to it by the estate of Felix Mendelssohn.)

Writing some sixty years later, Wilhelm Furtwängler gets it right when he points out that Beethoven "intended nothing more nor less than to indicate that the second fermata should be held longer than the first".[7] I suspect that Furtwängler knew this long before 1951, judging by his earlier performances of the Fifth Symphony, including his fine recording of the late 1930s with the Berlin Philharmonic. But then, some paragraphs later, Furtwängler also misses the point when he brackets the main thematic material starting in m.6 as follows (see Ex. 5), thus misrepresenting Beethoven's true phrase structure. This mis-

1. Today we have several ways of showing differing durations of pauses and fermatas, ranging from the shortest to the longest: 〉 , ⸵ , ⌒ , 𝄐 , as well as the by now well-established 𝄐corta and 𝄐lunga .

5. Felix Weingartner, *Ratschläge für Aufführungen der Symphonien Beethovens* (Leipzig, 1890), pp.64, 69.

6. See Schenker's devastating derogation of Weingartner's analysis in his *Beethoven: Fifth Symphony*, pp.31–35.

7. Furtwängler, "Beethoven und Wir," in *Ton und Wort*, (Wiesbaden, 1955); p.225.

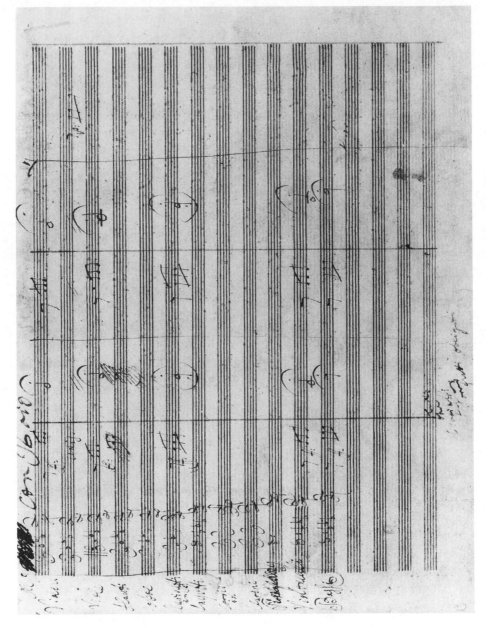

Plate I   Autograph manuscript of the opening of the first movement of Beethoven's Fifth Symphony

marking creates a five-bar phrase at the end of the example, which Furtwängler neglects to explain.[8]

Ex. 5

To return to Beethoven's text, if further confirmation be needed that m.1 is an upbeat measure, it can be found most compellingly in the two empty measures, mm.123–24 (Plate II) and even more emphatically in the last three measures of the movement, representing mm.1–3 in a four-bar structure, meaning, of course, that the final chord is a 'three' (not a 'one') (Plate III). This means further that, since in the notational conventions of the 18th and 19th centuries last and first measures of a piece had to complete the total structure—full circle as it were—the 'three' of the final bar here tells us that m.1 is a 'four,' i.e. an upbeat measure. Measures 123–24 confirm the same idea—otherwise why would Beethoven have structured in these two measures of rest—indicating unequivocally (a) that he was thinking in four-bar phrasings, and (b) that, m.122 being a 'one,' m.1 in the repeat (and, of course, also m.125) are 'fours.' Beyond these very telling structural signs, many additional confirmations are scattered throughout the movement.[9] And there is still further confirmation in the fact that all the strong *harmonic* structural points are in the 'one' position; that is to say, the various tonic positions in the over-all harmonic/tonal scheme and/or other major architectonic tonality-confirming structural points are also all 'ones.'[10]

What is fascinating about the opening motto in terms of harmonic implica-

8. It is possible that the bracketing in Ex. 5 was a printing error, for in the succeeding musical example of a slightly later passage from the first movement, Furtwängler's bracketing is correct. On the other hand, it seems to me from listening carefully to Furtwängler's 1930s recording that he interprets m.1 not as an anacrusis but a downbeat gesture. I must admit it is hard to tell with certainty because, being the first measure and the first sound produced, there is no rhythmic or metric reference point to cling to.

9. The three-note motto is found in mm.18, 22, 59 (horns); 179 (violins); 195 (wood-winds); 248, 303, 478 (trumpets and timpani), 491.

10. On the C minor tonic—mm.19, 254 (recapitulation), 304, 374, 423, 439, 484; on the dominant (V) positions as, for example, mm.26, 346, 399, or 471; in the Eb major 'second subject' episode—mm.60, 110, 118, 122; and various F minor subdominant (IV) episodes—mm.130, 196, 241.

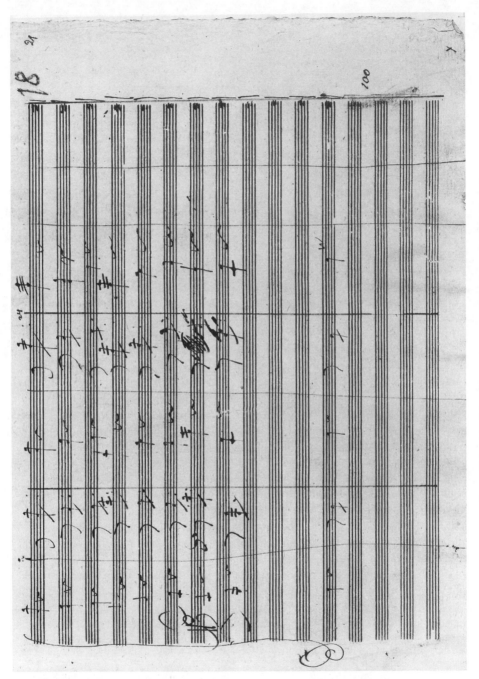

Plate II Autograph manuscript of mm.118–124 of the first movement of Beethoven's Fifth Symphony

Plate III   Beethoven, Fifth Symphony, m.472 to end of first movement

tions is that it is actually harmonically ambiguous. There are in the first five bars no unequivocally clear, explicitly stated harmonic specifications—to put it another way, no clarifying chords. We see simply unison octaves which theoretically could have any number of harmonic associations. The first pitches, G and Eb, could for example be heard in the key of Eb, and indeed are apt to be heard that way, since, again, there is no previous clear-cut harmonic reference point. I can well imagine that someone hearing those first two measures for the very first time could very likely think and hear the key of Eb. But we know, of course, that Beethoven's Fifth is in C minor; and after having heard the piece dozens or hundreds of times, we tend to hear those opening measures in that key, simply by prior association, by previous reference and memory. The next two pitches, F and D, are also harmonically ambiguous because they could be heard, for example, as belonging to the key of Bb, especially if one has heard the first two pitches (G and Eb) in Eb major. If Beethoven had on the other

hand written F and Bb in mm.3–5 , that

would have confirmed for us the Eb-ness of the first two bars. The F and D, however, strongly imply by prior association a dominant (G), with F its seventh, D its fifth, and thus set things up for the real beginning of the body of the movement in mm.6 and 7 in a very clear, unmistakable C minor.

I find it curious that Beethoven used only clarinets in the winds—no bassoons and horns. Given the thunderous *ff* dynamic with which he heard this motto, he might have considered the horns and bassoons as logical instruments to increase the power of the opening statement. But I suspect that he omitted the horns because the D in mm.4 and 5 is not a 'natural' note on the Eb hand-

horn—it would be a 'stopped' note, B 𝄞 —even though it was a rela-

tively accessible note and was frequently called for in the works of Mozart and Haydn and many other composers of the period. Beethoven used the horns very conservatively in his Fifth Symphony, resorting to altered notes only in three places: one note in the first, one passage in the third, and one in the fourth movement.[11] In any case, it is devilishly hard to make the two clarinets heard, especially with the large string sections of today. All one can ask is that the clarinet players give their all and in good acoustics their instrumental color might be heard—or at least sensed. (In recordings, with available microphone techniques, I do not see why the clarinets cannot be made audible. Yet on only one of the nearly seventy recordings analyzed here [Norrington's] were the clarinets discernible—and then only because they held their note longer than the strings!)

There are two additional performance problems in the first six bars that must

11. Here Weingartner errs grievously when he states (in his *Suggestions* p.65) that Beethoven "makes, especially in this symphony, more frequent use of stopped notes ["Stopftöne"] than elsewhere." On the other hand, Weingartner does suggest another—and reasonable—possibility for Beethoven's not using the horns in mm.1–5: in his words, "that he saved them for the later dynamically amplified appearance of this theme."

still be resolved. One is the problem of not having the first three notes sound
like a triplet (| 𝄽 ♪♪♪ | 𝅗𝅥 |). They surely will sound thus if the first note
is played with an excessive attack or accent. The way to avoid this is for all the
musicians to think and to feel a slight weight on the second note. For just as in
the hierarchy of a 4/4 bar, 'one' and 'three' have a greater weight than 'two' and
'four,' so in a four-note group of eighth-notes, 'one' and 'three' also carry more
weight than 'two' and 'four'—except that, of course, in this case Beethoven
doesn't give us the 'one.'

The second problem—and it is one of the more difficult baton-technical
problems in the symphonic repertory—is the need to go directly from m.2 into
m.3 and from m.5 into m.6. Under no circumstances must the conductor allow
extra empty bars or pauses here. We already know that Beethoven was quite
capable of writing empty measures when he needed or wanted them, especially
when it meant preserving structural integrity and logic (as in mm.123–24
or m.301 and m.389). Had Beethoven wanted a bar rest between the end
of the fifth bar D and the entrance of the second violins in m.6, for example,
he would have written as much. The fact that he didn't should tell us very
clearly that any extra measure or extra pause of any kind is not permissible here,
because it violates the internal structure and rigorous logic of the first movement's architecture.

Technically this means that here—and in all similar places in the movement[12]—the conductor must release the *ff* fermata note in m.5 simultaneously
with the downbeat of the next bar. But since this 'next bar' is in all instances
marked *p* and is, as we have seen, an upbeat gesture, the conductor's beat here
must also represent that *p*—in other words with a small light beat—and it
should express the upbeat feeling of the measure as well. An upbeat gesture, as
in the fourth beat of a 4/4 conducting pattern, works very well here. All this is
not easy and the main reason why the opening of Beethoven's Fifth is considered to be one of the most feared conducting challenges in the entire classical
literature.

It is sad to report that very few conductors in our recorded sampling—some
ninety-odd recordings—passed the test of these opening measures. Arthur Nikisch in the first complete recording ever made of the Beethoven Fifth—as a
matter of fact the first complete recording of any symphony—recorded in 1913
with the Berlin Philharmonic, manages the first measure quite well, but holds
the fermatas much too long[13] and at equal duration, thus ignoring Beethoven's

12. Measures 21–22, 24–25, mm.128–29, mm.249–50, 252–53, etc.; the reader may wish to refer
to the score for these examples.
13. Nikisch was here undoubtedly following Wagner's advice who, in his *Über das Dirigieren* (p.25)
fantasizes that the voice of Beethoven cries out to conductors from his grave: "Make my fermatas
long and terrifying! I did not use these fermatas lightly or as a moment of hesitation, before thinking
what to do next; rather . . . to throw into the violent and fast *allegro* figurations, when necessary, a
pleasurable or terrifying holding back." (Wagner actually uses the words "anhaltenden Krampf,"
literally translated an "impeding [or holding back] spasm [or constriction]"—surely a striking example of Wagner's sometime penchant for hyperbole.) "Thus shall the life of these notes be drained to

added measure and its meaning. Worse yet, he inserts a full two extra measures between m.5 and m.6. (It is well to add that after this somewhat erratic start, Nikisch settles down to what is in many ways a very respectable, at times even remarkable, rendition of the symphony.)

His performance raises the question of the length of the two fermatas. No one can, of course, prescribe their length precisely and with certainty; it is, like all such things, to some extent a matter of taste and feeling. But I would like to suggest two guiding principles: First, contrary to Wagner's fanciful suggestion and to Nikisch's implementation of that suggestion, the fermatas should not be very long—the second, of course, slightly longer than the first—so as not to impede the intended flow and energy of the opening statement. Let us remember that Beethoven's tempo marking is *allegro con brio*. It seems a little ridiculous to bring the motion of the music to a virtual standstill (as in Nikisch's performance) when the music has just barely gotten started. Relatively short fermatas, which keep the tremendous energy of the opening gesture fully charged, are recommended. Second, I strongly urge that the fermatas not be

---

their last drop of blood; thus do I hold back the waves of my sea and look into the abyss, or slow the passage of the clouds, scatter the trails of mist and look into the pure blue ether, into the radiant eye of the sun.*That* is why I use fermatas, as suddenly intervening, long sustained notes in my *allegros.*"

For all of Wagner's abject admiration and general understanding of Beethoven's symphonies, in this instance his imagination ran wild in a frenzied hermeneutic fantasy that bears no relationship to the substance of the written score. As Schenker in his biting comment on Wagner's poetic effusion puts it, referring as well to Schindler's oft-cited but dubious claim that Beethoven considered the opening motive of the Fifth Symphony to represent 'fate knocking on the door,' "even if we would like to think of Beethoven wrestling with fate throughout the [first] movement, then not only fate would be involved in this wrestling, but Beethoven himself, and not just Beethoven the man but, much more, Beethoven the musician. If Beethoven was really wrestling with notes, then no legends and no hermeneutic interpretations to explain this tonal world will suffice, if one fails to think and feel these notes as they themselves think [and feel]." (Schenker, *Beethoven: Fifth Symphony*, p.7)

Parts of the Beethoven Fifth had been recorded before 1920, mostly the *Andante con moto*, at the time the most popular of the four movements. These recordings, dating from as early as 1911, were made by recording company 'house' or 'resident' orchestras (Victor Concert Orchestra) and all-purpose groups such as Prince's Orchestra, but also by the New York Philharmonic, led by Josef Stransky. They show that these performances were innocent of any imagination or respect for the score, usually severely cut and mutilated to fit onto the single-sided wax discs of those days. Instead of string basses, tubas were used, and it is rather comical to hear them struggling with the famous thirty-second-note cello/bass passage at m.114 in the *Andante* movement. On some recordings it sounds like only two or three violins and one cello were used. The only performer who sounds truly professional by today's standards, indeed beautiful, is the then-young principal bassoonist of the New York Philharmonic at the time, Benjamin Kohon (in the Stransky recording). Intonation in these earliest recordings was usually pretty execrable, and the whole purpose of these recordings seems to have been to present Beethoven as a good 'tune' composer who could compete with the popular songs of the day and with Caruso's super-popular recordings.

It is most enlightening to compare these performances with Toscanini's 1921 recording of the last movement of Beethoven's First Symphony with the La Scala Orchestra, a stunning rendition, both interpretively and technically. How Toscanini must have worked those poor opera musicians over to achieve such an excellent rendition!

timed out in exact multiples of a measure's tempo, that is to say, for example, an exact three measures on the first fermata, four on the second.[14] The fermatas should be free in time, unpredictable in length, spontaneously and intuitively generated, so that m.3 and m.6 come somewhat as a surprise—perhaps even to the conductor. A predetermined, calculated duration of these fermatas undermines the spontaneous energy and drama of the music, and makes the opening motto sound ordinary.

If we have taken this much time and space to understand only the first five measures of the symphony, it reflects accurately the importance and difficulty of performance of this most famous of symphonic openings. And as previously indicated, our conductors did not for the most part stand the performance test very well.[15] (See Fig.1)

As for tempo, no one adheres to Beethoven's marking of ♩ = 108, except Norrington, Brüggen, Gardiner. The closest are Karajan and Dohnanyi (♩ = 104); Toscanini, Reiner, Steinberg, Carlos Kleiber, and Giulini (a surprise) with ♩ = 100. Furtwängler—another surprise—comes next, clocking in at 98, with Böhm, Klemperer, Kubelik, and Krips bringing up the rear with ponderous tempos of ♩ = 80 or below. A chart of the tempos taken by all the conductors whose recordings were sampled is shown in Fig.2.

Resistance to Beethoven's metronome markings is, of course, of long standing—virtually a venerable tradition—going back to Schindler's and Nottebohm's days, Schindler going so far as to make the dubious and unsubstantiated claim that Beethoven wanted the first five measures performed at a much slower tempo than the rest of the movement—he suggested ♩ = 126 (i.e., ♩ = 63)—beginning the real *allegro* only in m.6; while Nottebohm, more cautious, thought that Beethoven's "too fast metronomization" was "probably" the result of his having "determined the metronomic indications at the piano," thus arriving at markings "which he [Beethoven] could hardly have advocated for the

14. Bülow did something like this, we are told, calculating and maintaining an exact predetermined duration of multiple measures for the fermatas throughout the movement. Unfortunately, Igor Markevitch also suggests such an approach in his *Die Sinfonien von Ludwig van Beethoven* (Leipzig, 1983). Worse yet, he develops a whole new metric/structural analysis for the entire movement based on his strictly measured fermatas. Making the first fermata exactly three measures long, the second four measures long, he then reasons that these measure durations add up precisely to make neat four- and six-bar phrases, as follows (Ex. 6). This conception is based, however, on two quite erroneous premises: (1) it completely disregards the fact that the eighth-note figures are upbeat measures, not 'ones' (although Markevitch is even inconsistent on that point, because he positions the third group of eighth-notes in the last measure of a six-bar structure, which makes it an upbeat figure); (2) It makes the assumption by implication that fermatas should always be measured and strictly calculated—a totally untenable theory. Or did he mean that fermatas should be so controlled only in the first movement of the Fifth Symphony?

Ex. 6

15. For discographical details regarding all the recordings herein, see Discography, p. 549.

Fig. 1

positive	negative	positive	negative	positive
M.1 not tripletized	Fermatas same length	Fermatas differentiated	Extra mm. (mm.2–3 and/or mm.5–6)	No extra mm. added
Dorati	Ansermet	Abbado	Bernstein [3,4]	Ansermet
Harnoncourt	Boult	Ashkenazy	Böhm	Ashkenazy
Karajan	Brüggen	Bernstein	Boult	C. Davis
Markevitch	Gardiner	Böhm	Brüggen [5]	Dohnanyi
Masur	Giulini	C. Davis	DeSabata	Haitink
Scherchen	Harnoncourt	DeSabata	Dorati	C. Kleiber
Suitner	Hickox	Dohnanyi	Furtwängler	Kletzki
Szell	E. Kleiber	Dorati	Jochum	Kubelik
Van Otterloo	Klemperer [1]	Furtwängler	Knappertsbusch	Leibowitz
	Koussevitsky	Haitink	Mitropoulos	Mehta
	Kubelik	Jochum	Norrington	Mengelberg
	Leibowitz	Karajan	Nikisch	Muti
	Masur	Kempe	Ormandy	Ozawa
	Mengelberg	C. Kleiber	Reiner	Sawallisch
	Mitropoulos	Kletzki	Schuricht	Steinberg
	Norrington [6]	Knappertsbusch	Solti	Suitner
	Schalk	Krips	Stokowski	Szell
	Stokowski	Maazel	Strauss	Thomas
	Thomas	Mehta	Szell	Weingartner
	Van Otterloo	Munch	Toscanini	
	Wand	Muti	Wand	
	Weingartner	Nikisch		
		Ormandy		
		Ozawa		
		Sawallisch		
		Schuricht		
		Schwarz		
		Strauss		
		Suitner		
		Szell		
		Toscanini		
		Walter		
		Weingartner [2]		

1. Klemperer's undifferentiated fermatas appear to be of the Bülow type: exactly measured and predetermined.
2. Only in the repeat.
3. By adding the extra measure(s), these conductors completely destroy the four-bar-ness (*Viertaktigkeit*) of the motive.
4. Bernstein's insertion of these extra empty measures is more surprising than Toscanini's. Toscanini was, after all, not an 'intellect,' someone with a remarkably analytic mind; he was a musician with superior instincts and an uncomplicated, unegotistical view of music. But Bernstein was a kind of intellect who frequently thought in deeply analytic ways, and who knew, as his various television explications of Beethoven's work demonstrated, how tightly, how perfectly, how rigorously Beethoven constructed the first movement of the C minor symphony. And yet he seems to have been totally unaware of adding, willy-nilly, unwanted measures to Beethoven's strict form.
5. Only in m.480.
6. Norrington's second fermata is shorter (*sic*) than the first one.

Fig. 2.

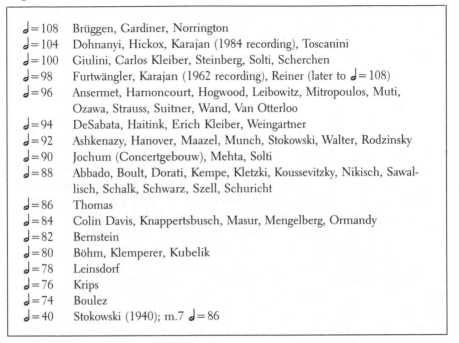

♩ = 108	Brüggen, Gardiner, Norrington
♩ = 104	Dohnanyi, Hickox, Karajan (1984 recording), Toscanini
♩ = 100	Giulini, Carlos Kleiber, Steinberg, Solti, Scherchen
♩ = 98	Furtwängler, Karajan (1962 recording), Reiner (later to ♩ = 108)
♩ = 96	Ansermet, Harnoncourt, Hogwood, Leibowitz, Mitropoulos, Muti, Ozawa, Strauss, Suitner, Wand, Van Otterloo
♩ = 94	DeSabata, Haitink, Erich Kleiber, Weingartner
♩ = 92	Ashkenazy, Hanover, Maazel, Munch, Stokowski, Walter, Rodzinsky
♩ = 90	Jochum (Concertgebouw), Mehta, Solti
♩ = 88	Abbado, Boult, Dorati, Kempe, Kletzki, Koussevitzky, Nikisch, Sawallisch, Schalk, Schwarz, Szell, Schuricht
♩ = 86	Thomas
♩ = 84	Colin Davis, Knappertsbusch, Masur, Mengelberg, Ormandy
♩ = 82	Bernstein
♩ = 80	Böhm, Klemperer, Kubelik
♩ = 78	Leinsdorf
♩ = 76	Krips
♩ = 74	Boulez
♩ = 40	Stokowski (1940); m.7 ♩ = 86

concert hall." Bülow, one gathers, must have also taken very deliberate tempos in this movement, because it is known[16] that he rearranged the whole movement in a 2/2 *alla breve*, thusly (Ex. 7), a tactic which almost certainly must have led to a slower tempo conception.

Ex. 7a

Having navigated the craggy reefs of Beethoven's opening signature motive, we can now continue our exploration of the first movement with the actual main theme at m.6. But immediately new performance/interpretive problems arise. For, on the assumption that we are now going to be dealing with four-bar phrases with a definite hierarchical division of each four bars into strong, less strong, and weak beats, it becomes imperative for the conductor, but more crucially the musicians, to know on which beat of a four-bar entity they find themselves.

The theory has often been advanced that the Fifth Symphony's first movement should be felt and performed in two-bar structuring. Four-bar entities, of course, divide easily into two-bar units and, the truth be told, it would in many

16. So described in Theodor Müller-Reuter's study "*On the Rhythmic Meaning of the Main Motive in the First Movement of Beethoven's C minor Symphony*," an article in an 1898 issue of *Musikalisches Wochenblatt*.

instances be impossible to tell whether a performance is being interpreted in fours or twos. Also, there are passages, such as mm.34–43, that perhaps really should be felt in groups of 'twos' (mm.34–37) and then in 'ones' (mm.38–43); similarly in mm.158–67 (all 'twos'). But the great danger in conceiving the performance of the first movement entirely in two-bar units is that it can easily lead to accentuations every alternate bar, which over the long haul becomes not only boring but destructive of the longer periodization spans, as, for example, Koussevitzky's recording demonstrates.

I would also suggest that a two-bar structuring would severely undermine the basic architecture and the very conception of the piece at the outset, for it would be incompatible with the four-bar-ness (the "Viertaktigkeit", as Schenker called it) of Beethoven's opening signature motive, which surely must be heard as a four-bar entity. In principle it is better to think of the movement as being in quarternary structuring, with the obvious binary phrasings (such as mm.34–37) coming quite naturally, and yet fitting into the four-bar periodization.

Certain is that the movement cannot—should not—be performed as an endless series of undifferentiated 'ones.' Even Beethoven's ingenious motivic concision and almost endless inventiveness in varying and shifting his little four-note motive around, cannot survive such cavalier and insensitive treatment. But how often the work has been performed exactly in such a careless, perfunctory fashion!

While the second violins may by now have been told, should they not have understood it that way, that their three G's in m.6 have an upbeat character, it is equally important for the violas to know that they are coming off a strong downbeat 'one' (confirmed by the clear establishment of the main tonality of C minor), and that the first violins are in a 'two' position, a weak beat, leading towards the C, a 'third' beat in the four-bar phrase. Such an understanding is not only applicable here, however, but *throughout the entire movement*. But since no orchestra of sixty to eighty musicians is likely to remember at every moment where they are in the four-bar structuring and since, as just noted, Beethoven constantly moves his motivic materials around, shifting them into unexpected places (beats), it is very useful to mark the orchestral parts with tiny but clearly visible brackets (⌐——) every four bars, thus clearly, visually delineating the architectural design, including, of course, the already mentioned deviations from the four-bar groupings.

It may be interesting and instructive for the reader to see what such an instrumental part would look like. Herewith three brief examples, all from the first movement: the first violins' mm.7–23, the cellos' and basses', mm.26–43 and mm.44–67.

I have also, when conducting this symphony, resorted to subtly beating a four-beat pattern (as in a 4/4 measure)—not all the time, but intermittently at certain points—to give the musicians another visual indication at a glance—a confirmation, perhaps—of where phrase-wise in the relentless flow of bars they happen to be at the moment. I emphasize the words "subtly" and "intermittently," for ultimately we must not subvert Beethoven's conception of the music of this movement on one structural level as being, essentially, in 'one.' To deprive it of that 'oneness' entirely would be to do severe damage to the feeling and expression of the music. It must retain that inexorable, relentless, to some extent 'driven' feeling and pacing which can only be achieved by beating in 'one.' The fact that conducting in 'one' is the most restrictive of all the beat patterns—obviously so, because the conductor can only show a constant series of downbeats (not even a 'two,' in turn necessitating a quick rebounding upward from the bottom of the downward beat to be able to descend again for the next downbeat)—is in itself a visual representation of the relentlessness of the music's motion. This then confirms for me how wondrous and correct Beethoven's conception of the first movement is in notating it in a fast 2/4, felt (and conducted) in one.

Following through on our established four-bar patterning, we arrive at another 'one' in m.19—it is well for the winds, timpani, and basses to know that in m.18 they are in an upbeat position—and thus a 'three' in m.21, the first violins' sustained fermata. (This is hopefully again not too long, in fact, one well-held bow.) This in turn puts us on a 'four' (upbeat) in m.22 and another 'one' in m.23,

Some may scoff at the next idea or consider it naive, but if the musicians all hear the Ab in m.22 as the minor ninth of the dominant G, and the F in m.23 as the seventh, it will make a tremendous difference in the rightness of the sound of these measures, not only in terms of intonation, but one will clearly be able to hear (and transmit to the listener, the audience) the *feeling* of the dominant in the basic tonality.

Here, however, we have a structural anomaly. If m.23 is a 'one' and, by analogy to mm.6–9, m.25 is a 'four' and m.26 another 'one,' then Beethoven ended up one bar short in completing a four-bar phrase. Was he simply compensating for the 'extra' measure in m.4 by now subtracting one? Perhaps; but it is much more likely that m.24 and its fermata are intended to comprise both the 'two' and 'three' of the present four-bar phrase, and that Beethoven was reluctant to expand the F to three bars ![musical notation]. Several possible reasons come to mind: (1) he did not want that fermata to be longer—and longer looking—than the previous one (mm.4–5); (2) it would have been the only such phrase and duration in the entire movement; (3) perhaps he saw the three measures (above) as a visually disturbing and ill-fitting deviation in the otherwise so consistently binary and binary-looking design; (4) perhaps he simply neglected (or forgot) to make such a change.

The approach to this first expressive and structural climax in the movement has led to many different (but mostly misguided) interpretations, ranging from holding the violins' sustained G excessively long (Karajan and Munch, for example) to making a noticeable, even huge, ritard in mm.19–20 into m.21 (Walter, Karajan, Jochum); or—conversely—actually rushing into m.21 (as Strauss does), and then slowing up m.22 (many conductors) and/or adding an extra empty measure between m.21 and m.22 (Nikisch, Furtwängler, Strauss, Reiner, Koussevitsky, Toscanini, Munch, Bernstein) and between mm.24 and 25 (Furtwängler, Jochum), or, as Brüggen and Harnoncourt and some of the other 'period instrument' conductors do, make a big diminuendo on the f-sustained G and many of the other first movement fermatas. In many recordings, the ten winds now additionally playing in m.22 are barely audible—in Giulini's recording totally inaudible—thus subverting Beethoven's idea of timbrally and dynamically reinforcing the strings.

In any case, by m.26 we are once again on structural terra firma, still on the dominant (G), the A^b's in the first violins and violas having, apart from being on weak beats, a melancholy *minore* feeling.[17]

I now point to the cellos and basses in m.28–9 to remind us that this motivic figure, derived from the main theme, of course, is situated on a 'three' and 'four' of the 4-bar phrasing (transcribed in metric reduction as ![musical notation]). It really does make a tremendous difference in the sound and feeling of the music, if the cellists and bassists know that they are to play those four notes in that metric frame of reference, not merely some isolated, unrelated notes flying by at about a half a second per bar. It makes a tremendous difference—as it does for any instrument (see 1.oboe, 1.clarinet, two horns in mm.32–33) whether one is playing the motive on 'three' or 'four' or 'one' or 'two', for the simple reason that each of these four beats has

17. For those who assume, or presume to know, that Furtwängler was constantly drawn to overly slow tempos—a myth which on close inspection is not entirely tenable—it might be well to realize that he is one of the very few conductors in my entire sampling who is virtually on Beethoven's metronome mark in m.26 (Furtwängler ♩ = 104 to Beethoven's 108).

its own 'personality,' as it were, its own function (and feeling) in the hierarchical scheme. Translated (again) into a 4/4 meter for easier accessibility, the four

(a) [musical notation] variants, should all be played differently, espe-

(b) [musical notation] cially in terms of directionality, (a) rebounding from a strong beat, ending on a weak beat; (b)

(c) [musical notation] starting on a weak beat, heading for a stronger beat; (c) somewhat like (a), but slightly 'weaker';

(d) [musical notation] and (d) somewhat like (b), but thrusting towards an even stronger, indeed the strongest (often to-

nality-confirming) of the four beats, a 'one.' If this kind of structural conscientiousness is maintained by the entire orchestra (and, of course, the conductor), Beethoven's first movement music will have a wonderful spontaneity and flexibility of feeling, a variety in pulse and flow, while the merely routine, unknowing, undifferentiatedly mechanical rendition will turn it into deadly boredom. This sort of awareness may be a lot to ask of musicians, but then Beethoven in all his great music always asks a lot, indeed pushes our human talents to their limits, as he pushed his own.

I must here speak of another performance practice question that one constantly encounters in the works of Beethoven, Mozart, Haydn, and many other composers of the 'classic' period, namely, the ambiguous notation of phrase-ending final notes, almost always written as quarter-notes (as those in m.29 in the lower strings, or the winds in m.33, the horns in m.35, the oboes and bassoons in m.37, and hundreds of similar places in this symphony alone). In almost all cases these were to be played as eighth-notes. The fact is that, as any reasonably comprehensive study and analysis of scores (printed and manuscript) of the period will show, it was at the time a notation and printing convention to place quarter-notes at the downbeat endings of phrases—or, to put it more precisely, to use that rhythmic unit which was contained in the time signature, therefore eighth-notes, not sixteenths, when the meter was defined in eighths; quarter-notes when these were specified in a time signature. Furthermore, the musicians of the time understood—and were expected to understand—this particular convention, and that they were to fit such release notes to the duration of the prevailing primary voice or melody. Thus, for example, the quarter-note in m.29 of the Beethoven Fifth's first movement is to be played by the second violins, violas, cellos and basses as an eighth-note, not as a full-length quarter-note (which a pedantically strict reading might indicate), if for no other reason—quite apart from the notation and performing conventions just mentioned—that the first violins' A♭ would conflict disturbingly with the C minor chord of the lower strings.

I should note in passing that many phrasings and bowings in the widely used Breitkopf and Härtel edition need some revision or adjustment, since they are in discrepancy with Beethoven's manuscript (as Schenker pointed out as early as 1925.) I will cite here only a few and let them stand for a host of others, these omission or errors being mostly a result of a combination of hasty engraving and inadequate proofreading on the part of the original editor and publisher (Ex. 8a, b, and c). The parallel passage (mm.138–42) *has* the slur in the bassoons. Other parallel passages of these phrasings (mm.138–42) also contain the slurs.

Ex. 8a (mm.15–18)   Ex. 8b (mm.28–29, 32–33)
Bassoons and Cellos   1st and 2nd Violins

Ex. 8c (m.34–38, 278–82)
1st Violins

Since mm. 34 and 38 are 'ones,' structurally speaking, it follows, if we maintain a strict sequential four-bar reading of the score, that m.42 will be another 'one.' But clearly it isn't; instead m.44 is (Ex. 9). This is confirmed by at least two facts: first, m.44 represents the arrival point, the peak, of the crescendo which began in m.34; second, m.44 clearly marks the unequivocal arrival and confirmation of the basic tonality of C minor, further confirmed by the next twelve bars' obvious division into three four-bar phrase units. If m.44 is a 'one,' then mm.42 and 43 are extraneous to the previously ongoing four-bar- patterning. They are therefore an extension, a stretching, of the four-bar phrase starting at m.38 to six bars, a device Beethoven uses many times in this movement (and in numerous other works, of course). Obviously, however, this six-bar phrase is still divisible into a binary format (three times two).

Ex. 9

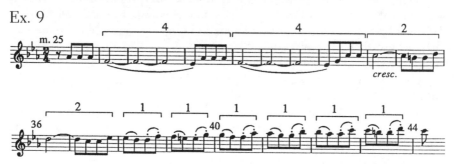

I call mm.34–43 a six-bar phrasing. But there are several other ways of looking at this passage, one of which is cited by Furtwängler[18] and was alluded to earlier: two four-bar phrases, then compressed motivically to two two-bar and finally further to six one-bar units (see the phrase bracketing in Ex. 9).

In this passage it is most important to observe Beethoven's dynamics, alas, so often totally ignored (especially by Bernstein, and even Toscanini, both of whom

18. Furtwängler, "Beethoven und Wir," p.228.

crescendo here too much too early). Measure 34[19] must start *p*—with no preceding premature crescendo—and lead through six successive *sf*'s to only one *f*,
which is followed by a (*subito*) *ff* nine bars later. The *sf*'s must be played in the
context of the rising over-all dynamic level, rather than, as is far too often the
case, as six hard-hitting equally loud *sf*'s. (For most musicians, unfortunately, *sf*
means a loud—or loudest—aggressive attack, regardless of the prevailing dynamic level; and many conductors, out of ignorance or timidity, fail to teach the
orchestra otherwise.) My suggestion is to mark the first *sf* (m.38) as *mpsf*, the
fourth one (m.41) as *mfsf*. There is nothing worse than a premature and exaggerated crescendo, for—as mentioned earlier—when it climaxes too early, the true
arrival point, which should be the real climax, becomes a mere anti-climax.
Furthermore, in this particular passage (mm.38–43) Beethoven achieves the bite
and increasing tension primarily *harmonically*—if only performers would be
aware of this miraculously dissonant progression (see Ex. 10), representing the
sf downbeats in mm.38–43). One way to make an orchestra appreciate the remarkable harmonic daring of this passage, is to play the chords (as in Ex. 10)
very slowly and well sustained. It is a startling ear- and mind-opening experience. If instead of the loud undisciplined bashing away at the six downbeats
which one usually encounters in this passage, players (and conductors) would
hear these extraordinary clashing dissonances and hear them as an intensifying
progression, Beethoven's intentions would be much better served!

Ex. 10

Speaking of dynamics, it should be no longer necessary to point out that in
Beethoven symphonies—Mozart and Haydn as well—trumpets and timpani,
and sometimes even horns, have to play at modified, that is to say, lesser dy-

---

19. Beethoven's original score indicates a crescendo in the horns in m.34 not to be found in the
first edition score and parts. Unfortunately, on quite a few recordings (notably those of Toscanini,
Muti, Mehta, Colin Davis, Gardiner, Leibowitz, Giulini, Krips, Norrington, Hogwood, Harnoncourt, and Carlos Kleiber) the horns here make an unpleasant out-of-context crescendo swoop:

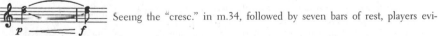 Seeing the "cresc." in m.34, followed by seven bars of rest, players evi-

dently did not realize that Beethoven's crescendo only *begins* here and then takes a whole ten bars
to reach *f*. That the conductors in question did not hear this dynamic exaggeration, probably even
encouraged it, is astounding to me.

Brüggen avoids the horns' *cresc.* in m.34, but has the oboes and bassoons make an even worse
completely out-of-context crescendo swoop.

namic levels than marked. Since Beethoven, unlike later composers such as Strauss and Mahler, even Schumann, used undifferentiated, vertically uniform dynamics, trumpet parts, if played at the absolute dynamic levels notated, would constantly predominate, overbalancing such less projecting instruments as flutes and clarinets. When this occurs it is especially deplorable, since the trumpets of Beethoven's day were by their nature limited to a few 'natural' tones, which rarely permitted them to participate in melodic lines or chromatic alterations, and thus relegated them to sustaining tonic and dominant pitches (an occasional third could sometimes be sneaked in). These, of course, when played too loudly are not only boring to hear, but severely distort the true picture of the music. It is distressing to hear in the dozens of sampled recordings herein analyzed how many, many times Beethoven's music is devastated by loud, overbearing trumpet and timpani playing, and—worse yet—tolerated not only by conductors but also by recording producers and engineers.[20]

While on the subject of balances and instrumental ensemble, we should note in mm.44–51 the somewhat unusual low-lying thirds of the two bassoons, three octaves below the flutes, a sonoric/harmonic touch which a conductor should try to bring out, or at least to not let it be buried and totally ignored. In those same measures care must be taken that the first violins, for the moment the only instruments playing Beethoven's main motive, be clearly heard. This can be easily accomplished if the woodwinds and horns hold to a *f* (not the usual *ff*), and if the violins make a slight compensating crescendo in mm.46–47 and 50–51, since on their lower strings they lose a certain degree of projection.[21] Even so, none of these balancing efforts should on the one hand diminish the

---

20. It is a sad fact that more often than not, when dynamic imbalances occur, the automatic solution is presumed to be that the 'weaker' instruments simply play louder. It seems rarely to occur to anyone that perhaps the 'louder' instruments *should play softer!* This kind of misguided thinking seems also to be behind the widespread penchant for doubling instruments, especially the woodwinds, in classical symphonies. Instead of the brass playing loud and conductors using large (or augmenting their) string sections as well as doubling the woodwinds, might it not be better to scale down the resultant inordinate volume levels by *reducing* the output of the brass and timpani, maintaining the woodwinds at their normal size and dynamic levels, and keeping the string sections at a size more common in Beethoven's time? It should be remembered that in addition to their pitch limitations, the trumpets of Beethoven's day produced an intrinsically softer, mellower, less projecting sound than the trumpets of today. They blended much more readily into the over-all texture. Similarly, the calfskin heads of classical timpani did not have the brilliance and impact of today's plastic heads, and therefore were not as obtrusive as they tend to be nowadays in classical symphonies. Such approaches would, by the way, come very close to 'period authenticity' without the necessity of resorting to actual period instruments.

The most grievous example of distorting Beethoven's music by way of uncalled for doublings and enlargement of orchestral forces that I ever had the displeasure to hear, occurred some years ago when Karajan visited Boston with the Berlin Philharmonic, performing the *Eroica* in that city's wonderfully responsive and sensitive Symphony Hall acoustics, using six trumpets, eight horns, enormous numbers of woodwinds, and, of course, the Philharmonic's entire string section (18–16–14–12–10). It was a truly painful and revolting aural experience! (Karajan also recorded Mussorgsky's *Pictures at an Exhibition* with twelve trumpets, ten trombones, and eight horns!)

21. It is remarkable—and a welcome surprise—that Schenker, who was, after all, not a conductor or a practicing musician, in his Fifth Symphony tract, offers exactly that advice (p. 17).

energetic effect of mm.44–51 by the winds playing too softly, lacking in energy, or on the other hand preclude the possibility of a dramatic sudden dynamic increase to *ff* (m.52) by playing mm.44–51 too loud. The two problems in this passage are demonstrated, alas negatively, by a number of conductors (Kletzki, Kubelik, Mengelberg, Dorati, Wand) including two world-famous conductors, Toscanini and Bernstein. Toscanini who, as I have mentioned earlier, allows a runaway crescendo starting at m.34 to peak much too early, then allows the first violins in m.44—and even more in m.52—to be quite overbalanced by the winds and timpani, almost to the point of inaudibility in mm.52–55. Bernstein's problem is that he is already so loud at m.44 that there is no room left to create the *real* climax (*ff*) at m.52. Conductors who solved the performance problems of this entire passage brilliantly are Erich Kleiber, Weingartner, Nikisch, Jochum, Karajan, Mehta, Schuricht, and Reiner.

It is of paramount importance to understand that Beethoven in his autograph consistently used the notation ♪ ♪. for his main motivic cell, not ♪ or ♩ ♩ ♩ ♩. Obviously this was meant to preserve the rhythmic/gestural integrity of the motive, even when several of these cells are linked together chain-like (as, for example, in mm.44–55). The Breitkopf and Härtel editor in the first edition, however, changed many of these measures to read ♩ ♩ ♩ ♩ (see mm.44 and 47, first violins, also mm.49–50). Beethoven tried to show the distinction between the main cell 𝄾 ♩ ♩ ♩ or ♪ ♪. on the one hand and a more linear variant, such as (m.38)

or (mm.102–104). It is important, therefore, that conductor and musicians be aware of these two distinct ways of phrasing, and keep them discrete, a not so easy task over the long haul of the entire first movement.

How terrifying and startling Beethoven's diminished chord in mm.52–55, with its cascading violin motives and thundering timpani, must have sounded to Beethoven's audiences in 1808! It still packs a terrific wallop nowadays when played correctly, but playing it 'correctly' does not include the timpani's drowning out the rest of the orchestra, which, unfortunately, is the case in Carlos Kleiber's generally excellent recording, in which the violins are virtually inaudible under the murderous *ff* of the Vienna Philharmonic's over-enthusiastic timpanist. On the other hand, Carlos's father, Erich, achieves the best balance in this entire passage (mm.44–56), with the marvelously calibrated sonorities of the Concertgebouw Orchestra.

Many conductors inadvertently (or perhaps consciously—if so, mistakenly) make or allow an accelerando with the crescendo starting at m.34. Bruno Walter, who started his recording of the Fifth Symphony with a sedate ♩ = 88 accelerated to a healthy ♩ = 96 by m.44.

The diminished-seventh chord in m.52 allows Beethoven to move to the most closely related key to C minor, namely E♭ major, and, having thereby signaled the end of the exposition, to move as well to his 'second subject.' Beethoven's transition to this point is as dramatic as it is consequent and succinct: two horns proclaiming an intervallically expanded variant of the movement's opening

motto (mm.59–62). Instead of the pitches , he now

gives us ... in a form rhythmically directly related to the movement's opening measures (Ex. 11). This moment represents, of course, a crucial juncture in the over-all form and continuity of the movement, and it has led many conductors to a variety of 'personal' but willful form-damaging

Ex. 11

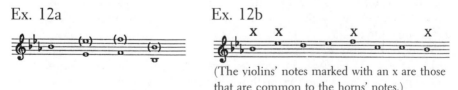

interpretations. It goes without saying that the second subject (starting at m.63) will want to be performed in recognition of Beethoven's own often articulated concept of "flexible tempo" or "tempo of feeling." Surely a degree of relaxation in the tempo is wanted here, or is at least possible. And yet the transition to such a tempo must be accomplished with subtlety—to come back to Wagner's cautionary word—"imperceptibly." For all its *dolce* amiability, the violins' motive or theme, soon answered in the clarinet, is still part of Beethoven's *allegro con brio*, with its connotation of a certain bristling energy and restlessness. Moreover, the relationship of this second subject to the main theme via the transitional link of the preceding horn call (mm.59–62) must be preserved and made audible, despite the dramatic differences between the horns and the violins in dynamic, in sonority, in articulation—e.g. *ff* to *p*, strong brass to gentle strings, *marcato* to *legato*. And this can best be achieved when both parts are expressed in not necessarily an identical tempo but a closely related one.

But how *does* the violins' theme relate to the horn fanfare? Clearly, in that the former is an elegantly embellished variant of the latter: the one, the horns': brash, eruptive, and vertical/harmonic/declamatory; the other, the violins': calm, conciliatory, horizontal/melodic/lyric (Ex. 12 a and Ex. 12 b), both lasting four bars and beginning on an 'upbeat' measure/gesture. Helping to make the transition from one theme to the other, the 'hammer blows' of the horns are softened by the quieting effect of the underlying 6/4 chord in m.63. The tempo modifi-

Ex. 12a   Ex. 12b

(The violins' notes marked with an x are those that are common to the horns' notes.)

cation called for in the transition from the horn 'fanfare' to the violins' second subject must be done subtly. My own experience in conducting the work has convinced me, that, if one holds to Beethoven's ♩ = 108 in the exposition, a subtle slowing to 100–04, possibly even to $\textsf{J}$ = 96, gets the desired results without any wrenching at the structural seams. Many conductors slow down too much here, causing the second subject to seem to belong to a totally different work. Such was evidently a well-established tradition, probably fostered by Bülow, and still adhered to by Nikisch in 1913—he slows to $\textsf{J}$ = 80 at m.63—but even by a 'modern' conductor like Carlos Kleiber who, strangely enough, slows down during the two incisive chords (mm.56–58) before the horns' entrance— a very peculiar effect. But the strangest, most unnatural interpretation comes from the baton of Richard Strauss, who precipitously drops about ten metronome points right at m.63. Karajan, oddly enough, *increases* the tempo (to $\textsf{J}$ = 104) in his 1962 recording, but hews more to the older tradition of relaxing the tempo (to $\textsf{J}$ = 94) in his 1982 recording.

The second subject carries with it certain very interesting performance and interpretational problems. One pertains to the long arching line of this passage which culminates in another secondary climax at m.94. In effect, mm.63–73 is a single line—a *Klangfarbenmelodie*—shared by three instrumental colors (timbres): violin, clarinet (mm.67–70), flute (plus violin) (mm.71–74). The three segments of this twelve-bar melody must be linked together in performance into a single line, which means that the players involved in these exchanges must be aware of this need—or must be made aware of it by the conductor. But mere awareness of the process may not be sufficient to achieve the desired result, for the violins must literally hand their four-bar segment over to the clarinet; the clarinet must in turn pick it up from the violins and hand it back to them and the flute—all this very much as in a relay race when the baton is smoothly transferred from one runner to another. The problem here is that, for reasons beyond my comprehension, musicians generally tend to shorten—chop off— final notes in phrases if they are followed by a rest. This is a world-wide bad habit, a disease most prevalent among string players, particularly violinists—who, of course, are those very players most likely to have to carry a tune or a melody in the classical/romantic literature. I can predict with certainty that, unless a violin section will have been trained otherwise, it will in any orchestra play m.66

not as , especially if

played with a downbow. What is curious—and distressing—about this careless habit is that, when such a final phrase note is *not* followed by a rest, as in m.74, for example, this rhythmic/dynamic note-dropping will not occur.[22] One asks oneself, what is the difference between m.66 and m.74—musically, structurally?

22. I hereby cite several such similar places in the literature where almost without fail players will unceremoniously, unthinkingly, before a rest, drop the last note of their phrase: Dvořák *New World Symphony*, first movement, mm.111–13, third movement, m.160, 170; Schumann Second Symphony, last movement, mm.65, 73, 77, 213, 225, 509 etc.; Berlioz *Corsaire* Overture, mm.31, 69–71(woodwinds), 153 (flute, 1.violins); Brahms First Symphony, first movement, m.252–58; second movement, mm.61–62; third movement, m.58 (woodwinds); Brahms Fourth Symphony, first movement, mm.157–164 (winds), mm.227–41; second movement, m.39; Mozart "Linz" Symphony, first movement, mm.109–10; last movement, mm.73–92 (strgs), mm.104–15; Mozart, "Jupiter" Symphony, first movement, m.5, 8; Menuetto, mm.4,8; and hundreds more.

None. Indeed, the need to give the second quarter-note in m.66 its appropriate full length is perhaps even greater there than in m.74 because the violins' B♭ has to be connected up with the clarinet's B♭ in m.67.

It should go without saying that the three-phrase segments involved in this tripartite melodic exchange should be played with the same (*p*) dynamic. Of the innumerable recordings I have sampled, most fail to achieve these phrase link-ups, most notably Bernstein, Koussevitzky, Gardiner, Harnoncourt, Toscanini, Ashkenazy, Böhm, Thomas, and Szell. They lose the long line, and instead of one long twelve-bar phrase, we are given three short four-bar phrases. Here again Erich Kleiber excels in his recording, also Furtwängler, Knappertsbusch, Schuricht, Mitropoulos, Jochum, Dorati, Brüggen in theirs (although Brüggen loses the long line completely in the recapitulation (mm.307–30).

The second typical problem in this passage—including now mm.75–93 as well—has to do with bowings (and slurrings).[23] Many different right and wrong bowings have been tried here over the many years of the existence of this work. In

mm.63–66, Beethoven writes  in the first

violins. (Note, by the way, the absence of any dynamic nuancing except for the initial *p*.) A few bars later (m.75), Beethoven begins four two-bar bowings in the first violins, in m.83 (now also incorporating the second violins and cellos) a three-bar bowing, then two more two-bar-ers, and then—surprisingly—a four-bar entity on a single bow (see Ex. 13). One is bound to ask, why the three-bar bowing in m.83–85, and indeed in mm.63–65 for that matter? And why the four-bar bowing at the end of the entire passage, just when, with the ongoing crescendo, a string player will almost *have* to use more bow to achieve the desired crescendo?

Ex.13

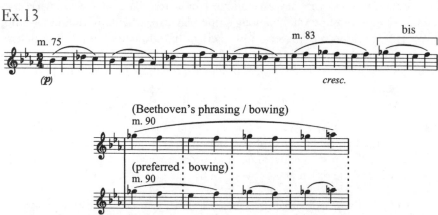

Schenker (p.10) makes as good a case for preserving Beethoven's designated bowings, at least in mm.83–93, as anyone to my knowledge has ever made. After comparing Beethoven's first bowing/phrasing version with his second 'corrected'

---

23. The ensuing discussion is of necessity somewhat technical in nature and is best understood by reference either to the score or at least to musical Exx. 13 and 14.

one (both in the autograph) in which the six measures of mm.88–93 are combined in one bow (not as in some editions, two and four), Schenker documents the reasons—that is, his rationale for Beethoven's reasons—for this unusual phrasing/bowing. It is deserving of serious consideration, even if in the end we may wish to differ with it. Schenker begins by noting that Beethoven retains the basic four-bar structuring throughout this 'second subject' episode, but untypically shifts the phrasing (and therefore the bowing) away from the four-bar periodization to the upbeat part of the structure—remember that m.64 is a 'one'—maintaining this cross-bowing until m.83. Schenker then suggests that, in order to prepare and anticipate the reconciliation of the phrasing with the underlying structure at the climactic

ff of m.94—as he puts it, to achieve "the metric readjustment" [*Ausgleich des Metrums*]—"it finally could be risked to not only move the head of the motive, G♭, to relatively strong measures in the grouping, like mm.86, 88, 90 and 92, but also to begin the bows on G♭ and thus reconciling them with the motivic kernel [*Motivkern*]. One can see from this that, to achieve this bowing, the combining of the first three bars 83–85 as a transition to the earlier bowing (mm.75–82) became necessary. But to do this four times in a row, in mm.86, 88, 90 and 92, certainly seemed disturbing to the master [Beethoven]; and therefore he felt compelled to indicate one bow [per bar] for the two bars mm.86–87 and then, from the middle of the [phrase] group, thereby also maintaining its equilibrium, the remaining six bars (mm.88–93) under one single bow. In order, on the other hand, not to jeopardize the crescendo effect because of a too-long bow, [Beethoven] divides the same line in the cellos into a two-bar and four-bar bowing and, additionally, adds the flute in m.91 [to the violins], joining the last three bars [of the phrase] under one slur. The irregular ten-bar-ness [*Zehntaktigkeit*] of this phrase stretching"—Beethoven's final phrase before m.94 is indeed another one of his stretching extensions of four bars into six—"as well as its strange bowing game [*seltsame Spiel der Bogenführung*] lend this passage an indescribably irresistible magic."

There is one slight flaw in Schenker's analysis, the result of overlooking one important point, namely, that what he called the "head of the motive" note (G♭ in mm.84, 86, 88, and 90) had already occurred earlier in the analogous D♭ of m.76 and the F of m.80. Since these two notes are also situated on a 'one,' a strong beat, Schenker's suggestion that Beethoven felt the need to *shift* the head note "to relatively strong measures" is misleading. No such shift was necessary, since the earlier head notes were also placed on 'strong measures.' Schenker was, of course, right to note Beethoven's shifting of the bowing in mm.83–89. But he probably should have questioned Beethoven's six-bar one-bow phrasing in mm.88–93, especially since he notes the different (and more logical) two-and-four-bowing of the cellos in the same phrase. (All editions after the first edition adopted this latter bowing for the violins as well.) But even the latter bowing—taking into account Beethoven's flute doubling an octave higher—

seems an odd way to deal with a crescendo, especially the last stages of a crescendo which is to reach not *f* but *ff*. I do not see how the result of Beethoven's
bowing could lend this passage an "irresistible magic." It could, in my view and
experience, only lead to a most ineffective crescendo, for I doubt that any violinist or cellist in the world can produce a crescendo in a four-bar single bow from
f (for that would be the dynamic for m.90) to *ff*.

   In order to determine the best bowing for the entire second subject passage,
one cannot start with its tail end; one must go back to its beginning, m.63. A
good bowing I have found—many years ago—for this passage (good in terms of
satisfying all of its multiple demands and implications) is the following (Ex.
14a). The assumption here is in part that the violins will use relatively short,
light bows in mid or upper position, not the usual full-length (and therefore
much too loud) bows[24] that have become the maddening norm in so many
orchestras, especially in the United States. But a simpler bowing, as in Ex. 14b,
works just as well as long as the violinists don't drop the last B♭.

Ex. 14a   Ex.14b

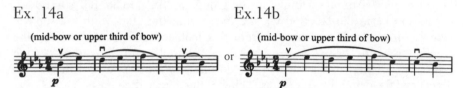

   The reasoning behind dividing the four-bar phrase mm.63–66 into three
bows—perhaps at first glance an odd choice—is that such a bowing satisfies
three performance conditions: (1) that the phrase begins with an upbeat feeling
(because of its 'weak' fourth-beat placing)—this is best achieved with an upbow;
(2) that the strong-beat character of m.64 be preserved (best achieved by a gentle
downbow); and (3) that the last measure (m.66) connects to the entering clarinet (best achieved by an upbow). If on the other hand one is intent on preserving Beethoven's three-and-one phrasing,[25] one obviously has only two choices:
down-up or up-down. Both carry risks, as any uneven bowing (in this case six
beats to two beats) does—although not insurmountable ones—being more difficult to balance out evenly. The former bowing (down-up) carries the risk of
the entrance at m.63 being too heavy and losing its upbeat character, while it
does work well for m.66. The other bowing (up-down) carries a different risk,

24. I do not know what bowings Karajan used in his various recordings, but suffice it to say that he
never achieved a real *p*, Beethoven's very special '*p dolce*.' But it is not only a matter of the dynamic
level per se; what really matters is the tone color, the sonority—a warm, relaxed, quietly singing, unedgy
sound—that can make the difference between making this passage sound magical or ordinary. This is
best achieved by using a light, gliding—very little vertical pressure—bowing. The entire passage should
offer a maximum in sonoric contrast to all the previous vigorous, eruptive, bursting sounds—as Furtwängler once put it: "like the turning of a gigantic hinge." I have already mentioned Bernstein's and
Toscanini's problems here. Günther Wand, in addition to allowing the players to drop their last notes,
thus annulling Beethoven's long-line *Klangfarben* exchange, also never achieves the real *p* and lovely
sound needed here. But two conductors who do are Erich Kleiber and Furtwängler.
25. For all we know, Beethoven was persuaded to use the three-and-one bowing because of the
repeated C's in mm.65–66, perhaps really preferring a single bow for the entire four bars.

namely, while it satisfies the anacrusis character of the phrase entrance, it makes m.66 much too heavy. And this is not desirable since both m.66 and its parallel m.74 are 'threes' and should have only a moderate weight, less than a 'one.' An upbow achieves this result more naturally.

But ultimately any one of the three bowings suggested here can be made to work well, as long as the players approach it with intelligence and sensitivity, making sure that the entire phrase is evenly balanced and preserves the integrity of its particular four-bar structuring ('four-one-two-three'). In this connection, we should observe that the entire 'second subject' episode has no dynamic nuancing; it is *p dolce* from m.63 until the crescendo of m.84. That doesn't mean, of course, that the passage should be played in some cold, mechanical, abstract, dynamically 'flat' manner; but it does suggest that the exaggerated heavings and roller-coaster dynamics one often hears here in performance

as, for example,   or   are quite unnecessary and, in peaking

on a 'two,' quite out of place.

To coordinate with the melodic lines, the second violins and violas[26], starting in m.63, and bassoons, starting at m.67, should also have four-bar phrasings/ bowings, as Beethoven's autograph clearly shows. Similarly the clarinet and flute should be aware of the upbeat character of the first measures of their phrase (m.67 and m.71 respectively), that is, that the phrase does not start on a 'one', but rather straddles the underlying four-bar structuring by being shifted one bar early. (How many times I have heard clarinetists and flutists play this phrase incorrectly, without any awareness of its true placement!). The same applies, of course, to the bassoons who, just looking at their bare part, have little way of knowing that their entrance in m.67 is in fact on a 'weak' upbeat part of the structure. Likewise, cellos and basses must play their motive going from a 'two' to a 'three,' which is quite different from the last time they played it in the mm.28–38 section.

At m.93 a terrible habit—a bad tradition and misguided interpretation—has crept into the performance of innumerable conductors: holding up the tempo for this one bar (what the Germans call *ausholen*), and at the same time staccato-shortening the final quarter-note, the A in the violins—a spurious addition not to be found in Beethoven's autograph or the first edition. Walter, Kubelik, Böhm, Szell, Mehta, and Carlos Kleiber, among others, are the conductors most guilty of this tempo distortion. The effect is ludicrous in all respects, for it interrupts the flow, the inexorable drive of the previous ten bars to the B♭ summit at m.64, manifested not only by the mounting crescendo and the insistent impetuousness of the

26. The vulgar swooping crescendos one hears on Brüggen's Harnoncourt's and Gardiner's recording (in mm.75 and 77, for example) are totally gratuitous additions, not indicated in Beethoven's autograph, nor in the commonly available score and parts.

melodic line, but the powerful harmonic progression (A dim.—Eᵇ minor—Cᵇ major—C dim.). The ineptness and banality of this idea can also be measured by the presence of the two eighth-notes in the violas and basses, which clearly signify a driving forward, not a holding back. Conductors who make this *ritenuto* probably think they are helping to clarify the form at this juncture (as if everything Beethoven does here harmonically, melodically, dynamically, structurally isn't already clear enough), little realizing that they are instead ravaging the form. It is a Bülowian arrogance, which assumes that a conductor knows better what Beethoven actually wanted, and that the (presumably) ignorant public needs to have these formal aspects blatantly pointed out to them.

In m.94 Beethoven reasserts the full four-bar structuring—Weingartner saw it as re-establishing the "unequivocal priority of the four-bar phrasing"—in blazing *ff* orchestral colors, constantly playing with the succinct motivic material, keeping it fresh and unpredictable, finally reaching a mighty Eᵇ major cadence in m.122. Contrary to what the Viennese novelist and composer E. T. A. Hoffman wrote in 1810 about Beethoven's Fifth, hearing in the first movement only a kind of "mechanical repetition of a four-note motivic cell," we can see the many fascinating ways Beethoven uses his main theme, which go far beyond mere repetition. The theme is constantly varied, in register and orchestration, and by continually shifting it around within the four-bar structuring, we hear and see it always in different contexts and juxtapositions. Equally astonishing is its sequencing, that is, the *manner* in which each motivic/thematic variant develops out of its predecessor. In each instance we sense that no other sequence of events seems to have been possible; and it is the utter naturalness, the organic inevitableness of this continuity which arouses our wonder. It is as if in each new motivic variant all its previous appearances are echoed. It is this extraordinary sense of inevitability—the feeling on our part as listeners that every moment of the music is the result of a natural, utterly logical progression to which there simply are no better alternatives—that we register with awe. As Bernstein once said, speaking of the first movement on one of his telecasts, it is music "that follows its own laws. And we can't resist it."

The Eᵇ cadence, at once powerful and stabilizing, provides the springboard for the music to return to the very beginning for a mighty 124–bar repetition, a repetition which is fortunately traditional nowadays but in earlier times was often omitted. Nikisch and Strauss and even Walter (recording in the late 1950s) forgo this important repetition. To disregard it is to destroy the intended proportions of Beethoven's flawless formal design. Because the movement's four sections—exposition, development, recapitulation, and coda—are virtually identical in length, the weight and thematic centrality of the exposition are seriously undercut when the movement is presented in four equal parts, ABA¹C. Repeating the exposition, thus strengthening its role as the fountainhead of all that follows, gives the movement its final needed balance and proportion (AABA¹C). One might also add that the exposition, with its 124 bars lasting little more than a minute—played at the *con brio* tempo, of course—is so short that there is no excuse for eliminating its repetition. And when it is done at a bright *con brio* tempo, it is a welcome return, not a boring repetition as when done too slow.

Proceeding now to the development section, we have already seen that m.125 is a 'four,' an upbeat gesture. But note that the strings' four-note response bounces, as it were, off of a 'one,' descending to a ('weak') 'two.' This string response to the winds has several functions; it expands, enlarges, the original motto, not only durationally but harmonically. The Db and C (along with the winds' Bb and G) relate back to the previous Eb cadence and at the same time serve to prepare and set up the new temporary tonality of F minor (m.130). This ingenious modulation can be capsulized as follows (Ex. 15). Although there is no E♮ in mm.127–28, one clearly hears the C dominant function here, making the F minor of m.130 inevitable.

Ex. 15

Note too that by expanding the original motive by one bar by virtue of the strings' response, Beethoven does not have to tamper with the four-bar structuring; everything is in place, the mm.127–28 representing 'two' and 'three.' These also relate back to mm.3–4, implying that the fermata must be held as long as the second fermata of the opening motto and perhaps even, because of the one-bar extension and the harmonic implications here, a touch longer—still, however, not so much as to interrupt the flow, the rhythmic undercurrent, of the music.

The development section begins innocently enough by restating the main thematic material, only now in F minor and subtly reorchestrated, adding clarinets to the earlier instrumentation. Two three-bar crescendos a little later are both followed by a *p subito*. What is unusual about that—and very exciting when performers are aware of it—is that these *p*'s come on a 'four,' the fourth beat of the four-bar phrase, not as a more ordinary composer might have done, on a 'one' (Plate IV, p. 141). It is curious that in almost all performances the second of these *p*'s (m.153) is never played as softly as the first one (m.145). While most conductors do observe the *p subito*, the flute and oboe tend to enter too loudly. The reason, I suspect, is that their entrance is preceded by thirty bars of rest. In m.153, all that these players see in their parts is *p*. But what kind of *p*? In relation to what? Since they don't participate in the previous three-bar crescendo, they are entering cold, as it were, out of the blue, out of context. A conductor must keep a sharp ear open to make sure that the flute in m.153 be no louder than the violas and cellos were in m.145 (the same notes transposed up a fifth).

The next performance pitfall lurks between m.158 and m.168—which, by the way, includes another one of Beethoven's six-bar add-ons. The temptation here is to make the crescendo too early. Beethoven indicates a crescendo only in m.166, but many conductors and orchestras start it half a dozen bars earlier. What difference does it make? All the difference in the world! When the passage is played as Beethoven wrote it, the sudden two-bar crescendo in mm.166–67 comes as a terrific surprise—taking us all the way from *p* or *pp* to a full orchestra *f* in a mere two bars. The effect is startling and truly Beethovenian. Making a gradual crescendo starting, say, in m.160, reduces the whole passage to something quite ordinary. So does not maintaining a real *p* (*pp* in the trumpets and timpani) in m.158, a carelessness of which a host of conductors is guilty. A tally of how various conductors fared on this passage breaks down as follows: those who held the crescendo back are Toscanini, Furtwängler, Klemperer, Jochum, Reiner, Muti, Haitink, Van Otterloo, Kempe, Ansermet, Ashkenazy, Dohnanyi, Böhm, Colin Davis and both Kleibers (father and son), Brüggen, and Gardiner; those who made (or allowed) a runaway crescendo are Nikisch, Weingartner, Mengelberg, Wand, Bernstein, Walter, Ormandy, Masur, Harnoncourt, and three of the 'early music' specialists, Norrington, Hogwood and the Hanover Band.

Measures 168, 172, 176, and 180 are all 'ones' in the four-bar structure, the eighth-note figure always on 'four.' Here Beethoven's dynamic indications are wonderfully explicit: *f*—but only *f!*—at m.168, *più f* at m.175 and a dotted line indicating a crescendo,[27] i.e. ever more *più f* (m.176–77), and finally arriving at a *ff*. The problem often is that the *f* at m.176 is too loud, thus making it virtually impossible to carry out Beethoven's graduated dynamic increase in mm.175–79.

The 'horn call' of mm.59–62 is now heard in the violins (in G major), and nine bars later in C major. At m.182 and m.190 new generally ignored performance problems arise. The vast majority of conductors pay inordinate attention here to the violins and the descending figure in the lower strings, while paying no attention to the remarkable things Beethoven is doing in the winds. On 90 percent of the recordings sampled, these wind interjections are either totally inaudible, just barely audible, or unevenly audible. And yet this is one of Beethoven's most daring and 'modern' ideas in the whole movement, the winds clambering up the range ladder in diminished fifths(!) (Ex. 16); moreover in a slightly truncated form of the original motive, reduced now from three

Ex. 16

m. 182

eighth-notes to two.We can see here, as I mentioned earlier, how each new musical thought flows out of some previous, sometimes immediately prior, idea. The first time this anapestic variant is used occurs a few bars earlier in mm.177–78. It suggests that Beethoven may have felt that, after nearly one hundred reiterations of the original three-note cell, it was perhaps time to vary it and try something different. In any case, the problem here (mm.182–94) is (a) to make the winds audible against the strings and (b) to match up the five wind groups dynamically: trumpets, bassoons, clarinets, oboes, flutes (bassoons, horns, clarinets, oboes, flutes in mm.190–94). Balance problems are exacerbated by the fact that these two passages

---

27. Beethoven's frequently used marking *più f* is nowadays often misunderstood. It does not signify a sudden increase in dynamic but is, rather, Beethoven's way of indicating a gradual crescendo, usually from *f* to eventually *ff*.

Plate IV Mm.145–168 of the first movement of Beethoven's Fifth Symphony

are lacking any specific dynamic indications, either in Beethoven's autograph or the two early editions (Breitkopf & Härtel, and Peters). Did Beethoven simply forget? Possibly; we'll never know. My suggested dynamics are *ff* in the woodwinds, *f* in the brass. (Woodwind doublings are *not* the answer—although I suppose it represents an easy, lazy way out of the problem.) All it takes to clarify this remarkable passage is (1) to rehearse the winds once alone, making them aware of the importance—and novelty—of the passage, and (2) to restrain the strings from playing too monstrously loud. Suddenly everything Beethoven wrote can be wonderfully heard. The only recordings on which this passage is fully mastered are those of Nikisch, Toscanini, Reiner, Szell, Mitropoulos, Krips, Ozawa, Mehta, and Harnoncourt. Some conductors on the other hand—Karajan, Furtwängler, Wand, Masur, Brüggen, Gardiner and Carlos Kleiber—seem to be unaware that there are any winds participating in these measures at all!

At m.196 begins one of the most remarkable passages in all of music. Starting in F minor with a variant of the mm.59–62 'horn call,' alternating winds and

m. 195

strings in heavy sustained chords (note the clash of the trumpets' C against the bassoons' Db in the second measure, rarely brought out in the seventy or so sampled recordings), Beethoven modulates his way through an amazing series of harmonies: F minor, Gb major, F seventh, Bb minor, Cb major, Db major, F# minor, D major—still later, after a *ff* main-motive interruption—D diminished sevenths. If that were all, it would already suffice to call the passage 'astonishing,' 'ingenious,' 'miraculous.' But there is much more; in these thirty-two bars Beethoven not only manages to incorporate, halfway through, a long fourteen-bar diminuendo from *ff* to *pp*, changing at the same time from two-bar alternations of winds and strings to one-bar alternations, but also manages to include an out-of-sync *five-bar* phrase. This five-bar phrase still comes as a surprise and a shock to listeners and performers,[28] even to this day when, after nearly two hundred years of hearing and performing this piece, one might have expected the surprise to have worn off.

But exactly where is this five-bar phrase? On that subject many great minds have disagreed over the years, and possibly there are two equally valid answers to the question. Let us examine the score. We know from all of our previous structural analysis that m.196 is a 'one.' If that is so, and if we follow through with more four-bar phrases, then mm.200, 204, 208, 212, 216, 220, and 224 will also all be 'ones.' And if all *that* is true, then the *ff* main motive bursting in at

28. How powerfully the four-bar structuring of the movement can take hold can be heard in a recording made of Bruno Walter rehearsing the first two movements of the Beethoven Fifth. As Walter, who after all had by that time conducted the symphony dozens if not hundreds of times, approaches the end of this passage, twice he exhorts the musicians to play the oncoming *ff*—only he does so *one bar too early!* The gravitational pull of the four-bar structuring had him momentarily confused, so that suddenly he could not deal with that five-bar anomaly.

Similarly, Toscanini in his 1952 broadcast and telecast performance of the Beethoven Fifth can be seen and heard to conduct the *ff* of m.228 one bar early!

m.228 has got to be a 'five.' But we know that this main motive always starts—
and *has* consistantly started—on a 'four.' Here too it *is* on a 'four,' but sitting *in
the place of* a 'five.' That m.228 is metrically and gesturally a 'four' is borne
out by the fact that mm.229–32 clearly comprise the original basic four-bar

structuring, , exactly as in mm.1–5, 6–9,

59–62, and many other places. Further reiteration and confirmation of this
structuring come a few bars later when the three-eighth-note motive *is* in its
right place, starting on a 'four' (m.240). Should we still doubt that evidence,
then we need only count through two further four-bar phrases to receive unequivocal confirmation of the fact, for in m.248 we come to a full recapitulation
of the opening of the symphony, fermatas and all, although quite reorchestrated.
In this interpretive version then, all flows naturally from the m.196 starting point
in eight four-bar entities, with the last one extended to five bars to accommodate
the eruptive anacrusis eighth-note motive (Ex.17).

Ex.17

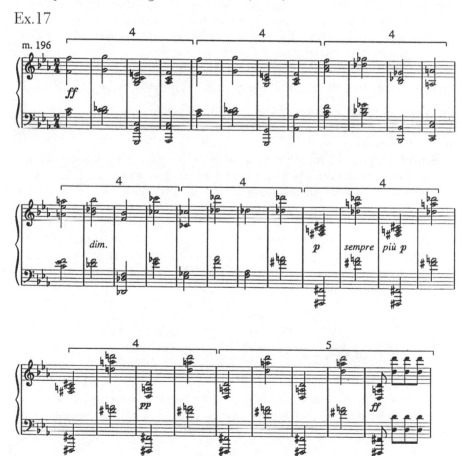

An alternate reading of this passage is offered by Schenker,[29] in which he contends that the initial four-bar phrasing is abrogated in m.209, turning "an originally weak measure into a strong one," in turn making m.210 a weak one. From there on (m.209) in Schenker's accounting there remain twenty bars up to and including the anacrusis measure m.228 (five four-bar units or ten two-bar units). This in turn means that Schenker's five-bar phrase comprises mm.204–08 (Ex. 18).

Ex.18

There is logic to both interpretations. In Schenker's version, the logic lies— so some would say—in the fact that the higher notes (in the woodwinds) are now on strong beats, the lower notes (in the strings) on weak beats. This *is* a possible way of looking at it. But I personally do not find this reasoning particularly compelling, because I see no inherent priority that upper notes (or woodwind notes) carry in terms of weight or strength of beat. Indeed, I could more readily present a counterargument that, in thousands of so-called oom-pah ac-

companiments in both classical and popular music  the

lower note contains such weight and strength. By that criterion Beethoven could just as easily have meant the lower string notes to be on strong beats, the winds on weak beats (as in my Ex. 17; see the phrase bracketing).

Schenker was also basing his explanation of this passage on a harmonic and melodic analysis by means of which he delineates the *Urlinie* (the basic line)— in effect the fundamental long-term harmonic rhythms—of the entire movement. It would go beyond the scope of this study to explore fully Schenker's approach to the movement's basic harmonic line, but in respect to the passage under discussion he argues that the B♭ minor chord of m.209 is part of a large harmonic rhythm that gives it more than passing structural importance. His diagram (Ex. 19), emended to show the bar numbers to which the pitches (the harmonic stations) refer, is correct up to the F (m.196) and possibly even

Ex. 19

the B♭ (m.209). It is thereafter that Schenker's logic breaks down, for Beethoven's amazing continuing modulation, plus the two *ff* interruptions surrounding the seven bars of quiet diminished chords, cannot be fitted into his scheme. It

---

29. Schenker, *Beethoven: Fifth Symphony* pp. 12–13,18.

is a long stretch from m.209 to m.249, with a lot happening which cannot be skipped over—which Schenker, alas, does—nor explained in terms of his schematic assumptions, a process he sees defined by what he calls the *Quartknotenpunkte* (nodal points of a fourth). (For more on this see Schenker's *Beethoven: Fifth Symphony*.) I think that Schenker was (untypically) for once seduced by the beautiful symmetry of his diagram (Ex.19) G / C / F, F / B♭ / E♭. He also was clutching at straws, I think, when he argued that Beethoven's diminuendo in m.210, followed one bar later by a diminuendo in the winds, was meant to indicate "thereby that the character [of m.210] was a weak beat compared to m.209." Apart from his inability to demonstrate that this is in fact what Beethoven's two 'diminuendo' indications mean, Beethoven would not have put a diminuendo sign into an empty measure. Thus there is no more significance to the diminuendo in m.211 than that it happened to be the first measure in which the composer could indicate such a diminuendo in the winds, the previous bar being empty; and, as I say, Beethoven would not have written a diminuendo indication into an empty measure (nor would any logically thinking composer).

Schenker's account of the passage also has the advantage—so some would argue—that the *ff* interruption in m.228 now fits neatly into a four-bar unit, and is therefore not as disruptive, not as disturbing, as in the other version. Ah, but that is precisely the point! Did not Beethoven, the 'titanic,' even at times 'demonic' composer, who gave us so many 'disturbing' moments, so many shockers in his music, want such a disruption here? And how better could he achieve this disruption than to lull us into a comfortable acceptance of the four-bar structuring, coupled with a calming diminuendo, and then, suddenly, hit us with a shattering explosion? I submit further that it would seem very unlikely that Beethoven would have stuck his five-bar phrase somewhere into the middle of this passage. He would, judging by his general life-long working methods, have put it either at the beginning or at the end.

Finally, one last argument for what I will dare to call the 'Schuller version' resides in the fact that Beethoven breaks up his four-bar phrases into two-bar units of alternating winds and strings. I cannot find any overwhelming reasons why that pattern, established for at least twelve bars (mm.196–207), should be abandoned in the next two bars and be broken into artificially. It is more in the spirit of the ever-inventive, ever-explorative Beethoven mind that the pattern-breaking would have manifested itself in the break-up of the initial four-bar structures, divided at first into instrumentationally and registrally differentiated *two*-bar phrases, then into one-bar units, as well as in the ever greater 'hesitance' and stretching out of the harmonic progression (initially two bars per chord, then, in m 215, six bars per chord, and finally, in m.221, seven bars).[30]

Unfortunately, most conductors and performers do not know *what* to make of this passage and more or less drift through it, not aware of where they are in the

30. Another totally different, and in my view unnecessarily fanciful, analysis of the passage (mm.196–232) is contained in an article by Andrew Imbrie, (" 'Extra' Measures and Metrical Ambiguity in Beethoven, in *Beethoven Studies*, ed. A. Tyson (New York, 1973), in which the author, one of America's very finest composers, not only disputes Schenker's interpretation but develops an

metric/rhythmic scheme of things. Most recordings that I have sampled follow neither of the above versions, but for no discernible or logical reasons turn the beats around in m.199, which they treat as a 'one.' But that makes a complete mess structurally, for now we have a three-bar unit (mm.196–98), presumably followed by seven four-bar units, but leaving us at the end with a left-over 'one-two' (mm.227–28). Muti, Leibowitz, Rodzinsky, and Boult are to be especially commended as the only conductors who did not turn the beat around at mm.199 or 201 (whether by design or by chance, I cannot say).

For anyone interested how this passage sounds (and *feels*) when played in what I believe to be the correct structuring, I strongly recommend Leibowitz's recording. In it one can hear what I consider to be perhaps the most remarkable aspect of this entire episode, namely, that if the weightier beats are maintained in the strings (with the woodwinds as a sort of subtle 'afterbeat') in mm.210–27, then one will hear the wondrous effect of this phrase feeling being *reversed* when in mm.233–39 it is the woodwinds' turn to have the 'strong' beats and the strings the 'weak' beats.

Measure196 has also been a favorite place for conductors to slow down in tempo, many of them inordinately. This makes little sense since Beethoven has already slowed down the motion of the music by composing the single longest passage in the whole movement entirely set in half-notes. Surely any further slowing down is not required.

Figure 3 (pp. 148–49) shows how most conductors have handled this tempo question.

---

elaborate and somewhat convoluted theory, which attempts to demonstrate that Beethoven did not conceive the first movement of his Fifth Symphony in a basically quaternary or binary structuring. Imbrie argues that many passages are asymmetrically constructed out of various combinations of threes, fives, and sixes. It goes beyond my intentions and the scope of this study to fully describe and then refute Imbrie's reasonings. Suffice it to say that the major flaws in his discussion lie in the fact that (a) he completely ignores the existence (or possibility of existence) of the five-bar phrase under consideration here; (b) although he presents counterarguments to most of Schenker's analysis, he accepts the one point where in fact, I believe, Schenker errs—his pivotal argument that m.209 is both a 'four' and a 'one'—and builds his entire theory around it. And in order to do *that*, Imbrie felt the need to completely 'restructure' the entire previous passage of mm.168–96. In his somewhat tortuous speculations Imbrie feels compelled to argue that mm.176–78 and 179–81 are three-bar units. This in turn makes mm.179 and 187 'ones'—the latter a real impossibility, since Beethoven would never have placed a first-inversion chord on a 'one', particularly in such a structurally critical juncture at or near the beginning of the development section—*and* it makes m.196 a 'two,' another musical impossibility, given the harmonic, dynamic, and durational weight Beethoven gives this measure. I find it curious that Imbrie would ignore and (by implication) not accept a simple five-bar unit at m.224—a nice touch of asymmetry, after all—but would instead invent a much more complex and less tenable hypothesis of metric organization to justify his assumption of "metrical ambiguity in Beethoven."

It is sad to report that Igor Markevitch, one of the 20th century's finest conductors and composers, in his aforementioned study of Beethoven symphonies seems to be totally unaware of Beethoven's five-bar phrase at m.224, for although he offers an exhaustive bar-by-bar periodization analysis of the entire movement (mostly very arbitrary, in my view), he goes into *no* detail regarding m.196–247, in effect glossing over Beethoven's most remarkable metric anomaly.

Peter Gülke in his critical report for the new Peters edition of Beethoven's Fifth basically follows Schenker's analysis of the passage. (See *Zur Neuausgabe der Sinfonie Nr.5 von Ludwig van Beethoven* (Leipzig, 1978), pp. 56–57.)

Strauss is truly 'odd man out' in the interpretation of this passage. Having ambled along at various speeds (𝅗𝅥 = 96 at the beginning of the movement, by m.14 up to 𝅗𝅥 = 104, at m.25 down to 𝅗𝅥 = 96, at m.63 𝅗𝅥 = 88, accelerating back to 𝅗𝅥 = 94 around m.94), Strauss suddenly pushes the tempo dramatically at m.196 to 𝅗𝅥 = 104, slams through the next eight bars at a terrific speed (completely surprising and disorienting the orchestra), then—just as perversely—slowing down by m.210 to a sluggish 𝅗𝅥 = 84. Measure 233 is even slower (𝅗𝅥 = 80), whereupon he again pounces onto m.240 at a full speed of 𝅗𝅥 = 104, a 24-point jump, with the orchestra again surprised and disorganized.

Perhaps the most difficult aspect of this passage (mm.196–239) strictly from the point of view of instrumental/technical control is the proper sustaining of all the half-notes, especially when the music breaks down into single alternating half-note durations at m.210. What one almost always hears in this passage, instead of 𝅗𝅥, is ♩. 𝄾 and ♩.. 𝄿, or worse ♩. 𝄾 and ♩.. 𝄿 (with diminuendo). To hold each half-note without any decrease of sound right up to the bar line—but not beyond it either—is fiendishly hard and takes terrific concentration and control. It is something musicians are almost never asked to do. My point is that each half-note must connect precisely with each succeeding half-note, neither falling short of it nor overlapping with it; and, of course, at its designated dynamic the note must not diminuendo. I know of only one way to achieve this result and that is for all the musicians to count in their mind two quarter-notes or four eighth-notes, as they hold their note. This may sound naive or simplistic to some readers, but it works. Indeed, it is the only way the full magic and originality of this passage can be realized. I have heard only three recordings, those of Krips, Schuricht, and Mitropoulos, in which the passage was played in a well-sustained manner.

Some conductors, like Mengelberg, Leibowitz, Mitropoulos, and Ormandy, ask the strings to play two successive down-bows in mm.198–99, 202–203, 206–207. While this may add a certain weight and power to the sound, it also is bound to shorten the duration of the half-notes (to | ♩. 𝄾 |), thus illogically differentiating the strings from the winds and destroying Beethoven's intended sustained continuity as well as his four-bar phrasing.

The next danger spot comes in or around m.245, especially m.248. Here many conductors make a greater or lesser ritardando (some as early as m.245), then even slower at m.250. They seem to have no idea that they have just come upon the recapitulation (Ex.20) and that one way of making that clear might be to *stay in tempo*. Doing so in fact produces an extraordinary impact on the listener. It is as if an object were hurtling along at full speed and suddenly

Ex. 20

Fig. 3

Conductor	Basic Tempo	Tempo at m. 196
Norrington	𝅗𝅥 = 108	𝅗𝅥 = 104
Gardiner	𝅗𝅥 = 108	𝅗𝅥 = 104
Brüggen	𝅗𝅥 = 106	𝅗𝅥 = 98
Dohnanyi	𝅗𝅥 = 104	𝅗𝅥 = 96
Karajan (1984)	𝅗𝅥 = 104	𝅗𝅥 = 92
Toscanini	𝅗𝅥 = 104	𝅗𝅥 = 98 (at m.200 𝅗𝅥 = 90)
Hickox	𝅗𝅥 = 104	𝅗𝅥 =
Giulini	𝅗𝅥 = 100	𝅗𝅥 = 82
Steinberg	𝅗𝅥 = 100	𝅗𝅥 = 88
Scherchen	𝅗𝅥 = 100	𝅗𝅥 = 96
C. Kleiber	𝅗𝅥 = 100	𝅗𝅥 = 92
Furtwängler	𝅗𝅥 = 98–100	𝅗𝅥 = 86
Karajan (1962)	𝅗𝅥 = 98	𝅗𝅥 = 104
Reiner	𝅗𝅥 = 98	𝅗𝅥 = 90
Hogwood	𝅗𝅥 = 96	𝅗𝅥 = 96
Ansermet	𝅗𝅥 = 96	𝅗𝅥 = 76
Mitropoulos	𝅗𝅥 = 96	𝅗𝅥 = 96
Wand	𝅗𝅥 = 96	𝅗𝅥 = 84
Leibowitz	𝅗𝅥 = 96	𝅗𝅥 = 96
Suitner	𝅗𝅥 = 96	𝅗𝅥 = 86
Harnoncourt	𝅗𝅥 = 96	𝅗𝅥 = 84
Muti	𝅗𝅥 = 96	𝅗𝅥 = 92
Van Otterloo	𝅗𝅥 = 96	𝅗𝅥 = 90
Ozawa	𝅗𝅥 = 96	𝅗𝅥 = 88
Strauss	𝅗𝅥 = 96	𝅗𝅥 = 104 (sic)
DeSabata	𝅗𝅥 = 94	𝅗𝅥 = 74
Weingartner	𝅗𝅥 = 94	𝅗𝅥 = 88
Haitink	𝅗𝅥 = 94	𝅗𝅥 = 86
E. Kleiber	𝅗𝅥 = 94	𝅗𝅥 =
Walter	𝅗𝅥 = 92	𝅗𝅥 = 92
Rodzinsky	𝅗𝅥 = 92	𝅗𝅥 = 92
Stokowski (1975)	𝅗𝅥 = 92	𝅗𝅥 = 80
Maazel	𝅗𝅥 = 92	𝅗𝅥 = 90
Ashkenazy	𝅗𝅥 = 92	𝅗𝅥 = 88
Hanover Band	𝅗𝅥 = 92	𝅗𝅥 = 88
Munch	𝅗𝅥 = 92	𝅗𝅥 = 92
Jochum	𝅗𝅥 = 90	𝅗𝅥 = 88
Mehta	𝅗𝅥 = 90	𝅗𝅥 = 86
Solti	𝅗𝅥 = 90	𝅗𝅥 = 84
Szell	𝅗𝅥 = 90	𝅗𝅥 = 84
Schuricht	𝅗𝅥 = 88	𝅗𝅥 = 84
Kempe	𝅗𝅥 = 88	𝅗𝅥 = 84
Abbado	𝅗𝅥 = 88	𝅗𝅥 = 88
Schwarz	𝅗𝅥 = 88	𝅗𝅥 = 86
Boult	𝅗𝅥 = 88	𝅗𝅥 = 84

Dorati	♩=88	♩=92
Koussevitzky	♩=88	♩=84
Kletzki	♩=88	♩=82
Schalk	♩=88	♩=80
Sawallisch	♩=88	♩=84
Nikisch	♩=88	♩=80
Ancerl	♩=86	♩=82
Thomas	♩=86	♩=88
Knappertsbusch	♩=84	♩=72
Masur	♩=84	♩=80
Colin Davis	♩=84	♩=82
Ormandy	♩=84	♩=84
Mengelberg	♩=84	♩=84
Bernstein	♩=82	♩=72
Böhm	♩=80	♩=76
Kubelik	♩=80	♩=76
Klemperer	♩=80	♩=70
Leinsdorf	♩=78	♩=84 (sic)
Krips	♩=76	♩=76
Boulez	♩=74	♩=72
Stokowski (1940)	♩=86	♩=88

slams into a solid wall (the *ff* chord at m.249). This ungentle simile is purposely chosen, for the effect here must be terrifying, dramatic, unpredictable—and yet, in retrospect, inevitable.[31] I would also submit that Beethoven was perfectly capable of writing '*poco ritardando*' (see mm.7, 17,51 etc. in the third move-ment, and of course, hundreds of places in other works). Not having put a *ritardando* at m.248 ought to clearly suggest that he didn't want one! Here Nik-isch's 1913 performance is rather peculiar. He rushes the tempo at first, acceler-ating dramatically at m.244, then puts on the brakes at m.248 for a huge ritard, followed by excessively long fermatas.

"As in all great masterpieces, so too in the Fifth Symphony, the recapitulation is governed by the law [*Gesetz*] of transformation: we see not a merely empty rattling off [*abschnurrende*] [once again of the exposition], but in various details a new life-infusing repetition." Thus Schenker, describing the reprise.[32] And in truth, Beethoven ingeniously reconstitutes the material of mm.6–21 into a bril-liant new variant: the previous violin and viola parts are condensed into just violins; bassoons alone carry the bass line (modified), previously also maintained by the cellos; the lower strings now provide "life-infusing" pizzicatos; clarinets

31. Furtwängler once wrote about this recapitulation that "it is never prepared for [*eingeführt*] in any way; it is so to speak suddenly there" (*Ton und Wort*, p. 244). Unfortunately, Furtwängler in his performances did not follow the implication of his fine insight, for he makes a sizable ritard in m.247—as a result of which m.248 is *not* "suddenly there"—followed by an even slower and more ponderous eighth-note figure in m.250 (♩ = 72).
32. Schenker, *Beethoven: Fifth Symphony*, p.13.

(and flutes) add the harmonies previously sustained by the strings; and, above all, a slender melodic line is added in the solo oboe. This oboe emerges gloriously—like a cocoon releasing a graceful butterfly—into a plangent solo mini-cadenza (m.268), marked *adagio*, replacing mm.21–24 of the exposition. Although Beethoven gives no indication in the entire oboe line (mm.254–68) of its gradual emergence, its transformation, into a full blown solo—I think because he could not find the notational means to do so (*più espr.* in m.262 might have done it, but that expression was not in common use in 1808)—I share with Schenker the idea that the oboe should very subtly and gradually in mm.262–67 begin to emerge into a leading position. Schenker puts it very carefully: "The oboe should make itself noticeable in m.262." My own suggestion is that the oboe begin in m.254 as if merely playing a harmony part, and then by the subtlest form of increasing expressiveness—perhaps even a subtle crescendo—starting in m.262, grow into the dominant role for m.268.

Beethoven's talent for constant variation continues to be lavished on the recapitulation. Besides re-orchestration in the passage beginning with m.269 (the parallel to m.25), m.273 signals a four-bar crescendo in thundering three-octave string unisons, bringing the music up to *f* at m.277, in turn placing the succeeding reiterative *sf*'s (mm.282–87 in a totally different (namely *f*) context than in the exposition (mm.38–43). A deft modulation at a key point (mm.300–02) moves the music to C major (formerly in E♭) for the return of the second subject, thereby already announcing the priority of the C tonality from here on out to the end of the movement.

A long-standing controversy exists regarding the use of horns instead of (or in addition to) bassoons in mm.303–306 (Ex.21), and the reasons why Beethoven

Ex. 21

m. 303

did not use horns here, electing instead to give the passage to the bassoons. (The sketches show that Beethoven originally had in mind adding cellos to the bassoons.) There are several possible reasons, foremost that the notes c^1, d^1 are bad, partially stopped, notes on the E♭ *Naturhorn* of the day, hence Beethoven's decision to give the passage to the bassoons. The alternatives would have been (a) to pull in a second pair of horns, pitched in C, which could have played those notes easily and well, or (b) to have the two horn players change to C crooks prior to this point. This might have been a possibility because Beethoven *does* change the horns to C in the second and fourth movements of the symphony. But changing to C horns for m.300 would have necessitated not using the horns for at least 50 to 80 of the previous measures, at best from m.254 on. For it took anywhere from 30 to 45 seconds for the players to change crooks. This would in turn have meant elimination of the horns in the climactic *ff* measures of 296 to 272. That was an option that Beethoven, I imagine, readily rejected. He also ruled out the other option—adding a second pair of horns for this passage—presumably because it seemed a little silly to drag in two horns for a mere four bars, having previously, we should note, decided for whatever reasons (economical, practical, balance in the brass, etc.) to limit himself in this symphony to two horns.[33]

The performance problem that is presumed to plague conductors here and why many choose to substitute horns for the bassoons (or add horns to the bassoons) is that nowadays and for some time in the past, the *ff* of the modern horns and the *ff* of the modern bassoons are far from equivalent. And those conductors who interpreted (and still interpret) this symphony as an expression of the titanic Romantic super-hero or in post-Wagnerian, post-Mahlerian terms, are, of course, bound to be disappointed in the relatively smaller sound made by the two bassoons. Their unthinking, automatic solution is to scrap the bassoons and substitute horns. I submit this is nonsense, and not at all as logical or necessary as such conductors think.[34]

Schenker had already urged the retention of the bassoons, noting that "most of the time horns are used instead of bassoons, very much in error [*zu Unrecht*],"[35] arguing that Beethoven knew the bassoon extremely well—"his early works show that"—and that he knew exactly what he was doing when he gave this passage to the bassoons. He suggests that instead of trying to play the three eighth-note G's with "the most blasting" [*schmetterndsten*] *ff*, the weight and emphasis should be placed on the three succeeding half-notes, especially c^1 (in m.304), a 'one' in the phrase structuring.

The solution to this alleged dilemma, once again, lies not in the pursuit of playing everything ever louder, but in—perish the thought!—playing something occasionally a little softer. This (radical) thought is particularly appropriate to Beethoven's Fifth Symphony, for it is perhaps, particularly in its first movement, the ultimate work of which one can say that it feels and sounds lean, strong, muscular—with no excess fat on its frame ('lean and hungry' might be even more apt); obese, flabby, ponderous, lazy, heavy music it is certainly not.

I further submit that the nature of the horns and bassoons of Beethoven's day offers us another clue as to how to reconcile this apparent—and it is only apparent—discrepancy. On the one hand, the natural horn of Mozart's and Beetho-

33. Having already used three horns in his Third Symphony and in two keys at times (two in E♭, one in C) it is curious that Beethoven did not light on that solution for his Fifth Symphony.
34. So ingrained is this tradition in conductors—I would call it another 'bad habit'—that Bruno Walter, for example, in the aforementioned recording of his rehearsal of the Fifth's first movement, every time he got to m.303, kept addressing the horns, speaking about the horns, calling for the horns, when all the time only the bassoons had been playing. After repeated attempts at this passage, and Walter still calling for the horns, the two horn players quietly began to play along with the bassoons. Poor Walter seems never to have noticed that it was *bassoons* who were playing the passage, nor noticed any of the ensuing interplay!

In his official recording of the Fifth with the Columbia Symphony Orchestra, Walter did indeed use the horns instead of bassoons.
35. Schenker, *Beethoven: Fifth Symphony*, p.19.

ven's day was an instrument with a pure, warm tone, light and buoyant in sound, projecting well because of its purity of tone, not its loudness. The bassoon, on the other hand, had a rather full, round, woodsy sound that matched the horn very well.[36] The modern double horns of today, depending somewhat on make and manufacture, are generally, in response to the demands of Strauss's, Wagner's, and Mahler's literature, capable of an enormous sound that in no way relates to sounds of a late 18th- early 19th-century horn. The bassoons, however, have acquired over the last century a more refined, a more cultivated sound, but not necessarily all that more projecting. Under the circumstances, to replicate the sounds Beethoven had in mind—the sounds of the instruments he heard and for which he wrote his music—all we have to do is tell the horn players in mm.59–63 not to play their loudest post-Mahlerian *ff*, but instead a cultivated tensile *ff*, in which rhythmic energy and articulation play a more important role than sheer amplitude and sonoric obesity. At the same time, if we can encourage the bassoons to give their all in mm.303–306, a viable dynamic relationship between the two passages will have been restored.

Again, Beethoven's extraordinary imagination and sense for constant variation is at work in the recapitulation of the 'second subject' episode (mm.306–46). Not only is it wonderfully re-orchestrated but, whereas the lyric four-bar phrase was heard three times in the exposition, it is now played four times, alternating between violins and flute. This expansion in turn allows Beethoven to play around with the cellos' and basses' ♩♩♩ by having them share it alternately with the timpani. At m.323 a different simpler harmonic progression (C^7-F-D^7–G) is substituted for the earlier one (E♭-C^7-Fm-D♭-E♭^{7}-A♭). Then, the earlier eleven-bar phrase, culminating in the dominant, is now expanded to fifteen (mm.331–45) (Ex.22). But while the latter is expanded horizontally, it is thinned out vertically, i.e. harmonically: the sustained wind harmonies that fleshed out the passage in mm.83–93 are now removed, leaving only bare-bones counterpoint, implying a series of diminished chords. Horns come to the rescue to help re-establish the clear dominant by supplying the seventh (F), just barely avoiding a rhythmic collision with the F♯'s in the melodic line.

Ex. 22

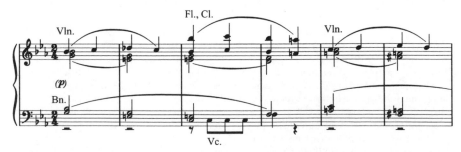

36. That is why, for example, Beethoven was able to use the bassoon in the great *Fidelio* aria, "Abscheulischer, Wo Eilst Du Hin" as, in effect, a fourth horn, or why Mendelssohn used the

Once again, in this entire episode (mm.306–46) the players must know where the four-bar structural points lie (mm.308, 312, etc., *not* mm.307, 311—which is the way one hears it played so often—mm.324, 328, 332, etc.), and the link-ups between violins and flute must be maintained to create long sixteen-bar or at least eight-bar lines.

Nikisch's and Furtwängler's and Schalk's link-ups are beautifully done; so are Erich Kleiber's, Giulini's, Mitropoulos's, and Walter's. But Bernstein's, Toscanini's, Weingartner's, Masur's, and Wand's are not. Strauss's interpretation remains the most peculiar, the most vagrant as to tempo variations. He takes the second subject (m.306) at a leisurely ♩ = 88, having come down from various faster tempos (♩ = 96, ♩ = 104), but by m.346 is up again at a hurtling ♩ = 112 (*sic!*), even faster than Beethoven's own basic ♩ = 108.[37]

---

bassoon in the company of horns in two famous passages: the trio of his Fourth Symphony's third movement and the *Nocturne* from the *Midsummer Night's Dream* music (which should be—but never is—interpreted as a horn and two bassoon *trio*, *not* a horn solo accompanied by two bassoons).

I was certainly disappointed that Harnoncourt chose to add horns to the bassoons in m.303, he who prides himself on performing Beethoven in the 'most authentic' way.

37. Earlier, in m. 266 before the oboe cadenza, Strauss makes an accelerando (*sic*) as he had done in the earlier analogous place, m.19. Knowing something about Strauss's attitudes in these matters, I have the feeling that he rushed these places simply to counteract the slowing down, the ponderous dragging, that most conductors indulged in in his time—and many still do today.

The A♮'s in m.340–46 not only lift the melodic line from its minor mood, but lift our spirits as well, as we listen to this powerful motivic expansion. It is as if the clouds are finally lifting and the clear blue sky of C major is now imminent. For it is those A♭'s that specifically allow the music to resolve to a joyous C major, for the first time in full force in the entire movement (not counting the brief intermittent C major measures in the 'second subject').

One of the more peculiar (and inept) interpretations of the climactic passage just before the development-extension (m.374) is Bernstein's. Having dragged along at a ponderous tempo for the entire movement—starting it at ♩ = 82, slowing it often to a really pathetic ♩ = 72 (only about 36 metronome points away from Beethoven's tempo indication!)—he suddenly lurches forward at m.362, jumping to—for him—a fast ♩ = 92. I can't tell from the recording whether this tempo leap is the result of an editing splice, or whether he suddenly remembered that the end of the movement was nearing, and that he'd better get the tempo moving for an exciting 'flash finish.'

Measure 374 has often been mistaken for the coda of the first movement. The coda does not arrive in fact until m.483, m.374 being instead in Beethoven's time a relatively new formal component of the sonata form—I am calling it development-extension—which one can find occasionally in some late Haydn symphonies, but which Beethoven really developed into major proportions in some of his symphonies, piano sonatas, and string quartets, and most importantly here (mm.374–482) in the first movement of the Fifth. It differentiates itself from the main development section in that room is made here for additional inventions, expressions, variants of material touched upon earlier that, in the composer's view, permits of (or demands) further elaboration and expansion. Thus Beethoven is intent here on giving all this accumulated material one more gigantic work-out, as if his initial ideas and vision had not yet been sufficiently thought through to their final conclusion. Thus the development-extension unfolds in three gigantic episodes, each of which rises from the tonic C and falls back to it (mm.374, 423, and 439). In the initial measures (mm. 374–481) of the development-extension, it is important that all instruments fully sustain their long half-notes: [musical notation: half-note tied, *sf*].

What one hears almost universally (especially with 'period instrument' conductors) is [musical notation with *sf* > *p f*] particularly in the strings, who generally spend most of their bow on the *sf*, leaving them with a *p* by the end (or even the middle) of the long note. This completely undermines and weakens the power and excitement of this climactic moment with its brilliant C major outburst.

The bright tonic major of m.374 is immediately converted into a powerful harmonic progression (Ex. 23), which Beethoven presently recycles, this time in the minor mode in the second episode (mm.427–29: I♭–IV^1–I♭^{1}–IV–V^2–I^1 etc.).

Ex. 23

In the meantime, one has barely recovered from the terrifying shock of the D♭ *ff* chord in m.382 (see Plate V) when an even greater surprise is in store: another completely unpredictable five-bar phrase (mm.386–90). Measure 390 is a 'four,' of course, and thus the inserted empty bar (m.389) is an almost freakish phrase aberration.[38] Almost as much of a surprise is the sudden eerie *p* reiteration of the four-note motive. This passage brought such consternation to early interpretors of the symphony, that they simply cut the empty bar out, compressing the phrase to its more comfortable and ordinary four-bar form. The elimination of m.389 can be heard on Nikisch's 1913 recording!

Measure 398 recapitulates the opening motive, this time without fermatas, and—*nota bene*—marked only *f*. Most conductors play this too loud, with great effect perhaps, but it makes the ensuing massive build-up and the ultimate release into *ff* at m.439 all anticlimactic. It is a shame when the cellos and basses fail to sustain the half-notes C–A♭–E♭–F etc., mm.423– 32, which

---

38. Beethoven's manuscript autograph already contains this 'extra-measure' phrase; that is to say, it was not one of the additions made a year later in 1809, and thus must have been included in the first performance. Still, it was not in his very first conception of that passage, as one can deduce from the manuscript which here shows some signs of struggle and indecision (see Plate V). Beethoven was in the habit of first ruling in the vertical bar lines, more or less equidistant—usually comprising six bars per page—and then filling in the staves with notes. The autograph shows that m.386, which initially had a simple rhythm of ♩ 𝄾 |, was sometime later divided into two measures and the present m.387 was inserted in that new space.

Based on my study of the manuscript, I believe that Beethoven originally meant to have two bars of silence here, before the *ff* outburst of m.390. Those two measures would have been, like mm.123– 24, 'two' and 'three' of a four-bar phrase, and thus the four measures comprising the final quarter-note D♭ chord through the eighth-note diminished-seventh chord motive four bars later were intended to recapitulate that earlier idea (mm.122–25). But then, having completed the page of score, Beethoven, perhaps feeling that the two silent bars were after all a little too empty, impeding the flow and energy of the movement, had the idea to insert a variant of the opening four-note call, but in *p*—an afterthought of the *ff* D♭ chord. Having done that, he then realized that the *ff* diminished chord, already written at that point, could not follow hard on the tiny *p* insert. It is at this point that he left the one empty measure already composed in place.

It is perhaps audacious of a mere mortal like me to second-guess an immortal like Beethoven, but it seems to me that in his revision his musical instincts provided him with the absolutely perfect solution. In any case, it is an astonishing thing to me that no other conductor or writer on Beethoven's Fifth Symphony has ever studied and analyzed how Beethoven came to write this five-bar phrase. It is sad to report that neither Igor Markevitch nor Peter Gülke, in their respective new critical editions of the Beethoven Fifth, make any reference to this remarkable five-bar phrase anomaly. Could they have been unaware of it?

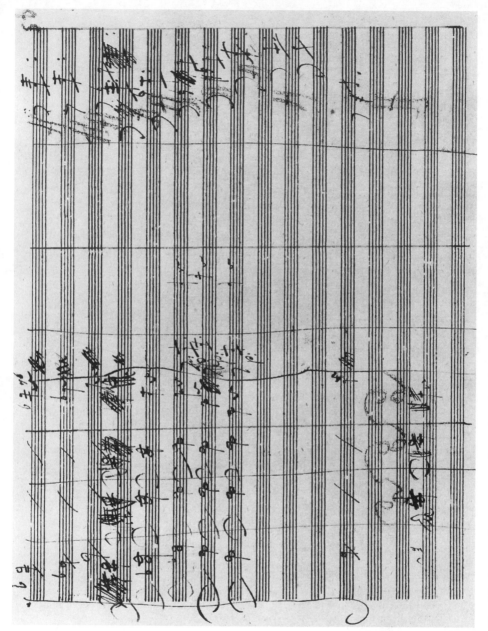

Plate V   Autograph manuscript of mm.383–390 of the first movement of Beethoven's Fifth Symphony

is often the case when the conductor suggests that the first six measures here be played all with down-bows. The result then is something like:

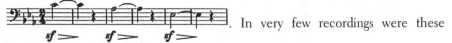 . In very few recordings were these

notes properly sustained, thereby seriously undermining the power and drive of this passage.

A crucial question, unfortunately not explicitly answerable, is whether the violins should have *sf*'s in mm.427, 429, 431, and 433. It is hard to know whether Beethoven (1) forgot to add them, (2) thought performers would simply assume them to continue every alternate bar for another eight measures, or whether (3) he intended some modifying effect from m.427 on. I rather doubt the last, and suggest the retention of the extra *sf*'s.

The viola line in m.423 (doubled in the violins an octave higher) is now restated *ff* and expanded, beginning in m.439, alternating winds (plus timpani) with strings, and heralding the second episode in the development-extension. Here again one cannot stress enough how important it is to understand, perform, hear, and *feel* Beethoven's four-bar structuring. Measure 442 is a 'four,' not a 'one' (as it is far too often played). These massive chordal structures, produced with an orchestra no larger than in many a Haydn symphony, are all the more overwhelming in their effect when delivered in one gigantic line, not just merely arbitrarily strung together. Beethoven's fertile imagination here pours forth a marvel of rhythmic/structural invention which operates on several levels. On the one hand, the earlier two-bar phrases, sitting atilt the four-bar infrastructure (on a purely rhythmic/metric level), are fragmented into various one-bar, two-bar, three-bar, and four-bar phrases (see Ex. 24). On another level—timbral-orchestrational—the passage breaks down into somewhat different phrase components, sometimes coinciding with the rhythmic/metric structuring, sometimes at odds with it. A third level—registral—intersects with the two other layerings, paralleling mostly the timbral layer, and frequently entailing gigantic registral leaps. The three levels of operation, in composite, produce a structural

Ex. 24

The example represents only the upper melodic line. The upper brackets represent various timbrally and registrally delineated phrase lengths; the lower brackets represent the basic four-bar metric structure.

polyphony that is reminiscent of some of Bach's more complex so-called mathematical fugues. It is precisely because of the enormous amount of variation and considerable orchestrational fragmentation in this polyphony of layerings, that the conductor and players must maintain a sense of the underlying four-bar infrastructure—musically anchored in it, as it were—lest the three-layered super-structure becomes a mere arbitrary thirty-bar jumble. Many conductors get bogged down in this fragmented texture, unable to maintain the relentless flow of the music. How exciting and right this sounds when fully comprehended can be heard on Mitropoulos's recording.

Beethoven returns to the opening motto of the symphony one final time (mm.478–82), more powerful and monumental than ever before, starkly 'harmonized' in bald thirds, seconds, and fourths, covering a tremendous range of six and a half octaves. The very low D of the basses (mm.481–82) must be fully exploited, requiring as many five-string or extension-equipped basses as possible. The most truly Beethovenian way of rendering this final return of the motto theme is *not* to back into with a ritard; it is hair-raising when approached absolutely in tempo. Unfortunately, most conductors abuse this passage, turning it into pompous bombast with their monumental ritards, often starting as early as m.476. The worst offenders here are Bernstein, Stokowski (1940), and Furtwängler, the latter's tempo in m.480 being an astonishing $\downharpoonright = 36$ ($\eighthnote = 144$)!

After this massive, shattering reiteration of the opening motto, reminding us of where this remarkable musical journey began 5½ minutes earlier, Beethoven gives us to believe that he will bring back still more of the exposition. Instead he finally comes to the coda. The main theme (of m.6) returns in m.483 as if in a dream, from far away, darkened by a somber open-string fifth in the cellos, and three winds—bassoons, clarinet, and oboe—weaving subtle *legato* garlands around the eighth-note motives. But before we can fully appreciate what is happening, Beethoven breaks in with a thunderous *ff*, and in twelve sharply articulated measures brings not only the movement to a climactic close but completes the entire cycle of four-bar phrases in m.502 on not a 'one' but a 'three,' leaving open the theoretical possibility that the entire cycle could start all over again with the 'four' of m.1.

After the rigorously perfected structuring of the first movement, the second movement, *Andante con moto*, with its lovely, song-like thematic material, provides a welcome contrast—perhaps even relief. Relief in the sense that, nestling between the relentless drive and *con brio* of the *Allegro* and the at times mysterious, even spectral, at other times rough and impetuous Scherzo (which is then extended without a break into the tumultuous triumphant Finale), another tightly constructed movement would have made the entire work almost beyond endurance, emotional as well as physical. As it is, the second movement, with its "tender melodic sentiment" (as John Burk called it[39]) and expressive simplicity, offers the perfect counterpoise to its mighty surroundings.

But even here, in this gentler, calmer music, there is an undertone of urgency, of tension, of motion—at least there *should* be in performance. This is clearly indicated by the words *con moto* in Beethoven's tempo heading and the metronome mark of ♪ = 92. And yet it is this underlying element (beyond all questions of exterior detail) which is most consistently ignored or rejected by conductors in renditions of this movement. In my sampling of recordings only Harnoncourt honored Beethoven's metronome marking, although a few others came within range: Leibowitz at ♪ = 88, Dohnanyi and Norrington at ♪ = 86, the two Kleibers and Mengelberg at ♪ = 84. Most conductors settled for a leisurely ♪ = 80, including Toscanini, Suitner, Reiner, Karajan, Bernstein, Weingartner, DeSabata, Ashkenazy, Strauss, Boult, and Wand. All the others 'interpreted' the movement in sluggish tempos around the low or middle 70s, with Walter, Krips, and Solti on the lugubrious side with an *adagio*-like ♪ = 66, and Stokowski with a 'schmaltzy' ♪ = 60. Gardiner, on the other hand, drives the music too fast, most of the time around ♪ = 100.

Some readers will immediately protest: 'Oh, Schuller, here, you go again with Beethoven's damnable metronome markings. We don't even know whether they are authentic; they're probably a mistake. Anyway, it's too mechanical and academic to follow rigidly those metronomics.'

Several responses come to mind. First, we don't *know*—meaning *know for sure*— that Beethoven's metronomics are *in*accurate and *not* what he really intended. Second, working within a composer's metronomic indications does not *necessarily* lead to 'mechanical,' 'academic' performances, as a number of inspired recordings that respect composers' tempo markings can attest.[40] 'Mechanical,' 'academic' performances result from bad conducting, not 'incorrect'—or for that matter 'correct'—tempos. Third, why is it that the protesters of Beethoven's metronome indications protest selectively; why do they accept many—most of the slower ones— and ignore the fast(er) ones?

Fourth—and this is the most important point—if conductors (and other performers) wish to ignore Beethoven's metronome markings, so be it. But how can

39. John N. Burk, *The Life and Works of Beethoven* (New York, 1935), p. 279.
40. Very few recordings, obviously, of *this* movement, since none (except Harnoncourt's and Leibowitz's, perhaps) respect Beethoven's metronome markings. But there are any number of recordings of various movements and works by any number of composers in which, contrary to the prevailing 'tradition' to ignore the metronome markings, some conductors have adhered to them and produced performances that are anything but 'mechanical' and 'academic.'

they dare to ignore his tempo headings, in this case *andante con moto*, not just *andante*. (Note also that in most cases Beethoven's metronome markings are confirmation and illustration of his tempo headings.) In Italian *andante con moto* clearly means 'in a walking tempo *with motion*.' It seems to me that the message thereby given is unequivocal and should be heeded, especially when it is a great master like Beethoven who is instructing us.

I want to make it clear that I attach primary importance to the *verbal* tempo heading, which in the case of most great composers is remarkably precise and refined, and in my view to be trusted as much as or perhaps even more than the metronomizations. This approach also allows us to deal effectively with those composers—like Brahms, Schubert, Debussy, and Strauss—who rarely or never used metronome markings. On the other hand, where the metronomics corroborate the verbalized tempo indications, as in the case of Beethoven's "*Andante con moto*, ♪ = 92," it seems to me we ought to be doubly eager to respect those indications. Indeed, this movement, when played at or near the designated tempo, reveals a very different character and feeling from what one usually gets in the typical conventional performance. It is, in fact, not a 'slow movement' at all, which seems to be the standard interpretation, I suspect, on the basis that 'this is a symphony, and it must therefore have a slow movement.'[41] It is merely a 'somewhat slower' movement, not only in that its tempo is slower than that of the first movement (from 108 per beat to 92), but for much of its duration its beats (eighth-notes) are not further subdivided into faster rhythms. Thus Beethoven achieves a significant degree of relaxation by both slowing down the tempo and adhering generally to the broader rhythmic units. Seen in this light, Beethoven's tempo is already sufficiently differentiated from its surrounding movements as to require no further 'improvement' from performers.

But further, of even greater import is the fact that Beethoven's *andante* theme in its melodic contour outlines a pitch progression from C via D♭ to E♭ (points x,y,z in Ex. 25). The tempo must not impede or imperil one's perception of— i.e. one's ability to hear—the melodic (and implied harmonic) *motion* contained in this thematic line. In that connection, the *f* dynamic in m.7 underscores

Ex. 25

41. It is not often enough remembered that Beethoven's initial intention was to call this movement *Andante quasi menuetto*. And conductors who think of this movement as a slow movement should be reminded that Beethoven's Seventh Symphony doesn't have a 'slow movement' either. But then such conductors are not likely to heed Beethoven's *allegretto* (♩ = 76) tempo in that case either, evidenced by the fact that if there is a tempo marking that is more abused than Beethoven's Fifth Symphony *Andante*, it is the *Allegretto* of the Seventh Symphony.

the arrival point (E♭) of the melodic/harmonic progression. And how often this remarkable *f* is ignored or suppressed or otherwise adulterated! Let me add lest I be misunderstood, I am not suggesting that the *f*'s in mm.7,9,11 are hard-hitting, aggressive *f*'s, but rather firmly expressive, lyric *f*'s, to be played with a certain warmth and 'cordiality'—in effect a *forte dolce*.

Indeed, tempo and dynamics are the two elements most often disregarded in performances of this movement, this despite the fact that these are the most unusual and original aspects of the piece. In many recordings, for example, the theme in the violas and cellos is played *mf* (or *mp*) with a thick, heavy sound—what one critic has called an "industrial strength" sound—that, when also played at too slow a tempo, completely falsifies this theme's discreet artfulness and simple elegance. It is not for lack of craft or harmonic sensibility that Beethoven chose not to harmonize the theme in conventional four-part voicing. It is precisely its Handelian or Haydnesque two-line simplicity that gives this statement its utter nobility. It is simple two-part counterpoint, elevated to the level of a deeply felt lyric theme. All that is needed in performances is a light, discreet *p* sound, rhythms that are clearly and accurately articulated (within the slurs), no extraneous dynamic nuancing (like ≤ ≥)—and Beethoven's perfectly chosen notes will do the rest. But only six conductors in our sampling honor Beethoven's dynamics: Haitink, the two Kleibers, Strauss, Gardiner, and Brüggen.[42] Toscanini almost does, producing an elegant, noble *p* in the first six measures, a healthy *f* in m.7, but then disbelieves Beethoven's sustained *f* and diminuendos to the next *p*. Others come close, delivering a fine *p dolce*, but are seemingly afraid of Beethoven's sudden *f* in m.7, reducing it to *mp* or *mf*. This includes Furtwängler, Boult, Böhm, Suitner, and Solti. Others, like Maazel, Masur, Colin Davis, Ansermet, allow a kind of creeping crescendo in the first six bars, especially where there are two bows per bar (m.4,5—three in m.6) which again works to annul the sudden *f* of m.7. Even worse are those who feel the need for a thick, heavy syrupy *mf* or *mp* in the first six bars—Weingartner, Wand, Masur, Stokowski, Ashkenazy, Krips, Karajan, Walter, Giulini, Bernstein (Bernstein's and Stokowski's are the thickest)—for, by going in this direction, they completely vitiate Beethoven's intended and surprising dynamic contrast in m.7.

All such 'interpretations' undermine Beethoven's remarkably refined, original, and sophisticated use of dynamics in this movement, flattening out the music's extraordinary dynamic contours to the point of blandness (see Exx. 26a,b below). For example, the three important E♭'s in mm.7, 9, and 11 are all differently set. Measures 7 and 9 both have *f*'s followed by *subito p*'s in m.8 and m.10. However, m.9 slurs into m.10, while m.7 does not into m.8. The difference is subtle,

---

42. But Brüggen is otherwise disappointing in this theme for, although he starts the violas and cellos at ♪ = 92, by the fifth measure he has dropped to ♪ = 76. It is hard to tell from the recording whether the musicians in the orchestra pulled him back, preferring the more leisurely tempo, and he simply followed them, or whether Brüggen purposely or inadvertently changed tempo after one or two bars. Brüggen's performance of the entire movement is quite erratic in terms of tempo, vascillating often between a low of ♪ = 64 to a high of ♪ = 84. It never does achieve the original Beethoven tempo of ♪ = 92.

but significant—and, when done right, perceptible and meaningful. Measure 11
is different again, in that here the *f* does have a diminuendo. Do the conductors
who make their diminuendos in mm.7 and 9 really think that Beethoven didn't
know what he was doing, or simply forgot the diminuendos? Or to put the question another way: if Beethoven writes a diminuendo in m.11, why does it not
occur to these conductors that mm.7 and 9, where there is none, ought to be
played without diminuendo? Do they not see the beauty and originality of this
subtle differentiation? And why do they overlook the fact that there are at least
five other instances of one-bar *f* to *p* diminuendos in the movement (mm.37,
60, 86, 195, 219)? Proof enough, I should think, that when there is no diminuendo indicated, we performers ought not to make one.

These may seem like simple, even naive, questions to some. But they are not
merely hypothetical or rhetorical; they are based on hard performance evidence.
And the fact that such questions have to be asked, reflects the sad state of the
aesthetics (and ethics) of conducting. It also reflects the wide latitude in musical
self-discipline between conductors and orchestral musicians. The latter would
be severely chastised by most conductors if they allowed themselves a vagrant
diminuendo or an arbitrary ritardando or any other willful deviation, and yet a
conductor is allowed such liberties, whether taken consciously or inadvertently.[43]

It is amazing how many distortions and deviations conductors are capable of
visiting upon this simple, innocent theme. Perhaps it is its very simplicity that
disturbs conductors, who feel they have to 'dress it up,' 'improve on it.' For
example, Nikisch virtually recomposes mm.7–11 (Ex. 26a).

Ex. 26a

Ex. 26b

Walter (this time with the Columbia Symphony) recomposes it another way
(Ex. 26b), having also in the first six measures gotten the musicians to play sixty-

43. A telling example of this double standard can be heard on a recording of Bruno Walter rehearsing the first two movements of the Beethoven Fifth with an unnamed orchestra (actually, the Los
Angeles Philharmonic). The violas and cellos play [♪] in m.7. Walter stops

fourths (instead of thirty-seconds), probably because his tempo ($\flat$ = 66) was so slow that the thirty-seconds, when played correctly in that tempo, seemed too leisurely and pedantic, even to him. It seemingly never occurred to Walter that the way to fix this problem was to speed up his basic tempo, at which the thirty-seconds would have been just right. It is also amazing how many conductors — among them Strauss, Koussevitzky, Ansermet, Karajan, and (the usually meticulous) Erich Kleiber — have asked for the C in m.1 to be played *tenuto*, when Beethoven has clearly marked this note with a *staccato* dot — a note, in other words, to be gently lifted.

Just as amazing — and disturbing — is the fact that scarcely anyone plays the pick-up notes ( ) in mm.8 and 10 *p*; they are usually rendered at various louder dynamics, mostly *mf*. Have not any of the many conductors who ask for this deviation (or allow it) realized that this upbeat gesture in *p*, followed immediately by an expressive *f*, is very dear to Beethoven's heart, as witnessed by the fact that it not only recurs (in a modified form) in m.57 and m.59, but is reiterated twice more in even more dramatic dynamic contrasts (*pp* to *ff*) in mm.28–29, 77–78?

The next performance problem appears in the woodwind phrase, mm.11–15. It is not known to me who started the bad tradition of separating mm.11–14 into four disjunct one-bar phrases, making caesuras at the end of each bar. It was probably Bülow, who was fond of making such phrase separations and sprinkling his performances with innumerable tiny pauses — *Luftpausen* in German. Weingartner describes this in *Über das Dirigieren*,[44] and chastises Bülow for it but then — in the woodwind phrase here under discussion — himself makes precisely such phrase separations in his own recordings. Many of the earlier conductors imitate this bad tradition — Nikisch, Mengelberg, Jochum, Koussevitzky, Reiner, Walter, Szell, surprisingly Toscanini, and DeSabata, Kempe, Solti, and even 'modern' conductors like Gardiner, Harnoncourt, and Norrington — a tradition which, I am happy to note, has for the most part been stamped out in more recent times. Note the beautiful line Haitink, for example, achieves here.

Speaking of this descending line, it is rarely realized by flutists, clarinetists, and conductors that the D♭ and B♭ in mm.12–13 are written-out appoggiaturas. Had Beethoven wanted to merely imitate mm.7–10, he would have written

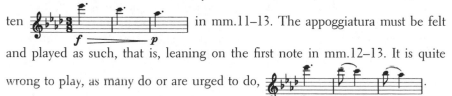

and played as such, that is, leaning on the first note in mm.12–13. It is quite wrong to play, as many do or are urged to do,

At the end of this phrase (recurring three more times in mm.60–64, 195–99,

and tells them to play *f* "without a diminuendo," then sings the passage, making a big diminuendo himself — exactly what the musicians had played in the first place. Walter seems to have been totally unaware of his own contradiction.

44. (1905), p. 38; (1913), p. 34

219–23), the second clarinet should play its rhythm (♩. ♪) clearly against the flute's rhythm (♪♪♪), i.e. not converted into sixteenth triplets, as many conductors have demanded or allowed (for example, Furtwängler, Mengelberg, Boult, Koussevitzky, Maazel, Szell, Solti, Masur). Other conductors (Giulini, Ansermet, Klemperer, Steinberg, Gardiner, Brüggen—even the usually exemplary Haitink) allow this little note to be swallowed, becoming virtually inaudible, especially in the several parallel viola/first bassoon passages.[45] If conductors question Beethoven's rhythmic notation here (♩. ♪ against ♪♪♪), they are advised to note that Beethoven in this entire movement is constantly juxtaposing these two rhythmic cells.[46] Obviously all cannot be accidents or mistakes.

The last note (m.15) of the woodwind phrase is often held too long, most commonly as | ♪ ♪♪ ↾ | , or even | ♩ ↾ | (as with Reiner, Karajan, Masur, Giulini, and Walter). Now this may seem like a 'musical,' 'sensitive' way of ending the phrase, rounding it off, as it were, to link up better with the incoming strings. But it is wrong; it is wrong (a) because Beethoven is absolutely consistent throughout the movement in ending all his phrases with an eighth-note (mm.19, 20, 31, 199, 242, etc.—even m.22, after the previous sixteenths); and (b) because the rests in this music are not arbitrary, accidental gaps to be somehow filled in; they are not 'empty' moments, but an important and integral part of the music. The silences in rests are of vital importance in all great music; they are the places where the music breathes, and where it flexes itself. They are also often the windows into the structure, into the 'building,' of the music. These windows must not be boarded up; silences must not be devalued. The temptation to elongate final phrase notes before a rest is all the greater when the tempo taken is too slow: it obviously makes the rest also too long, and there is then the greater temptation to fill in that rest.

The next phrase (in the strings, mm.15–19) generally fares not much better in performance. The tendency here is for the violins to crescendo on the three anacrusis notes in m.15, particularly when using an up-bow. An even greater temptation exists in m.16, where many conductors and string sections are wont to make a premature crescendo. But the originality of Beethoven's conception lies precisely in the fact that this phrase crescendos dramatically from *p* to *f* in *one* measure, not the two or two and a half measures most performances offer out of sheer laziness or inattention. Again, Beethoven's creative imagination is at its most vivid in the way he has enriched this phrase with dynamic variety and contrast (Ex. 27).

Ex. 27

45. Measures 18, 67, 202, 203.
46. See mm.26, 33, 158–62, 238–39.

While the second movement of Beethoven's Fifth is more spacious in its design and flowing melodic lines, less rigorous in its patterning and less involved with minute—one might almost say mosaic—construction than the first movement, it is nonetheless planned out in a marvelous, grandly logical form. Beethoven may not have called the movement 'Variations'—it is not strictly speaking in a conventional 'theme and variations' form—but it is nevertheless essentially variational in conception, as the formal analysis in Fig. 4 easily demonstrates. It should be noted that what I have called the 'theme' is itself

Fig. 4

Theme			Var.I			Var.II		
A	B	C	A^1	B	C	A^2 A^3 A^4	B^1	C^2
mm.1–22	mm.22–31	mm.31–49	mm.49–71	mm.71–80	mm.80–98	mm.98–123	mm.124–147	mm.147–166

Var.III			Coda
A^5	A^6	A^7	B^2
mm.166–184	mm.184–205	mm.205–228	mm.229–247

divided into three thematic segments, and that B is itself but a variation of A by way of both contraction and expansion (see Ex. 28), and C a variation of B

Ex. 28

(mostly by way of orchestration and transposition (from A♭ to C major). Also notice that in what I have called Variation I, only the first part of A is significantly altered, while B and C remain virtually unchanged (with but minor rhythmic alterations). Variation II undergoes more extensive modifications. The first part of A is varied three times in three immediately contiguous variations, featuring in succession violas/cellos (m.98), first violins (m.106), cellos/basses (m.114). B and C now also undergo substantial alterations, most dramatically C, which is stretched from its binary (two-bar) structuring to a ternary (three-bar) phrasing. In Variation III, Beethoven takes us to A♭ minor (the only time in the movement) and at m.184 into a grand canonic interplay between upper strings and woodwinds. A^7, which in part relates to B^1 in Variation II, is unique in the movement's over-all scheme in that it is broken up into two tempos, a *più moto* and a *tempo primo*. The coda's B^2 is initially yet another variant, a new fusion, of the original A and B.

That is the grand scheme of the movement in its largest, boldest outlines. On a smaller, more detailed, phrase level, we see construction in long eight-bar phrases—long by comparison with the four-bar structuring of the first movement. But we also find a number of seven-bar phrases, even one eleven-bar phrase. But most of these asymmetrical phrasings are the result of contraction (*Verkürzung*) or expansion (*Dehnung*). This freer, irregular, more flexible structuring also contributes to the music's sense of looseness, of spontaneous, almost improvised invention.

In m.23 two performance problems arise. One is the balance in the two woodwind pairs, specifically the balance of the second players with the first players. Only rarely does one hear these second players—no matter what orchestra—match the firsts dynamically in true balanced duets, a fact borne out by virtually all the recordings I have sampled. Somehow this passage (mm.23–26) is generally interpreted as 'solos' for first clarinet and first bassoon, on the one hand because these players usually take the initiative in that direction, and on the other hand because very few conductors hear the resultant imbalance and therefore see no reason to make a correction.[47] It also results from a widespread tendency among a majority of conductors to conduct and hear only 'the melody,' rarely hearing and balancing the harmonies. Of course, when the four woodwinds *are* blended on a recording into a well-balanced quartet, it is not always possible to ascribe this success to the conductor; it may simply be that the respective second players instinctively knew that they should balance with the firsts, and the conductor, aware or unaware, accepted this gift. It may also be that the conductor actually asked for the instruments to balance, but without being present at the rehearsals, it is impossible to deduce from a recording who should receive the credit for the right balance. In any case, only in very few performances were these balances right, those of Haitink, Krips, Carlos Kleiber, Solti, and Bernstein.

Another problem in this phrase results from a mistake in the printed score and parts, as compared with Beethoven's manuscript. In mm.23 and 25 (as well as in m.72 and m.74) the woodwinds' slur should end with the third eighth, while the violins' slur as printed all three notes on one bow. Beethoven was—again subtly, imaginatively—confirming the differentiation between the woodwinds' quarter-notes and the violins' eighths in m.24 and m.26. In this connection, we should note that Beethoven has no crescendo indications in mm.23–26, either per two-bar phrase or over-all in these four-bar phrases. Yet the vast majority of performances contain such crescendos—trivializing the passage with an obvious emotionalism that is far removed from the almost unearthly stillness and held-back tension the phrase has when played without crescendoing—particularly when then followed by the powerful *ff* outcry of mm.29–31.

Finally, some conductors (Strauss, DeSabata, Boult, Toscanini, Norrington, and

47. This is a good example of why I described the ear as the servant of the mind, of the intelligence, in Part I of this book. No matter how physiologically sharp the ear is, it cannot (will not) hear what it is ignorant of, what the mind has not told the ear to hear.

Schuricht) like to move the tempo up a notch or two at m.23, mostly, I suspect, so that they can broaden the tempo again with the *ff* in m.32 for greater 'effect.'[48] In my opinion Beethoven creates enough contrasts in texture, dynamic levels, and orchestration to make any adjustments in tempo quite unnecessary. Indeed, as I have already pointed out, since these two phrases comprise identical thematic and harmonic material, it ought to be our obligation to present it in the same tempo, precisely to let all other Beethoven-inspired variations and changes be clearly, un-distractedly heard.

As mentioned before, the sudden dynamic contrasts of mm.7, 9, and 11 are reintroduced in mm.28–9, this time not from *p* to *f*, but *pp* to *ff*, a stunning effect rarely rendered correctly, alas. In m.30 (and the analogous m.79) the staccato dot on the dotted sixteenth is an engraving error, not contained in Beethoven's manuscript. But the two succeeding eighth-notes *do* have staccato dots, while the notes in mm.32 and 34 do not; and thus the latter should be played in a well-articulated but sustained manner. This is necessary to mention since, once again, numerous conductors who have not trusted Beethoven's notation, have caused the brass, oboes, and low strings to play *staccato* in this triumphant passage (Nikisch, Mengelberg, Boult, Thomas, and Walter among them).[49]

---

48. Strauss even crescendos through the four measures 23–26, only to correct himself with an exaggerated *pp* at the end of m.26. A word on Strauss as a conductor, especially of other composers' works, may be appropriate here. The received wisdom about his conducting has always been that he was a musician given to fast, bright, no-nonsense tempos, to inexorable tempo steadiness and control, to a certain 'coolness' of expression, espousing in general the 'new objectivity' of the 1920s.

The evidence of his recordings, however, tells us that this is all a myth, a myth probably promulgated as much by himself as by observers or admirers. His recordings show that he was in fact a highly erratic and willful interpreter, especially in matters of tempo. I can think of only two conductors who could outdo Strauss in tempo deviations: Stokowski and Bernstein, and perhaps we can add Mengelberg.

The second movement of Beethoven's Fifth is a striking example of Strauss's wayward way with tempos. In this movement alone I count fourteen major tempo changes, as the following table shows. (The other movements are not much steadier).

m.1	♪=80	m.124	♪=84	m.196	♪=78
m.23	♪=90	m.132	♪=96	(m.205	♪=106)
m.32	♪=82	m.141	♪=106	m.219	♪=70
m.50	♪=86	m.148	♪=76	m.229	♪=78
m.72	♪=90	m.191	♪=84	m.245	huge ritard
m.81	♪=82				

(Beethoven's metronome marking is ♪=92.)

49. To give them the benefit of the doubt, they may have been influenced by the timpani part, since it is difficult (though not impossible) to play long-sustained notes on the timpani. German timpanists generally play with a dry hard sound—more so in earlier days—and it may be that the conductors just mentioned, facing that reality, felt that the brass should then match the timpani. I should mention in this connection that most printed scores contain an error in the timpani part: the *sf*'s in mm.35, 36 should be placed on the first beat, not on the third (similarly in mm.84, 85).

I caution here against another bad tradition, namely, that of playing m.31 in a 'stop-and-go' manner, that is, a big cadencing slow-down on the first two eighths of the measure, then a fresh pick-up in the brass on the third eighth. It is a much more exciting realization of Beethoven's intentions here to keep the tempo moving in m.31, to *connect* (not separate) the two C major *ff*'s (mm.30/31 and m.32), and to clarify and maintain the thematic link between m.23 and m.32. (Listen to Dohnanyi's recording to savor the full effectiveness of this.) As already mentioned, many conductors like to broaden the tempo dramatically at m.32, some (like Karajan, Knappertsbusch, Suitner, Ashkenazy, Colin Davis, and Hogwood) even—quite unmusically—already at m.29 or m.30; while others prefer to noticeably brighten the tempo. Those who broaden at m.32 argue that doing so heightens the 'majestic' effect of the passage. These include, apart from the conductors just mentioned, Strauss, Bernstein, and Solti (the last-named, already at a dangerously slow basic tempo of $\flat = 68$, slows to a ponderous 62 at m.32). Those, on the other hand, who press forward here believe a faster tempo helps to heighten the effect of the bright C major sung forth by the brass and timpani, giving the passage its 'necessary' urgency. (Included, as might be expected, are Toscanini, DeSabata, Ansermet, and Stokowski). As suggested, neither approach is valid or 'necessary.' Beethoven has composed enough contrast and drama into the passage to make any obvious additional 'improvements' quite superfluous.

The two crucial performance problems the conductor must address in mm.32–37 are (1) the over-all balance and (2) the rhythms in mm.33, 35, 36. We should recall that the brass instruments of Beethoven's Vienna in the early 19th century were not as brilliant and powerfully penetrating as the brass instruments of today. It was therefore not unreasonable—certainly not a case of 'bad instrumentation'—on Beethoven's part to pair the oboes here with the brass and timpani, given also Beethoven's customary habit of almost always using uniform dynamics for all instruments in a given measure. But a good balance in mm.32–37 can be easily managed if the conductor admonishes the brass not to play their absolute loudest (which a *ff* seems always to signal to most brass players) and, on the other hand, to encourage the oboes to give their all. Oboe players, especially seasoned ones, long ago having given up trying to be heard here, are prone to 'take it easy' or in some cases not play at all, just appear to be playing. This problem reaches really ridiculous proportions when conductors (Karajan, for one) have the idea of doubling the brass in this symphony. Then truly all hope is lost for the oboists.

The second problem here is differentiating clearly between the two rhythmic layers: the triplets in the upper strings and the martial dotted sixteenth- thirty-second rhythms of the rest of the orchestra. This usually takes a little rehearsing (as well as the conductor's firm hand), and is best achieved when the two disparate rhythmic forces do not listen to each other, but simply maintain their own appropriate rhythms. It is also worth mentioning that, again, as in mm.23–26, the pairs of winds (oboes, trumpets, horns) should balance dynamically. It detracts tremendously from the magnificent full effect of this music when the second players, who after all supply the all-important harmony, are weaker than the firsts. In the innumerable recordings I have analyzed, only in a very few were the instruments properly matched and balanced here.

Conductors should know that the diminuendos in the brass in mm.37–38—innocent looking enough in the score—are very hard to produce without losing the intonation. It is up to the players, of course, to master this problem, but for the conductor—particularly the inexperienced conductor—who might be quick to berate his players, it is well to know that a long-sustained, even diminuendo from *ff* to *pp* is one of the most difficult things (technically) to achieve on wind instruments. A smile of encouragement works better here than a frown. The players will in fact appreciate it enormously if the conductor indicates his awareness of the technical difficulties here.

The next phrase (mm.39–48) is one of the most magical moments in all of music, a harmonic progression of such daring (for its time) as only a Beethoven (or a Mozart) could have produced. But the full effectiveness of this passage depends most crucially on being played not only with a true *pp* but a soft velvety sonority. Only a quiet, warm *sul tasto* sound will ensure the hushed rapt mood and the beautiful stillness of this extraordinary passage—the feeling that the music is motionless but not inert. Very few conductors seem to know how to achieve this special mood, but Nikisch, Strauss, Reiner, the two Kleibers, Knappertsbusch, Karajan, Schuricht, Dorati, Dohnanyi, and Haitink are notable and welcome exceptions. (Incidentally, Harnoncourt here slows down to a static $\flat$ = 60, down 32 points from his basic (good) $\flat$ = 92.)

In mm.48–49 care must be taken that the three E♭'s (viola/cello, bassoons, clarinet) are produced with equivalent *f*'s. On most of the recordings sampled the bassoons were substantially weaker (softer) than the other instruments.

Since mm.49–61 are a variant of mm.1–11, it stands to reason that the gratuitous interpretational liberties taken by certain conductors there will be perpetrated again. These include a heavy, fat *mp* or *mf* for the violas and cellos (rather than an elegant, elegiac *p*); the creeping crescendo which annuls the *subito f* in m.56; the softening of the strong dynamic contrasts in mm.56–61: all to be rigorously avoided. What is new and problematic in this passage—the problem will come back again in mm.98–105—is the pizzicato accompaniment. The danger here is—and dozens of recordings prove this—that the violin pizzicatos will be too soft, even barely audible, relative to the bass's pizzicato, unless the violinists are cautioned to play a little louder, conceptually, say, *mp*, and with what string players call a 'full' or a 'deep' pizzicato. This is, of course, not a problem limited to this passage in Beethoven's Fifth; it can apply to literally thousands of pizzicato passages in the literature, simply because we are confronted here with an unalterable acoustic phenomenon: namely, that the projection of a pizzicato depends in large measure upon the thickness of the string which is plucked, and thus thicker, heavier strings, as on a bass or cello, will project more effectively than a violin string. Even on a violin the lowest (and therefore the thickest) G string will produce a louder pizzicato than the upper (thinnest) E string. To put it another way, the same amount of energy in plucking a string will inherently produce a bigger, more projecting pizzicato on a bass string than on a violin or viola string. The lesson to be drawn from this is that as a rule, especially in a *p* dynamic, violinists should play a little louder and bassists a little softer to achieve the desired result of a vertically balanced pizzicato throughout the string section. Conductors should hear this imbalance when it occurs, but in my experience most in fact do not. And the recordings sampled in this analysis bear out my point. Only very few of Beethoven's beautifully chosen violin pizzicato notes in mm.49–56 can be heard, whereas the basses' pizzicatos are uniformly well represented.

A few score misprints in this passage must be mentioned. The clarinet's slur should stop with the E (written F#) in m.53. Both bassoons should play in unison from the second sixteenth in m. 57 to the downbeat of m.59. In m.57 the

last three sixteenths of the violas should read as follows: , similar

to the analogous passage in m.8. Though not strictly speaking a misprint, the woodwinds in m.59 should be marked in score *and* parts as follows: the flute beginning *p*, the others (oboe, clarinets, bassoons) *mp* beginning at the last three sixteenths.

Many conductors, for reasons that I cannot fathom, have in the past made a big ritard in m.60 and then, even more ridiculously, an *a tempo* in m.61. Measure 60 being a variant parallel to m.11, the beginning—not the end—of a four-bar phrase, there cannot be any justification for distorting Beethoven's line, as, for example, Colin Davis, Mengelberg, Furtwängler, and Klemperer have done. (But even more shocking are the two bad, blatantly audible splices in Solti's Vienna Philharmonic recording at mm.55 and 61—again the work of some famous, well-paid recording producer!)

Whatever has been said regarding mm.11–38 is obviously applicable as well to mm.60–87. Again, special care should be taken not to swallow the G's in the second violas and first bassoon at the end of m.67; nor should this last beat diminuendo into m.68. The problem of balance and continuity mentioned in connection with m.48 recurs in m.97, but in a different version. Here the clarinets are apt to sound weak relative to the tripled bassoons, violas, and cellos.

In mm.98–104, if the violas and cellos play a true light *p dolce*, and the violins' pizzicatos—especially the sixteenths, these perhaps even with the slightest accent—are rebalanced, Beethoven's intentions will be fully realized.

One of the symphony's most heavenly passages, mm.105–14, is unfortunately rarely rendered in a 'heavenly' manner. And again it is simply a matter of conductors and players not following Beethoven's explicit instructions. While almost everyone makes the *subito pp* at m.105—only Strauss, Koussevitzky, Walter, and the early music specialists Norrington, Hogwood, and the Hanover Band do not (*sic*)—the results in mm.107–14 reveal in proportions of about three to one a cavalier disregard of Beethoven's score. In column I of Fig. 5 are listed those conductors who have inadvertently or deliberately defied Beethoven's cautionary marking, *sempre pp*, in m.107; in column II are those relatively few who have

Fig. 5

Column I		Column II
Hickox	Boulez	Nikisch
* Abbado	* Schalk	Furtwängler
Strauss	Mitropoulos	Karajan
Weingartner	* Schuricht	Reiner
Walter	Krips	Boult
Toscanini	Böhm	Solti
Munch	* Leibowitz	Carlos Kleiber
DeSabata	Ansermet	Maazel
* Muti	* Knappertsbusch	Haitink
Giulini	Kubelik	Dorati
Ormandy	* Dohnanyi	Masur
Jochum	Colin Davis	Sawallisch
Kletzki	Ashkenazy	Schwarz
Bernstein	Wand	Suitner
Ozawa	* Kempe	
Stokowski	* Klemperer	
Koussevitzky	* Steinberg	
Mengelberg	* Thomas	
Erich Kleiber	Mehta	
Szell	Norrington	
Gardiner	Hogwood	
Harnoncourt	Hanover Band	
Van Otterloo	Brüggen	

respected and understood his marking. In the latter group the results are truly beautiful; in the former, ordinary at best. (Asterisked names in column I indicate conductors who maintained a reasonable *pp* at m.107, but then allowed—or urged—the violins, seconds as well as firsts, to creep up dynamically, to crescendo in mm.110–14) (see Plate VI).

At mm.123–124 the fermata should not be held overly long—Wagner's previously mentioned fantasy admonition notwithstanding. Most conductors take a slower tempo here, mainly for a kind of *misterioso* static effect, a superficial 'profundity' which is both unnecessary and misplaced. How this passage, including the clarinet and bassoon solos and the ensuing woodwind quartet, can sound when taken at very close to Beethoven's intended tempo can be heard to wonderful effect on Dorati's recording with the Royal Philharmonic Orchestra (London). Dorati, himself a fine composer, was one of the few who, respected a composer's score, and though he was not regarded by the critical and conductorial fraternity as a 'great classical' conductor—being stigmatized early in his ca-

Plate VI Mm.105–118 of the second movement of Beethoven's Fifth Symphony

reer as a "ballet conductor" and a "specialist" in Bartók, Stravinsky, and modern ballet scores—he was actually one of the finest Beethoven conductors of our time.

The woodwind quartet passage at m.132–46 loses a lot of its intended effect when the four woodwind players, especially the flutist, indulge in a kind of fancy wandering rubato, as if Beethoven had written

 Such

players—and the conductors who allow these indulgences—do not seem to understand the special nature and character of this passage (mm.132–43). What is special about it is that it is harmonically stationary: B♭ major stretched out through six measures, followed by six measures of E♭. Though the passage is harmonically stationary, Beethoven provides more than enough interest by two means: one, the copious use of seventh and ninth degrees throughout, and the other the simple but here ingeniously applied device of contrary motion between the two instrumental pairs (flute/oboe and two clarinets). What all of this means is that any fancying up of the passage with rubatos and gratuitous crescendos/diminuendos—as one almost always hears here—is unwarranted. A pristinely 'simple' rendition of the passage is wanted, precisely because the music is, as it were, locked into two tonalities (B♭ and E♭), each stretched out into a kind of six-bar fermata. And, as I say, Beethoven has provided enough other highlights to prevent the passage from getting boring or uneventful or static.

Indeed, one piquant touch hardly ever realized is the extraordinary cluster-like

clash of notes at the beginning of m.139 . This is never heard because the flute and oboe tend to make a diminuendo in m.138 , thus vitiating and defusing the intended 'dissonantal clash' in m.139.

It is amazing to me that not a single conductor recording the Beethoven Fifth ever bothered to consult the original manuscript or the facsimile published in Germany in 1942. Most reprehensible perhaps is Karajan's, Furtwängler's, and other Germany-based conductors' negligence in this regard. Beethoven's manuscript lies in the Staatsbibliothek in Berlin, and yet in the (at least) four recordings Karajan made of the Beethoven Fifth, he apparently never saw fit to study the original manuscript, not only in respect to this passage in the *Andante* movement but in respect to the several dozens of other errors in the endlessly reprinted first edition. Equally disturbing is the disregard of Beethoven's intended phrasing by the Hanover Band, who proclaim loudly in their CD booklet that they played from edited parts made to correspond to Beethoven's autograph. (There is mighty little evidence of this in their recording.) The effect of this passage when performed as Beethoven intended is totally different from

what is almost always heard. It gains in clarity and firmness, avoiding the sentimental mood of a *Romanza* that the usual (wrong) phrasing imparts to this
passage.[50]

Measures 145–46 are interpreted by many conductors in a *tenuto* manner.
There is little justification for this. Beethoven did not use the short-cut notation

♪. found in most available scores, writing instead ♪ ♪ ♪ ♪ ♪ ♪. But the lack

of staccato dots in mm.145–47 (in the autograph as well) has led many pedantically minded conductors to interpret these ten woodwind sixteenth-notes in a
tenuto manner. But coming from the previous staccato sixteenths (in the strings
as well), this sudden tenuto makes no musical sense. An interesting alternative,
assuming that the incoming horns should sound a little broader and heavier in
their *f*, is to have the woodwinds in mm.145–46 progress gradually from a staccato to a *tenuto*. Many fine woodwind players over the years have done this
instinctively; it goes logically with the crescendo.

In the context of the consistent binary or four-bar structuring of this movement, mm.148–53 constitute an interesting anomaly: two three-bar phrases (see
Ex. 30). In many recordings (and performances) one can hear the relative discomfort of the musicians at the intrusion of this 'foreign element,' many

Ex. 30

musicians, of course, not realizing that the sudden three-bar phrasing is the cause
of their unease. A conductor's comment here, urging the musicians to feel—perhaps even enjoy—the non-conformity of this passage goes a long way towards making it sound right and 'comfortable.'

The next passage in the strings (mm.158–66) is one of those rare moments that
is almost always played correctly (except, obdurately—instead of 'authentically'—
by the Hanover Band and Hogwood's Academy of Ancient Music players). Nonetheless it is worth taking note not only of Beethoven's detailed dynamic markings
but his phrasing/bowing as well: three short light bows in m.162 and m.163, followed by one long bow over two bars in mm.164–65. And, again, the *più p* here (in
m.161) does not mean a sudden *subito p*, but rather a softening to the *pp* of m.162.

The key of A♭ minor in m.167 has led many a conductor to turn this passage—yet another variant of the opening theme—into a funeral march. How
this can sound much more in keeping with Beethoven's intentions—let us remember that he originally intended to call this movement *Andante quasi men-*

---

50. The new Gülke edition (Peters) is correct in this respect.

uetto—can be heard on Dorati's aforementioned recording: buoyant rhythms with a sense of underlying urgency in the strings' repetitive pizzicatos, and the single oboe notes poking whimsically through the texture.

The crescendo found in many editions in the flute part in m.177 is spurious. There is none in Beethoven's autograph. The entire passage should be played without crescendos or dynamic swells, maintaining (even in the high register of the flute) a pure simple *p*, followed similarly by the violins. The crescendo starts only with the entering cellos and basses in m.181.

In mm.182–83 care should be taken that the violas and second violins hold their dotted eighths just the right duration (Ex. 31). Held too long, they interfere

Ex. 31

with the incoming thirty-second-note run; held too short, they fail to make the necessary connection to those incoming scales.

A big balance problem has plagued performances at mm.185–94 from time immemorial. (Berlioz already complained about it in his *A Critical Study of Beethoven's Nine Symphonies* [English edition, New York, 1912].) Far too often the wonderful canon between strings and woodwinds, two beats apart (not three beats or one bar!) is left unrealized (Ex. 32). Most conductors are so busy con-

Ex. 32

ducting the strings here that they quite neglect the three-octave woodwind counter-line. Since the violins already outnumber the woodwinds (four to one) and are hardly in danger of not being heard, it would behoove a conductor to give his full attention to the woodwinds. Matters are made worse when the sustained brass and timpani as well as the highly rhythmic lower strings play their less important non-melodic parts too loudly, too heavily. What is truly astonishing—and depressing—is the way Beethoven's contrapuntal intentions here are so roundly ignored, even on recordings where it is obvious that the woodwinds are doubled, proving by the way that the mere doubling of woodwinds in *f* passages does not necessarily guarantee proper balances.

Only a relatively few conductors have managed this passage (mm.185–94) successfully, notably Nikisch, Jochum, Krips, Erich Kleiber, Giulini, Bernstein, Mitropoulos, Thomas, Suitner, Haitink, Colin Davis, Reiner, Szell, Solti. Renowned conductors who failed in this passage to the point where the woodwinds are virtually inaudible are Weingartner, Furtwängler, Toscanini, Walter, Kubelik, Gardiner, Brüggen, and Harnoncourt. Amazing!

Equally amazing is how many conductors fail to discipline themselves and their musicians to produce an appropriate *p* in mm.199–204, a passage essentially the same as mm.15–18 and 64–68, but with the crucial exception that this time the earlier crescendo is completely withheld by Beethoven.

Beethoven's *più moto* tempo (m.205) is ♪ = 116, a 24-point increase over the basic *andante* tempo. But here again, most conductors keep a considerable distance from Beethoven's tempo, in fact more or less to the same degree they deviate from the basic tempo. The range of deviation is astonishing, all the way from Weingartner's rather fast ♪ = 124 to Colin Davis's and Giulini's ♪ = 86, a mere 30 points off the mark! Giulini jumps to this (for him) relatively lively tempo after grinding to a virtual standstill in the previous string passage with a deadly ♪ = 54—no *con moto* there! Jochum hits it pretty much on the nose with a lively ♪ = 112, and others, like Furtwängler, Ansermet, Strauss, Toscanini, Dohnanyi, and Dorati, come close enough (between ♪ = 104 to 108) to capture the spirit and intent of this light-hearted episode.

Measures 210–13 have been adulterated in every conceivable way. Even though Beethoven's grace notes in the oboe are clearly before the beat, and the G♭'s have staccato dots, any number of conductors insist on such variants as

𝄾 [♫♪] (Böhm, Furtwängler, Wand, Haitink, Krips), 𝄾 [♫ ♪] (Mengelberg, Weingartner), 𝄾 [♫ ♪ *3*] (Maazel), 𝄾 [♫♫] (Erich Kleiber, Nikisch),

𝄾 [♫♪] (Jochum, Ansermet). In the meantime the poor first bassoonist is left to fend for himself. Beethoven neglected to put staccato dots for the bassoon's notes—he is obviously partnered with the oboe—with the result that everything has been tried, from ♪ to ♪ (Kubelik, Dorati) to [♩♩] (Bernstein)—as well as the correct one: ♪.

One of the most abused passages in the entire movement is the recapitulatory phrase at m.218 (and onward to the coda at m.229). Most conductors want to make a huge tempo expansion here; and if they have somehow managed the woodwind phrase without too much distortion, they will surely want to make up for that in the next string phrase and pull it completely asunder. (The worst offender in this latter respect is Knappertsbusch, who actually inserts a huge fermata of silence [*sic*] of almost an entire measure's duration before the last sixteenth of m.226.) This approach to both phrases clearly is wrong and self-indulgent. For one thing, apart from the fact that the score clearly states *Tempo I* at m.218, dragging the tempo here pretty much destroys any possibility of hearing the woodwind phrase as Beethoven's final recapitulation of m.11, now ingeniously modified—surely modified enough by Beethoven to not warrant still further digression—by a remarkable pyramidal pile-up of an E♭ chord (bassoon, second clarinet, first clarinet, oboe in succession). Second, since the *Tempo I* of m.218 follows a *più moto* (marked ♪ = 116), it ought to be obvious that an excessively slow tempo at m.218 is unjustifiable, not only in terms of simple tempo contrast, but because it also completely defies the tempo relationship between the two passages that Beethoven had in mind: ♪ = 116 to ♪ = 92, a modest relationship of 5:4. The extent to which a tempo distortion of this passage destroys Beethoven's conception and structure can perhaps be best measured by the fact that, for conductors who have already done the whole movement at a slower tempo than indicated and without Beethoven's *con moto*, going even slower here removes the passage entirely from consideration as part of this quasi-*menuetto* movement. It becomes, against all logic and sense of continuity, an *adagio*, many conductors—Karajan, Furtwängler, Knappertsbusch, Szell, Ashkenazy, Thomas, Maazel, Bernstein, Colin Davis, Giulini—ending up anywhere from between 12 to 20 points metronomically below *their* already slow *Tempo I*, and Mengelberg off by an incredible 26 points. (A bad editing splice on the second beat of m.218 further mars Karajan's 1982 recording.) Others, such as Jochum, Kubelik, Ansermet, Walter, took such a slow tempo to begin with (♪ in the 60s) that they could not—or dared not—go even slower at m.218. (One conductor, Masur, actually went faster at m.218.) The worst offender is Böhm, who makes a *huge* ritard at mm.218–19, then suddenly quite irrationally does an *a tempo* at m.220. The best performances of this section are those of Dorati, Dohnanyi, Mehta, Reiner, Toscanini, Erich Kleiber, Wand, and—virtually perfect—Haitink.

The note-by-note pyramid in mm.218–19 is often poorly performed; it needs to be perfectly balanced in the four separate entrances. The printed score is perhaps a little at fault here, since these two measures are somewhat carelessly or ambiguously marked. In the commonly available scores, a diminuendo wedge in the bassoon part is entirely missing, while the others' dynamics are incorrectly lined up, making it unclear where the diminuendo is in fact to start. Beethoven's manuscript is clear, however, in that the ——— signs (except for the bassoon's, which *is* missing) start on the third eighth of m.218, extending for a total of four beats. This leaves matters still a little unclear in respect, for example, to the oboe's entrance (should it be *mf* or *mp*), also the first clarinet's. One possible version of clarifying Beethoven's intentions might be the following (Ex. 33a). This, however, is more difficult to realize than the version usually preferred (Ex. 33b), because it is hard enough to get all three woodwind players—the two

Ex. 33a

Ex. 33b

clarinets and the oboe—to match up their attacks and dynamic when they are to be all equal; it is even harder to calibrate the three entrances in such a way as to create an over-all diminuendo from *f* to *mp*.

As for mm.224–28, once again Beethoven has provided enough contrast with the three previous occurrences of this phrase—dynamically, harmonically, and orchestrationally—as to not require any additional deviation from the text. Even so, distortions of this passage abound so plentifully as to have become a virtually irreversible tradition. That it was not a tradition necessarily handed down from 19th-century interpretational practices is shown by the fact that, for example, Mengelberg slows the tempo to an incredible 50 to the eighth-note, while Nikisch keeps the tempo beautifully flowing. Conductors who over-sentimentalize this phrase seem not to realize that it is, as it has always been in the three previous incarnations, a variant of the immediately preceding woodwind phrase, only this last time heightened in intensity by the means mentioned above. This suggests—or should suggest—that m.224 is not some brand new episode in the movement, that it should not break the bounds of relationship to its predecessor phrases, and that it is not an excuse for a willful, self-indulgent, emotional exploitation.

It is endemic to all unwanted, unintended tempo changes that they require readjustment, or at least raise the question of how to continue: in the same tempo, in a new one or in the originally intended one? Orchestra musicians know that every time a conductor deviates from the tempo he has to 'fix it' again, and that this is always problematic, creating in fact unnecessary problems which would not have arisen if the tempo had been kept steady in the first place. The enormous tempo stretchings in most performances at mm.224–28 are no exception. For they exacerbate the question of what tempo should be taken for the coda, beginning m.229. The answer, of course, is the 'right tempo,' namely that of Beethoven's original *Andante con moto*, perhaps a tiny bit slower, more relaxed, in a sort of dreamy, reminiscent mood, as the clarinets and bassoons nostalgically harken back to earlier main theme statements. Instead, however, many conductors use this passage for further self-indulgence, trying to make the passage either slower than anyone else has ever done it, or softer, or more legato. The extremes here are represented by Knappertsbusch and Ashkenazy, at a creeping ♪ = 50 and ♪ = 56 respectively (only 42 and 36 points away from the intended tempo!) and, on the opposite tempo scale, Thomas, who actually beats his own *Tempo I* by four points (♪ = 84). The slow tempo at m.229, already requiring an adjustment from the previous six to ten bars, now prompts most conductors to make an accelerando with the crescendo that starts in m.235. Such an accelerando—one of the most durable of bad habits in any case—becomes quite unnecessary, when a better tempo is maintained in the first place, as Haitink's, Reiner's, Leibowitz's, and Dorati's recordings demonstrate.

One of the most problematic passages in the entire movement is mm.240–41. Here the convention in Beethoven's time of notating all instruments at the same dynamic level obscures the primary thematic line which ascends through the orchestral range from the low basses, cellos, and violas to the violins, and in the third beat of m.241 from the violins to the woodwinds (minus oboes) to create the following over-all line (Ex. 34). The only way to achieve the intended

Ex. 34

effect is to ask all the players not involved in this arpeggiated line to lessen their *f* slightly, especially the brass and timpani, at the same time asking the players who do participate in the primary line to bring it out. It is astonishing how few recorded performances came even close to realizing this passage correctly, how few even understood that there was a problem here. The only conductors who get it right are Nikisch and Weingartner; with some others—Jochum, Walter, Klemperer, Maazel, Mehta, Haitink, Dohnanyi, Furtwängler, Suitner—coming close (Furtwängler, and Suitner, for example, only lose the woodwinds' part of the line).

I am not partial to big portentous ritards in the last six bars of the movement, although I realize they are very popular. Firm *ff* chords in the strings, well-sustained E♭'s in woodwinds, a resolute tempo, and the final two bars played with great finality, are all that is required to bring the movement to a stirring close.

Any reader following my tabulation of the various and sundry performance vagaries visited upon the *Andante* movement, will have noticed that they represent in total a much greater latitude in interpretation compared with the first movement. This is undoubtedly accounted for by the structural, formal difference between the two movements: the one tightly organized, precisely constructed out of minute motivic cells, the other conceived in longer melodic spans and subtle variational procedures, which in themselves provide more room, as it were, more opportunities to take liberties. The reader may remember that Beethoven and Wagner, but even Mattheson and Junker, already referred to the phenomenon that a slow movement inherently offers and suggests opportunities for tempo and other interpretational modifications that a tightly argued *allegro* simply does not. This is not to suggest that this condition justifies the taking of more liberties in the slow movements of composers' scores; it simply states the fact that such a tendency exists, and leads me to suggest that conductors ought to be even more on guard against arbitrary indulgences in slow(er) movements than in faster movements. For let it be stated for the record that, predilections, tendencies, and willful traditions aside, there is absolutely no categorical reason why or proof that slow(er) tempos can't be maintained at a steady pace, as, of course, any number of fine performances by the best conductors can attest. Stated this way, it may sound to many a reader as an obvious truism, hardly worth mentioning. Yet the evidence that many conductors consider slow movements fair game for egotistical musical self-gratification is overwhelming, and is in itself a devastating comment on the generally deplorable state of conducting as an interpretive, re-creative art.

We turn now to the third movement of Beethoven's Fifth Symphony and return also to the stricter, tighter type of structural organization of the first movement, as well as its conception in single-measure-per-beat notation and four-bar phrase structuring. And once again, as we shall see, the four-bar structuring is not unequivocally displayed in Beethoven's notation, which fact has unfortunately led to innumerable misinterpretations of the music. But even more grievous are the common ignoring and rejection of Beethoven's tempo and metronome marking, *Allegro* ♩. = 96. It really baffles me why conductors are so reluctant to honor Beethoven's metronomization here, when they are perfectly content to honor it in some of Beethoven's other symphony Scherzos, such as the First ($\mathbf{J}$. = 108), the *Eroica* ($\mathbf{J}$. = 116), the Seventh ($\mathbf{J}$. = 132), all much faster than the third movement in the Fifth. Is it because the music is in C minor? Is it because the opening phrase is *legato*, rather then *staccato?* Is it because the two opening phrases are marked *pp?*

Perhaps it is for all of these reasons. But if so, none of them is justified. It is, after all, a Scherzo—although curiously Beethoven did not call it thus in his manuscript score—and in its relationship to other Beethoven scherzos it is only slightly slower than many of them. Even so, the quarter-notes are still mighty fast ($\mathbf{J}$ = 288).

The worst consequence of taking a slower than indicated tempo is that, as we shall see abundantly below, the Scherzo in most performances turns out to be slower than the last movement. This makes no sense whatsoever, by any possible reasoning. It is not only inaccurate but perversely disrespectful of Beethoven; it is furthermore irrational in that it tends to respect the metronome marking of the fourth movement ($\mathbf{J}$ = 84), but not of the third. What kind of logic is that? Apart from Toscanini, whose tempo is a perfect 96 to the dotted half-note, only a few conductors in our huge sampling come close to Beethoven's desired tempo: Norrington (94), Strauss (92), Hogwood, Muti, Dohnanyi, and Reiner (90), Karajan, Dorati, Masur, DeSabata, the Hanover Band (all at 88), Weingartner, Erich Kleiber, and Leibowitz (86), Brüggen and Suitner(84). Most conductors settle for a comfortable, leisurely $\mathbf{J}$. = 80, which, as I have pointed out, if maintained throughout the Scherzo, will make it slower than the Finale. Even slower, around 70 and 72 (more than 20 points off from Beethoven's mark) are Knappertsbusch, Walter, Solti, Krips, Kletzki, Van Otterloo, and (surprisingly) Haitink. Boulez takes a ponderous $\mathbf{J}$. = 66! Gardiner is once again on the fast, nervous side ($\mathbf{J}$. = ca.104).

But let us assume we have accepted Beethoven's tempo indication; we still need to determine where the four-bar phrase structuring falls. Where is 'one,' for example, in the opening phrase? Well, it is not m.1, although many conductors and musicians have too easily assumed that to be the case. As in the first movement, we have to peruse analytically virtually the entire movement to determine with certainty where the four-bar periodizations lie. Once again it is not all that obvious, the movement being notated in all look-alike single bars. We have to use the same analytic tools as in the first movement: tell-tale major phrase junctures, confirmation through tonality or tonic anchoring, evidence of consistency in the phrase structuring, etc. By such means we discover that the first four notes are in an anacrusis position, and the first 'one' bar is m.2, thus making mm.1, 5, and 9 'fours,' i.e. upbeat measures (see Plate VII). This in turn means that the fermata measures (mm.8, 18, 52) are 'threes,' a logical place for them, being on dominant (V) positions harmonically. These fermatas are preceded in the previous measure by a *poco ritard.* This is where the performance and interpretation problems begin, for most conductors simply have ignored— and continue to ignore—Beethoven's word *poco.* Most conductors make huge

Plate VII   Beethoven, Fifth Symphony, opening of the third movement

ritards here, not only defying Beethoven's admonition of *poco*, but also bringing the piece, the motion and flow of the music, to a virtual standstill when it has barely gotten started. Worse than that, most conductors ignore the score further by starting the ritard as early as the fourth or fifth measure. If one does start the ritard early, then inevitably it will result in a *molto ritard*. If conductors would but consult Beethoven's manuscript, they would see that he originally had the *poco ritard* in m.6, but then on second thought specifically moved it one bar later to m.7. This ought to tell conductors something, unless they simply want, once again, to assume that Beethoven didn't know what he wanted. The worst sinners on this point are Nikisch, Mengelberg, Furtwängler, Krips, Koussevitzky, Böhm, Stokowski, Munch, Ansermet, Muti, and Harnoncourt. Excellent, on the other hand, are Strauss, Maazel, Steinberg, Dorati, Karajan, Jochum, Schuricht, Dohnanyi, Ozawa, and, as might be expected, Toscanini.

In most available scores and parts m.13 has a *sfp* in the cellos and basses. But this is incorrect; Beethoven's manuscript has *sf* ⊳──, implying that m.14 is once again *pp*. We should also note in passing that in this passage (mm.9–18) Beethoven used another one of his phrase stretchings, extending the expected four-bar line, starting at m.10, to six bars. For if m.6 is a 'one,' then m.16 must also be a 'one,' and that means in turn that m.15 is a 'four' type, in this case, however, transformed by the addition of two measures into a 'six.' By that reasoning m.13's *sf* ⊳── falls on a 'four,' and functions therefore as a kind of off-beat syncopation. It is very effective that way, a little expressive surprise in the phrase structuring early on in the movement. But it is also possible, I suppose, to think of the six-bar phrase divided in two three-bar entities, in which case m.13 would be a 'one.' Whichever choice one makes, I think it is important that *a choice be made*, that the cellos and basses feel and perform the phrase with some unified conception, rather than just playing the *sf* in some arbitrary uninformed way.

Measures 1–18 constitute the introduction to the actual body of the movement, which begins with the horns in the anacrusis m.19, that measure being a 'four.' This needs, alas, to be stated and emphasized, for too often this music has been played as if m.19 were the head of the phrase, i.e. a 'one,' thus being thereafter one measure off in the phrasing. Such a misinterpretation also ignores the fact that the three horn notes in m.19 are a close variant of the opening motif of the symphony (Ex. 35), also, as we know, in an anacrusis position. It is

Ex. 35

therefore important for the horns to know that m.25 is a 'two,' a weak 'beat' in the four-bar phrase, not the accented 'one' so often heard. The same caution applies to the upper strings and woodwinds in mm.33 and 37. Care should also be taken that the bass line starting in m.27 is well sustained by the cellos and basses, in contrast to the preceding seven measures.

Since a four-bar-ness (Schenker uses *Viertaktigkeit*) functions crucially throughout most of the movement—there are again, as we shall see, a few interesting exceptions—it is imperative that conductor and players understand this,[51] but more than understanding it intellectually, *feel* it and make it felt—and perceivable—to the listener. The difference between the right and the wrong phrasing can be heard in many recordings; for example: (correct) Toscanini, Furtwängler, Haitink, Jochum, Reiner, Dorati, Mehta, Dohnanyi, Ozawa, Muti, Gardiner; (incorrect) Weingartner, Klemperer, Suitner, Knappertsbusch, Walter, Ormandy, Giulini, Solti, Thomas, Stokowski,[52] Ashkenazy, Boulez, Harnoncourt, Brüggen.

Measures 40–45 is another stretched phrase (by means of repetition). Another kind of stretching, although this time without disturbing the four-bar periodization—a melodic stretching—occurs in mm.54–60. This was necessitated by the need for the music to return to the tonic key of C minor, having landed in B♭ minor in m.45 and m.53. The intervallic melodic twists and turns in mm.54–60, already of course partially anticipated in mm.11–14, represent Beethoven's ingenious way of moving in a minimal amount of time from B♭ minor to a basic G(major) pedal in m.60. Beethoven accomplishes this by traversing a brief cycle of fifths: B♭ -F-C-G (see encircled notes in Ex. 36). It is this very widely (and

Ex. 36

same as mm. 10-13

wildly) skipping line—it must have sounded very 'modern'and strange to listeners in 1808—which prompted Beethoven to cast it in long slurs. His intention, I am sure, was to subdue the twisting contours of the line and its *pp* dynamic by an eerily smooth *legato*. Most performances, however, break the strings'

51. It must be pointed out that again, as in the first movement, unwary orchestra musicians are not likely to be able to deduce from merely looking at their individual parts where the four-bar periodization falls. Unless they sat down with the score one day and figured out the correct periodization or have been told by a good conductor how the phrases go, they are unlikely ever to have thought about the subject. Many horn players, for example, see a *ff* in m.19 and simply assume from that scant information that it is the 'downbeat' beginning of the phrase.

52. Stokowski's recordings of Beethoven's Fifth are complete musical travesties. Apart from many typically Stokowskian phrasing and dynamic excesses—many of them the result of the most absurd technical electronic alterations (equalizing, dialing entire sections or individual players into inaudible oblivion, others into exaggerated prominence)—Stokowski indulges in some bizarre re-orchestrations. He doubles the clarinets in mm.38–41 (and the oboes in mm.90–93) in the third movement with muted *(sic)* trumpets; in the last movement he doubles the piccolo runs in m.329–32, 346–49 with a flute (not another piccolo), and has the horns play most of the last thirty measures an octave higher. But then, Kletzki, not to be out done by the Beethoven re-arrangers, has his first trumpet in the Czech Philharmonic play a high C in m.415.

phrasing and bowing, thus undermining—not underscoring—the intended effect. It is also important for the strings, especially the entering violas, to know that m.57 is a 'four' in the phrase structure.

Another six-bar phrase extension occurs in mm.66–71, making the violins', horns', clarinets' thematic entrance in m.71 an upbeat measure (similarly that of the trumpets and woodwinds in m.79). The six-bar extension in mm.92–97 parallels the one in mm.40–45, in turn making m.98 a 'one.' Measures 97–100 is the sixth time that we have heard this phrase,[53] each time, however, heading off in a different direction. This time it leads to a major thematic excursion—a kind of miniature development section—which will play a most important role and undergo a most remarkable transformation in the recapitulation (mm.141–235) of the Scherzo, after the Trio.

Because conductor and musicians are usually unaware of the four-bar phrasing, the entire section mm.101–40 is more often than not played incorrectly. For if viewed only from the individual parts—flute, oboe, first violins, cellos, for example—the phrase mm.101–104 would seem at first glance to start on a 'one' (Ex. 37a,b). The legato phrasing in the flute and cellos underscores that impression, as do the *f*'s in the cellos in mm.105 and 109. And this is indeed

Ex. 37a

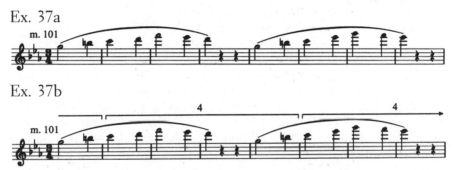

Ex. 37b

how it is played in the majority of cases,[54] including, alas, by many renowned conductors like Kubelik, Solti, Dohnanyi, Ozawa, Walter, Klemperer, and the English 'authenticists.' Those interested in hearing how this section sounds when played with the right phrasing might listen to the recordings of Karajan, Jochum, Haitink, Dorati, Toscanini, and Reiner. Karajan and the Berlin Philharmonic manage this entire passage especially well. Listen to how the cellos not only play mm.105 and 109 with an elegantly expressive *f*, but how they feel it as a 'four,' an effect akin to a syncopation, an accent on a weak beat (very much like the 'fourth beat' accent in m.13). Again, Beethoven's phrasing/bowing for the cellos (and bassoons), starting in m.115, would lead many to assume that m.115, is a 'one' (Ex. 38a). But it is not; it is a 'two' (Ex. 38b). The important

53. Measures 1–4, 9–12, 58–61, 97–100, as well as mm.45–48, 53–56, the latter two in transposition.
54. Flutists especially love the incorrect phrasing because it allows them to crescendo into the high F and G in m.103 and m.107 respectively, a crescendo they would be obliged to avoid in the correct phrasing, where the high note is on a 'two' and should not be emphasized.

Ex. 38a

Ex. 38b

Ex. 39

'ones' coincide with the tonic C's in m.118 and m.122. Similarly, m.131 surely looks like a 'one' in all the parts—three sturdy tonic-dominant chords—and yet it too is a 'two,' the C minor chord in m.132 then being a 'three' and the next main-theme repetition starting (correctly) on an upbeat measure. This periodization requires that the three chords in m.131 not be played as loud as possible—which most orchestras love to do, thinking it is the arrival point of the phrase and the peak of the previous nine-bar crescendo—but lead instead to the weightier 'three' in m.132, followed than by a real *ff* in m.133.

When orchestras play m.133 erroneously as a 'one,' they end—if indeed they are feeling four-bar phrase entities at all—on a weak 'four' in m.140. That is, of course, impossible: neither Beethoven nor any other composer of the period would ever have ended a movement or a major section of a movement on the weakest beat of a phrase. And one can hear and feel the embarrassed hesitation and obvious discomfort of players in performances in which m.133 and m.137 are perceived as 'ones.' Measure 140 therefore is a 'three,' completing the first part of the Scherzo (Ex. 39), just as the first movement of the symphony also ended on a 'three.'

By rights, one might argue, Beethoven should now have added an empty measure before attacking the Trio of the Scherzo. But evidently he felt the need to plunge right on into what must have been a most shocking surprise to musicians and listeners in 1808: the rambunctious, galvanic outburst in the cellos and basses, turned even more surprisingly immediately into a 'proper' little fugato.

In the meantime—before we leave the Scherzo proper—we should not fail to appreciate the delicious piquant dissonances with which Beethoven spices up much of this 'development section': the woodwinds' A♭'s, for example, in mm.111–14, rubbing not only against the G pedal points in horns and bassoons, but against the G-E♭'s in the strings, engaged at the same time in an interplay with the timpani whose G's intermittently also clash with the A♭'s of the wood-winds and violas (see Ex. 40). The irregularity of periodization we have glimpsed

Ex. 40

at the end of the Scherzo proper, in effect a seven-bar structuring (mm.134–40)—or viewed another way a three-bar structuring in mm.138–40—is contin-ued in the Trio, the first section of which can be seen as consisting of two six-bar phrases (mm.141–52) (Ex.41), followed by a five-bar and a three-bar

Ex. 41

unit (mm.153–57 and mm.158–60, respectively). It is as if Beethoven wanted to shake up the previous existing structure, pummel it into different unpre-dictable shapes—of course, only on the surface. For underneath, these ir-regular phrases still combine into a binary-based over-all period of twenty bars.

The just-mentioned six-bar phrasings of the Trio can also be thought of as being each divisible into three bars (twice), or—more remotely—into two bars three times. I lean strongly towards the former conception for a number of rea-sons, primarily harmonically oriented. It becomes clear rather quickly from a harmonic analysis that, just as the Scherzo is basically in C minor, so the Trio is primarily in the dominant, G, leading eventually with the return of the Scherzo back to the tonic C minor. But within this larger harmonic scheme one finds smaller harmonic groupings that clearly help to define the phrase structuring. The schematic (Fig. 6) of the first section of the Trio displays its harmonic functions in detail, revealing some very fascinating patterns. We see that the first three measures, set in G (with its ancillary subdominant C), are exactly mirrored, although transposed up a fourth, in the second six-bar unit, both segments then appearing in the next unit in a contracted form (mm.141, 142, 147, 148 combined selectively, as can be clearly seen in the trumpet and bassoon parts, in mm.153–56). The extra measure 157 (in D) was needed to bring the progression back to the temporary tonic G; thus the five-plus-three structuring. We can see also that the three-bar subdivision of the first

Fig. 6

measure	141 142 143 144 145 146	147 148 149 150 151 152	153 154 155 156 157	158 159 160
key a)	G C G C modulates to I IV I IV	D G D G circles to stay in V I V I around	G C D G D I IV V I V	G G G I I I
tonality b)	V I V I	II V II V	V I II V II	V V V
periodization	⌞ 3 ⌟ ⌞ 3 ⌟ ⌞___ 6 ___⌟	⌞ 3 ⌟ ⌞ 3 ⌟ ⌞___ 6 ___⌟	⌞ 2 ⌟ ⌞ 3 ⌟ ⌞__ 5 __⌟	⌞ 3 ⌟ ⌞ 3 ⌟

The higher harmonic positions [a)] are to be understood as in the key of the Trio: G (dominant). The lower harmonic positions [b)] are to be understood as in the key of C, the Scherzo's primary key.

six-bar phrase links up naturally with the Scherzo's final three measures, whose $\frac{\text{C-G-C}}{\text{I-V-I}}$ cadence is immediately reverberated in the Trio's $\frac{\text{G-C-G}}{\text{V-I-V}}$ (I-IV-I).

A similar, closely related harmonic pattern can be seen in the second part of the Trio, now returned to four-bar phrasings. The eight measures 161–68 are in G, although in a sense just barely so, since the first four measures are hinged to the seventh and ninth degrees of G—a third inversion, as it were. Now the following four-bar patterns evolve (Fig. 7), which turn out to be harmonically expanded and transpositionally modified variants of mm.153–56, melodically/thematically also

Fig. 7

measure	169 170 171 172	173 174 175 176
key tonality a)	G C G C I IV I IV	C F C F IV VII IV VII
b)	V I V I	I IV I IV
periodization	⌞___ 4 ___⌟	⌞___ 4 ___⌟

referring back to the Trio's first part. As in that section, so here now in the second section, the eight measures 169–76 are immediately mirrored in contracted form in mm.177–80 (G-C-C-F), after which a prolonged G pedal point (of twelve measures) sends the music back to C major. A repeat of the second section, modified primarily dynamically by a prolonged diminuendo but still principally located in G major leads to a return of the Scherzo proper (C minor).

As can be seen from even this relatively cursory survey of the Trio's larger harmonic rhythms, it is remarkably strict in its functional organization. Like boxes within boxes, harmonic progressions and their thematic/melodic counterpoints constantly rotate around the various tonic axes, with the largest 'box'—in G major—containing all the smaller ones.

I have presented this much harmonic analysis—there is much more that could be said on this subject—to show how the Trio's harmonic functions do in fact determine its over-all phrase structuring and in particular the 6-6-5-3 partitioning in the first section.[55]

The dramatic intensity of the Trio's by now famous cello-and-bass passage has led many a conductor to exaggerate it dynamically, urging the players, either by gesture or verbal bidding, to play as loudly and roughly as possible. The entire section is marked only *f*, in contrast to the *ff* in the Trio's second ending (m.159), a dynamic that is rarely observed (or observable, since the previous passage is so often thundered forth in a pulverizing *ff*). When played too loud, the risk is that in mm.141–42 the players' bows will indiscriminately hit adjacent strings, impairing the note clarity of the passage, which is hard enough to produce cleanly on the lowest strings of the bass at a fast tempo. It also makes the open strings in the cellos (circled in Ex. 41), especially the bright A in m.144, spurt forth in an out-of-context, edgy, explosive way.

Two more problems must still be dealt with in this section (applicable also to mm.194–97). One is the need to keep the bass line fully prominent throughout, particularly in mm.158–60. This is important in order to fully re-establish and confirm the key of G, not only chordally/harmonically but melodically/thematically. The problem in this regard is twofold: as just mentioned, if the rest of the orchestra, including timpani and brass, is playing *ff*, the cellos and basses will surely be outbalanced and inaudible. Second, it is an established (though generally little regarded) fact that the vast majority of listeners associate melody and theme almost exclusively with the upper and middle registers of our hearing range. To put it another way, generally people do not expect to hear anything thematic, melodic—or important—in the bass register. Their ears are more likely to follow an upper register line, even if it is of lesser consequence. Thus in this case, the average ear will follow the violin line, beginning in m.154, and be completely distracted from following the all-important bass line, unless conductor and orchestra explicitly treat it as the main voice. It is therefore especially crucial here that the timpani not play too loud, or let its notes ring too long.

The second problem is one of bowing. In mm.158–60 the upper strings must use the following bowing ♩ ♪ ♪♪♪ ♩ ♪ ♪♪♪, for the bowing often used, namely, ♩ ♪ ♪♪♪ ♩ ♪ ♪♪♪ results in the three eighth-notes sounding like a triplet or, in effect, turning these 3/4 measures into 6/8's. As for the tempo of the Trio, everything conceivable has been tried over the years. A few conductors (including Solti, Thomas, and Carlos Kleiber) like to go faster at m.141, ostensibly producing 'greater excitement'; most like to go slower,

---

55. Some interesting anomalies and ambiguities remain unexplained. For example, is m.142 really in C? I believe so, because the measure's last note is an F. Had Beethoven changed it to F♯, the whole phrase would have a totally different feeling, much more in G major. But then, how do we explain the C♯ in m.148; and why did Beethoven use it here when he did not use it eight bars later in exactly the same situation in the first violins? And does that C♯ make m.148 fall more in a D tonality?

thinking of a weightier, heavier effect—a stereotype often, alas, associated with double basses. Some, of course, believe in staying at the same tempo, as in the main part of the Scherzo. The first (faster) approach can lead to a lack of pitch clarity (already at a premium on the lower bass strings), if not actually a bit of a note scramble. The second approach (slower) can lead to an overly ponderous, pachydermatous effect that is quite foreign to Beethoven's music, which, even at its weightiest, has a lean tensile strength that is closer to Haydn, Handel, and Bach than to Bruckner and Wagner.

In the end, any of the three tempo approaches work, this being one of Beethoven's more indestructible and memorable passages, that is, as long as the tempo modification is within reason, within feel-able range of the original tempo.[56]

On a par with the originality of the irregular phrasing in the first part of the Trio, all couched in a mini-fugato, is the radicality of the motivic fragmentation in the beginning (mm.162–65) of the second part (Ex. 42). This unusual passage

Ex. 42

has given players problems for generations, it being difficult to play these short bursts of phrases *without* the expected downbeat notes at mm.163, 167. The rhythmic instability implied here can in fact be destabilizing, even disorienting. In performances and recordings one often can hear the G's in mm.163, 165 come in early, rushed. More than that, however, it is very hard to prevent the passage from sounding like this: ♪ as innumerable recordings attest. (I have often jokingly called this the first true 5/8 in classical music.) An accent on the final F of m.162 and m.164 can be avoided by making a slight, subtle diminuendo on the last two or three eighths.[57]

56. The most perverse interpretation of the Trio on record has got to be Stokowski's, not only because he took a sluggish tempo of ♩. = 68, but because he made the poor cellists and bassists of the London Philharmonic Orchestra play the entire passage *on the string!* So did Scherchen.

57. Several conductors have added to the 'irrationality' of this passage by imposing even more irrational interpretations on it. For example, Reiner, whose recording of the Fifth is in almost all respects exemplary, nonetheless had the bizarre idea to insert a fermata over the rests of mm.163, 165, thereby delaying the incoming G's by a whole bar and making the rests equal to about five quarter-beats, and thus drastically—and unforgivably—recomposing Beethoven's music.

The other notable performance aberration in this passage is one perpetrated by Kletzki, who conducted it in such a way as to result in the rendition in Ex. 43. I can imagine the musicians' consternation encountering this 'interpretation' at their first rehearsal with Kletzki, and their

The second section of the Trio (mm.162–97), developing along the lines of an expanded variant of the first section, entails two more six-bar phrasings (mm.182–87, 188–93) over a G pedal point. But these 'sixes,' unlike those earlier (which divided into two 'threes'), seem to partition into 'twos' and, in the second group into reiterated 'ones'—all of these, needless to say, still easily containable in the movement's generally quaternary periodizations. More important, however, is to not neglect the little chromatic alterations Beethoven inserts in the seconds, violas, and cellos, in mm.189–91, as well as the interesting bassoon and trumpet parts in mm.192–93.

In the third part of the Trio, Beethoven returns to the beginning of the second section, but this time leads the music in an entirely different direction via an eighteen-bar diminuendo. The whole passage, which eventually moves from the strings to a flute-dominated woodwind septet,[58] functions as a transitional link to the recapitulation of the Scherzo (m.236).

At this point we, incidentally, find further proof (if such be still needed) of the particular four-bar structuring I have suggested as the basis for most of the movement. The key is m.236, which is not contained in the original theme statement. Beethoven had to add a measure if he was interested in preserving the four-bar structuring at the joining of the Trio to the Scherzo recapitulation. Having arrived at m.234, had he now simply repeated the opening phrase literally, he would have ended up with the following phrase (Ex. 44), including a three-bar unit (bracketed). The insertion of the one bar of dotted half-note C

Ex. 44

restored the four-bar symmetry. This then provides further proof that (a) Beethoven was constructing the movement in quaternary phrase units, and that (b) m.237, like m.1, is a 'four,' an 'upbeat' measure, in the structure.[59]

Ex. 43

subsequent head-shaking in disbelief as they struggle to play the passage well *incorrectly*—not an easy thing to do.

58. I suggest that two-bar slurs be added to the woodwinds in the four bars mm.214–17, to help maintain the *pp* dynamic and to relate more easily to the violins' bowings/phrasings.

59. I am, of course, aware of the numerous discussions and arguments that have raged over the question whether or not Beethoven intended a repetition of the entire Scherzo *and* Trio at m.236, a debate first initiated by Schenker nearly seventy years ago.

Actually, the discussion goes back to the third and fourth decades of the 19th century, when writers such as Fétis in Paris and various musicians and conductors, in Germany including Mendelssohn, began to question two extra measures that had mistakenly been left in score and parts, mm.238–39. (These are no longer in present-day scores.) As Schenker first pointed out, further

At m.231 many conductors start a ritard, in the mistaken notion that this is necessary to prepare the return of the Scherzo. Beethoven was quite capable of writing *ritard* or *poco ritard* in m.231 or m.232 if he had wanted to (he did so only twelve bars later). What he really intended (or at least *possibly* intended) was that the Scherzo should suddenly reappear, out of thin air, as it were—unheralded, in the nature of a surprise.

What follows now for the next 138 bars is surely one of the most astonishingly visionary and, for its time, innovative musical creations. The Scherzo is recapitulated but transformed into a spectral, skeletal shadow of its former self. And when this recapitulation has run its course, the Scherzo does not come to a close, as in Beethoven's first four symphonies, but, as in the Sixth Symphony (composed virtually simultaneously with the Fifth), a coda-transition leads directly to the fourth movement, which is an even more striking flight of fancy, an almost motionless, music, as if even the ghost of the Scherzo had now died, only to be revived 42 bars later with the glorious full-of-life C major brilliance of the Finale.

The entire recapitulation and coda-transition contains numerous performance pitfalls. Again, of primary importance is an awareness at all times of the periodization. Being a recapitulation of the Scherzo, though varied, it leans on the exposition with similar, at times identical, phrase structuring. Fig. 8 shows the relationships between exposition and recapitulation. The first hint that the Scherzo's recapitulation will not be merely an identical repeat, as in Beethoven's earlier symphonies, comes in m.238, when the initial *legato* is changed into a *spiccato* and a few winds replace the earlier mixed sonorities. The transformation to the skeletal apparition of the former Scherzo is fully accomplished in m.245 when, all flesh

---

confirmed by some Beethoven biographers—Nottebohm, Thayer, for example—Beethoven sometime early on dropped the idea of a complete recapitulation of the Scherzo *and* Trio, which, however, through a series of mishaps and misconnections—with publishers and copyists—never became unequivocally clarified, not in the autograph or in the printed materials derived therefrom.

The controversy has continued into our own time, with no one side having succeeded in establishing an unequivocal victory pro or con. A considerable literature has developed over this question, most notably by Heinrich Schenker (1925), Willy Hess, "Die Teilwiederholung in der klassischen Sinfonie und Kammermusik" (in *Die Musikforschung*, Vol. XVI, 1963); Walter Riezler, *Beethoven*, (Zürich, 1944); Robert Simpson, "The First Version of Beethoven's C minor Symphony" (in *The Score*, No.26 (1960); C Canisius, *Quellenstudien und satztechnische Untersuchungen zum dritten Satz aus Beethovens c-Moll Sinfonie*, Diss. Heidelberg, 1966; Peter Gülke, *Zur Neuausgabe der Sinfonie Nr. 5 von Ludwig van Beethoven* (Leipzig, 1978); Igor Markevitch, *Die Sinfonien von Ludwig van Beethoven* (Leipzig, 1983).

Having read all these materials, especially those arguing for the complete repetition, I am still not completely convinced that this represents Beethoven's absolute or final intentions; there are too many imponderables and factual lacunae to be fully persuasive, although Gülke's summation of the matter—and conclusion in favor of the repetition—is impressive.

Under the circumstances, barring new reliable documentation, and claiming no irrefutable knowledge on the subject, I believe it is unnecessary to re-argue the case here. I refer the reader to the sources above and suggest that conductors make up their own minds based on a thorough analysis of the relevant disputations.

Fig. 8

Recapitulation		Exposition
m.236–44	=	m.1–8
m.245–54	=	m.9–18
m.255–80	=	m.19–44
m.281–323	=	m.97–139

stripped from its bare note bones, the music is reduced to the spooky sounds of a staccato bassoon and pizzicato cellos. Thinly piping woodwinds alternating with pizzicato violins render what is left of the tune, while nervously twitching grace notes in violins and violas add an unearthly, eerie touch. The grace notes can be played before or on the beat,[60] as long as they are played very, very fast. They must sound disembodied, though of course clearly audible.[61]

Note that the dissonantal interplay of mm.111–14 returns in the parallel place mm.295–98, this time a single horn providing the tiny discordancies. It is also worth noting that when played in the four-bar phrasing I have suggested, the pizzicato grace notes in the first violins (mm.300, 304, 308, etc.), which are virtually impossible to play without a tiny accent, fall—appropriately—on relatively strong beats ('threes') of the phrase units. (They fall on a weak 'two' the way the piece is often *incorrectly* played.) Very few conductors get this ghostly Scherzo recapitulation right, either in its phrase structuring or in the special mood and spell it casts. Conductors who have done especially well here are Reiner, Karajan, Furtwängler, Haitink, Wand, and—absolutely uncanny in the dance-like swing and shadowy unreality he imparts to the music—Carlos Kleiber. That must be heard to be believed.

A brief word about Carlos Kleiber is perhaps in order, even if this book is not about individual conductors, but rather about specific works in specific performances. Kleiber is so unique, so remarkable, so outstanding that one can only describe him as a phenomenon. This does not mean that he is a 'perfect' conductor—perhaps no one can be that—but he has so many extraordinary attri-

60. Conductors, musicologists, music historians have argued for generations about the placement of Beethoven's grace notes (not only here but, for example, also in the slow movement of the *Eroica* and in a dozen other pieces). Nobody can be absolutely certain of Beethoven's wishes in this respect, or even whether he was entirely consistent in his use of grace notes and whether their usage might have varied in different musical contexts. All that is certain is that in Beethoven's time, notational conventions required that grace notes associated with a first beat in a measure be placed at the beginning of that measure, not before it. This is, of course, in discrepancy with grace-note placement in relation to other beats, where they usually are placed, vertically seen, *before* the beat. This notational discrepancy leaves the question unanswered and ambiguous. In any case, in this Scherzo passage I prefer to play the grace notes on the beat, for the pragmatic reason that, based on my own experiences with this passage, if the grace notes are played before the beat, they tend (a) to be played too slowly, and (b) as the passage proceeds, the grace notes, especially in the violas where the grace notes are quite awkward technically, tend to come progressively earlier and slower.
61. What does not seem to make much sense is what Ashkenazy does with the Philharmonia Orchestra, which is to have the violins play the grace notes *before* the beat, the violas *on* the beat.

butes that make him a great and important representative of the art of re-creation and performance.

To begin with, his conducting technique—that term is almost a misnomer because his manual and gestural talents go way beyond 'technique'—is not only astonishingly expressive, but amazingly free and flexible. In its variety of gestural choices—mostly concise, clear and to the point (although sometimes a bit outrageous)—his conducting is virtually improvisatory. That is to say, he almost never conducts anything, any passage, any musical idea, gesturally the same; his work is never pre-programmed. Also, as a 'switch-hitter,' he can at will do things with either or both hands—and with an infinite variety of shadings. In these respects he is like a great jazz improviser who has endlessly diverse ways of expressing himself, even time and time again on the same subject. With all of these gifts Kleiber is not only an inspiring leader/conductor, but a hell of a good show.

As a musician/philosopher Kleiber is in the grand lineage of 20th-century German conductors—a sort of combination of Furtwängler/Erich Kleiber/Fritz Busch—who above all goes for the grand line, the large shape, and the clarification of inherent structure(s). Not that he, like some conductors of that tradition, ignores details, but he is not obsessed with them and therefore is able to see the forest in spite of the trees. Indeed, I find that he is not (or elects not to be) conscious of every detail in a score, even details which others would rightly consider very important. Like many maestri, he is selective in what he chooses to point out (by his gestures) to the orchestra and the audience. And like many others, he has some questionable musical habits, such as almost always crescendoing too much too early—he loves to drive orchestras to a climax—conducting too often only the primary melodic or thematic lines, frequently neglecting to make the orchestra really play the softer dynamics, occasionally indulging in unnecessary over-conducting.

But whatever he does, he does with such consummate control, gesturally and intellectually, and with such a joy of music-making, ranging from complete confidence-building relaxation in front of an orchestra to passionate, almost ecstatic outbursts, that one can only be compelled to admire in awe—even if one does not always agree with every aspect of his performances. Kleiber is a virtuoso in the best sense, a virtuoso with a mind.

Tempo in the Scherzo recapitulation is critical, perhaps more so than in any other part of the Scherzo. I say this because this skeletal music, consisting more of silence than actual notes played, requires, virtually by definition, a certain minimal motion and pulse, just to hang together, as it were. A slow tempo of ρ. = 64, such as Böhm takes here, or even a leisurely one such as Walter's ρ. = 72, simply cannot work; nor does an overly hasty rushed-through tempo, such as Strauss's ρ. = 104.

The final six-bar phrase (mm.318–23), which would normally have ended the movement or led to a coda, ushers in the aforementioned transitional passage (m.324), leading directly to the fourth movement. Its uncanny stillness is achieved by the simplest of means: motionless sustained strings, in this case an oddly spaced and voiced A♭ tonality, without the fifth (E♭) 𝄢 ⎯⎯⎯ marked

ppp, the only occurrence of this dynamic in the entire symphony,[62] and a softly pulsating timpani pattern which recalls in summary the several rhythmic variants of the main theme used earlier in the recapitulation of the Scherzo:

♩♩♩♩|♩♩ ♩ | ♩♩ ♩♩♩ ♩ | ♩♩♩♩|♩♩♩♩|. The timpani's note is C, emphasizing the third of the key of A♭, adding to the suspenseful, unreal atmosphere. At the same time that C, reiterated over fifty measures, secretly anticipates and leads to the brilliant C major—the "éclat triomphale" (as it has been so aptly called by Gülke)—of the Finale. Two performance problems must be mentioned here. Care should be taken that the last notes in the first violins in mm.341, 343, 345, 347, 349 not be dropped, either dynamically or rhythmically. One often hears the passage played as follows: 𝄞 or 𝄞. This has the negative effect of destroying the long melodic sequential line that Beethoven has created here, in which the silent beats are just as important an element of the music as the audible notes. Making the final notes in each three-note group too short tends to chop Beethoven's 28-measure-long line into too many two-bar segments, rather than one long arching line.

The other problem is really more of a question, and has to do with Beethoven's unusual phrasing (and, presumably, bowing) starting in m.352. There is a certain metrical/structural ambiguity in this passage, made all the more tenuous by Beethoven's uncommon bowing/phrasing pattern (four-, six-, five-bar). It appears that Beethoven was operating simultaneously on three levels of perception: (1) the ongoing four-bar infrastructure—between m.324 and the entry of the final movement there are eleven four-bar units plus one (extended) six-bar unit; (2) the sequential pitch contouring which appears to fall into five three-bar groups; and (3) the aforementioned bowing pattern of four-six-five. Thus a multi-layered structure evolves—a 'polyphony' of perceptual levels, as it were—as shown in Ex. 45. There are two more sub-patterns embedded in the pitch

Ex. 45

62. I am convinced that Schubert was influenced to use the special *ppp* in the last movement of his *Octet*, mentioned earlier (see p. 26, fn. 15), by Beethoven's example. In both pieces the dynamic is used exclusively at the one point that represents the dynamic nadir of the movement, building from that point to a critical juncture—in Schubert's case the recapitulation, in Beethoven's case the brilliant fourth movement.

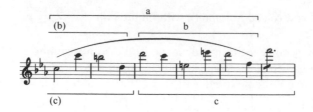

Bracket a, ⌐————a————⌐   delineates the bowing pattern;

bracket b, ⌐————b————⌐   the sequential pitch contouring;

bracket c, ⌐————————⌐   the underlying four-bar metrical structuring.
                 c

contouring (bracket b), outlining two ascending scalar patterns, which fall into a five-plus-four module, as shown in Ex. 46. Given this complex three-layered

Ex. 46

construction, I can only believe that Beethoven's bowing is intentional; and it is, of course, entirely practical at the *ppp* dynamic level. So perhaps the passage is not so much ambiguous as it is unusual and original, and ought to be performed with an awareness of its multi-layered patterns.

With the end of the transition, we have come to one of the more remarkable moments in the entire symphony: the burst of radiant C major that constitutes the sudden arrival of the Finale movement with an overwhelming release of energy that has been pent up for nearly two minutes (and nearly 175 measures), ever since the middle of the Trio.

Although nearly every conductor understands and feels the momentousness of this juncture, where the third and fourth movements meet and where the Scherzo erupts into the great Finale, very few seem to comprehend the intrinsic relationship between these two movements on even the simplest and most basic terms: namely, that the Finale is, as conceived by Beethoven, in its tempo, its pulse and beat, slower than the Scherzo. It is beyond my ability to understand why the vast majority of conductors insist on doing the reverse: conducting the Finale in a tempo faster than the Scherzo. Actually, the problem is not so much that the last movement is played too fast, but that the Scherzo is played too slow. Beethoven's metronome markings make the intended relationship of the two movements very clear: Scherzo, 96 (to the dotted half-note); Finale, 84 (to the half-note).[63] But even if we choose to disregard the metronome indications, it should be self-evident that a Scherzo, particularly a Beethoven Scherzo, is inherently faster than a Sonata-Allegro Finale movement. And yet this tempo relationship is reversed and perverted by virtually all conductors—with a very few rare exceptions—and worse, it seems that conductors never even give the matter any thought at all. It is one thing to perform a piece of music or a movement at a slower or faster tempo than intended by the composer, but it is quite another matter to reverse the *roles* and functions of two movements.

This confusion of tempos is made all the worse when conductors additionally make a huge ritard (not indicated by Beethoven) in the final measures of the Scherzo going into the Finale, some conductors even ritarding beyond the tempo they take for the last movement. In my sampling of recordings the only conductors whose Scherzo tempo was faster than that of the Finale and close to Beethoven's tempo were Hogwood, Hickox, Brüggen, Norrington, Toscanini, Karajan, Dorati, Steinberg, and Strauss, with Toscanini absolutely on target in both tempos (96 and 84), and with some of these it was because their Scherzo tempo was so slow that the Finale almost *had* to be faster. All the rest set the last movement's tempo either the same as the Scherzo's or faster, with Muti, for example dropping a staggering 24 metronome points.

But the worst habit of all—again, a tradition which many conductors seemingly inherit unquestioningly or are afraid to oppose—is holding the tempo back considerably in the first two to four measures of the Finale and then letting (or making) it propel forward from m.5 or m.6 on. The most preposterous version of that idea I have ever encountered is Ozawa's, who starts the last movement at $\d = 62$, twenty-two points below Beethoven's tempo designation, but in m.3 jumps 30 points to $\d = 92$, now 8 points *above* Beethoven's intended tempo. The silliness of this 'interpretation' can perhaps best be underlined by two pet

---

63. We can see this relationship at another rhythmic unit level even more dramatically displayed: Scherzo—$\d = 288$; Finale—$\d = 168$. This is one point in regard to which the usually infallible Schenker errs. In comparing the two tempos (Scherzo and Finale), he suggests (p.69) thinking of the last four bars of the Scherzo as "slightly faster" ("etwas beschleunigt") than the quarter-notes of the Finale. Perhaps Schenker's error is merely semantic or inadvertently ambiguous, but on the face of it he is saying that the dotted half-notes of the Scherzo ($\d. = 96$) are slightly faster than the quarter-notes ($\d = 168$) of the Finale. This is, of course, mistaken, because 96 is, of course, slower than 168 and, in any case, can hardly be described as "slightly" faster. I think Schenker may have wanted to suggest feeling the *quarter-notes* of the Scherzo as faster than those of the Finale. But even in that case the word "etwas" (slightly) is misapplied, for a drop from $\d = 288$ down to $\d = 168$ can hardly be called 'slight.'

I am also concerned by Schenker's implication—he refrains from saying so explicitly—that the Finale should be felt (and conducted?) in 4/4. Some conductors (myself included) have tried this and one can say unequivocally that it doesn't work—at all. One must be grateful to Beethoven for marking the movement $\d = 84$ (not $\d = 168$), with the clear implication and suggestion, despite his C (not $\mathbf{\phi}$), that the movement be felt and conducted in two.

phrases orchestra musicians privately, derisively, apply to such an (mis)interpretation (Ex.47).[64] Scherchen indulges in the same tempo distortions.

Ex. 47

Many conductors who start the movement at a slower (presumably 'stately,' 'majestic') pace and then accelerate, arriving at full tempo usually around m.16, who also take the repeat, find that by the end of the exposition they have gathered so much momentum, that at the repeat they are forced either to suddenly jam on the tempo brakes to reach the same slower tempo taken the first time, or to stay at full speed, which, of course, makes a mockery of the repeat. The point is that all such tempo twistings and deformations as described here are not only totally unnecessary but significantly detract from and undermine the real intended effect, which is powerful, thrilling, and majestic enough just as Beethoven wrote it. What can be more exciting and breathtaking than the outburst of triumphant C major with its simple powerful, elemental theme, following minutes of mysteriously spectral, stifled tension!

In view of the extraordinary popularity of Beethoven's Fifth Symphony and its overwhelming impact upon audiences over more than a dozen decades, it is interesting to learn that many early listeners found much of the last movement wanting in various respects. Some found it "commonplace" and "overly noisy," others thought it "blatant" or "vulgar," expecting, one suspects, a polite Mozartian Rondo. Ludwig Spohr complained about the "disreputable" sounds of the trombones and piccolo, little appreciating the fact that these instruments and the contrabassoon were used here for the first time in a symphony. Berlioz found the Finale "repetitious." Even as late as 1890, the critic-musicologist Hermann Kretzschmar could not resist pointing out that he considered "the themes [of the last movement] simple to the point of triviality."[65]

The opening theme of the Finale may indeed be 'simple,' but it is also, because of its simplicity, remarkably strong and compelling. Furthermore, underneath the apparent simplicity there lie some fascinatingly original and complex

---

64. Incidentally, the German musicologist and conductor Peter Gülke has pointed out the interesting relationship between the opening main theme of the last movement and the woodwind theme of the *Andante con moto*.

65. Hermann Kretzschmar, *Führer durch den Concertsaal* (Leipzig, 1890).

structural interplays. Our understanding of this music must begin with the knowledge that most of the movement is constructed in two-bar phrases and measure groups, which are extended frequently to three-bar units (very much like the fours in the first and third movements extended sometimes to sixes). These two-bar units are fused into larger phrase and structural entities, as for example mm.1–6, mm.7–12, mm.13–18 (all three times two measures). What is unusual in these phrase constructions is that the longest and therefore weightiest rhythmic values, dotted half-notes, fall on the second, fourth, and sixth measures of the first six-bar phrase, thus falling on the 'weaker' measure of each two-bar unit. This is somewhat unusual but also complex in the subtlest of ways, in that the longer rhythmic value produces a kind of syncopation, not only because the dotted-half-note is longer than the previous two notes, but because the G in m.2 is also the highest note in the thematic line, having risen from the tonic C. This 'syncopation' and subtle shift of weight reverberates throughout the first part of the exposition, in different rhythmic configurations. For example, the inherent basic syncopation in mm.14,16 (♩ ♩ or ♫ ♩), itself a reference to mm.1–2 (|♩ ♩ |♩· 𝄾 |), recurs again in mm.18, 19, and in diminution in mm.20, 21,[66] as well as in the accompanimental figures in the bass line (♪ ♩) and some of the winds ([𝄾·] ♪ ♩).[67] This veritable barrage of syncopations and offbeat accentuations culminates in the powerful descending unison passage in mm.22–25  . This figure in turn will, somewhat smoothed out rhythmically and dynamically, play an important role in the second part of the exposition, returning as  . It is because of all these intricate

---

66. One interesting confirmation of this inherent feeling of syncopation is the fact that virtually all orchestras automatically play—and this means feel—mm.20,21 as syncopated, playing with an accent or *sf* on the second and fourth beats, a feeling, of course, emphasized by timpani, cellos, basses, and contrabassoon. This cross-accentuation is usually so strong in most orchestras that I have often as a conductor been moved to tell the players who play on the first and third beats to accent *them*, to restore some degree of beat equilibrium to these measures and to avoid a result such as:

67. Who, by the way, among conductors has ever bothered to distinguish between the two accompanimental rhythms between cellos/basses and timpani in mm.20, 21 (♪ ♩ and 𝄾 ♪ ♩)? I myself never had the time to sort these rhythms out in rehearsal until a few years ago with several orchestras, including the Spokane Symphony and the Cologne Radio Orchestra. When the instrumentalists involved became aware of and actually heard the rhythmic differences in those two measures they were amazed not only at Beethoven's ingenious inventiveness, but that they had never in the several hundred times they had all played the piece noticed this rhythmic differentiation, and that no conductor had ever pointed out this detail before.

rhythmic and thematic/motivic relationships that a rock-like steadiness of tempo is very much needed, any tempo deviation undermining Beethoven's structural framework and rhythmic detailing.

To add to the complexity of this initially "simple-" looking music, the phrasing takes an interesting turn in m.18, which functions in two roles. It is at once the sixth measure in the third six-bar phrase (mm.13–18) and the first measure of two two-bar units (mm.18–19, 20–21). This elision in the periodicity explains why there is an uneven number of bars in the first phase (mm.1–25) of the exposition.

It used to be popular among conductors to broaden the tempo measurably at m.26, and worse, to prepare for this broadening with a substantial ritard in mm.24–25, as one can hear in Mengelberg's recording, for example. It seems to me quite unmusical to impede the exhilarating rush of sounds Beethoven has created in mm.22–25, just to achieve some rather obvious grandiose effect at m.26. Fortunately, judging by recordings of more recent vintage, most conductors no longer indulge in this particular distortion, sensing that it is more effective to keep the established momentum going across the structural seams at m.26. Many conductors do, however, still lean towards a heavy ponderous sound at m.26.[68] This is particularly inappropriate once one realizes that m.26 is not some major arrival point but a surging continuation of the movement's first theme statement, leading on towards a much more important juncture, namely, the arrival of the dominant in m.45.

A caution needs to be raised for mm.26–27, where too often the horns, enthusiastically seizing upon the first good little tune they have been offered by Beethoven, blast in with an enormous ff that completely overbalances the woodwinds that also have this phrase.[69]

A serious balance problem arises in mm.28–29 and 32–33, in which the majestic rising arpeggiated figures in the bass instruments are usually drowned out by timpani and brass if care is not taken to avoid this. Interestingly, the bass figure, lasting *three* measures (mm.28–30), continues beyond the timpani roll, overlapping with the continuing theme in horns and woodwinds. The effect is very odd indeed when, in poor balancing of this passage, the bass instruments are inaudible in mm.28–29 and then suddenly emerge for the final five beats of their figure, which, of course, makes no musical sense at all but has been tolerated time and time again by both conductors and recording producers. It does not seem to have occurred to many conductors that that bass figure is the 'response' to the 'call' of the winds in m.26, to which it must therefore be con-

---

68. One of the worst distortions of this phrase occurs on Abbado's Vienna Philharmonic recording, where the maestro in m.27 stretches the tempo *enormously*—in effect making a fermata in the middle of the measure—to the obvious consternation of the Vienna musicians, who come completely apart rhythmically. (It is even more shocking that a major record producer would allow such a mishap to remain uncorrected and to appear on a professional recording by a world-famous conductor and orchestra.)

69. Dangerous enough with two horns, one can imagine how this passage sounds when with Karajan or Klemperer conducting, the horns are doubled.

nected musically and equalized dynamically. Of the innumerable recordings sampled, in only a small number was the proper balance achieved and the entire bass line heard, most notably those of Toscanini, Jochum, Reiner, Dorati, Haitink, Solti, Gardiner and Nikisch.

Measures 26–33 comprise two four-bar groups (or, if you will, four two-bar units, depending upon which instruments one is looking at), followed in m.34 by a seven-bar phrase, which is in effect another one of Beethoven's stretchings of material, in this case the immediately preceding material of mm.26–33. Both main ideas in those measures are extended, the violins' melodic line being a variant of the winds' theme at m.26, including the unusual phrase/bowing

Ex. 48

articulation, and the bass instruments continuing in eighth-note motion, m.34 being analogous to m.30 (see Plate VIII). These two elements, supported by an added contrapuntal line in the violas and first bassoon (in inversion to the violins), are spun out into a modulatory bridge leading to the secondary dominant D major in m.41, which in turn sets up the mandatory move to G (m.45).

At first glance one may be puzzled by the existence of an irregular seven-bar phrase (mm.34–40), especially if one is aware of the recapitulation of this material, mm.240–49, where it reappears in regular even numbered multiples of two-bar phrase units. We will return shortly to that recapitulation, one of Beethoven's most extraordinary inspirations. For now it will suffice to explain that the seven-bar phrase in question was necessitated by the modulatory move to D major (m.41), a progression that could not easily have been achieved in six measures given the starting point of C major (in m.34). This can be readily seen, for instance, by eliminating m.40, which would cause a very crude and inept move to m.41 (from m.39). The 'extra' measure in the phrase is, in fact, m.38, a G 6/4 chord, which had to be added in order to get to the D major arrival point of m.41 as soon as feasible, as the progression in Fig. 9 shows. Measure 38 is a sequential repetition of m.37, the latter's notes being of necessity transposed

Fig. 9

mm.34–35	mm.36–37	m.38	m.39	m.40	m.41
C	D^7	G$^{6/4}$	D	G$^{6/4}$	D

up one step to extend the rising melodic progression from e^2 to f#^{2}, begun in mm.34–37. The passage has thus a kind of self-fulfilling inevitability about it, including the seven-bar phrasing. What makes it even more interesting and is in a way the reason for the irregular phrase length, is the fact

Plate VIII   Beethoven, Fifth Symphony, fourth movement, mm.31–39

that m.35 is a one-bar rhythmic contraction of the two-bar winds' theme

at m.26. Instead of ♫ Beethoven permutates it to

♫ , a kind of doubling up of the phrase. Note also

how m.39 is a rhythmicized variant of m.37 (Ex. 49). Be it noted that this one-bar mutation does not destroy the basic two-bar phrasing (mm.34–35, 36–37), for it is ingeniously encapsuled in it. In other words, the two bars contain both the original thematic form and its contraction (see Ex. 50). In the end it is this one-bar thematic variant which in turn allows Beethoven to move with it to G major in m.38 and then contrast the phrase even further rhythmically in mm.39–40.

Ex. 49   Ex. 50

Two errors in all scores and parts, one quite serious, the other less so, must be mentioned here. One concerns the second violins, which from the fourth beat of m.34 on should be played one octave higher, as is made unequivocally clear in Beethoven's autograph. (The same is true of mm.240–43.) The other error occurs in m.32 in the bass, which should start this measure with the low contra C (see m.238 for comparison).

I have gone to considerable pains to detail the structural analysis of this passage for two reasons: first, the hope of alerting conductors to the remarkable polyphonic, harmonic, and structural originality of this passage, and to the fact that care must be taken to make all these elements function correctly together; and second, because, as already alluded to, this passage undergoes an even more ingenious, almost miraculous transformation when encountered again in the reprise (see below, pp. 214–15).

The next episode (mm.45–63), on the dominant, seemed not to raise any terribly serious performance problems in the majority of recordings sampled. Minor shortcomings do occur, of course, such as the failure of most performances to observe the special *pp* in m.54 (as compared with the previous *p*'s in mm.46,50). Also, since most conductors generally concentrate only on the highest-lying and most obviously prominent melodic lines, the four-note cello

phrases (the first one in mm.45–47: ♫) are often neglected. That is unfortunate, since this motive takes on considerable prominence in the development section (see mm.106–08 (contrabassoon, cellos, basses) or mm.112–14 (trombones, bassoons) etc.).

Most conductors and orchestras pick up considerable speed at mm.45—although some, such as Carlos Kleiber, Abbado, and Kubelik actually slow up there *(sic)*—and then, at m.58, increase the momentum even more for the strings' sixteenth-runs. This, of course, makes for a superficially very exciting effect for audiences, but is an unnecessary 'improvement' of Beethoven's music. What bothers me about such rushing of the tempo is that in most cases it is inadvertent, more a matter of lack of control of the tempo than some well-considered musical intention. A steady, unrushed tempo is actually even more powerful and dramatic in effect. I should also mention that in many performances and recordings the violins in mm.58–59 are partially or totally covered by the over-played *ff* of the rest of the orchestra, especially the brass and timpani. (In the Ansermet, Mengelberg, Norrington, Hanover Band, Hogwood, and Colin Davis recordings the strings are either virtually or, in some cases, totally inaudible, especially in mm.58,59.)

Before we leave this section (mm.45–57), we should take note of the recurrence of the quasi-syncopation we found in the opening of the movement, that is, the longer note value (mm.46,50) falling on a weak measure in the four-bar phrasing. This idea returns many more times in the Finale, not only in this particular form (| ♩ ♩ | ♩. ⁏ |) but in the development section in a rhythmic variant, as in mm.106–108, 132–34, etc. (Ex. 51). It is well for conductors and their orchestras to appreciate this unusual phrasing and to make

Ex. 51

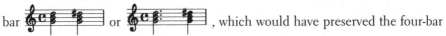

it subtly felt whenever, in the course of the movement, it occurs.

Let us also note in passing the five-bar phrase mm.53–57, another one of Beethoven's phrase stretchings from four to five measures, caused by the interpolation of the G augmented chord (m.57). That is to say, had Beethoven not felt the need for that transitional chord, going instead from a pure G major triad to the C major of m.58, the extra bar would not have been needed. There is also the possibility that Beethoven could have dealt with the present two bars in one

bar  or , which would have preserved the four-bar structuring. But the 'dissonance' of the augmented chord to Beethoven's ears required the phrase expansion and extra spacing.

However unproblematic and reasonably well-played the section just discussed seems to be, the next episode (mm.64–71) is quite the opposite. It has almost never been played correctly on any recording that I know of, nor have I ever heard it 'interpreted' properly in concert. It is difficult to understand why Beethoven's dynamics (various *fp*'s and *p*'s) are summarily ignored or rejected. Almost everyone, including the celebrated Beethoven'specialists' Furtwängler, Karajan, Klemperer, Walter, Masur, and even Toscanini, simply play the whole eight-bar passage at a mediocre *mf*, neither *f* in the *fp*, nor *p* in between. The only conductors who respect Beethoven's dynamics are Reiner, Dorati, Carlos Kleiber, and Jochum, although the last allows too much crescendo in mm.65, 67, and 69–71 towards the *f* of m.72, which then annuls the *p*'s . There are two

viable interpretations of Beethoven's *fp*'s here. One is to take the marking literally, that is a true *f* followed instantly by a true *p*, that *p* continuing through the
next seven beats (*without* crescendo). Proof that Beethoven is serious about this
p can be seen in the cello part in m.65, as well as the first violins. I would add,
that this *fp* should be treated as an expressive (singing) *fp*, not an aggressive,
hard-hitting one.

The other sometimes offered interpretation is to consider the entire eight bars
at a basic *p* dynamic level, with the *fp*'s treated as accents in *p*, just as one might
interpret *sf*'s in a *p* context. But the problem is that these are *not* *sf*'s; they are
fp's, a distinction Beethoven makes frequently in his music and one which we
as performers ought to honor. Additionally, I should point out that all generally
available editions contain serious phrasing and bowing errors in this passage.
According to Beethoven's autograph score, the following is the correct setting
(Ex. 52)[70]

Ex. 52

A partial reiteration at m.72 of the previous eight measures leads to the repeat
of the exposition (m.84). The descant violin figures of mm.65–71 are now
echoed in the piccolo—one piccolo against the entire orchestra. This certainly
presents a balance problem, but not an insurmountable one. It is important that
the orchestra keep to only a *f*, not just to allow the piccolo to be heard, but
because Beethoven follows this *f* six bars later with a *più f* and three bars after
that a full *ff*. Unfortunately in most of the recordings sampled the orchestra is
too loud at m.73 and the piccolo is barely audible or, in some cases, totally
inaudible—understandable perhaps in a performance, but there is no excuse in
a recording.

70. The *sf* in the first violins in m.68 that one finds in most editions should be eliminated; it is an
error and not to be found in Beethoven's autograph. This misprint was pointed out as early as 1925
by Schenker and mentioned with some frequency in subsequent writings on Beethoven's Fifth. It is
thus all the more amazing still to find this errant *sf* in many performances, including one in early
1994 by Kurt Masur and the New York Philharmonic.

This is perhaps the appropriate occasion to lament the fact that the instruments Beethoven adds to his basic 'classical' orchestra in the Finale—piccolo, contrabassoon, and three trombones—are woefully neglected in most performances and even most recordings. This is unforgivable, especially on recordings. Here Beethoven has the innovative daring to employ these instruments, as it turns out for the first time in any symphony, and for all one can hear of them on most recordings, they might just as well have never been used by Beethoven. This is true, surprisingly, even of the trombones, an inexcusable lapse I ascribe as much to the recording producers as the conductors (although the latter presumably have the final artistic approval of the recorded performance and should object to the muzzling of these instruments). But worse yet, there is not a single recording on which the contrabassoon can be heard at all, at any time. This presumes the ironic, almost humorous situation that in all these recordings a contrabassoonist was hired, was paid for his services, and not one note of his playing is heard. He or she might just as well have stayed home. It is almost the same with the piccolo, which on many, many recordings is often either inaudible or barely present. All this is particularly deplorable since these instruments were meant to enrich Beethoven's instrumental palette, adding three new and unusual colors to the collective orchestral sonority. This is especially true of the trombones, a brilliant powerful addition to Beethoven's four brass (two trumpets, two horns). Since trombones are anything but weak non-projecting instruments, it is amazing to me that on so many recordings the trombone color is either virtually or totally hidden. It is a color which conductors should relish to exploit, but for some inexplicable reasons mostly don't.

We should also note—and feel and make audible, i.e. comprehensible to a listener—that m.71 (and m.64) is the first instance where the longer note value (♩.) occurs on the 'strong' measure of a two-bar or four-bar phrase. These notes should be well sustained in a melodic sense, that is, 'dropped' neither rhythmically (to |♩ ♪ ↱ ♩ |) nor dynamically (⊃⊂). Both defects can be heard on most recordings.

The *più f* on the second beat of m.77 has given many conductors interpretive problems. There are in fact two theories current as to the meaning of this *più f* (and many similar instances in Beethoven's oeuvre). Some musicologists and historians claim to know that Beethoven used *più f* to indicate a crescendo. I have already referred to one such instance in the Fifth Symphony's first movement, mm.175–79 (although there the additional perforated lines extending out from the words *più f* tend to make this interpretation more reasonable). The other theory claims, of course, that *più f* can only mean a sudden dynamic jolt, since Beethoven regularly used the designation "cresc." or "cresc. . . . . . ." to indicate a gradual dynamic increase, and that the two notations cannot possibly mean the same thing and are not interchangeable. But I still tend to agree with the first formulation: a graduated dynamic increment. The problem some conductors and orchestras have at m.77 is, however, not of such a subtle theoretical/ interpretive nature. Their problem is more mundane in that they allow the phrase at m.71 to be excessively loud, after which they naturally find it impossible to increase the volume at m.77 (either as a sudden *più f or* as a further crescendo). Various circumventions are then resorted to, such as changing Beethoven's *più f* to a *subito p* and then crescendoing to the *ff* at m.80—a really tawdry, banal effect (this was a favorite of Ormandy's)—or, like Böhm and Reiner, resorting to making an accelerando (*sic*). None of these 'solutions' are necessary or justifiable; Beethoven's score tells us precisely what to do and is eminently realizable with some care, some rehearsing.

The measures directly prior to the first and second endings, and the second ending itself, (m.85) *do* take some special care in terms of ensemble balance to realize Beethoven's intentions. The problem is to bring out all the moving eighth-notes in mm.80 through 89 against the sustained sounds of the brass and the sixteenth-note barrages of the timpani. These arpeggiated eighth-note figures must be well heard and projected, for they provide the rhythmic momentum, the surging energy, that propels the music in the first instance back to the recapitulation and the beginning of the movement, in the second instance towards the development section.

Beethoven here solved a difficult problem most effectively in his typical direct, unhesitating fashion, which solution, however, if not performed with understanding, that is, with a deliberate judicious pacing, can sound too abrupt and a little awkward. I am referring to the fact that Beethoven, having maneuvered his music to the key of F minor (m.80), and having also arrived at a point where the next imminent formal sections would be a recapitulation of the exposition and, the second time, the development section, had to prepare simultaneously for both eventualities, in the one case returning to C major, in the other going on to E major (m.86) and thence to A major (m.90). I am certain—and the sketch books and autograph confirm this—that Beethoven struggled considerably to arrive at this remarkable twin solution, and having succeeded, I think it is mandatory for us performers to respect his decision of a first ending and an exposition repeat.[71]

Many performances contain a self-inflicted problem around m.82. Conductors who have slowed up the tempo either at m.26 or m.45 and now, eyeing the return to the repeat or the plunge into the development section, suddenly realize that they must increase their pace to arrive at a tempo identical or at least similar to the one they had taken at the beginning of the Finale. This readjustment often creates ensemble problems and rhythmic raggedness in the orchestra, as can be heard on Bernstein's, Jochum's, and Steinberg's recordings. Con-

71. I recall a seminar on Beethoven symphonics, held by Igor Markevitch at my invitation at Tanglewood in 1982, at which in answer to a question on whether to honor repeats in classical (but especially Beethoven and Brahms) symphonies, Markevitch suggested that if the composer has written an extensive first ending with new material not otherwise represented in the work and/or has evidently spent considerable effort on redirecting the music back to a repeat, then that repeat should be honored. It is an interesting and entertainable proposition, quite apart from questions of formal balance and proportions. It is all the more disappointing to read in Markevitch's Beethoven symphony studies *Die Sinfonien von Ludwig van Beethoven* (Leipzig, 1983), that he strongly advocates eliminating the repeat in the Fifth Symphony's Finale (p. 307).

versely, many conductors who make the repeat cause serious rhythmic/ensemble problems when, having pushed the tempo forward before the first ending, suddenly put on the brakes for the return to the opening of the movement.

The sudden movement to E major in the second ending (m.86) is one of the most exciting moments in the Finale, coming as it does as somewhat of a harmonic surprise. Here it is important to bring out, as mentioned earlier, the eighth-note figurations, not only in the bass instruments but in the continuation of the line in the second violins (m.88) and the woodwinds (m.89). What must be clearly audible to the listener is the following total line (Ex. 53). Admittedly,

Ex. 53

the completion of this line, especially in m.88, is somewhat under-orchestrated. However, some subtle rebalancing and dynamic reduction to *poco f* or *mf* in all the sustaining upper-range instruments and the brass, can readily achieve the appropriate realization of these four measures. (One can even borrow a few first violins and/or violas to add to the second violins in m.88.) Unfortunately more than half of the recorded performances indicates that the conductors didn't even know there was a problem here. Those who did very well here are both Kleibers, Kletzki, Ormandy, Toscanini, Mehta, Dorati, Krips, and above all Haitink and Maazel, while the famous Beethoven conductors Furtwängler, Karajan, and Szell deserve only a 'fair' rating. Solti in his Vienna Philharmonic recording projects the bass instruments well, but loses the second violins and woodwinds.

The development, initially featuring only a transposition of the exposition's second episode (m.45), soon reworks and extends much of the earlier material, at times by inversion (see oboe and flute in mm.96–99) or by modulation (eventually to the key of Bb, m.106), and seems to represent relatively few performance problems. But one might watch that no crescendo occurs in the ascending woodwind line in m.102, for it will undermine the effect of the rather sudden *f* in m.106. I also don't understand why Gardiner has to change the strings' figures in mm.100–101 and mm.104–105 to separate bows, when Beethoven's slurred triplets really work very well—indeed better.

In the meantime the four-note cello motive I mentioned earlier (in connection with mm.46–48) is now elaborated and sequentially extended (mm.91–93) and distributed variously among the strings, eventually to take on primary importance in m.106. For here this bass motive is now elevated to the function of theme and accorded further elaboration, expanding it eventually from its customary two-bar shape to its current four-note configuration, but stretched to three-bar units (mm.107–109, 110–12).

Many cautions are in order here, the first one regarding dynamic levels. As so often in Beethoven's mature works, he applies his favorite climax-generating device: *f* via *più f* to *ff* (mm.106–122–132). The danger is that the various brass and timpani entrances, if louder than a cultivated *f,* will prematurely drive the whole dynamic level up, making a *più f* and a climactic *ff* very difficult or impossible. The worst offender is often the timpanist who, having rested for 28 measures and spying the *f* at m.104, is eager to participate in the build up and crashes in with a thunderous *ff* roll. (This disturbance can be heard on countless recordings.) But the biggest balance problem occurs in mm.118–21, in which the trumpets enter thematically (aided and abetted by horns and timpani) and are apt to drown out not only the other thematic lines (violins, mm.118–20 and violas/cellos, mm.120–21) but also the triplet figures in the woodwinds. These last are very important because it is now their turn to be transformed into primary material, triumphantly leading, in alternations with the brass, to the climactic majestic dominant pedal point of m.132. The staggered canonic layering and structuring that need to be clearly preserved in performance can be appreciated graphically in Ex. 54.

Ex. 54

I have never understood why the violins' phrasing/bowing in mm.106–107 and 109–10 (as well as the cellos' in mm.112–13 and mm.115–16) as shown in most available editions, including the first edition and Beethoven's autograph, is universally ignored and rejected. If the reason is that it is 'impractical' to play five beats in *f* on one bow—which is not really true—then let us at least preserve Beethoven's sense of *legato,* so important to contrast with the staccato alternate measures (108–109, 111–12 etc.),

and not play . A compromise of with *sustained* quarter-notes would

serve Beethoven's intentions much better. The oboe triplets in mm.113 and 116 are almost never heard, but that is through sheer neglect. Since they continue the previous strings' triplets gestures and are also the link to the triplet lines in mm.118–21 (and beyond that to mm.122–31), they *are* important and can with a little care in balancing be made quite discernible. A serious error exists in the contrabassoon, cello, and bass parts in mm.118–19: the D♭, analogous to the woodwinds in m.106 and the second violins in m.112, should be tied across the bar line to the whole note.

Two problems stand out in the next section, mm.122–31. One, simply a bad habit rather than a real problem, is the failure to sustain the dotted half-notes in the strings. In almost all orchestras this passage, either solicited by or tolerated by their conductors, is rendered as if written [♩ music notation] or [♩ music notation] or, worse yet [♩ music notation]. The English 'authenticists' and Harnoncourt are particularly negligent in this respect. Indeed, they make a stylistic (bowing) fetish of these diminuendos. The other problem is to create a viable balance in the *più f* between woodwinds and brass (and timpani) in their triplet alternations.

Balance problems continue to plague most of the next section (mm.132–52), but all can be dealt with successfully in intelligent rehearsing and subtle dynamic readjustments without excessive doublings or orchestrational retouching. The woodwinds' theme, beginning m.132, at first in octaves, then in thirds—remarkably positioned in four octaves (Ex.55)—can easily be brought into prominence (without resorting to doubling) by judicious modification of the strings.

Ex. 55

Let us remember that rhythmic energy and intensity are just as effective—perhaps even more so—as sheer volume and mass of sound. A healthy, cultivated *ff* in the brass (at m.136), added by Beethoven to the woodwinds as a timbre augmentation (not as superseding the woodwinds), will help to keep all elements of the passage in balance and under control.

For whatever it's worth, I should like to add my own voice in objection to doubling the woodwinds at m.132 with the horns, unfortunately a long-standing tradition, first proposed (but later retracted) by Weingartner and since then favored by many conductors (Karajan, Böhm, Koussevitzky, Szell, DeSabata, Abbado, even Erich Kleiber). Such doubling and 'fixing up' of Beethoven is both misguided and unnecessary, as Weingartner himself realized later in life when he conceded that "after all, Beethoven had known best."[72] In this instance add-

---

72. Weingartner, *Über das Dirigieren.*

ing the horns in mm.132–36 destroys and distorts Beethoven's original and much more interesting conception. As Schenker once put it: "It is part of Beethoven's art [and genius] to be concerned *even* in a full *ff* with [instrumental] shadings, while simultaneously working towards a dynamic intensification by giving the theme initially only to the woodwinds and adding the horns [and trumpets] only in m.136. It is therefore *ff*-inimical (*ff-widrig*) when a conductor brings the horns in already in m.132."[73] I would add that introducing horns at m.132 thickens the texture in an anti-Beethovenian way, and will certainly cover over and obscure the unusual, fascinating low voicing in thirds in the bassoons.

If the orchestra players are aware of the harmonic content of Beethoven's terrifying dissonance at m.141 (Ex. 56), especially the A♭ and F players whose notes clash with the bass G, the effect will be much more overwhelming than that produced by sheer loudness. It should be remembered that in music only that which is truly *heard*, i.e. that which the musicians actually hear and feel—and only that—will project in performance. One more caution: m.142 is *not* louder than the previous ten measures, as Beethoven's cautionary *sempre ff* ought to make very clear. (Nor should one follow Ormandy, Bernstein, Furtwängler, and Walter, who added *subito p*'s either at m.142 or m.146–47, followed by vulgar super crescendos.)

Ex. 56

Many conductors, preparing for the recapitulation of the Scherzo, mistakenly start jockeying around for an 'appropriate' tempo to accomplish that transition, thereby upsetting the orchestra's accumulated momentum and causing ensemble and rhythmic irregularities (hear the recordings of Munch, Mehta, and Colin Davis). But even worse is the ritard some conductors make—another deplorable tradition—in the final three bars (mm.150–52) before the Scherzo's return (notably Szell, Ashkenazy, Ansermet).

Beethoven's Finale development section is one of the master's most extraordinary creations, but as we have seen it is also fraught with many problems if not performed as Beethoven intended and so notated. Three conductors have excelled in recordings of this passage, not only in terms of *textual fidelity* but in the drive and excitement they bring to the music, mostly *by virtue of* textual fidelity. They are Reiner, Toscanini, and Furtwängler. Reiner drives the music relentlessly forward, considerably exceeding Beethoven's basic tempo—by m.132 he has reached ♩ = 106 (*sic!*)—and yet maintaining, as almost only he could in that generation of conductors, a mesmerizing control both over the music and his musicians. It is high musical drama at its best. Toscanini and Furtwängler achieve almost the same kind of musical excitement but, interestingly, by quite different means: Toscanini with his typical hard-driving, somewhat muscular, tensile approach, Furtwängler by his (for him also typical) unique gift to create long lines, cohering great architectural spans of time, combined with a remarkable depth and warmth of sound. For those who like to keep track of various conductors' metronomic inclinations, let me just add that Furtwängler,

73. Schenker, *Beethoven: Fifth Symphony*, p.69.

too, gathers terrific tempo momentum here; by m.145 he is at $\sphericalangle$ = 98 (again considerably *above* Beethoven's suggested tempo).

From the earliest days of the Fifth Symphony virtually everyone, even some of the work's detractors, has considered the recall of the Scherzo in the middle of the Finale one of Beethoven's most original and irresistible strokes of genius. And so it is, but much less so when the tempo relationships between the Finale and Scherzo are disturbed or, worse, even reversed. Great confusion reigns here at the juncture of the two movements in most recordings, with few conductors returning to a faster beat ($\sphericalangle$. = 96), and indeed most remaining at an equivalent tempo or, even more strangely, slowing to a leisurely un-scherzo-ish pace. As in the Scherzo proper, many performances have the beat turned around, erroneously taking mm.160, 164, 168, etc. as 'ones' in the four-bar structuring. Where the 'ones' are is, once again, vitally important to know, because the long oboe line (eventually joined by flute and bassoon) is in its phrasing and pitch moves out of phase by one bar, that is, in subtle syncopation with the basic four-bar periodization. This is not something one could readily deduce from viewing the oboe part isolated from the score. Seeing a phrase such

as , almost anybody would assume a

four-bar or six-bar phrase-unit with the respective first measures falling on a structural downbeat. Such, however, is not the case, m.173 being a 'one.' Yet most oboists play the line—and most conductors *allow* them to play it—that way. Someone who has not heard the passage played the right way can hardly imagine what a difference it makes. The fact is that the entire Scherzo recapitulation, starting at m.153, consists of twelve four-bar groups plus, at the end, one six-bar extension.

To further clarify the 'correct' interpretation, especially by the oboe, we have to understand the function of this Scherzo interpolation, the reason for Beethoven's insertion of it. Most ordinary composers would have gone directly from the sustained chords of mm.150–52 (perhaps adding one more similar measure to preserve quaternary structuring) to a recapitulation. But Beethoven was not an 'ordinary' composer. His inspiration was to bring back the Scherzo—at least part of it—to create a parallel to the situation that exists prior to the arrival of the fourth movement. He felt the need to re-create that same mood, making the thus delayed reprise of the Finale all the more powerful and overwhelming— again a moment of tremendous release of pent-up, suppressed energies. The eerie stillness, the uncanny staticity of the original transition passage must be replicated here. Only this time, instead of a quietly pulsating timpani and a shadowy, twisting violin line, it is the thin line of a single oboe that represents the only slightly quasi-melodic movement in this music. Therefore the oboist must play with a minimum of expression, holding back, suppressing all overly emotional utterance, creating a sense of motionlessness (yet with a beautiful tone, of course)—not the often heard emotional, rhapsodic, indulgent display that many oboists and conductors offer here. It is in that sense that I would reject the impulse to ornament the oboe's line with extraneous dynamic nuances, such as those recommended by Markevitch and many other conductors in, for example, mm.177–79 . The fact is that, in the midst of this harmonically and melodically virtually motionless realm, the very move to F♯, only a slight half step from the surrounding sustained G's, already contains all the 'expression' that is needed.

I never cease to be surprised and amazed at the poor intonation displayed by most oboists in this solo passage, especially in the first four bars (mm.172–75). Doubters of this statement should listen to the recordings of Karajan, Muti, Mehta, Abbado, Colin Davis, and Kubelik, all presiding, needless to say over first-rate, world-famous orchestras. Similarly, it is rather shocking on how many recordings the clarinets in mm.164–67 are out of tune, invariably quite sharp.

As in the final measures of the first transition passage, so here too no ritardando is necessary or wanted, especially since the final six-bars (mm.201–206) are already in effect a written-out *ritenuto*, a 'stretching' of the phrase.

All truly great music is marked by two extraordinary qualities: one is its memorability—it etches itself into our memory. The other is its sense of inevitability, that amazing and amazed feeling we have as listeners that, regardless of how original, how unpredictable and surprising a certain musical idea (or a movement, or a whole work) may be or may appear to be at first hearing, in retrospect we perceive it as the best and only possible idea appropriate for that work; it was, in short, inevitable. Such thoughts come to mind in contemplating the astonishingly unpredictable, even radical idea of recapitulating the Scherzo in the middle of a Sonata-Allegro fourth movement. For, as extraordinary as this idea may be—one cannot think of any composer of the time who could have had such an inspiration—we *hear* it as if it had been the most logical thing to do, as *inevitable*. As Schenker once put it: "What logic in these occurrences, which appear to drive where they themselves are driven!"[74]

Everything that has been said about the first 33 measures of the Finale will apply to the identical segment of the recapitulation. At the 34th measure (m.240) the music takes a remarkable and decisive turn, a turning point impelled by the need for the second theme of the recapitulated exposition to be set in the tonic key of C. Thus Beethoven seized the opportunity to modulate to the subdominant F and thence to the dominant G. If we retrace his compositional steps, we can see that Beethoven, realizing that he would need more maneuvering room to accomplish this modulatory process (see Fig. 10), was inspired at the same time to invert structurally part of the earlier passage— another extraordinary stroke of genius.

74. Schenker, *Beethoven: Fifth Symphony* (Vienna, 1925), p. 65

Fig. 10

Comparison of exposition (mm.34–44) and
recapitulation (mm.240–53)

Exposition (11 measures)	C	D⁷	G⁶ᐟ⁴	D
	mm.34–35	mm.36–37	m.38	m.39

Exposition (11 measures)	G	D⁷
	m.40	mm.41–44

Recapitulation (14 measures)	C	F	G	C
	mm.240–41	mm.242–43	mm.244–45	mm.246–47

Recapitulation (14 measures)	G	C	G⁷
	m.248	m.249	mm.250–53

What is remarkable and was (as far as I know) at the time unprecedented in the symphonic literature, was Beethoven's idea of turning the earlier music of mm.34–40 registrally upside down in its entirety in mm.240–49 (see Exx.57a and b). What

Ex. 57a

Ex. 57b

was originally in the bass is now in the highest register piccolo); what was on top as the primary melodic line is now in the bass; and the viola counterline in the tenor range is now in the woodwinds, distributed over three octaves.

But there was one problem with Beethoven's idea: he didn't have quite enough instruments (even in his expanded orchestra) to realize all of it fully and easily. One piccolo in its medium register can hardly replicate a whole section of cellos and basses, coupled with contrabassoon. Because this six-bar phrase (mm.244–49) is such a remarkable invention, it is one of the very few places where I would in a live performance suggest a minor instrumental adjustment to provide Beethoven's intentions with a viable acoustical realization. Apart from assuring that the orchestra not exceed the implied f dynamic, the piccolo part should be played by two piccolos (marked $f\!f$), the second piccolo to be played by the second flutist. (Doubling the piccolo with a flute is also possible, but it is not as good an idea because the flute in its highest register has a quite different and more penetrating sound which, compared with the thinner piccolo timbre, would be intrusive and out of place.) Additionally, one might then borrow one of the oboists or clarinetists to fill in for the vacated second flute, more likely a clarinet since the clarinet's notes are doubled exactly in the second violins and can thus be spared. In order for the piccolo doubling to be feasible, the second flute must quickly switch to piccolo in m.236, play the second flute part in mm.250–52 on piccolo, and switch back to flute in the ensuing measures of rest.

I would not think that this bit of cosmetic instrumental realigning would be necessary in a recording, for there are easy electronic means to give the piccolo its needed presence. Despite this, on only eight out of the nearly ninety recordings sampled can the piccolo be clearly heard (Muti, Jochum, Dorati, Suitner, Thomas, Haitink, Krips, and Carlos Kleiber). On a dozen others it is barely discernible, if one listens *very hard*. On the rest (some 50-plus recordings) it is literally, totally, completely inaudible! The most bizarre case is that of Stokowski whose dialing madness, as mentioned before, causes him to virtually 're-invent' Beethoven's entire symphony. In the passage in question, instead of helping the piccolo electronically, he makes a hash of this entire episode by first dialing the violins way up in mm.240–43, then way down in m.244, simultaneously raising the cellos and basses tremendously, in the meanwhile suppressing entirely not only the piccolo part but the important woodwinds counter-line. At m.250 suddenly all instruments burst forth $f\!f$, sounding almost like a bad editing splice (perhaps it was).

Again, everything that has been said about mm.11–71 should be applied to mm.250–80, with only the additional comment that the first clarinet and first bassoon in mm.263–64 in the recapitulation have replaced the two horns of mm.55–56 including the written pp (not p or $m\!f$!), a fact not often realized by the respective musicians and most conductors. Special care must also be taken that the horns observe religiously Beethoven's p *dolce* in mm.273–80.

We come now to a passage in which I believe Beethoven did make a 'serious mistake,' probably an inadvertent omission. Except for some orchestrational modifications, mm.281–88 is an exact transposition of mm.72–79. Somehow, however, Beethoven, in writing out the transposition, forgot about the piccolo runs that are such an important part of the original passage. The piccolo staff in the autograph in the respective measures is blank. Since Beethoven in his manuscript scores did not bother with rests in empty measures, we cannot be absolutely sure that he left the piccolo out intentionally. Had there specifically been rests in mm.282–85, it would have clearly indicated Beethoven's intention to eliminate the piccolo in the recapitulation. But I feel certain that Beethoven simply forgot to include the piccolo. (More composers make most of their mistakes and omissions in recapitulations, usually, in haste, taking certain obvious reiterations for granted.) What surprises me more is that, to my knowledge, not *one* writer on Beethoven's C minor Symphony has ever commented on or questioned this curious omission. For my part, I have added the missing piccolo notes (by transposition up a fourth) in my performances of the symphony for many years, often to the amazement of piccolo players who had no idea, of course, that an important piccolo passage might have been left out.

At m.289 we encounter another one of Beethoven's stretched phrases, this one extended to five bars. Interestingly, from a purely technical/theoretical point of view, Beethoven did not have to resort to a five-bar phrase. The modulatory progression contained in mm.289–93 (C#° over F-Bb$^{6/4}$-F°-F#°) could easily have been accomplished in four measures, but evidently Beethoven felt that the move to the C$^{6/4}$ pedal point would thus be too abrupt and inserted the extra measure (m.293).

This progression and its arrival at the dominant (G) pedal point produce exactly the same feeling one gets at the arrival point of a cadenza in a classical concerto. Beethoven's re-working of the second exposition theme, now at a tremendously high level of intensity, feels indeed like cadenza material.[75] It is seething with rhythmic excitement (waves of triplet eighths), crowned with triumphant theme proclamations (first in the strings, then in the brass), and leads, almost as expected, to a climactic subdominant, dramatized by second beat *sf*'s (mm.308–11), and thence to a series of dominant–tonic chordal exchanges that in almost any composer other than Beethoven would have signaled—indeed would have *been*—the end of the movement.

But Beethoven is not through yet. Having just recycled the second main theme of the exposition, he realized that he could turn once more to other primary thematic material, not yet fully exploited. The winds' theme of mm.26–27 is now revived, although not as there stated, but as it had appeared in slightly altered and contracted form in the bass instruments in mm.246–47. This is undoubtedly why this theme is now given to two unison bassoons, immediately discreetly succeeded by two horns. As simple and as uncomplicated as this sounds, it is, strange to say, one of the most maltreated passages in the entire symphony. For reasons I cannot understand, the dynamics here are consistently ignored by virtually all players and conductors, more often than not completely

75. The *sf*'s for brass and timpani in m.296 are not authentic; the dynamic here should be a simple *f*. Also it should be obvious that, after the *sf* of m.294, the succeeding dynamic is *f* (not *ff*), followed soon by Beethoven's favorite crescendo strategy, *più f* and then *ff*.

reversed. Granted that on purely acoustical terms the difference between a *ff* of bassoons and a *p* in horns will not be as dramatic as the bald dynamics might indicate. On the other hand, with a little effort Beethoven's contrasting dynamics *can* be achieved. What is so disturbing—and so inexcusable—is that, judging by the recordings sampled, most players and conductors don't even try to achieve any meaningful dynamic contrast. On almost all recordings one hears a kind of tepid *mp—mf* in the bassoons and anything but an echoey *p* in the horns, in many instances in fact a coarse *mf* or *f*! Every dynamic except the right ones can be heard:

Bassoons		Horns
f	—	*mf*
mp	—	*mp*
mf	—	*mp*
p	—	*p*
mp	--	*mf*
p	--	*f*

Not a single recording manages to represent Beethoven's intended dynamics correctly, although on some performances the players seem at least to be trying and in some instances come fairly close (Muti, Colin Davis, Haitink, Krips, Toscanini, and Jochum, although Jochum's recording is spoiled by a crudely edited splice at the horns' entrance).

At m.329 (and m.346) two performance problems—one of them merely another bad habit—mar most of the recordings. The vast majority distort Beethoven's rhythm | ♩.   ♩ | as |♪ ♪ ♪ ♪ | or |♪·   ♪ |, primarily in the strings, although often the winds seem to be influenced by them as well. This rhythmic misrepresentation makes the four bars, mm.329–32, sound choppy and disconnected. Furthermore, if Beethoven had wanted a caesura before the fourth beat (which, be it noted, *is* staccato by way of contrast), he would have written a rhythm (like ♩ ♪) to achieve that effect. Only a few conductors—Reiner, Koussevitzky, and Karajan (whose penchant for *sostenuto* playing is well known)—manage to sustain these chords properly, thereby achieving a grand four-bar line rather than four chopped apart one-bar units. The other problem is that the lower three or four notes of the piccolo in mm.329–332 are hardly ever heard. It is true, of course, that notes within the staff on a piccolo have very little projection, but again, that is an acoustical limitation which could certainly be mitigated audio-technically in a modern recording.[76] Stokowski solved the problem in his typically impulsive way by add-

---

76. One occasion when a player might have wished *not* to be heard at all occurs on Steinberg's recording with the Pittsburgh Symphony, where the piccolo player entered two bars early (in m.323 instead of m.325). It is amazing to me that this error was simply left on the recording.

ing a flute (in the higher octave) and dialing both instruments up to levels that are twice as loud as the entire rest of the orchestra!

One final word on the piccolo runs in mm.329–32, 346–49: they are to be tongued, not slurred, not only because they are so written by Beethoven, but because they will project better when tongued.

During the *sempre più allegro* (meaning accelerando) between mm.352–61 the conductor must convert gradually to one beat per bar. It is best to be in 'one' by m.357, even though the accelerando continues for another four measures.

A number of performance problems haunt the coda (m.361), marked *presto* o = 112. Beethoven's metronome marking, which once again almost no one observes, is perfect, producing when adhered to a triumphantly tumultuous, ecstatic, spine-tingling excitement that is simply overwhelming and irresistible.[77] A favorite tempo for many conductors is o = 100, many others well below that in the 90s, and a few laggards like Klemperer, Walter, and (surprisingly) Dorati at 88, 84, 88 respectively. Only a few conductors, Szell, Norrington, and Gardiner among them, manage Beethoven's 112. However, Szell slows up significantly (to 104) at m.389, a tradition clung to by many conductors, particularly of the earlier generations or those who studied in the earlier German tradition (Ormandy, Steinberg, Ashkenazy, Mehta). Mehta starts the *presto* at o = 100, slows to 96, then 92 and finally to a rather ponderous o = 88. Other conductors accelerate somewhere before m.389: Ansermet, for example, also Böhm, Haitink, Weingartner, Reiner. Strauss is once again the oddest case of all, starting the *presto* at 100, accelerating twice to 104 and 108 but in between also twice *reducing* the tempo noticeably, in effect zig-zagging through *five* different tempo changes in little over thirty seconds of music. Furtwängler who, as I have mentioned before, is often thought to be a slow-tempo conductor starts the *presto* at a lively o = 104 and before long has pushed the tempo to o = 120, even *beyond* Beethoven's 112. Furtwängler's coda is an exhilarating, transporting experience, as is Toscanini's, all in the same tempo (o = 106), relentlessly and excitingly driven, and despite some ensemble raggedness overwhelming in its impact.

In a way more problematic, because they are harder to correct, are the common performing bad habits, such as not really playing *fp* (emphasis on the *p*), or dropping the dotted half-notes in mm.361, 363, 365 etc. These notes should be fully sustained; and one 'trick' to get both the almost lyrical sustaining of the *presto*'s two-bar phrases and to assure a real *p* effect, is to tell the strings (and later in m.369 the woodwinds) to play basically *p* and make accents (in *p*) on the downbeats of the alternate measures. The timpani and brass *f* punctuations provide the necessary energy accents.[78] In practice, very few orchestras (and

---

77. I have often wondered why almost all conductors are remarkably eager to pounce on the fastest *prestissimo* tempo possible in the coda of Beethoven's Ninth Symphony, and yet resist and argue against a similar approach in the Fifth Symphony.

78. Let us note in passing that cellos and basses are here playing the same figure they played so often in the first movement. The tempo is virtually the same, only the notation uses larger rhythmic units.

conductors) are able to maintain a basic *p* for the first sixteen measures of the *presto*. This is not easy, at this speed and with the constant interruptive *fp*'s, which tend cumulatively to spiral the dynamic level upward. But even when the woodwinds enter in m.369, the over-all dynamic level should not increase.

Although many have criticized what they consider to be an excessively long and repetitious ending—"interminable C major chords," "a miscalculation by Beethoven"—I maintain that his coda and ending are perfect, especially when done at his tempo. (It *does* become laborious and boring when conducted at a ponderous tempo of, say, ♩ = 88, some 24 points off the mark.) A work of such monumental scope, of such formal complexity, which has been pauseless since the beginning of the Scherzo, *needs* a monumental ending. In this respect it is very much like Mahler's Third Symphony, which has an even longer (D major) coda, primarily because it is an even longer symphony than Beethoven's Fifth. And perhaps we also need to be reminded that the last forty-odd measures are not as devoid of interest as some would have us believe. Unfortunately most performances (and recordings) make the coda sound emptier than it is. Apart from the ecstatically rising melodic line echoing the Finale's opening main theme (at m.389), Beethoven has a number of trump cards up his sleeve, cards which are, however, never played by most conductors. For example, how many times has anybody been able to hear the canonic response in the cellos, basses, and contrabassoon in mm.390–400, a dramatic counter-line that, when projected properly gives a tremendous lift to this passage. Alas, it is usually buried—inaudible—under a barrage of timpani and brass (especially when the brass are doubled).[79] Similarly, the majestically rising figure in the lower strings and contrabassoon in m.427 is almost always (unnecessarily) covered by the surrounding din. On most recordings even the trombones are not allowed or encouraged to shout forth their triumphant harmonies.

If everything Beethoven put into his coda is brought out, there is no need to doctor up his orchestration, as Kletzki does, by having the first trumpet play a high C in m.415; or as Stokowski does when he has the horns play a whole octave higher (except for the high G's) in mm.403–14 and again m.419 to the end.

One final word on the subject of revising, supplementing, retouching Beetho-

---

79 Berlioz complained about this passage in his *Mémoires*, writing: "There is along with the final statement of the theme a canon in the bass instruments in pitch unison at a distance of one bar, which would give this melody renewed interest if it were [in fact] possible to hear the imitation of the winds. But unfortunately the whole orchestra is at the same time playing so loud, that [this canonic imitation] is inaudible." I can only conclude that Berlioz heard a typical poorly balanced rendition, such as one can still hear almost any day in our own time and, alas, on most recordings.

I was particularly disappointed in Gardiner's, Harnoncourt's, and Brüggen's recordings at this point (m.390), for I would have thought that they and some of the other 'period instrument authenticists' would have been more successful in realizing Beethoven's remarkable canonic imitation at mm. 389/90—99.

ven's orchestration and the doubling of instruments. That final word will not be mine, but Schenker's, who put the case for honoring Beethoven's text—and genius—as well as anybody ever has. "All tricks of reinforcement and doubling of instruments which the composer himself did not foresee, will prove to be superfluous for any conductor who really understands how to read this score. The score will always prevail over all such conductors who, for the sake of one cheap effect or another, feel the need to paint the score over with a stylistically inappropriate instrumental rouge."[80]

The recent (in certain circles) much-touted recordings of various Beethoven symphonies performed on 'period' instruments, led by Roger Norrington, Christopher Hogwood, and Monica Hugget are of such questionable musical quality and in any case so utterly removed in conception and performance practice from all the other 'normal' recordings, that I feel compelled to deal with them separately, all the more so because the three leaders and their orchestras (the London Classical Players, the Academy of Ancient Music, and the Hanover Band) exhibit an almost identical approach in their 'interpretations.' Furthermore, there has been such an avalanche of hype promulgated by these 'authenticists' in the last decade or so—many of their claims totally spurious and chimeric—that I feel compelled in the context of this book to offer a serious challenge to their cultish cant (and their supporting record companies), based on the hard evidence of their recordings.

But before I can elaborate on the specifics of their recorded performances, a whole range of fundamental assumptions and pretensions must be dealt with—without, however, turning this study into a disputation on the early music movement (far too much attention has been paid to it already, at least in regard to its encroachment on the 'classical' and now, 'Romantic' repertory). The three just mentioned 'authenticists' have spread such a bewildering amount of confusion, invoking various (untenable) theories, premises, and pseudo-historical assumptions, that the mind boggles as to where to begin to expose and refute these claims.[81] But perhaps one way to start is to detonate their most cherished claim,

---

80. Schenker, *Beethoven: Fifth Symphony*, p. 69. Having studied the Fifth Symphony autograph, I am convinced that the timpani part in the last measure should simply have a trill, meaning a fast roll. I believe that Beethoven initially started to write a thirty-second-note tremolo, changed his mind, altering it to a pure roll, but forgot to cross out the tremolo. As it stands in all editions it makes little sense, and at the fast tempo cannot be played as notated.

81. Lest I be misunderstood at the outset of this particular discussion, let me point out that I have been a supporter of historically and musicologically informed performances of 'early music' for many decades, going back to the early days of such pioneers as Arnold Dolmetsch, Noah Greenberg, George Malcolm, and Robert Donington; later heroes of mine were David Munrow, (occasionally) Nicholas Harnoncourt, and, of course, Charles Mackerras, Raymond Leppard, and (more recently) John Eliot Gardiner. I will be immodest enough to mention a few of my own activities in the realm of 'historically informed' performance, such as 'authentic' renditions of Ars Nova repertory in the early 1950s, what I believe to be the first performance in the United States of Monteverdi's *Orfeo* with the full authentic 42-piece instrumentarium specified by Monteverdi, and—closer to the subject of this book—the performance of Beethoven symphonies at Beethoven's metronome tempos and his specified dynamics, for example, as long ago as the early 1960s.

that they are giving us—finally—the first opportunity ever to hear Beethoven's symphonies in "a form which" he himself "would recognize."[82] The arrogance of this assumption is staggering, for it suggests that these particular 'authenticists' (a) know precisely what Beethoven performances were like in his own time (even under his own direction), and what therefore "he would recognize," and (b) that all previous interpretations and performances of these works by many generations of conductors and orchestras were bound to be 'inauthentic.' There is the further assumption that the allegedly authentic performances on period instruments we are being given are by virtue of that fact alone inherently better or preferable to anything done on modern instruments. They also claim (or try to create the impression) that their performances are based on textual fidelity with, again, the implication that textual fidelity is a subject that has now come up for the first time, and that they are the rescuers of Beethoven's symphonies from textual infidelity and the romanticist 'improvisers'.

Taking these points one by one, in reverse order, we have to begin by reminding these folks that the textual fidelity revolution was initiated by Toscanini early in this century, and that virtually every conductor since then has to one degree or another been influenced by Toscanini, and that any number of 'modern' conductors (Haitink, Carlos Kleiber, Abbado, Dorati, Dohnanyi—just to name a few) are, as far as the text, the score, goes, much more 'authentic' in their performances than anything Norrington, Hogwood, and the Hanovers have yet produced. (We will return to this point again.)

The assumption that a performance being on period instruments of itself defines that performance as somehow 'authentic' or 'better' or 'preferable,' is so lacking in logic—and so full of chutzpah—as to be laughable. To put it another way, the fact that an orchestra is playing on 'period instruments' in and of itself does not guarantee that that performance is somehow 'authentic' or adheres to the philosophy of 'textual fidelity.' A performance on 'period instruments' may indulge in as many interpretive aberrations—and often in fact does—as that by an orchestra playing on 'modern' instruments.

On still another point, one can safely assume that every conductor, from Wagner and Bülow to Karajan and Thomas—not to mention such superior Beethoven interpreters as Erich Kleiber, Schuricht, Reiner, Weingartner—has claimed fidelity to Beethoven's intentions. The implied notion, spawned by these latter-day 'authenticists,' that only those who perform Beethoven on period instruments could possibly have the appropriate insights into Beethoven's intentions and that their revelations have become possible only in most recent times, is nothing short of ludicrous. Further, the idea that these performers would somehow know what Beethoven would recognize or not recognize, and that the mantle of authenticity can now be assumed because one has (allegedly) irrefutable evidence as to orchestra sizes, performance venues, payrolls, reliable eyewit-

82. The claim of Horace Fitzpatrick in the sleeve note for the Hanover Band's recording of Beethoven's First Symphony, (Nimbus CD 5003).

ness accounts, and other 'quantitative' data, provides a neat way of avoiding all the really relevant and difficult 'qualitative' performance and interpretation questions. It is interesting to note that the so-called 'evidence' is selectively respected or alternatively disregarded when it suits these authenticists' own intentions, such as, for example, 'conducting' a Beethoven symphony from a fortepiano, when there is ample and conclusive evidence that Beethoven often conducted premieres of his works in the modern sense of conducting, that is standing in front of the orchestra, waving his arms and gesturally representing the music without benefit of a fortepiano.

Hogwood does indeed make the claim (through his sleeve-note writer for his recordings of the first two Beethoven symphonies) that the ideal, true Beethoven symphony performance can only be achieved with a conductorless orchestra, specifically one led as in the old days from the keyboard (even though there are no keyboard parts in Beethoven's symphonies) with some help from the concertmaster/mistress. This is, of course, to begin with a smokescreen, an attempt to hide the fact that Hogwood is nonetheless 'conducting' the ensemble, making all the decisions as to tempo, dynamics, phrasing, balances etc. or more the case, as we shall see, failing to do so, even as he conducts. Second, his notion that conductor-less-ness by itself will remove generations of stylistic encrustations that have accumulated over the years as the legacy of conductors not sanctified by the early music movement is hogwash, mainly because Hogwood (and his two major confrères) simply substitute their own interpretations and modifications for those of their predecessors. Hogwood's performance of the Fifth is hardly the pure, pristine, freshly heard Beethoven he claims to be resurrecting from the past, but is instead, with his idiosyncratic deviations from the text, as vagrant an interpretation as any in the record catalogues.

Third, Hogwood claims (again through his sleeve-note spokesman) that a maestro-less orchestra, led from the keyboard, cannot indulge in the "wider variety of nuance and tempo modifications" that Hogwood is ostensibly trying to supplant. "The old system inevitably necessitated a constant pulse," resulting in "uncomplicated rhythmical performances."[83] Apart from the fact that this reveals that Mr. Hogwood seems never to have heard a Toscanini Beethoven performance, one needs to ask what is so good about an "uncomplicated rhythmical Beethoven performance"? Could it be that Hogwood has never heard of Beethoven's beloved "elastischer Takt" and the tempo rubato, not only in Beethoven but in Mozart and Haydn? Could it be that he has not even read his Dolmetsch and Donington?[84]

Similarly, the authenticists' claims of authenticity on the assumption that their performances are based on the text turn out be equally fallacious, since in general they fail to follow the text any time it is convenient for them not to do so — as we shall see anon. Furthermore, insofar as many of their claims to 'textual

83. Sleeve note for Symphonies 1 and 2 by Beethoven, Christopher Hogwood and Academy of Ancient Music (Oiseau—Lyre CD 414 338).
84. Arnold Dolmetsch, *The Interpretation of the Music of the XVII and XVIII Centuries* (London, 1915, 1944); Robert Donington, *The Interpretation of Early Music* (New York, 1974).

fidelity' are, as it turns out, based on evidence *external* to the text, and insofar as their performances are not inspired, hard-won interpretations based on a painstaking consideration of the score as a *prescription*, but instead dull run-throughs, non-interpretations, which regard scores as *descriptions* of a work, their understanding of 'textual fidelity' is exactly the opposite of mine and that of most of the values expressed in this book.

Last, even if one were able to claim possession of absolutely reliable knowledge in regard to original intentions and practices—Beethoven's, his musicians', the effects of various personal and instrumental capacities and limitations—are we to assume that such intentions and practices are imperishable, immune to reconsideration and revision?[85]

Given all these erroneous assumptions and spurious premises, it is not surprising that the performances of the three authenticists under consideration here are riddled with deficiencies, misinterpretations, bald disregardings of the text, and, in addition, something not found on any of the other 60 to 70 recordings, a terrible unBeethovenian sound, ranging from scrawny to coarse and grating— sounds one sincerely hopes Beethoven would in fact not be "able to recognize." Especially annoying is the sound of the Hanover Band's performance that was recorded in some cavernous hall or church whose excessive reverberation completely defeats and counteracts the essentially chamber music approach they have taken, even to the extent of reducing the string section to six-five-four-three-two players.

Now to a few performance specifics. The opening measures of Beethoven's Fifth Symphony's first movement are variously mishandled, Norrington making the second fermata actually shorter than the first, as well as 'embellishing' both fermatas with huge diminuendos and adding two extra empty bars before m.6; Hogwood getting the fermata lengths right, but allowing the three eighth-notes to sound like triplets; the Hanovers also making a sizable diminuendo in mm.4–5 in the strings, which, however, the clarinets do not make, thus causing them to protrude incongruously at the end of the measure. (If I have earlier complained about not hearing the clarinets in the opening measures in any recording, I was not expecting or hoping to hear them in this inept way.)

---

85. It might be well for our three 'authenticists' to ponder the words of the very wise (and not ambitiously self-promoting) late Robert Donington, who in his invaluable *The Interpretation of Early Music* (p.38) writes: "A merely tacit assumption that early methods, instruments and techniques are superior for early music ignores the possibility that there might be exceptions to that basic truth. This now and fashionable habit of mind is indeed sounder than its previous opposite, and gives better results; but it still flies somewhat in the face of probability. In the course of musical history, there must, we should suppose, have been some flaws upon which we have made genuine improvements." Donington then adds "And in fact there were many. An uncritical assumption that whatever is old is best is no more reasonable than an uncritical assumption [of the opposite]." (The last three words are my paraphrase of his implication.)

Mackerras gives these thoughts a slightly different twist, as quoted in *Charles Mackerras: A Musicians' Musician* by Nancy Phelan (London, 1987): "Although we should try to learn as much as possible about how eighteenth-century musicians performed, we should not turn our knowledge into an inflexible dogma, but use it to vitalize our modern performance."

As for tempo, Norrington is, of the three, generally the most respectful of Beethoven's metronome markings, while Hogwood and the Hanovers, despite all claims to textual fidelity, conveniently ignore them, settling for a leisurely $\circ$ = 96 and $\circ$ = 92, respectively, in the first movement. This is a particularly annoying example of how cavalierly these musicians treat Beethoven's score and intentions when it suits their interpretational purposes. The Hanovers are particularly high-handed in their dismissal of Beethoven's tempos and metronome marks—flatly stating in their liner note that "these tempi are not at all suitable"—suitable to whom, to what?—and that the "authenticity" of their tempo conception is based on the notion (unsubstantiated, of course) that "Beethoven inherited the conventions of tempo in use during the latter part of the eighteenth century," and that "the evidence"—but what evidence?—"of these conventions and the technical demands of the instruments of the time combine to suggest that fast movements were played slower . . . and slow movements faster." (Note the cleverly hedging word "suggest.") At another point their decision to opt for "a late 18th-century concept of tempo" rather than Beethoven's markings of 1817 is justified by the opinion that it "solves certain problems of both technique and ensemble." [86] What problems, pray tell? Apart from the unsubstantiability of these arguments, it is a gross example of historical revisionism to claim that Beethoven adhered to "18th-century conventions" when all sorts of unquestionable evidence, not the least a break-through work like the Fifth Symphony, shows him to have been engaged in supplanting those conventions. The conventional tempos of the past were specifically singled out by Beethoven as outdated and belonging to "the barbarous period of music" (as he put it in a well-known letter to Ignaz von Mosel), welcoming therefore the help of Mälzels's metronome in more precisely defining the new tempo possibilities. (I would not make an issue of the Hanover Band's transgressions in regard to tempo—for as we have seen the vast majority of conductors have likewise ignored or rejected most of Beethoven's metronome indications—were it not for their claims of authenticity, their callous misrepresentation of incontrovertible evidence, and their particularly arrogant dismissal of Beethoven's tempo indications.)

Common to these three recorded performances is the general lack of dynamic (and therefore timbral) contrast, the result of a tendency to favor higher decibel levels. Norrington and Hogwood never get down to a real p at m.6, and the former's Classical Players start in m.14 the crescendo Beethoven has in m.18. (So much for textual fidelity!) The Hanovers spoil the first phrase by lopping off every long note value (half-notes) to ♩ ♪ on the, again, chronologically irrelevant evidence of a 1774 violin method of an obscure German composer and theorist, Georg Simon Löhlein (the name misspelled in their sleeve note), indicating how rhythmic values were played at that time ("shortened slightly in

---

86. Horace Fitzpatrick, sleeve note (Nimbus CD 5003).

order to separate them"). I say irrelevant because, again, it stems from "the barbarous period of music" thirty or more years before the composition of the Fifth Symphony, whose conventions Beethoven was drastically overhauling if not rejecting outright, irrelevant also because Löhlein was a minor theorist whose writings are hardly representative of Haydn and Mozart's era (let alone Beethoven's), and whose theories are in many respects not corroborated in any of the major performance practice tracts of that period (such as Leopold Mozart's *Violinschule*, the famous Quantz and Carl Philipp Emanuel Bach, *Methods*); and whose obscure 1774 treatise was selectively chosen as 'evidence' because it happened to suit the Hanovers' peculiar purposes. In any case, this form of 'articulation' featured in all the Hanovers' violins and violas creates a chopped-up, dissected phrasing that completely negates both the four-bar structuring of the music and any sense of a complete musical statement in mm.6–21 or any other phrase (or period structure).

In the upcoming fermatas (mm.21 and 24), Norrington and the Hanovers indulge in supposedly 'authentic' dynamic nuances: fade-to-nothing diminuendos, down-and-up ($(\gt\lt)$) swoops. (In many places later they impose the so-called *messa di voce*,[87] an up-and-down swell, especially disturbing in the timpani rolls—again a misguided stylistic application which was primarily, as the name implies, a vocal effect, rarely applied to instruments, and one which flourished in the Baroque era, certainly not in Beethoven's time. Many of these dynamic effects, especially in Norrington, are, to make matters worse, electronically manipulated which makes them painfully 'artificial,' rather than 'authentic.' (Hear this, for example, on the repeat of mm.4–5.)

The vulgar dynamic swoops in the horns ♩ ♪ (m.34–38) and woodwinds (mm.36–37), so disturbing in many of the 'normal' recordings, are here (in Norrington and Hogwood) so exaggerated as to become a tasteless mannerism or fetish. At m.44 and, again, m.52 there is real trouble in these performances. With the Hanovers and Hogwood the violins are nearly covered in the first instance, and totally covered—inaudible—in the second (m.52). Norrington achieves a reasonable balance, but distorts the four-bar phrases at mm.44 and 48 with four-bar diminuendos.

In the 'second subject' episode none of the three performances achieves any interconnection, musically implicit or technically explicit, between the various links of the long melodic chain (except that in the Hanovers' recording the extreme reverberation causes a certain amount of acoustic overlap). Hogwood's performance is, in addition, particularly offensive in its extreme dynamic exaggerations in Beethoven's foreshortened two-bar phrase groups (as in mm.75–76, for example) which are performed as [musical notation, *p* ⟨ *mf* ⟩ *pp*]. Beethoven's wonderful *subito p*'s in mm.145 and 153 are seriously compromised in all three

87. Spelled *mesa di voce* in the Hanover Band's sleeve note.

recordings, while the ingenious dynamics of mm.158–67 are pretty much ignored, and played as *mp* to *mf* at m.158 and a much-too-early crescendo. So much, again, for authenticity and textual fidelity!

In mm.182 and 190 the wind counterfigures I singled out previously come off reasonably well (primarily as a result of the much-reduced string sections), although by no means ideally, the articulations and dynamic levels in the various wind sections being somewhat less than uniform. In mm.196–239 there is little attempt to sustain Beethoven's note values, all under the assumption of Baroque practices. The Hanovers are particularly annoying here, reducing all ♩.'s to ♩.. in the winds, but to ♩. in the strings. Later, when Beethoven abbreviates the instrumental alternations to single measures, the strings give us no more than a chintzy quarter-note.

The big climactic moments (like mm.228–32 or mm.240–52) in all three recordings sound harsh, strained, and uncultivated; the trumpets are particularly blary throughout. The excessive reverberation on the Hanovers' recording produces other unpleasant side effects, such as the trumpets' loud D in m.232 bleeding over into the soft sounds of m.233. At times the overwhelming decibel levels, overbearing reverberation, and general dynamic boisterousness of the Hanovers' recording make the Beethoven Fifth sound more like Mahler's Eighth.

Balance problems continually plague all three performances, as for example in m.354 where the violins disappear completely under a barrage of timpani and crude brass. At the climactic D♭ chord of m.382, Norrington treats us to one of his manneristic down-and-up inverted *messa di voce* effects. None of the recordings manages to come even close to a *p* in mm.387–88, normally one of Beethoven's more heart-stopping moments. Unwanted diminuendos or excessive *sf*'s variously spoil the development-extension section, topped in Norrington's recording by a crazily noisy, all out of proportion, timpani swell in the fermata of mm.481–82. Finally, none of the performers shows any respect for Beethoven's exquisite coda *pp*'s at m.483.

The second movement doesn't fare much better. Again, the question of tempo is resolved in unanimity by all three 'authenticists' by disregarding Beethoven's ♪ = 92, settling for a more "suitable" ♪ = 86. The first theme statement is marked by a uniform ignoring of the *p dolce* dynamic—the violas and cellos playing with a non-*dolce*, rough-hewn tone somewhere in the *mf* range—and is all gussied up with extraneous crescendos and diminuendos. The dynamics in mm.7–12 are also pretty much ignored, with no contrasting expressive *f*'s (some of them also eroded by long diminuendos) and with the *p* pick-up notes in mm.8 and 10 all played too loud. Again these are 'sins' committed as well by other conductors in other recordings, but given all the claims and hype regarding 'authenticity,' one has a right to expect something a little less shoddy.

Hogwood, like many conductors who start with a relatively moving tempo at the beginning of the movement, also (like them) slows considerably to an *adagietto* ♪ = 60 during the woodwind phrase (mm.11–15) and the ensuing string passage. In the meantime the Hanovers somehow produce an enormous mysterious E♭ wrong note in the accompanying harmony of m.16. Again in all three recordings the written *p*'s and *pp*'s are as scarce as hen's teeth.

Beethoven's wondrously mystical strings-and-bassoon sequence at m.39 is platitudinized in these performances, in the Hanovers' case by a cold non-vibrato, in Hogwood's by an unpleasant stringiness of sound, in Norrington's a paralyzing dullness. Similarly, the problems that surfaced in many recordings in the variational episodes of mm.50 and 98, are not dealt with much more successfully; essentially all three ensembles disregard Beethoven's *p dolce*, callously crescendoing into m.56, thereby of course precluding any of the composer's intended dynamic and textural contrasts. Dynamic control is obviously not of much interest to these authenticists. This becomes painfully evident in the crucial typical Beethovenian dynamics of m.105 (*subito pp*) and m.107(*sempre pp*), roundly ignored in all three renditions—similarly in the breathtaking misterioso passage of m.161–66.

One of the more difficult passages in the *Andante con moto* movement is the episode at mm.114–23—difficult in terms of balance and articulation. But no one has distorted this passage more thoroughly than our trio of authenticists, imposing all sorts of gratuitous yo-yo like dynamic effects on the accompanimental repeated sixteenths. This displays a degree of arrogance vis-à-vis Beethoven that even a Stokowski or a Bernstein would not have mustered.

The Hanovers claim that they recorded their performance using "edited and revised parts based on the autograph and the first edition of 1809." They must not have looked very closely, for many passages that are in discrepancy with the generally available Breitkopf or Peters editions, are performed in the traditional (incorrect) manner.

Skipping over many, many further details to some of the grosser textual and stylistic misdemeanors perpetrated by the three groups, let me cite the opening of the Scherzo which is played with an astounding coarseness and, as usual with these ensembles, with a total disregard of Beethoven's dynamics. Worse yet, Hogwood's tinny fortepiano doublings, as he leads from the piano bench, are painfully audible here. The difficult scuttling cello and bass passage in the Trio is a grand mess in all three recordings, especially that of the Hanover Band. Norrington's performance is technically cleaner but so decked-out with gratuitous nuances (see Ex. 58)—talk of a "wider variety of nuances"—as to make one think of Schumann or César Franck. I must cite also the wondrous

Ex. 58

bridge passage leading to the Finale, which all three groups perform with a degree of insensitivity and a blatant disregard of Beethoven's text, including *mf* timpanis, that I find simply astonishing.

In the Finale some of the worst offenses include the Hanovers' adoption of an absurdly fast tempo for the opening ($\d = 96$, compared with Beethoven's $\d = 84$)—although they soon retreat to a saner and safer $\d = 88$; Hogwood doesn't seem to know that the violin *sf*'s in mm.68 and 277 are spurious; none of the important piccolo parts are bought out in any of the three recordings, least of all the crucial one in mm.244–50 (they seem as unaware of the problem as all the other conductors); the coda's *presto* is messy with Hogwood, but surprisingly good with Norrington who not only hits the tempo ($\o = 112$) right on the nose, but manages the basic *p* dynamic level very well. This is surprising because Norrington, like the others, produces a generally rather loud last movement with almost no contrasting soft nuances, perhaps trying to live up to his sleeve-note which calls the Finale of the Fifth "unremitting loud" (*sic*). The vital canonic counter-line in the bass instruments in m.390 is totally obscured by brass and timpani in all three recordings.

In Norrington's defense, his writing or speaking about his involvement in the 'period-instrument' movement is considerably more rational than some of his recorded performances. He admits in his interview with Jeanine Wagar in her *Conversations With Conductors* (Boston, 1991) that he sometimes wonders, upon hearing "a modern orchestra playing superbly, probably with some control over vibrato, with an awareness of phrasing," why he "bothers with old instruments." He goes on to say, rightly, "in the end I don't think that early music is about instruments; it's about music." He suggests it is more a matter of creating a historically informed performance," one where the "playing [is] relevant to the music." In the same interview-conversation, speaking on the subject of rubato and tempo fluctuations, Norrington asserts that in Beethoven he doesn't change tempi," doesn't "feel it necessary" to do so, whereas in later Romantic-era music (Mendelssohn, Brahms) he does feel "the need to do it." He further points out quite correctly that in Beethoven's time a) in orchestral music—what Norrington calls "public music"–tempi were not changed because, among other things, "that would have been a very difficult thing to do" in that "there weren't any conductors around, in the modern sense" to direct the tempo fluctuations; and b) that there was very likely a considerable difference in the way "public music" as opposed to music played in private salons—sonatas, small intimate chamber pieces—was performed, in the latter case obviously more freely. Norrington is again historically/philosophically on target when he says that this difference is one "we no longer observe today, because all of our music has become public, but it's a very important difference."

I could continue with this dreary recital of the essential fraudulence of the authenticists' claims and pretensions. But perhaps the point has already been sufficiently made to debunk the pretenders to the throne of authenticity and historical fidelity—they are the new emperors without clothes—and to show that these conductors (and their performances) are less faithful to Beethoven than those whom they claim to replace. The hype and controversy that have surrounded (and supported) their efforts have unfortunately obscured the outstanding contributions of some of the finest conductors of the past (or of the present, for that matter). They have also, alas, managed to obscure the longstanding good works in 'historically informed' performances of Charles Mackerras (remember his *Messiah* of 1966), Raymond Leppard, and more recently, John Eliot Gardiner.

Beethoven: Seventh Symphony

It is debatable whether Beethoven's magnificent Seventh Symphony has fared better or worse in the hands of most conductors than his Fifth Symphony. It is perhaps in some respects a less vulnerable work; in the sense that it is for the most part less rigorously constructed than the Fifth (especially in the first and third movements) and therefore allows for slightly more interpretational maneuverability—it is also less vulnerable to damage. Somehow the apparently more spontaneous, more intuitively realized rhythmic characteristics of the Seventh Symphony have led most interpreters to rein in their interpretive ambitions. Rhythm is, after all, the most powerful, most compelling, and at the same time the most readily understandable feature of a musical work. It speaks more directly to the musician—and, of course, the listener—than do, say, harmony, instrumental timbre, or even melody (some of the other tools of the composer). And if Wagner's almost universally accepted characterization of the Seventh Symphony, especially its last movement, as the "Apotheosis of the Dance," is not particularly useful conductorial-interpretive advice—it strikes me as rather non-specific—it *has* in general caused attention to be focused on the rhythmic/metric aspects of the work, its elemental and linear drive, and has in turn discouraged the kind of tempo divagations so routinely visited upon the Fifth Symphony.

On the other hand, in several other basic respects the Seventh is maltreated at least as much if not more than the Fifth. Beethoven's tempos of the second and third movements, especially the latter's Trio, are as cavalierly rejected as any in the entire symphonic literature. In those two movements Beethoven's unique and remarkably precise phrasing-articulations are universally ignored; reduced to merely convenient, mediocre, non-committal approximations, thus losing their truly Beethovenian singularity. And if uncontrollably loud timpani, urged on or permitted by conductors, is a widespread problem in the Fifth Symphony, the brutalization of timpani parts in the Seventh is even more ram-

231

pant. The destructive impact here of this particular offense is even more serious, because beyond anything in the Fifth Symphony, Beethoven put more of his primary thematic material in the bass range, where an overly loud, competing timpani will do the most severe damage, obliterating those important bass lines. The ensuing analyses of recordings will show how often the criminally loud and over-recorded timpani parts obscure entire sections of music where the main discourse is being—or supposed to be—carried in the cellos and basses. I cannot think of another major symphonic work in which this abuse is so common or so damaging.

Because many of the conductorial misdemeanors encountered in the recordings of the Seventh are in nature the same as those we've encountered in our examination of the Fifth Symphony, I will not be as exhaustively detailed in the analysis of the Seventh's nearly fifty sampled recordings. On the one hand I will emphasize those digressions that are the most grievous, and on the other hand those most injurious to the special, perhaps even unique, qualities of the Seventh Symphony.

Although, the tempo markings of the Seventh Symphony's first movement— both introduction and main body, the *Vivace*—are in general more respectfully treated than are most of Beethoven's metronome and tempo markings, there is still enough divergence on the part of conductors as to the 'appropriate' tempo to warrant discussion here. The difference in tempo interpretation between this movement and so many other Beethoven symphony movements, but especially the Fifth, is one of degree. Whereas in the Fifth—and indeed in, for example, the Trio of the Seventh's Scherzo movement—most conductors are as much as 30 points off the indicated metronome marking, in the introduction of the Seventh, the margin of deviation is considerably less, more like 15 points. Interestingly enough, the tendency to disbelieve Beethoven's $\boldsymbol{\lambda} = 69$ *poco sostenuto* and to take a slower tempo was more common among the earlier generations of conductors—Furtwängler, Stokowski, Casals, Boult, and some of their imitators, and yes, even Toscanini—than it is today among the present generation of established maestri. Figure 1 shows the range of tempos chosen, with Barenboim and Casals the slowest and Weingartner and Batiz on the fastest track.

The strangest tempo conception of this movement I have encountered is one proffered by Arnold Schönberg in his *Structural Functions of Harmony*.[1] Although, as far as I know, he never actually conducted the Seventh Symphony, he seems to have been convinced that the score's $\boldsymbol{\lambda} = 69$ should have been $\boldsymbol{\lambda} = 69$. In a footnote to a harmonic analysis of the opening of the Seventh's first movement, Schönberg states: "I am convinced that [$\boldsymbol{\lambda} = 69$] is a misprint. Evidently the two episodes on mediant and submediant [mm.42–52 and mm.24–34] have a march-like character." Schoenberg then backs away from his 'conviction' a little with: "If $\boldsymbol{\lambda} = 69$ seems too fast, I would suggest $\boldsymbol{\lambda} = 52$–54. Besides if one of these masters [Beethoven] writes sixteenth-notes, he means it;

---

1. Arnold Schönberg, *Structural Functions of Harmony* (London, 1954;) p.171.

Fig. 1

♩=52	Barenboim, Casals
♩=54	Furtwängler, Toscanini, Previn, Sanderling
♩=56	Bernstein, Kubelik, Norrington, Maazel
♩=58	Boult, Ferenczik, Solti, Collegium Aureum, C. Kleiber, Celibidache
♩=60	Dorati, Colin Davis, Stokowski (1959), Keilberth, Brüggen
♩=62	Walter, Kletzki, Stokowski (1928), Dohnanyi
♩=64	Thomas, Steinberg
♩=66	Mengelberg, Abbado, Fricsay, Ashkenazy, Böhm, Jochum
♩=67	Karajan, Masur
♩=69	Haitink, Klemperer, Muti, E. Kleiber, Ansermet, Leinsdorf
♩=70	Cantelli (Philharmonia)
♩=72	Harnoncourt, Gardiner, Mehta, Szell
♩=76	Weingartner
♩=78	Batiz

he means not eighth-notes but fast notes, which will always be heard if the given metronome mark is obeyed."

That is debatable, of course, especially in the case of Beethoven. And yet it is a curious but intellectually intriguing notion which relates interestingly to the Brahms First Symphony's *poco sostenuto* introduction. For, as the discussion on that music's tempo shows (see p.280), there is considerable internal evidence that Brahms's *poco sostenuto* could also be in a much faster tempo than has been traditionally and universally accepted. If Beethoven's *poco sostenuto* were, like Brahms's, an afterthought relating to the movement's main *Allegro* tempo, then Schönberg's idea would not be so far-fetched. Schönberg's idea also begins to have some merit—especially in his secondary suggestion of a compromise ♩ = 52–54 tempo—when we realize that the final six bars of the introduction, (which, I must confess, I have often intuitively felt are interminably slow, too elongated and fragmented, especially when done *slower* than ♩ = 69), if executed at Schönberg's suggested tempo relate more organically and naturally to the upcoming *Vivace*. The silences between the reiterated E's relate better to the *Vivace's* ♩. ♪♪♩ rhythm, while the metronomic relationship between the two tempos, ♩ = 52 – ♩. = 104—a classic 1:2 ratio—also seems not beyond the realm of consideration.

I personally cannot envision doing the Seventh's introduction at Schönberg's fast tempo—I would also suggest that in the case of Schönberg's supposition, Beethoven might have used a time signature of ¢—but I may, after all, also be completely brainwashed by the traditional (Beethoven's) tempo conception heard since my early childhood. Still, I find it an intellectually intriguing idea, probably to be explored further.

Beyond the tempo question, a number of other interpretive temptations arises, plaguing many performances. Beethoven's pillar-like chords every other measure (mm.1–7) are clearly marked with a staccato dot. And yet many conductors (Previn, Harnoncourt, and Sanderling, for example), wishing to show how 'deeply' they feel Beethoven's music and how 'profound' these chordal eruptions are, insist on making them into long sustained chords—Colin Davis pushing things to an extreme by making his orchestra (the London Symphony) play Beethoven's ♩ as ♩ ♪. Perhaps some of these conductors cannot reconcile short chords with a *poco sostenuto*, not realizing that any slowish tempo can have within it sharply articulated, staccato chords. Indeed, this is clearly Beethoven's intention: to demarcate the lovely lyric four-note theme first heard in the oboe and three times varied in other instruments, with incisive, almost shocking interruptions. Of course, reasonable men and women can argue about how short is short; there are degrees of staccato. I merely suggest that *some* kind of staccato needs to be heard here, not ♩ or ♩ ♪. Leibowitz unfortunately carries things to the opposite extreme, making the four chords excessively short—actually staccato sixteenths(!)—and unnaturally sharply articulated.

I find particularly annoying the stretching of (ritarding in) every second measure in the opening (mm.2, 4, 6) by many conductors. Their idea, I suspect, is to extract some additional expressivity out of those measures, on the supposition that distorting/stretching a phrase makes it somehow more 'profound,' more 'emotionally compelling.' I acknowledge that the subtlest flexibility of tempo might not seem harmful, but when conductors (Ashkenazy, Böhm, Furtwängler, Ferenczik) stretch the measures under discussion to 5/4 or 4½/4 it goes beyond the bounds of taste and artistic discipline, seriously distorting Beethoven's form and continuity—and this at the very outset of the movement. Tempo distortions of this sort and this magnitude prevent the all-important initial expository material from being heard properly.

The next problem many conductors create occurs in m.4 when they allow the oboist, who admittedly has a gently wavy expressive phrase, to 'do something with it,' 'make something out of it'—an understandable desire, except that if overdone, as it most often is, it obscures the four-note *main* motive now in the unison clarinets. The same problem would arise as seriously in m.6, except for the fact that the horns, now carrying the four-note motive, are in a relatively more projecting register than the clarinets are in m.3 and m.4; and therefore the horns are not as easily obscured as the clarinets. In fact, it is not easy to get the horns to play as softly in m.5 as the clarinets in m.3, which by all rights they should. For our modern 20th-century all-purpose horns this is a problem that needs special attention, in a way not true for the early 19th-century natural horns, crooked in A. On Reiner's Chicago recording these balances are all perfect.

The crescendo that almost always develops in m.5 and m.6 is often continued by many conductors in m.7 and m.8, perhaps because the music now modulates fairly quickly—in just two bars—from D major through G and C to F major, and they are inadvertently led to express this dynamically when in fact Beethoven intends that only the harmonic movement and the resultant quietly moving voice-leading be the sole expressive medium.

A major interpretive transgression occurs with disturbing frequency at m.10 (or sometimes m.9), namely, a substantial change in tempo—when none is indicated—in most cases slower. This tempo mutation is then just as surely followed by a compensating accelerando in m.14, helplessly responding to that most primitive of urges of pushing the tempo forward, because Beethoven happens in that very same measure to be asking for a tremendous crescendo—from *pp* to *ff*. Many conductors indulge(d) in this tempo aberration to the point where for many years it was a hard and fast tradition. Major offenders in this 'tradition' were (are) Furtwängler, Celibidache, Mengelberg, Walter, Stokowski, Jochum, Ferenczik, Fricsay, Masur, Ashkenazy, Mehta, Sanderling, Böhm, de Burgos, Leibowitz—even Carlos Kleiber; while those who held the tempo, to allow the wonderful contrast between the first nine measures' half-notes and the next thirteen bars' sixteenth-notes to have its full effect, were (are) Reiner, Cantelli, Karajan, Haitink, Ansermet, Solti, Szell, Muti, and above all, Gardiner and Abbado, whose recordings of the Seventh are altogether splendid, perhaps, all in all, the finest recordings of the work.[2]

In contrast to those who take a slower tempo at m.10, a host of other conductors rushes the tempo of the incoming sixteenths—if not there then surely in m.14. Toscanini, Weingartner, Steinberg, Casals, Bernstein, Keilberth, Kletzki, Boult, Kubelik, Erich Kleiber, Colin Davis, and the Collegium Aureum belong to that group, while Gardiner rushes noticeably but only in m.14. Previn is odd man out here, as he holds stubbornly to his already very slow (too slow) tempo. Then there are conductors (Dohnanyi, Barenboim) who rush the tempo even before m.9 and m.10, Dohnanyi, for example, as early as m.3. Klemperer, on the other hand, slows down substantially at both critical junctures, ending in m.15 with a ponderous tempo about 15 points below Beethoven's metronomization.

Incidentally, it makes no sense to treat the sixteenths of mm.15–22 any differently than those of mm.10, 12, and 14. Or are we to believe that if the music is soft (*pp*), we should adopt a slower tempo, and if the dynamic is *ff*, we should go faster?!

The next big problem—one to some extent of Beethoven's making, but certainly not unsolvable—occurs in mm.15–22, and even more so in the parallel section mm.34–41. Given Beethoven's powerfully sustained harmonies in the woodwinds and brass in mm.15–21, especially the trumpets' *ff*, if literally main-

2. Gardiner's recording is, except for a half dozen interpretive lapses (mentioned in the text), all in all remarkably good—I should add 'correct'—and provides an excellent example of how a correct, respectful-of-the-score disciplined performance can be exciting, passionate and, in many moments, even revelatory. I disagree with Gardiner's strong feeling that such a performance can only be achieved on period instruments. (It is not the instruments by themselves that produce the 'right' sounds; it is and should be the ears of the performers and their sensitivity to *how* the instruments sounded in Beethoven's time that can produce the 'right' sounds; and the 'right' players can do that on any instruments, modern or 'period.')

tained, the second violins' four-note theme, echoing the oboe's opening statement, can hardly be heard. It is indeed difficult to bring out the second violin
line  sufficiently—especially, being in
the upper middle (or if you will, lower upper) register, since it is less projecting
than the first violins' four notes two bars later—unless one asks the winds, particularly the trumpets and horns, to back off a bit after their attack. 'A bit' usually
does it. It is also not necessary—indeed not advisable—for the woodwinds to
play their loudest *ff*. However it is done, it is urgent that the second violins (in
mm.15–16 and mm.19–20) be heard as the primary voice, toward which end
Beethoven's *sfs* (four per phrase, one for each note), if vigorously attacked, can
be most effective. Reiner and Abbado are about the only conductors on recordings who get this passage really right. (Carlos Kleiber also does it beautifully
on the German Unitel video, less so on his Vienna Philharmonic recording.)

I would also like to point out another aspect of this passage (mm.15–22) that
is almost always overlooked. The sixteenth-note scales, usually treated in distinct,
unconnected one-bar units, are actually composed in two bar phrases, starting
in the lower register and climbing each time into the upper range. Example 1
illustrates what I mean. The scalar link-ups at mm.16, 18, 20, 22, not immediately discernible from the instrumental parts, need to be specially rehearsed, and

Ex. 1

in effect the violins have to take over the ascending lines from the cellos
and basses, while the latter have to be aware of passing on their line to the violins.

An awful misdeed occurs in m.21 in two recordings: Stokowski's 1928 re-cording and Bernstein's (with the Vienna Philharmonic). In that measure one can hear a painfully dissonant C being played, this in a G major seventh chord. On Stokowski's recording it sounds as though the bassoons are the offending instruments (were their parts wrong, written in tenor clef?), while on Bernstein's recording it sounds like two horns playing a quite powerful 🎼. I know from many personal experiences with Stokowski and many examples in his recordings that he rarely paid attention to such 'minor' details as wrong notes. But what is truly inexplicable—and shocking—is that the recording producers and engineers did not in both instances hear the offending notes.

It is perhaps understandable why oboists in mm.23–28 (and flutists in mm.42–47) consider this lovely tuneful episode their private solo. With its gen-tly swaying contours, it is one of Beethoven's most graceful and attractive melo-dies. But I suggest there is another way of looking at these two passages. Based on the study of the autograph score, I believe what Beethoven really had in mind in mm.23–28 was a well-balanced wind sextet of two oboes, two clarinets, and two bassoons, with the lead voice, of course, in the first oboe; similarly, in mm.42–47, where it is a flute-led wind septet intermittently fleshed out with divided violas. This represents a quite different conception of these two passages, but one which I think is more faithful to Beethoven's intentions, and at the very least, represents a logical alternative interpretive option. Once heard played in this fashion, I think many conductors will be convinced of its interpretive viabil-ity. The one recording on which these passages were performed in the manner suggested here was Dohnanyi's.

I am passing over the fact that many conductors and players ignore the special *pp* in m.29 (and m.48) in the second theme, most conspicuously, Celibidache, Reiner, Jochum, Mengelberg, Ashkenazy, Mehta, Dohnanyi, Brüggen, Gardi-ner, Harnoncourt, Strauss, de Burgos, Leibowitz, Kletzki, Ferenczik, Batiz, the Collegium Aureum (to name just a few)—Gardiner has a poor splice here—and, most grievous, Stokowski in his 1928 Philadelphia recording, in which he manages a big fat *mf (sic)* in m.29[3]; also the fact that most conductors crescendo too much and too early[3] between m.29 and m.34 (and in the parallel passage between m.48 and m.53), never noticing Beethoven's important distinction be-tween the two passages: the crescendo in the C major section (m.30) is initially held back, beginning only three and a half measures later and is therefore a rather abrupt crescendo; whereas in the F major section (m.48), it begins in the second bar and is thus stretched across four full measures.

---

3. Oddly enough, Stokowski in his 1959 Symphony of the Air recording, went overboard in the opposite direction, in m.29 producing a just barely audible quadruple *p*. Stokowski could be a man of extremes.

   In my hundreds upon hundreds of pages of copious, detailed handwritten (yellow legal pad) notes on the more than 430 recordings (over-all) studied and analyzed for Part III, the crescendos made too much and too early were *so* frequent—tantamount to an epidemic—that I was forced to resort to the abbreviation "tmte" in my notes to cope with the cataloguing of this interpretational abuse.

More important is what happens in mm.34–41 — or what *doesn't* happen in the majority of performances and recordings. If the analogous earlier passage (mm.15–22) was difficult to bring off, the later one is even more problematic. Never one to merely repeat a musical idea verbatim, Beethoven adds another thematic layer in m.34, by way of canonic imitation. As Ex. 2 shows, the

Ex. 2

woodwinds imitate the strings fairly strictly[4], a factor that must be made audibly clear. And yet, in most recordings, judging by the aural evidence, this canonic layering is left unattended to, or at best left to chance. It is in fact not easy to make the fullness of Beethoven's structuring here absolutely clear. Only a few conductors achieved it: Reiner, Celibidache, Klemperer, Kletzki, Boult, Jochum, Erich Kleiber, Casals, Sanderling, Leibowitz, Furtwängler (in his Vienna Philharmonic recording), and above all, Abbado; while those who seem most ignorant of what is at stake here include Walter, Bernstein, Masur, Colin Davis, Keilberth, Harnoncourt, even (surprisingly) Weingartner and Gardiner. Many other conductors, Carlos Kleiber, Stokowski, Haitink, Toscanini, Muti, Ansermet, Furtwängler (in his Berlin recording) and Cantelli, come fairly close to realizing the full extent of the structuring. The problem is that most conductors get so involved in conducting the sixteenth-note scales that they tend to forget all about the canonic byplay and fail to hear whether it is properly balanced or not.

Let us move on to the *Vivace*, and here we encounter, apart from the tempo issue (which I will not detail again at this time), many performance problems, the two most crucial of which, and the most universally ignored by conductors and orchestras, are 1) the slipping rhythmically into a 2/4 (instead of Beethoven's 6/8), and 2) Beethoven's subtly varied articulative differentiation of the basic rhythmic cell, the famous 'Am–ster–dam' rhythm ⌐⌐. As for the latter, hardly anyone seems to notice that Beethoven uses three clearly and consistently differentiated versions of this rhythmic figure: (*a*) ♩. ♪♪, (*b*) ♪ ⅞ ♪♪, and (*c*) ♩. ♪♪. The most that ever happens is that occasionally, by accident, this or that player or section will happen to play the rhythmic figure as written by Beethoven. Mostly this occurs with the version (*b*), the more commonly used version of the three. Generally, most conductors and performers either fail to

---

4. Beethoven ingeniously eschews an absolute pitch imitation at points *a* and *b* to avoid, at point *a*, a pitch not accommodatable in an E minor chord and, at *b*, the bland doubling of the violins' pitches.

notice the different notations or assume that they are meaningless: Beethoven's carelessness or pointless idiosyncracy.

In point of fact, these articulative distinctions are anything but meaningless. Beethoven is astonishingly clear and consistent about his intentions. To begin with, the flute and oboe have the cell type *a*, the sustained version, at mm.63–66, and quite logically so, since those measures represent a transition from the sustained reiterated E's at the end of the introduction to the main theme of the *Vivace* in the flute (m.67). The first appearance of one of the staccato versions, cell type *b*, occurs at m.85. Cell *a* returns immediately thereafter in a *ff* restatement of the main theme. This is especially crucial to render correctly in the cellos and basses, whose harmonic foundation in mm.89–96 needs to be well sustained to support the powerful theme above it. In m.98 the cellos and basses turn melodic with an important counter-response to the violins a bar earlier: again most effective when sustained, that is, played 'melodically' not merely 'rhythmically.'

Let us next notice the distinction made between the more sustained violins at mm.111–14 as well as m.118 (cell type *a*) and the lightly skipping woodwinds (cell type *c*) in mm.116–17. The interested reader, with score in hand, can continue to trace the progress—and process—by which Beethoven varies the three cell types, brilliantly fitting them into different musical and expressive contexts. I shall single out several unique instances: (1) the very precise, cogent—and therefore crucial—differentiation Beethoven makes in the *pp* C major section at the beginning of the development (m.181), where after four bars of the sustained type, the rising figures, in canonic succession (cellos/basses—first violins—second violins—oboe-flute/bassoon) are set in the lightly skipping *c* type; (2) a little later, m.222 and onward, Beethoven mixes the two types, *a* and *b*, the former reserved for primary thematic material, the latter given to the secondary responses (see Ex. 3). Notice that the viola participates in both types. It is essen-

Ex. 3

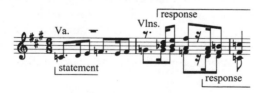

tial to insist on these articulative distinctions,[5] for only thus can one bring clarity and sense to the rather complicated polyphonic and fragmented texture of this brief episode, which, to complicate matters, is set in three three-bar phrases and one five-bar extension;[6] (3) whereas the different articulations in the section just described were handled successively (sequentially), in the next episode (mm.236–50) they are presented simultaneously. Beethoven gives the winds the sustained

---

5. Only a few conductors and orchestras succeeded in making these distinctions, most notably Abbado, Masur, Stokowski (in his 1959 recording), Kubelik, Harnoncourt, and Gardiner.
6. Measures 222–24, 225–27, 228–30, 231–35.

version, the strings the skipping staccato version. The roles are reversed in mm.256–57, 260–61, where the winds are staccato and the strings sustained. A similar contrast is made at m.278 and onward between the violins (sustained, melodic) and the cellos/basses (staccato, accompanimental). It should be clear from even these few examples that these articulative markings are anything but meaningless or arbitrary, and that they ought to be observed strictly and conscientiously.

As for the matter of maintaining a 6/8 rhythm throughout the *Vivace*, that is, of course, not easy; but it *can* be done as a very few recordings show (Carlos Kleiber, Tate). Most performances, as demonstrated by the recordings, slip into a 2/4 at some time or other. If not done in 2/4 ♩. ♪♪ , the rhythms most readily encountered are ⌐5¬ ♪♪♪ or ⌐7¬ ♪. ♪♪. , which, ironically, if Beethoven (or any composer for that matter) had actually written that, there would be (have been) the loudest howls of complaint for writing such a 'complicated' rhythm. The fact is that rhythmic training, rhythmic accuracy, rhythmic discipline are in extremely short supply, in the present day as in the past. It is a subject glossed over in music schools and conservatories and treated very casually in professional life. It so happens that to play a true 6/8—a dancing, swinging 6/8—is one of the most difficult things in music to do. The two rhythms shown above, plus the one often encountered in the Beethoven Seventh (2/4 ♩. ♪♪ ♩. ♪♪), are very easy to slide into, unnoticed except by very sharp minds and ears. The problem in the Seventh's *Vivace* is exacerbated when the tempo is taken too fast. Even at ♩. = 104 it is hard to maintain a true 'Am-ster-dam' 6/8 rhythm; but the faster one goes, the harder it is to distinguish between 6/8 ♩. ♪♪ and 2/4 ⌐5¬ ♪♪♪ or ♩. ♪♪♪ , because the actual distance between the respective third notes, already close at ♩. = 104, becomes virtually immeasurable. But, of course, it *can* be done correctly. Therefore it is doubly inexcusable when conductors (and orchestras), taking a too-slow tempo, still don't keep to a true 6/8 and slip into a 2/4 or the '5' or '7' variants shown above. Most conductors (and, alas, musicians) don't even hear these rhythmic differentiations, as is easily demonstrated by the recordings of the Seventh. The following table, representing only a small sampling, shows how conductors and orchestras fared in this matter.

correct 6/8	incorrect only intermittently	incorrect often		incorrect most of the time
C. Kleiber	Brüggen	Celibidache ⎴ (despite their		Beecham
Tate	Jochum	Harnoncourt ⎵ slow tempos)		Toscanini (NBC)
(almost all	Masur	Abbado		Monteux
the time)	Norrington	Böhm	Leibowitz	Casals
	Reiner	De Burgos	Paray	Hogwood
		Dohnanyi	Strauss	Walter
		E. Kleiber		Gardiner

To return to the beginning of the *Vivace*, some strange things go on here in certain recordings, quite apart from the fact that most conductors ignore Beethoven's *p sub* at m.67. For example, Carlos Kleiber starts the *Vivace* somewhat under tempo but accelerates into the main theme to $\bullet$ = 112 (faster than Beethoven's $\bullet$. = 104). I'm sure Kleiber was trying to make a kind of transition from the slow introduction gradually to the full *Vivace*. Furtwängler does something quite the opposite (as does Boult—did they both get this idea from their mentor Nikisch?). Furtwängler speeds up at m.59, trying, I assume, to get a leg up on the upcoming *(vivace)* tempo, but then makes a substantial ritard in mm.64–66. The two ideas together make very little sense. In the Collegium Aureum's recording an errant horn enters in m.65 (on a low A, as I recall), and in m.63 the rhythm is exceedingly messy. The group makes up for that in a way by bringing out the wonderful low bassoon thirds (a somewhat unusual voicing for Beethoven) in mm.75–80.

Celibidache's tempo for the *Vivace* is a pitiful, pedestrian $\bullet$. = 72 (as compared to Beethoven's 104), the slowest on record. A word about Celibidache is *à propos* here. Since he has not recorded commercially since the late 1940s and indeed has refused to record on a variety of philosophical, commercial and technical grounds, the numerous LPs and CDs issued in the last two decades are all live concert performances, in most cases recorded under less than ideal if not downright poor technical conditions. Celibidache has managed through the years to create a mystique about himself—almost as a kind of cult figure— and has apparently legions of admirers. This admiration, however, seems to be totally subjective, indicating no awareness of Celi's (as his musicians call him) erratic interpretive behavior, his performances ranging from the sublime to the perverse. A man of extraordinary passions and, at the same time, of keen intellect, his work has vacillated between the brilliant, searching, and deeply moving on the one hand, to the coarse and perversely idiosyncratic on the other. Within the same piece he may at once be maniacally insistent on textual fidelity and arrogantly dismissive of the score, reconstructing and recomposing it at will. Celibidache is a kind of Roumanian Stokowski, both conductors representing a very, very rare combination of genius and charlatanism.

To his many admirers I would like to suggest listening seriously to just one of his 'recordings': the Beethoven Seventh, recorded in 1964 with the Stuttgart Radio Symphony Orchestra—a fine orchestra, by the way, typical of the many good German radio orchestras of the '60s who dutifully followed Celi's every interpretive twist and turn.

Celibidache's interpretive willfulness exhibits itself most prominently and consistently in the realms of tempo and dynamics (in many ways, of course, the central themes of this book). In matters of tempo Celibidache can be incomprehensibly arbitrary, inconsistent, and illogical. Consider his tempos in the Seventh as compared with Beethoven's.

The two really gross misinterpretations are found in the first movement's *Vivace* (so draggingly slow and ponderous) and the Scherzo's Trio (see Fig. 2). How any man of Celibidache's intelligence can indulge himself in such distortions of the composer's intent is hard to fathom. Since it cannot be ignorance of the facts, it must be ego and arrogance. On the other hand, as wrong as

Fig. 2

	Celibidache	Beethoven
I. Poco sostenuto	♩ = 58	♩ = 69
Vivace	♩. = 72	♩. = 104
II. Allegretto	♩ = 56	♩ = 76
III. Scherzo	♩. = 132	♩. = 132
Trio	♩. = 44	♩. = 84
IV. Allegro con brio	♩ = 66	♩ = 72

his tempos may sometimes be, he is at least—unlike many other conductors—consistent with them, rarely changing or losing control of his preferred tempo once established.

Equally shameful is Celi's callous ignoring of Beethoven's dynamics. Everything in this Beethoven Seventh performance is squeezed into a boring middle range: *mp* to *mf*. There are no *ff*'s and there are no *p*'s or *pp*'s, and certainly no distinctions between the latter two, especially as in the Scherzo or, say, at m.29 in the first movement. He does not even notice Beethoven's *fff* in the last movement, nor his special *pp* in the fugato of the *Allegretto*.

It comes as a complete surprise, then, that in other matters Celibidache is remarkably conscientious and intelligent. Take, for example, the articulative rhythmic distinctions Beethoven makes (see Ex.3) throughout the first movement. Celibidache is among a rare few who insists in this regard. Since he takes such an impossibly slow tempo in the *Vivace*, one can hear very clearly the Stuttgarters' successful contrasting rendition of these rhythms, although the slow tempo completely vitiates and undercuts Beethoven's intentions in other ways.

Even more remarkable is Celibidache's handling of the *Allegretto*'s famous theme. He is the only one who honor's Beethoven's phrasing/articulation, including the full *tenuto* of the first beat. Celibidache is also one of the few who brings out the woodwinds *sf*'s in the beginning of the Finale and throughout the movement.

His fans will claim that his finest insights outweigh the tasteless indulgences. I am not so sure. In any case, his performances are filled with irrational, willful but passionately felt interpretations, side by side with genial inspirations, all in bewildering juxtapositions that reflect no sense or known logic.

Almost everybody does well in the *ff* re-statement of the first movement's main subject (m.89), except when the timpanist is allowed to play too loudly. Unfortunately, this happens on far too many recordings and is especially damaging in passages like mm.97–100, where an overly loud timpani can easily obscure the important answer of the cellos and basses, responding to the violins, clarinets, and bassoons. This obtrusive timpani playing in Beethoven (and Brahms) symphonies seems to be a particular problem with German and German-based conductors, and, more likely, German timpanists. The German school of timpani playing is addicted to the use of very hard sticks, producing a hard, dry sound. In my view it is a rather 'military' and insensitive sound, an inflexible approach to timpani playing; and while it can produce under certain circumstances a welcome pitch and sonic clarity, unfortunately it also lends itself too readily to out-of-context, overly obtrusive playing. Many Seventh Symphony recordings are marked by such insensate, clangorous, obstreperous timpani-playing, most notably Karajan's 1977 Berlin Philharmonic recording. That timpani player's relentless bombardment of the timpani part is truly astonishing; it would be ludicrous if it weren't so distressing. He plays as if he were a soloist in the piece, and treats the symphony as a 'Concerto for Timpani'—and Karajan tolerates it.

If some readers like more thunderous timpani-playing in their Beethoven Seventh, they can indulge themselves to the utmost in a number of recordings, including those of Klemperer, Fricsay, Masur, Casals, Barenboim, Muti, Colin Davis, Keilberth, Harnoncourt, the Collegium Aureum, and above all in Batiz's (one of the worst offenders) and in Furtwängler's Vienna recording, in which the timpanist plays so loud that even the high horns, playing *ff* in mm.89–100, are completely covered. If the reader on the other hand is looking for recordings in which the balances are correct and the timpanist is an intelligent sensitive member of the over-all ensemble, he/she should turn to those of Toscanini, Cantelli, Reiner, Haitink, Previn, Sanderling, Kletzki, Kubelik, Tate, and, to a lesser extent, Abbado, Carlos Kleiber, and (most of the time) Gardiner.

A small point but nonetheless worth mentioning concerns mm.107–18. Here Beethoven ran into an interesting problem with the flute part. In the previous six measures the flute had been partnered with the oboe and bassoon, initially answering the first violins and then joining the violins in octave unison (in m.106). But at that point Beethoven realized that continuing to double the violin part in the flute at the octave would be impossible because of the flute's then limited high range. Accordingly he gave the flute, starting in m.107, for three beats parallel thirds to the violins. For these thirds to be heard against the whole orchestra, including brass and timpani, it is best to have both flutes double the part.

In mm.120–22 (and the parallel passage mm.332–34) it is a neat helpful trick to have half the second violin and viola sections play their sixteenth-note runs separate bow, for played entirely slurred they are easily covered by the rest of the orchestra.

I have always found the six-bar passage of mm.124–29 to be one of the most exciting in the entire symphony. For most music lovers and musicians, I suppose it has lost its fascination, simply through familiarity. But we should bear in mind that there is nothing like this in any music that I know of prior to this moment. The climactic powerfully reiterative double notes in the winds, followed by a dramatic sheer two-and-a-half-octave drop to the low strings, is, when seen and heard in historical perspective, one of the most daring leaps of the imagination devised until then. Unfortunately the passage loses some of its power and drama when wind players, out of sheer inattention, rush the eighth-notes and let up dynamically on the second in each pair of notes, i.e. ♫ ⁷ ♫ ⁷. Only when both notes are played in perfect time and made equal in strength, can the

full effect of this extraordinary passage be appreciated, as can be heard so perfectly on Toscanini's New York Philharmonic recording.

Two problems arise regularly in the wonderfully inventive ten-bar passage of
mm.141–51. As the crescendo mounts, starting in m.146, if it is also driven too
much too early, the little three-note figures ♪♩ ♩ in the woodwinds, especially
in the clarinet, bassoon, and oboe parts (mm.147–49), are apt to be drowned
out entirely by the encroaching strings. Conductors (and players) should take
note that Beethoven writes not *cresc.* but *cresc. poco a poco*, a marking which is
used to indicate a very gradual crescendo. In innumerable recordings conductor
and orchestra have already reached full *f* or *ff* in m.148 or m.149, thereby completely obliterating the single woodwinds trying desperately to be heard. The
other problem concerns the little three-note figures themselves which, especially
in the cellos and basses, are rarely played rhythmically correctly. What one hears
far too often is ♪♩ instead of | 𝄾 𝄾 ♪♪♩ 𝄾 |. This is particularly dis-
turbing when the woodwinds play correctly (which they generally do in most
orchestras), while the low strings play the sixteenths too fast, like grace notes. I
have also heard on far too many recordings what appears in Ex. 4b:

Ex. 4a

Ex. 4b

The effect of overly quick sixteenths, especially in the lowest register of the
basses, is that of a series of pitch-less grunts—hear it yourself on, say, Barenboim's recording—an effect Beethoven surely *did not* have in mind.

The bass parts in mm.152–61 (and the analogous recapitulation section,
mm.364–71) are to this day extraordinarily difficult; one wonders how Beethoven
dared to write something so audacious in 1812 (along with the even more challenging passage in mm.278–99). Even today these passages are rarely played cleanly
and even more rarely well recorded, especially when a booming timpani runs constant interference in the overly reverberant halls so often used for recordings nowadays. How clear and clean these passages can sound can be heard to wonderful
effect on Abbado's remarkable Vienna Philharmonic recording.

Abbado's performance is also exemplary in that astonishing C major passage
which initiates the development section (mm.181–94), as is Haitink's (with the
London Philharmonic) and Gardiner's. Other conductors and orchestras can learn
a lesson not only in what a real orchestral *pp* is, but what a feeling of suspense and
mystery such a *pp* can convey. Moreover, Abbado and Haitink make the earlier–
mentioned canonic byplay crystal clear, avoiding all crescendos as the figures rise
in pitch, the unwanted crescendos that most orchestras seem unable to resist.

There is a strange scriptural anomaly in mm.205–206 and mm.211–12, which has confounded conductors, probably since the Seventh Symphony's publication in 1817. The second violins' rhythmic figure ♪ ♪♪♪♪ ♪ ♪♪♪♪, first encountered in m.201, continues through mm.205–206, while in the parallel place five bars later, the second violins are silent. It seems strange that Beethoven would have added the violins to the winds in one place and not the other. Given that premise, one has the choice of either eliminating the second violins in mm.205–206 (save on the downbeat of m.205) which Strauss, for example (in his early 1926 recording) and Reiner do, or inserting a similar rhythmic figure (on the pitches E/G♯) in mm.211–12. Through the years various conductors have taken one or the other of these options, but most conductors—many of them perhaps even unaware of the 'apparent' discrepancy have—simply gone with the printed score and parts.[7]

Ex. 5

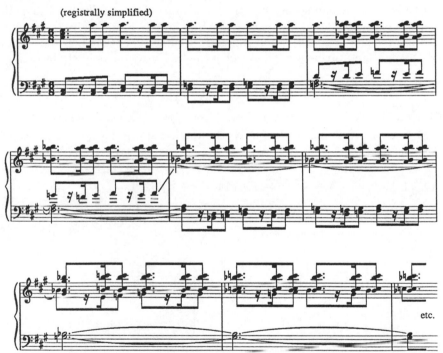

(registrally simplified)

etc.

7. A possible explanation once suggested to me—in fact by one of my brightest conducting students—for the presence of the second violins in mm.205–206 and their omission in m.211–12 relates to the winds' instrumentation in both passages. Largely because of the horns' note limitations, Beethoven was unable to create the kind of middle-low-range voicing in the bassoons and horns in mm.205–206 he could so readily get five bars later, namely [notation]. In mm.205–206 the A horns of Beethoven's day could not play the low G (seen now in the second violins) or

Among the many daring passages in the Seventh Symphony, none is more radically innovative than that remarkable chain of minor and major second dissonances near the end of the development section (mm.236–49) (Ex. 5). Oddly, these bracing, almost terrifying dissonances are frequently softened by conductors, especially conductors afraid of strong dissonances or conductors fearful of their audience's negative reaction, conductors who rarely conduct any 'dissonant' contemporary music. In my view Beethoven's dissonances, very modern sounding even today when forcefully projected, must be given full intense harmonic expression, as opposed to indulging in the 'too-much-crescendo-too-early' approach, which so many conductors adopt here and with which they try to impress and overwhelm their audiences. Notice the rhythmic distinction between the two motivic cells—one with sixteenth-note rests, the other without—a differentiation which almost nobody makes, except notably Celibiache.

Two truly troublesome spots occur at mm.256–57 and mm 260–61, the problem being to make audible the descending string patterns. At fault in most performances and recordings are the brass and timpani, especially the latter. All that needs doing is to ask the brass to back off a little on their sustained notes and the timpani to soften its roll slightly. On almost all recordings the strings (Ex. 6) are quite inaudible; or to put it another way, on only six recordings

the 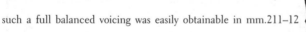 of the bassoons. The latter could, of course, have played that desired low G but that would have thrown out of balance Beethoven's full, evenly registered three-octave voicing. Whereas

such a full balanced voicing was easily obtainable in mm.211–12 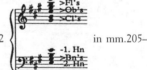 in mm.205–

206 the best Beethoven could have done is

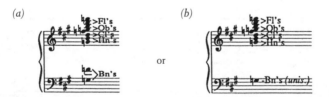

In option *a*, I suspect, he would have considered the low G too weak in just one bassoon (whereas it is quite strong on the open G string of the violins), while in option *b*, he would have considered the gap between ⟨♮⟩ and ⟨♮⟩ and the lacking ⟨♮⟩ a serious imbalancing of the chordal voicing.

Thus, according to this reasoning, Beethoven needed the help of the second violins in mm.205–206, but not in m.211–12. I should point out that, in any case, this apparent anomaly is not an editor's or engraver's error; the passage is rendered as it appears in Beethoven's autograph.

Ex. 6

(out of 70-odd) is this passage well balanced, with the strings heard properly: Haitink's, Kletzki's, Klemperer's, Reiner's, Sanderling's, and especially Previn's. Muti's is the worst, for not only is the timpani much too loud but Muti has the two trumpets make vulgar crescendo swoops in m.257 and m.261, none of course indicated by Beethoven.

Skipping over many performance problems (of the type already discussed elsewhere) and much of the recapitulation, I come now to one of those passages whose correct expression depends on the full realization of the tiniest, most minuscule details. Tiny details—a shift of one note in an unexpected place, a harmonic/melodic clash that may flash by in a hundredth of a second, an unusual voicing almost hidden in a complex texture, unusual octave displacements—these are almost always the hallmark of the work of great composers, and Beethoven contributed his full share of such important minutiae throughout his entire *oeuvre*. Curiously, the more uncommon, the spicier the effect, the more conductors tend to ignore or suppress it. This is sad, since it is often (as I have previously pointed out) that particular, unique, unprecedented flash of invention, of inspiration, that represents an important break-through, and places its composer uniquely above all others of his time and moves the language of music forward in some significant and unexpected way.

One such passage is that between m.358 and m.363, in which in each measure tiny dissonantal clashes occur between the first and second violins (an almost identical thing happens in an earlier parallel passage, mm.146–51, but there the 'clashing' notes are an octave apart and therefore not as obvious in their shock effect). Reduced to a single-line reduction (Ex. 7), it is easy to see

Ex. 7

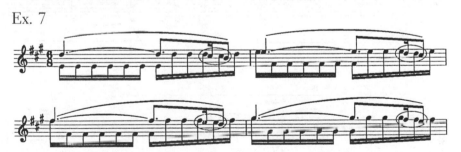

how unusual and how daring Beethoven's idea is (the clash points are circled in the example). Before Beethoven, as far as I can see or imagine, only Mozart or Bach—possibly Haydn—could have conceived something so bold, so 'wrong' according to all the textbooks and yet sounding so natural and wonderful. The problem is that the majority of conductors and, I'm sorry to say, most orchestral violinists, seem to be totally unaware of how unusual this passage is. They ignore it, suppress it, neutralize it, emasculate it. A lot of this comes from the baleful, boring habit of conductors conducting the first violins (or other lead voices) to the neglect of other equally important and interesting parts. The only way in which what Beethoven has written here can be fully realized is when both violin sections play at equal strength, above all at the clash points. Simple awareness will generally produce the right result. The only recording on which this wondrous effect can be savored is Dorati's.

Another point of awareness concerns the passage mm.376–82 (and its earlier parallel, mm.164–70), in which there occurs a canonic exchange of ideas between first violins/woodwinds and cellos/basses and in which each four-bar phrase is divided into primary and secondary elements. As reduced in Ex. 8, one can readily see that the first two measures (mm.376–7) in the violins and woodwinds contain primarily thematic material (bracketed $\overset{a}{\frown}$), derived from the flute's initial theme statement (mm.68–88), while the next two bars (mm. 378–9) are simply the bass line (bracketed $\overset{b}{\frown}$), transferred to the violin and woodwinds from the cellos' and basses' first two bars. It thus becomes clear that the violins' and woodwinds' phrase is not a full four-bar phrase but rather one broken into two components: primary (thematic) and then secondary (accompanimental), the reverse for cellos and basses. This makes a beautiful symmetrical structural design, diagrammatically represented in Fig. 3, one

Ex. 8

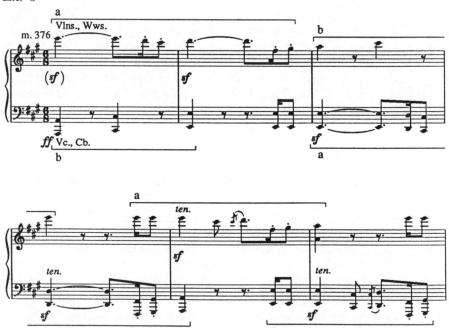

Fig. 3

which can (and should) be realized aurally/acoustically. This is accomplished by even the merest awareness on the part of the conductor and players of what is at stake, i.e. that the three notes, A - C# - E, in the violins and woodwinds (mm.378–9) are merely a transported bass line, not part of the melody, and must therefore be somewhat underplayed, while conversely the cellos and basses must bring their primary thematic measures to the fore. (Note also the emphasizing *sf*'s in the thematic segment not present in the accompanimental notes.) In that way Beethoven's simple but potent canonic/contrapuntal interplay can be made unequivocally audible and fully representative of both the idea and its notation. If on the other hand the timpani pounds away at full tilt, especially in a reverberant, boomy hall or recording studio, then the cellos'/basses' canonic answers are simply wiped out, and Beethoven's beautiful aural design is completely defeated.

By the simplest of moves—an unadorned half-step modulation downward— Beethoven transports us to the key of A♭ (in an A major piece!), but just as quickly by a magical four-step progression (A♭ - C6/4 - F - A6/4) brings us back in just eight bars (mm.391–99) to the central key. To underscore the mystery— and audaciousness—of this passage, it is absolutely necessary to play the basic

rhythmic figure ♩. ♪♪ ♩. ♪♪ *tenuto*, i.e. melodically—again one of those articulative distinctions Beethoven took pains to make but generally ignored by most conductors.[8]

Most composers, having gotten back to the central key, would now have gone quickly and directly to the movement's final denouement. But Beethoven has one more ingenious move up his sleeve, an idea he stretches and prolongs agonizingly slowly (mm. 401–22), almost to the point of being unbearable, when it simply *has* to erupt and be released into the final coda. It is kept harmonically ambivalent (although the implications of the movement's tonic key and its dominant are vaguely discernible) by a low-register, grinding, repetitious, but constantly growing chromatic figure in violas, cellos and basses

, repeated no fewer than nine times,

8. It is wrong for the horns and answering woodwinds in mm.399–400 to play

 instead of as they brazenly do in Erich Kleiber's, Jochum's, and Mengelberg's performances, even more lately with Carlos Kleiber (evidently a Concertgebouw horn tradition).

and doubled up rhythmically the tenth time—twenty-two bars in all. It is like
something slowly turning on a spit, rotating again and again on its own axis, but
growing and evolving eventually to gigantic proportions.

This chromatically twisting, writhing bass line—we will see other examples of
this in the last movement of the symphony—is set against the starkest of back-
drops, again harmonically ambiguous or vacillating: one a five-octave sustained
pedal point E (Ex. 9a) in the winds, and (Ex. 9b) a stationary melodic line

Ex. 9a   Ex. 9b

pivoting around an E axis. This amazing structure is sustained for, as I have
said, twenty-two measures, in a gigantic crescendo, with instruments gradually
being added and the upper-range figures becoming more and more rhythmicized. Listen to how beautifully Reiner and Gardiner control the dynamics
here (mm.401–22), restraining their musicians from crescendoing too early!

Many conductors through the years have made the bass players play this en-

tire section an octave lower, i.e.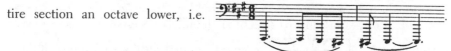

Though I am second to none in my love of deep low-register sounds,[9] and am
not only constantly demanding the full range down to low C and B in bass parts
in my own compositions, and also find myself occasionally tempted to add such
low-octave doublings wherever appropriate in the classical and Romantic works,
the Beethoven passage under discussion here is one in which, I submit, it is not
appropriate to add a lower octave. There are two reasons. One is that Beethoven
in the Seventh Symphony is very consistent and logical in his use of the low,
fifth bass string and, as far as I can tell, very clear about where he wants low
notes and where he doesn't. Below-E notes are conspicuously scattered throughout the work,[10] and one must assume therefore that when Beethoven keeps the
basses in unison with the cellos, he must mean it. It is rash simply to assume
that Beethoven forgot, or that the missing lower octave is a mistake. I believe
the second reason the lower bass octave in mm.401–22 is not wanted has to do
with the way Beethoven has structured the passage in terms of the chosen over-
all range—five octaves, as mentioned—within which three separate layers of

9. I was, after all, the first to write a *Contrabassoon Concerto*, and to write in 1947 a *Quartet for
Doublebasses*, and have written voluminously for the contrabassclarinet.
10. See for example mm.40–41, 77, 141–46, 366–74, in the first movement; none in the two middle
movements, but many places in the Finale, including that remarkable twisting, writhing pedal point
at the end of the development section (mm.298–328).

music function. I am convinced that Beethoven did not want his bass line too far removed from the lowest pedal point note , in the second horn. Finally, as regards this particular passage, it is tragic on how many recordings the timpani (in mm.421–22) completely obliterates the final two bars of the bass line (Ex. 10); most brutally in Casals's and Ferenczik's recordings.[11]

Ex. 10

I cannot resist pointing out another one of Beethoven's miraculous canonic interplays (mm.427–31), in which the low strings at first imitate the high strings but halfway through roles are reversed, and now the upper strings mimic the lower strings (Ex. 11a, simplified and reduced—(the diagonal lines show the

Ex. 11a

canonic interconnections). We might also note what an interesting composite rhythm results from these interlocking lines, especially in mm.429–31 (Ex. 11b).

Ex. 11b

A final comment on the first movement: I wish there was even just *one* recording (or, for that matter, one performance of those that I have heard in all

11. But the worst offenders are Mengelberg and his 1940 Concertgebouw timpanist, not only in this passage but in all the places—in all four movements—where the timpani plays.

my years of hearing the Beethoven Seventh) in which the horns are heard
as well in m.447 as in mm.442, 444, 446. It is obvious that high horn parts project more readily than middle-register ones; but this imbalance—a normal
acoustic phenomenon—can easily be adjusted by any horn player if he/she will
play, for example, in the descending passage just referred to (mm.446–47), a
little softer in the high register and a little louder in the middle register. Cannot *one* pair of horn players achieve—or *one* conductor ask for—such rebalancing?

In the Seventh's second movement, the famous *Allegretto*—the movement that
had to be immediately encored at the Symphony's premiere in 1813—there are
two major performance/interpretation issues rarely adhered to: the basic tempo
(Beethoven's metronomization is ♩ = 76), and the phrasing and articulation of
the movement's basic thematic/rhythmic cell (mm.3–4).

Dealing with the first issue, we find ourselves once again in a situation where
90 percent of all conductors ignore or reject not only Beethoven's metronome
marking but his Italian tempo and character designation as well. They evidently
distrust both and are sure they know better than Beethoven what he wanted.
They are usually about 20 metronome points off the mark, slower that is. Oh
yes, it is well known that Beethoven is reputed to have said at one point that
perhaps he should have named the movement *Andante quasi allegretto*. But
ultimately, though he had ample opportunity to change the tempo designation,
he did not. Also, why assume automatically—as I have pointed out in respect to
so many of Beethoven's tempo settings—that the markings are wrong? Why not
give them a try; and why question the *Allegretto* here and not the tempo and
Allegro con brio of the Seventh's last movement? It all makes very little sense,
and ultimately is so disrespectful of Beethoven and his intelligence.

As a matter of fact, Beethoven's *Allegretto* ♩ = 76 works wonderfully well.
Instead of the usual funereal, drudging-along, heavy-footed kind of affair, the
movement becomes light and airy, friendly and positive. At too slow a tempo,
the movement's extraordinary repetitions—a single two-bar rhythmic pattern
dominates the entire movement—become overbearing, if not boring. In Beethoven's tempo, it achieves, in the right conductorial hands, a wonderful flow and
a unity of conception instead of the usual drawn-out, belabored effect. How
right and beautiful this music can sound at Beethoven's tempo—or close to it—
can be heard, fortunately, on several recordings, most notably on both Erich
and Carlos Kleiber's, Dohnanyi's, Szell's, Karajan's, Reiner's, Harnoncourt's,
Gardiner's, and the Collegium Aureum's.

The second issue, the phrasing and articulation of the two-bar cell, seems to
be even more resistant to correct interpretation/realization. And yet, Beethoven's
marking is so clear and unequivocal, and moreover reiteratively consistent
throughout the movement. Here is the two-bar phrase as notated by Beetho-

ven: ♩ ♪♪ | ♩ ♩ |. One can see immediately that Beethoven took a lot of care in refining his notation, in specifying a particular articulation and expression. Most ordinary composers would have just left it at ♩ ♫ |♩ ♩ |; instead, Beethoven provides five phrasing indications for five notes. Yet for all his pains, most conductors ignore or—possibly—misunderstand them. The information is very clear: the first note is to be held *(tenuto)*, the next two notes are to be played *staccato* and the final two notes slightly lifted, slightly longer than the previous two eighth-notes. The ideal bowing for rendering that phrasing is: |♩ ♫ |♩ ♩ |.

I suspect that some of the confusion about the phrasing derives from the fact that many conductors (Solti, for example) think the *ten.* *(tenuto)* refers to the entire phrase and all of its repetitions, resulting in something like |♩ ♫ |♩ ♩ |. But that is a misreading of the notation, since Beethoven is absolutely consistent throughout the movement in marking that first note *ten.*, in whatever instruments, always followed by the two succeeding eighth-notes with *staccato* dots. It cannot be that the entire five-note pattern should be played *tenuto*, because that totally contradicts the *staccato* and *semi-staccato* markings that follow. Besides, to play the entire phrase with its several hundred repetitions *tenuto* would result in an appalling lack of variety, which a great composer like Beethoven would never have tolerated or wished for. Even in those recordings (and performances) in which there is an attempt to realize Beethoven's notation—generally rather half-hearted—the serious problem that invariably remains is that the first quarter-note is not held long enough. It generally ends up being around three-quarter length, as if Beethoven had written |♪. ♫ | or |♪.. ♫ | or |♩ '♫ |. It is invariably bad, careless bowing habits that cause this performance flaw; the note needs to be fully sustained, held right *into* the second beat. Only in that way is Beethoven's remarkable conception—three totally different articulations in but two measures—given its full due.

To conclude on this particular matter, the correct varied articulation, presented at the right tempo, will result in a wonderfully buoyant, flowing, almost swinging feeling in this movement, as one can hear on Carlos Kleiber's recording (Cantelli and Abbado get the right articulation/phrasing, but their tempo is a slowish ♩ = 63, while Harnoncourt at a tempo of ♩ = 69 gets two of the three articulations correct, but adds all kinds of annoying dynamic nuances and unwanted accents.)

If these basic aspects of the *Allegretto* movement are well attended to, the whole movement falls rather readily into place, presenting not too many other problems. Of those that remain, I should like to mention only the following. Between m.27 and m.42 there must not be any crescendo, accidental or otherwise, while the *cresc.. poco a poco* starting at m.51 must be handled very judiciously, that is, very gradually, spread over twenty-four bars. Many conductors pounce on this crescendo with such vehemence that they arrive at a full *f* or even *ff* after only eight measures—the usual 'tmte.' The worst offender in this regard is Mengelberg, who perpetrates a huge crescendo *even before* m.51, arriving there at a full tilt *f* or *ff*, maintaining it for the next eighteen bars or so, *and then* makes a diminuendo (!) just where Beethoven writes *più f*. Unbelievable!

In the magnificent climax of this expository phase of the movement (mm. 75–98), the three layers of activity (see Ex. 12) have to be well balanced against

Ex. 12

one another. It is very important that all three structural components be equally audible. What very often is obscured or underplayed is the rhythmic understructure (C in Ex. 12) with its bounding triplets in violas, cellos, and basses. Here again a pounding timpani can do enormous damage, making the important triplets, the brand-new element in Beethoven's traversal of the exposition, virtually or totally inaudible, as several recordings, most notably those of Mengelberg, Toscanini, Muti, Fricsay, and Ansermet, reveal.

The mention of the term 'exposition' serves to remind us that the form of the *Allegretto* movement is in itself one of Beethoven's miracles of invention, a wonderful new hybrid combining elements of sonata form, variation, rondo, chaconne, and march (or processional). It is variational because its main theme, a sixteen-bar phrase, is frequently varied although, as has often been pointed out, not thematically and not in the usual Beethovenian developmental sense but rather more limitedly through varied instrumentation and rhythmic ornamentation (as, for example, the added triplets in mm.75–99). It is a kind of rondo form since the A episode returns always after the interpolated secondary segments, three more times (four if one counts the coda as well). Thus the following quasi-rondo ABACABAA form evolves (Fig. 4):

Fig. 4

As Rondo		As Sonata form
Introduction	mm.1–2	Introduction
A (main theme)	mm.3–101	First subject
B (secondary episode)	mm.102–49	Second subject
A (main theme, varied)	mm.150–82	First subject
C (development)	mm.183–213	Development
A (main theme, varied)	mm.214–24	Recapitulation (varied)
B (secondary episode foreshortened)	mm.225–46	↓
A (main theme, fragmented)	mm.246–54	Coda
A (main theme as coda)	mm.255–75	↓
Closure	mm.276–78	Closure

It is in sonata form in that the movement has a first and second subject (the A major section), a development section (in this case a fugato), and a recapitulation and coda. It is also a chaconne in the sense that the ground figure or theme is subjected to a constant ostinato-like repetition, continually varied and gradually building to climactic peaks, as traditionally many chaconnes (and passacaglias) do. It is obviously also a march or processional, with its steady gait and relentless pulse. That is sufficient reason why conductors should not make any of the ritards that have become, alas, so fashionable and traditional, particularly the ones in mm.101 and 224 (hear Strauss's exaggerated pullback here), 148–49, 243–48, and especially the utterly vulgar one at m.213). (For an example of that at its worst, listen to Maazel's Cleveland recording.)

I should also like to point out that, just as the *Allegretto* is radical and novel in its innovative form, so it is radically innovative and novel in its fundamental conception and construction. For as far as I can tell, it was the first time in the history of Western music (or at least since the Ars Nova and the early Renaissance) that a composer had written a major work which managed to almost do without one of the most essential elements in the creation of a piece of music: melody. The emphasis in this movement is primarily on the element of rhythm, supported and colored by harmony and timbre (instrumental color), but virtually eschewing melodic invention. Some at first blush may consider that a heretical statement, but if one thinks about it more one may find it difficult to defend the movement's first subject as a great melody, or any melody at all.

It is on the other hand a remarkably striking fusion of rhythm and harmony, made all the more remarkable by the use of soft, darkly colored, low strings. A more truly melodic invention enters at m.27, the beautiful, almost Schubertian counter-melody in the unison violas and cellos; and the A major episode with its serene, sunny, step-wise woodwind line leans even more in the direction of melody. But even there the rhythmic and harmonic elements remain more or less in the forefront, or, to put it another way, support what little true melody there is to make it into a more fulfilled, a more complete musical statement. That aspect of the piece, along with its daring repetitiousness, has undoubtedly contributed to the success of the movement which has mesmerized audiences ever since its premiere. For audiences by and large react more strongly and positively to rhythm and repetition than to any other stimuli.[12]

The A major section (m.102) is one of Beethoven's loveliest lyric creations, even as the somewhat ominous pulse of the main theme rhythm continues (now in low string pizzicatos) and rippling triplets in the violins embellish the winds' long serene lines. Not in the realm of major interpretational problems but nevertheless small unnecessary annoyances, are the frequent disregard by clarinetists and second hornists of Beethoven's explicit phrasing articulations. It is incomprehensible to me why so many players in those two positions disregard Beethoven's staccato markings in mm.118–19 and 120–21, respectively, clearly intended to contrast with the legatos three bars later, when the flute, oboe, and bassoon in the virtually identical figures in mm.139–42, almost always play staccato. It makes no sense!

Another all–too–obvious performance detail, which however, for all its obviousness, is largely ignored by performers and conductors is to be found in the descending A minor scale in mm.144–48, distributed across three instrumental sections. In at least half the recordings of the Seventh Symphony, musicians, conductors, recording producers and engineers did not hear (and failed to correct) when players end their part of the scale with a long eighth-note, which, of course, distorts and disturbs the intended effect of one long four-octave, evenly articulated scale. If the three winds and violins in m.145 and m.147, respectively, fail to shorten their last notes to at least that of the ongoing triplet eighths, those elongated notes will bleed over into the continuing scales, making a clumsy, untidy, and unmusical effect. Among the few recordings that handle this scale passage well is Jeffrey Tate's with the wonderful (and highly disciplined) Dresden Staatskapelle and de Burgos's with the London Symphony Orchestra.

Starting at m.150, Beethoven recapitulates his main subject—varied of course—using one of his favorite devices, namely, turning the original music upside down. The lead line of m.3 is now in the cellos and basses, while the lovely secondary line, first encountered in m.27 in the violas and cellos, is now transferred to the upper register and given to three woodwinds. In addition, Beethoven mixes rhythmic figurations with almost Brahmsian daring; quarters and eighths in some of the strings (pizzicato), spiccato sixteenths in the others, and occasionally triplet eighths in the winds. But unfortunately those triplets are

12. Not until almost a century later did a composer take that idea a considerable step farther. In 1909, Schönberg, in the third movement of his *Five Pieces for Orchestra*, wrote a piece which all but eliminates rhythm in addition to melody (or theme) from the creative process. It is a movement which consists entirely of harmony and timbre, and yet is self-sufficient in making an astoundingly complete, if relatively brief, musical statement.

all too often played incorrectly, in the majority of performances only approximated and, slipping inadvertently into duple patterns. The recordings that got this right, that is, where one can really hear all three rhythms (as in m.156, for example) clearly and distinctly in their composite complexity, are those of Cantelli, Dorati, Carlos Kleiber, Haitink, Solti, Klemperer, Masur, Steinberg, Sanderling, Karajan, and Ferenczik, while all the others are incorrect, or rhythmically ambiguous, especially those of Casals, Muti, Fricsay, Jochum, Keilberth, Harnoncourt, Gardiner, and, surprisingly, Toscanini.

The fugato (mm.183–213) serves, as already mentioned, as a development section; at the same time it is yet another variation on the main theme first subject. One might call it a fugal variation. Apart from being an exemplary precision-controlled fugato, simple in the extreme and a triumph of economy and balanced proportions, it is even more remarkable for being held dynamically at bay for 26 measures—almost 45 seconds—until it is allowed to burst forth into a climactic recapitulation of the main theme in full regalia. The fugato is thus a very special, even unique moment in the movement, like a precious exquisite jewel encased on either side by the richest of embroidered frames. Therefore the secret of performing this fugato as intended is to play it in its entirety *pp*—not *mp* or *p*—but *pp*. Unfortunately, to judge by all the recorded evidence, this is rarely done. It is also imperative that the fugato be felt in Beethoven's 2/4 time signature, not the 4/8 to which it so often deteriorates in performance. In 4/8 the fugato sounds painfully pedantic and schoolmasterish. Here again tempo is a very crucial consideration, for at too slow a tempo it almost cannot help but be in 4/8 and thus stiffly academic. The only recorded performances that render the fugato as the special moment it is, are those of Cantelli, Toscanini, Erich Kleiber, Jochum, Walter, Kletzki, Abbado, Muti, Gardiner, and, above all, Solti. The last named's realization is my favorite. It is almost ghostly in its quietude, as if heard from a distance, a faraway long-ago remembrance—almost as if it were not a real part of the movement. Quite the opposite occurs in some of the worst performances of the fugato—those of Stokowski, Casals, Maazel, and Bernstein. They literally excel in their disregard of Beethoven's *pp*'s (and reminding *sempre pp*'s) and in their stiff, stilted rhythmic interpretation.

In the final moments of the movement all the problems attendant to the proper phrasing and articulation of the main thematic material come home to roost. On recording after recording the different wind sections or choirs play the basic rhythmic/thematic cell in any fashion they choose to or by chance happen to land upon. What is crucial in Beethoven's *Klangfarben* conception of this coda is that the tone colors change while the phrasing remains constant, binding the fragmented phrase segments together again. Seemingly, the conductors in question had nothing to say on these matters or perhaps did not even hear the lack of uniformity. The most flagrant abuses here occur on the recordings of the Cleveland Symphony (with both Maazel and Dohnanyi), Mengelberg, and Steinberg. Even Carlos Kleiber, who does so beautifully with the main theme at the beginning of the movement, at the end allows the winds too much phrasing inconsistency.

Certain aspects of the final five bars of the *Allegretto* movement have puzzled conductors for generations, and do in fact present a number of unusual performance problems and questions. First, there is the mixture of pizzicatos and arcos in mm.273–76. It is known that Beethoven was uncertain himself during the composing of the piece about the choice of pizzicato or arco and precisely when these should occur; there is even evidence that Beethoven's original idea was to retain the pizzicato which starts way back in m.255 through to the end. In any case, he ultimately committed himself to the ending we all know, the ending in all the available published scores and parts.

I must confess that since my earliest acquaintance with the work, even as a teenager, I have always found the sudden return to arco in m.275 in the violins strange and unconvincing, and have often been tempted to restore the pizzicato (and on several occasions have actually performed the ending that way—but not without a certain guilt feeling). In listening to the dozens of recordings in my research for this book, I was fascinated to find that at least four conductors have also insisted on that final pizzicato: both Kleibers, Strauss, and Klemperer. The problem of realizing these measures is complicated by the sudden appearance of a *f* in the strings in the third-last measure, for the question immediately and naturally arises: is that a *subito f*? Or are we to crescendo into it, on the assumption that Beethoven forgot to put in the crescendo? Why is the *f* there? Well, it is clear to me that the *f* got there once Beethoven had decided that he wanted to close the movement with the same wind chord with which it opened, which starts *f* and diminuendos to *pp*. But that still leaves the other two questions unresolved, and I'm afraid we will never have the absolute answers to them. As a result, conductors have through the years come up with all sorts of different solutions. The most favored solution is to make a crescendo into the *f* of m.276, many times starting it as early as m.273 in the cellos and basses.[13] To me that is the more obvious and ordinary remedy; I think the *subito f* effect in m.276 is much more interesting and more in line with Beethoven's constantly evident sense of surprise, his avoidance of the obvious, his vivid imagination for the untried and the unusual. I am also unwilling to assume that Beethoven 'forgot the crescendo.' There is scarcely any evidence that Beethoven forgot details; on the contrary, his manuscript scores are marvels of scrupulous, meticulous attention to details and minutiae of notation. The signs of struggle and tormented revisions in his scores provide further proof of that.

The final wind chord (as well as its partner at the very beginning of the movement) is rarely performed well or correctly. First of all, it seems to be very hard to tune, the problem notes here being the C's in second oboe and first bassoon, the minor thirds of the chord. There are only a handful of recordings out of the fifty-odd in which these chords are in tune. Second, hardly anybody seems to appreciate the unusual voicing of this A minor chord. It is a 6/4 chord, with the second horn in the 'bass' position. Apparently very few conductors

---

13. Incidentally, many conductors (such as Stokowski, Carlos Kleiber and Klemperer) had the violas, cellos, and basses play pizzicato in m.276. Since these are conductors who assume Beethoven's arco in the violins to be correct, it is difficult to fathom why they don't assume as much for the remaining strings.

know this or allow this unusual voicing to be heard. In virtually all recordings the second bassoon's A is far too prominent, while the second horn's lower E is subdued, suppressed, or played with very little presence.

Perhaps the strangest, and most bizarre rendition of the movement's ending—amounting actually to an almost complete re-composing of it—is Stokowski's (in his 1928 recording). After a huge ritard in m.271, Stokowski changes the violas in m.274 to arco *(sic)* but then—perversely—back to pizzicato in m.276. This is followed by a huge 'sexy' ritard in m.275, and to cap matters, the first violins are forced to hold out their final A with the winds in a long protracted diminuendo, his final tempo at the end being somewhere around ♩ = 26!

In the *Presto* movement, a Scherzo in everything but name, there are fewer performance problems—at least until one gets to the Trio—than in the Seventh's other three movements. It is not as encumbered with misguided traditions; even in the question of its basic tempo (again with the exception of the Trio section) conductors seem to find Beethoven's *Presto* and metronomic designation of ♩. = 132 much more acceptable than in the Scherzo of the Fifth, for example. They at least do it in some sort of fast tempo, if not exactly Beethoven's 132 marking.

Since this is again a movement to be conducted in 'one,' the question of the phrase structuring may arise. Here too, unlike Beethoven's Fifth Symphony, the Seventh's Scherzo offers very few complications or irregularities. For the most part it falls neatly into four-bar patterns, with the occasional six-bar extension interpolated along the way. Two of these are interesting. Whereas Beethoven in most of his fast tempo music forms his six-bar phrases by adding two measures to the end of a four-bar entity, in this Scherzo movement he twice (twelve times counting all three Scherzo sections and repetitions) attaches the extra measures at the front end of the four-bar unit. This occurs at the very beginning of the movement and in its recapitulation, m.89. The usual Beethoven 'sixes' on the other hand always occur in the rising phrase first heard in mm.11–16.

The phrase and period structuring here is so natural and regular that very few interpretive problems can arise—the kind of problems that abound in the first and third movements of the Fifth Symphony. But there is one section in which the phrase periodicity is not as obvious as it might seem when glancing at the instrumental parts, and is in fact so ambiguous and complex that no one can be sure what Beethoven actually had in mind. The problem begins in m.63, a 'three' of a four-bar phrase, where the oboe and bassoon overlap with the outgoing four-bar phrase of the strings. There is no way an oboist and bassoonist can discern from merely their own parts whether the note D (in m.64),

m.64

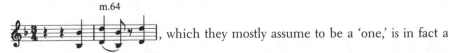, which they mostly assume to be a 'one,' is in fact a

'one' or a 'four.' And if it *is* a 'one,' then is it a 'one' in a five-bar phrase? Or is it the 'one' of a series of four-bar phrases that Beethoven has shifted, as Ives might have done, one bar earlier in the over-all structural pattern? And if that is the case, when and how does Beethoven realign the winds—a flute and second bassoon are added a little later—with the prevailing four-bar periodicity? There seems to be no unequivocal answer. The three possibilities, all viable options for a performance interpretation, are outlined in Exx. 13 a,b,c. This first option

Ex. 13a

is hard to bring off for the oboe and bassoon, since the B♭'s in m.66 land squarely on the tonic note—B♭ is the temporary key center of this section—and in both parts that measure feels more like a 'one'—mind you for the oboe and bassoon, not for the strings. In that case too, mm.63–64

 would be anacrusis notes to the F 'down-

beat' of m.65, which might feel strange at first try, but can be made to work. Or is the phrasing therefore polymetrically displaced as in Exx. 13b and 13c? In the former instance (Ex. 13b), we would have in the woodwinds one of Beethoven's

Ex. 13b

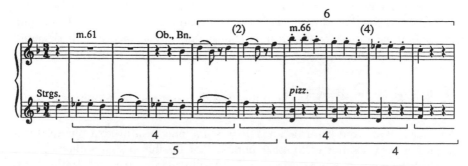

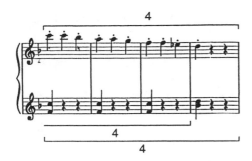

Ex. 13c

six-bar phrases, of the type peculiar to this movement (as already mentioned), and indeed a phrase identical, except for dynamic variants, to the first six bars of the Scherzo. This seems quite logical and is indeed how most oboists/bassoonists instinctively interpret the phrase, with perhaps a slight extra pulse on m.66. But if their phrase is in fact a 'six,' and it is followed by four 'fours,' then to become reconciled with the strings and their undisturbed four-bar patterning, there would have to be a 'three' at the end (mm.86–88). In that optional interpretation, the wind solos would be one bar out of phase with their string accompaniment, which is the way the passage is most often played, consciously or inadvertently.

Another possibility, represented by the lower string brackets in Ex. 13b, would be that the strings start with a five-bar phrasing, then align themselves with the winds, continuing in 'fours,' until they too would have to do a 'three' to arrive correctly at the *f* in m.89. This, however, is a fairly remote solution, since the harmonic structuring (alternating B♭ and F chords in four-bar units) is so strongly anchored in those keys that it would seem downright peculiar to change that to

$$\overset{1}{\raisebox{0pt}{ρ}} \quad \raisebox{0pt}{$\mathrm{\textit{~}}$} \quad \raisebox{0pt}{$\mathrm{\textit{~}}$} \quad \Big| \overset{2}{\raisebox{0pt}{ρ}} \quad \raisebox{0pt}{$\mathrm{\textit{~}}$} \quad \raisebox{0pt}{$\mathrm{\textit{~}}$} \quad \Big| \overset{3}{\raisebox{0pt}{ρ}} \quad \raisebox{0pt}{$\mathrm{\textit{~}}$} \quad \raisebox{0pt}{$\mathrm{\textit{~}}$} \quad \Big| \overset{4}{\raisebox{0pt}{ρ}} \quad \raisebox{0pt}{$\mathrm{\textit{~}}$} \quad \raisebox{0pt}{$\mathrm{\textit{~}}$} \quad \Big| \cdot$$

B♭   B♭   B♭   F

In the third option, shown in Ex. 13c, the first wind phrase would be a 'five.' Measure 69 would then be a 'one,' which would work well with the ongoing four-bar string structuring, but would also mean that the winds would have to feel mm.70, 72, 74, 78, 82 as 'twos' in a four-bar phrase. That seems rather difficult to do, in view of the fact that in these measures the winds start on the high note of a descending line after a two–beat rest, and these initial high notes sit squarely on the tonic and dominant positions. It also goes against the phrasing as Beethoven first and most often presents this theme, where the bar in question here (m.70, for example) is always a 'three,' much more structurally akin to a 'one.' On the other hand, though difficult, this interpretation is not impossible; it just takes a little extra effort, and indeed produces a very interesting and in its own way quite logical effect, very different from what one normally gets to hear in this passage. It would mean that the oboe and bassoon (and later the flute) would have to play mm.70, 74, 78, 82 relatively lightly and probably feel a slight crescendo (part of the over-all *crescendo poco a poco*) in each descending line, feeling a stronger pulse or weight on mm.69, 73, 77, 81. Again, the advantage of this is that in that phrasing the winds and strings are on common metric ground.

While we are on this passage, I should mention that it is very important that the oboe and flute know that the oboe has the lead through mm.74–47, relinquishing it to the flute in m.78. To judge by the recorded evidence, this is very rarely understood.

To return now to the opening of the Scherzo movement, two performance misdeeds are immediately committed in the first ten measures by most orchestras and conductors. Very few orchestras attain a true *p* in m.3, which is after all three dynamic levels below the initial *f*. There should be a dramatic drop in intensity and dynamic level, not the half-hearted *mp – mf – poco f* one hears most of the time. The sudden drop to *p*, after the boisterous brass and timpani-laden opening *f*, is so typically Beethovenian—something he undoubtedly learned from Haydn, the master of this kind of surprise—that it is to severely misjudge Beethoven's sense of humor and miss the whole point of this particular movement to not observe this *subito p*. Just as severe a misinterpretation is the universal bad habit of failing to hold out the dotted half-notes of mm.6 and 10. On recording after recording one can hear every possible distortion of this duration from ♩ through ♩ ♪ and ♩ to ♩ ♪ ⅞. The conductor may have to tell the winds not to breathe in m.6—it being quite unnecessary; as for the violins, they are in a perfect position, up-bow that is, to hold the note through. The beauty of this sustaining approach [14]—it is what Beethoven wrote, after

14. As far as I can tell, it is to be heard correctly on only two recordings: Dorati's and Ashkenazy's.

all—is that it turns the whole opening passage into a wonderful eight-bar phrase, mm.3–10 (ten bars if you count in mm.1–2). All too often in this movement, as in so many classical movements notated and conducted one-to-the-bar, performances achieve a kind of breathless, choppy, disjointed feeling. It is important for all concerned to produce long multi-bar phrases—eight-, twelve-, sixteen-, even twenty-four-bar phrases—the underlying four-bar infrastructure notwithstanding. This approach can be very important in the aforementioned winds-and-strings episode (mm.61–89), which really sings and swings when it is played as one long twenty-eight-bar musical thought. (Yes, reader, it *can* be done!)

Since sudden dynamic contrasts are one of the major compositional tools Beethoven consistently uses, it is surprising—and disheartening—to discover how few conductors and orchestras pay any attention to these markings. In most recordings the sudden *p*'s in this Scherzo are treated very casually, more in the region of *mp* or *mf*; on some they are ignored entirely. Nor are Beethoven's little *pp* echoes, scattered throughout the exposition always attended to. What I find particularly unpleasant are the accents that many orchestras make, especially the strings, as in mm.37–40. My sense is that such four-bar phrases, especially when they are set in *pp*, should be played very smoothly, letting just the subtle bow changes articulate the individual bars. The tied-over note in the first bassoon, horn, and violas in mm.41–43 and 37–39 is another clue that a *legato* unaccented approach is the right one. By the same token, care must be taken in violins and violas not to drop the eighth notes in mm.44 and 60, as if Beetho-

ven had written a diminuendo, , a common fault in many

recordings. These eighth-notes are, as we can see in the opening measures and passages like mm.25–28 (where incidentally, oddly enough, they are usually played correctly), an integral, not to be dismissed part of the primary thematic material.

That Beethoven attached great importance to the sudden *p* in m.3, and that it wasn't some idle, accidental effect, is confirmed by the sudden dynamic change in the recapitulation of this theme in m.93, this time not dropping to *p*, but intensifying to its opposite, *ff*. Again, sad to report, this startling effect is largely ignored in the vast majority of recordings and performances, although here and there an occasional timpanist enjoys hitting his drums a bit louder in that measure. The point is that most conductors ignore or are unaware of this important dynamic indication.

I cannot resist expressing my wonderment at Beethoven's ingenious choice of

pitches for the timpani part in the Scherzo: . Since the Scherzo proper is in the key of F, the timpani's F was a logical, even inevitable choice. Normally the other note for the timpani in an F major piece would have been

C[15]. But since the Trio was going to be in D major, Beethoven had to consider the choice of a second timpani pitch in that tonality. His decision was A, which, as the third in F major, could be used fairly effectively in that key as well. Furthermore, Beethoven at that moment of decision already knew that some of the actual Scherzo would also be in A major (see mm.17–28), and that his timpani F/A's could possibly even occasionally follow the contours of his thematic material (see mm.140–44). The A, being the dominant of D, could of course serve him well in the D major Trio section; and indeed the climactic passage of mm.207–21 demonstrates some of the most exciting and powerful use of the timpani in the entire classical repertory.[16]

The rest of the Scherzo section usually comes off rather well, except on one recording (Maazel's) I heard the violin trills in mm.117, 119, 121, etc. rebowed

as  instead of Beethoven's

It is in the Trio where the most severe interpretive problems and bad tradition again rear their ugly heads. Once again, it is the fundamental question of tempo which is at the heart of the interpretive dissention. Yet I fail to see why there should be any question as to the appropriate tempo, namely Beethoven's, and why anyone should question it in the first place. Beethoven is very clear about it, both in the Italian tempo designation and the metronome marking, which latter, as I have suggested before, confirms the former. *Assai meno presto* in Beethoven's very good Italian quite clearly means "very much less fast" or, in better English, "considerably less fast." It does not mean *andante* or *adagio* or

15. We have to remember that in the late 18th and early 19th century all composers were limited to two timpani—or thought they were, until Berlioz came along with his *Symphonie Fantastique* (using four timpani) and *Requiem* (using sixteen timpani). In addition, the instruments of the day could only be retuned very laboriously; the chain tuning and the pedal timpani would not come along for another half-century. Of course, some composers, notably Graupner, Fischer, Molter, Druschetzky, Salieri, and Spohr, had occasionally already written for multiple timpani—sometimes as many as seven or eight. But Beethoven apparently was either unaware of such earlier experiments (unlikely) or simply felt no need to follow in such footsteps.

16. In that connection, I am opposed to revisions of Beethoven's timpani parts, as many conductors and timpanists have done, making use of the fully chromatic modern timpani. While that is great fun to do—to line up the timpani parts with the bass parts, to fill in timpani notes in sections where, because of some temporary modulation, the potential or desirable timpani notes were simply not available in Beethoven's time—I oppose such an approach (a) because it usually amounts to an almost complete rewriting of Beethoven's music; (b) because Beethoven did his best to compensate for the timpani's limitations and did so ingeniously, often adjusting other pitches to make up for those deficiencies; and (c) how is one to know where to stop rewriting, how far to go in modernizing the timpani parts. As they exist they are an inherent part of Beethoven's conception. His timpani parts are not mere orchestrational add-ons, and I am willing to bet that in many, many cases the limitations of the timpani influenced the course of his composition. To tinker with that is, to my mind, impermissible.

'slowly'! No, it simply says "considerably slower." Now, I submit that dropping from M.132 to M.84 *is* considerably "less fast", "considerably slower;" constituting, in fact, a one-third drop in tempo.

I don't know who started this deplorable tradition of performing the Trio in a slow, ponderous, bombastic, dragging manner; perhaps it was Bülow. I doubt that it was Wagner or Habeneck or Mahler or Seidl. In any case, it is wrong, if for no other reason than that in some of the plodding tempos many conductors take, the Trio is no longer part of a Scherzo. (Even Bruckner's Trios are not taken this slowly!) The other immediate problem is that, taken at too slow a tempo, the entire Trio becomes a series of chopped apart two-bar phrases, rather than the eight-bar sentences Beethoven composed. Under such treatment the Trio loses its grand line, its breadth, its nobility. Even Furtwängler, the master of the grand eternal line, could not at his tempo of $\sqrt{} $. = 46 (50 in some recordings) maintain the long sweeping arching lines that Beethoven created in the Trio.

As shown in Fig. 5, the range generally goes all the way from $\sqrt{} $. = 44 to the lower 70s. The favorite tempo appears to be $\sqrt{} $. = 54—a 'mere' 30 points below Beethoven's intended tempo—the '54' a tempo shared by a half a dozen or so conductors. Toscanini's, Reiner's, Norrington's, and (surprisingly) Ashkenazy's stand out as the only ones either on or close to Beethoven's mark.

The ritard almost all conductors make in the second ending (four bars before the Trio) is a dead give-away that they are also going to ignore Beethoven's Trio markings. Note that there is no *rit.* indication in the second ending. Since Beethoven was quite capable of writing a ritardando—although he did so sparingly

Fig. 5

♩. = 44	Celibidache
♩. = 46	Furtwängler, Batiz, Mengelberg (slows down later to 38), Strauss
♩. = 48	Böhm, Weingartner, Ferenczik, de Burgos
♩. = 50	Stokowski (1928—in his 1959 recording he had advanced to 56), Ansermet, Fricsay, Sanderling, Jochum
♩. = 52	Colin Davis
♩. = 54	Erich Kleiber, Keilberth, Kubelik, Klemperer, Casals, Haitink, Mehta
♩. = 56	Walter, Abbado (speeds up to 66 later)
♩. = 58	Bernstein, Previn, Kletzki, Masur (speeds up to 64 later), Harnoncourt
♩. = 60	Barenboim
♩. = 62	Collegium Aureum
♩. = 64	Steinberg, Carlos Kleiber, Maazel
♩. = 66	Solti, Maazel, Boult, Cantelli, Brüggen
♩. = 68	Thomas (slows down to 60 later), Karajan, Muti, Leinsdorf
♩. = 72	Dohnanyi (slows down later to 64), Gardiner, Szell, Leibowitz
♩. = 74	Dorati
♩. = 76	Reiner
♩. = 80	Ashkenazy, Norrington
♩. = 84	Toscanini, Reiner

in his symphonies [17]—when we see none written, we ought to assume that there was not meant to be one. I am also convinced that Beethoven meant the sudden appearance of the Trio to be a surprise, undiluted and unrestrained by a ritard, a tempo drop that should be suddenly experienced at full speed.

The other problem encountered frequently in the Seventh's Trio is that it is played too loud and with too heavy a sound. The violins' sustained pedal point high A's, marked *p*, and the lack of low-register notes (at least until the entrance of the second horn in m.181), offer a clue that the whole Trio should be played with a light, airy sound, in a gently swaying tempo. Like the fugato section in the second movement, the Trio, except for its powerful *ff* climax, should be heard as a quieter, serene interlude, somewhat isolated from the driving Scherzo sections; and maintaining a real *p* will contribute immeasurably to that effect. Another common abuse is that the swells ($\prec$ $\succ$) in the first four bars are imposed on the next two measures as well (not so indicated by Beethoven, of course). Among other vulgarisms visited upon the Trio are the ritard directly before m.207 (the big D major *ff*) and one just before the return of the Scherzo. Here some conductors lose complete control of themselves. Mengelberg, in his performances, for example, after having slowed to ♩ = 38 around m.193, used to make a considerable accelerando with the second horn in mm. 199–204 and then pull on the tempo brakes again with a huge ritard in mm.205–6. Another ritard would come at m.221, slowing down further after that, and coming to a virtual $\frown$ standstill in m.235. (All of this is documented on his 1940 Concertgebouw recording.) Furtwängler even outdoes Mengelberg by slowing down to ♩. = 17 (♩= 51) before the return of the Scherzo. Other major self-indulgers in excessive ritards are Strauss, Böhm, Muti, Fricsay, and Sanderling.

Instead of worrying about a ritard in mm.205–206, what conductors should concentrate on is bringing out the startling dissonances in mm.201–204, especially the one in mm.203–204. Here a diminished chord in the woodwinds clashes with the stationary A's of the violins and the second horn's G♯ - A (See Ex. 14). The B♭ of the second clarinet is the most crucial note and needs to be brought out. Instead it is more often than not suppressed, quietly

Ex. 14

17. Beethoven's chamber music, especially the string quartets, his *Lieder*, his choral works, and, of course, the opera *Fidelio* make much more use of ritardandos and accelerandos, and a general flexibility of tempo than do his symphonies. I believe the reason is that in the latter he was dealing with much more tightly constructed forms, the unity and rhythmic energy of which he wanted to preserve at all costs. It is also true that while Beethoven's interest in freer tempos, *rubatos*, more frequent changes of tempos, developed later in life as his creative vision expanded and became more elaborate, he never lost his abhorrence of "slighting form for the sake of content," as Frederick Dorian put it so eloquently (Frederick Dorian, *The History of Music in Performance* (New York, 1942)).

buried somewhere in the chord. The few conductors who brought this daring, typically Beethovenian harmonic idea to the fore are Abbado, Harnoncourt, Casals, Walter, Toscanini, Leinsdorf, Sanderling, Kletzki, Thomas, Steinberg, Boult, and Dorati.[18]

Before going on to the rest of the movement, I must still allude to an apparently tiny notational detail, which is, however, of striking importance, simply because Beethoven uses the marking in question in a very special way. I am referring to the horns' *fp* in m.222. Perhaps horn players, looking merely at their part, may be partially forgiven for not understanding that this *fp* is not the usual one—although if they used their ears and were a bit curious they might be led to at least ponder the question. But there is no such excuse for the conductor who has the full score in sight and should in the total context of the passage in question be able to understand—and then elicit from the performers—the particular meaning and feeling of this *fp*. As Ex. 15 shows, the horns are

Ex. 15

merely a continuation of a D major triadic arpeggio, which Beethoven uses to both return to the repeat of the previous 34–bar section and lead to the 14–bar transitional postlude that concludes the Trio. The horns obviously have the role of continuing and picking up from the trumpets this three-octave arpeggio. This means in turn that the horns must take up the trumpets' line as collaboratively, as efficiently, as possible in respect to both tone and dynamic, so as to give the impression of *one* single descending seven-note line. If Beethoven had intended this entire two-bar line to be played *f* or perhaps with a gradual diminuendo, there would be less of a problem; but since he decided upon the second half of the arpeggio to be played *p*, he created a more interesting and more complicated performance problem. The link-up between the trumpets and horns must be accomplished with great care and discretion. What typically happens instead is that the trumpets inadvertently make a slight diminuendo in m.221, while the horns after two beats rest, burst in with an unthinking ordinary *fp*—in, mind you, a favorably projecting register compared with the outgoing trumpets—and the line is broken into two parts rather then one phrase unit. It is therefore essential that (a) the trumpets not diminuendo at all; (b) that the horns pick up exactly at the dynamic level of the trumpets—easily controlled by the ear; and (c) continue their part of the phrase in *p*, this necessitating playing the first note in m.222 in a manner shown in Ex. 16. This is not easy, especially at a fluent tempo of ♩ = 252 (or even at a modified slower tempo), but it definitely can be done with a little effort and care.

18. But perhaps I am giving too much credit to the conductor, for it could also be that it was some enterprising, harmony-conscious clarinetist who on his own brought out the desired dissonance.

Ex. 16

The form of the Seventh's Scherzo movement is quite unusual and was, as far as I know, a wholly original invention of Beethoven's. Following up on his various experiments with the Scherzo of the Fifth Symphony a half a decade earlier, he expanded the typical Minuet or Scherzo form, as developed and refined by the Mannheim school as well as Haydn and Mozart, to a much grander scheme. The Scherzo proper is here played three times, with its own internal repeats, and the Trio appears twice, also with its own repeats. After the third Scherzo, Beethoven teases us briefly with a third Trio, but after four bars—with tongue in cheek—breaks it off, and with the laughter of five short incisive chords ringing in our ears, closes the movement.

As one might assume, Beethoven was not content to repeat the Scherzo sections verbatim. He finds a most ingenious way of subtly varying the material. After restarting the Scherzo in its second appearance exactly as the first time, Beethoven suddenly surprises us with a *subito p* where previously there had been a vigorous *f*. From that point on (m.260) the bulk of the Scherzo, some seventy measures, is kept at *p* or *pp*, all the previous contrasting *f*'s suppressed. This entire *p* section should, on the face of it, not offer any serious performance or interpretive problems. Yet it is amazing—and discouraging—to hear how few conductors and orchestras seem to be able to carry out Beethoven's wishes. The worst offender, once again, is Mengelberg, who plays the entire *p* episode at least *mf*, more than once even reaching *f*. In effect he ignores Beethoven's idea entirely. (The same orchestra, the Concertgebouw, does very well with this same movement twenty-six years later under Erich Kleiber's excellent direction. What a difference a conductor can make!) The conductors (besides Kleiber) who have realized Beethoven's idea perfectly are Abbado, Szell, Dohnanyi, Kletzki, Ferenczik, Gardiner, Harnoncourt, Brüggen and above all—spectacularly—Toscanini. Much less successful or downright bad are most of the other recorded performances, most notably those of Strauss, Celibidache, Casals, Masur, Fricsay, Karajan, Solti, Mehta, Maazel, Leibowitz, de Burgos, and Thomas (whose English Chamber Orchestra players constantly strive for and are allowed to reach *mf*).

The final would-be recapitulation of the Trio near the very end of the movement is even more mishandled by the majority of conductors than in its two previous incarnations. Instead of realizing that Beethoven's joke consists precisely of having the Trio reappear in true 'peek-a-boo' fashion for only an instant, they manufacture some elaborate drama (or melodrama) to justify doing the four bars at even slower than snail's pace. Wrong! The joke's success depends entirely on the sudden unexpected reappearance of the same already twice-heard Trio, as if to elicit in the audience a groan of 'Oh, not again?!', when of course, just as abruptly, Beethoven lets us know he was 'just kidding.'

For the record, once again Mengelberg is the worst offender. His tempo for the last Trio appearance is ♩. = 30 (some *assai meno presto* tempo!), which he follows with a long fermata in the second measure, then ritards further down to ♩. = 23. Strauss's ritard is almost as bad.

With that sad commentary on the sometime state of the art of conducting, we must turn to the fourth movement of Beethoven's Seventh. There are lots of performance/ interpretation problems here, but I shall try to restrict myself to the most grievous conductorial misdemeanors. Again, my emphasis will not be on obvious generalities, but rather on those aspects of the work that are special, unprecedented, original, or perhaps even unique to Beethoven or this symphony. Of course, nothing in Beethoven's great works is ever ordinary or unspecial. Every measure of the Seventh's Finale bristles with striking ideas, remarkable solutions to new unprecedented problems, eye- and ear-opening revelations, ingenious revitalizations of old formulas—too many to treat in this context as exhaustively as they deserve.

Let me assume that we all know by now that the Seventh's Finale is a most extraordinary exploration of the potential expressive power of rhythm and meter, combined with a lively tempo, to move and excite the human spirit. But how does this manifest itself in particular? Many pieces, say by Haydn or Mozart— or even Bach and Rameau—had involved rhythmically exciting, driving Finales. What was new, what was so special about Beethoven's Seventh Finale? More than anything else, even more than the sheer scope and relentless energy and dynamic intensity of the movement, what gives it its emotional power is, I believe, the unsettling element of syncopation, a kind of shifting the beat into unexpected places, all of this—the unpredictable, the unknown—balanced with absolute formal control with the predictable, the known. These essential hallmarks of the movement are all enunciated at its very opening. As one follows the course of the movement one will encounter again and again the following

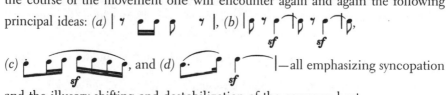

principal ideas: (a) | ♪ ... (b) | ... (c) ... and (d) ... |—all emphasizing syncopation and the illusory shifting and destabilization of the common beat.

Almost all conductors do well with Beethoven's tempo of ♩ = 72. Only a few of the earlier generation, like Furtwängler, Klemperer, Boult, Mengelberg, Böhm, and Erich Kleiber, tended to take slower tempos, but not by much (the slowest was Klemperer's at ♩ = 58), while some of the present maestri, like Muti, Ashkenazy, and Dohnanyi, tend to err on the fast side, again by relatively little—Muti, for example, ♩ = 78. But tempo is not a major point of discussion; other matters are. Already in the initial theme statement (mm.5–12) the *sf*'s in the cellos, basses, and brass (Ex.17) lose much of their punch and feeling of syncopation in most recorded performances, for the reason that the players do not feel the weight and pulse of the downbeats in each measure sufficiently to

Ex. 17

act as the counterpoise to the second-beat *sf*'s. Musicians and conductors some-times tend to forget that a syncopation can only sound syncopated if it is in reaction to a strongly felt beat. When a normally weak beat is stressed, that stressing can only be perceived as such when the normally strong beat is also felt in all its full strength and weight; otherwise the stressed weak beat has noth-ing to be overstressed against and is likely to sound as an ordinary strong beat. This is what happens when the aforementioned instruments play mm.5–12 as in Ex. 18, that is, with a quick decay of the sound to some substantially lesser

Ex. 18

dynamic, thereby turning the strong first beat into a weak beat. With that all sense of syncopation is gone and the strong beat has simply shifted over to the second beat.

In order for Beethoven's reiterated syncopations to work, it is necessary for the players to do two things: (1) allow no diminuendo from the *sf* second beat to the next 2/4 downbeat, in other words to maintain the basic *ff*, the *sf* being merely a more pronounced, more aggressive attack on top of the *ff*; (2) stop the note on the first beat, although tied, with an audible, perceptible, and rhythmi-cally precise release—not some vague indiscernible disappearance of the note. The only way I can think of notating this is as follows:

etc. The sustaining and clear release of the sound are more difficult to achieve when both the stressed weak beat and the strong first beat are the same pitch. When the pitch changes on the release note, as in mm.11–12 or mm.13–20[19] in the cellos and basses, it automatically becomes more audible—as listening to any recording of this passage will attest.[20]

---

19. The bar count I have used in the Finale is based on the principle that first and second endings in the many repeats in this movement are designated by the same measure number. Thus the first ending is m.12a, the second ending is m.12b. The next phrase thus starts with m.13, not m.14. The first and second endings at the end of the exposition are designated mm.122a-26a and mm.122b-26b.

20. This is a problem encountered in a wide range of musical repertory. I cite only a very few examples: Beethoven *Leonore* Overture N° 3, strings, mm.444–51; Tchaikovsky Sixth Symphony, third movement mm.265–69, 301–308; Brahms Fourth Symphony, third movement mm.11–15.

But Beethoven in many places in the Finale is not content to have one syncopation per measure. In mm.5–19 and many other subsequent passages he superimposes not only another syncopation but another *order* of syncopation. Having already included syncopation at the quarter-note level, Beethoven adds syncopation at the eighth-note level as well (in the woodwinds). This results in a double syncopation, the second one being a syncopation within and against the other.

In simplified notation the result is: etc.

Here too, the woodwinds must produce clear audible releases on their tied-over notes, especially when pitches stay the same (as in the flute parts in mm.5–12). Unfortunately, hardly anybody seems to be aware of this most extraordinary effect of double two-level syncopation; for on all but a few recordings (Carlos Kleiber, Harnoncourt, Cantelli, Bernstein, Kletzki, Jochum, Leinsdorf, Dorati) the woodwinds are either virtually or totally inaudible. So much for respect of Beethoven's wonderful idea!

Equally wonderful are Beethoven's viola parts in this movement—viola parts which, however, are hardly ever heard because they are un- or underappreciated by conductors (who are conducting the melody anyway, as if that were necessary), or because timpani parts, played too loudly, drown out what the generally very hard-working violas are playing. Look at the wonderful viola lines, the exquisite note choices, in passages like mm.5–16 and mm.155–62.

I am, after some fifty years of listening to the Beethoven Seventh, pretty tired of hearing only the high horns in mm.26–29. As a first horn myself years ago I remember how much I enjoyed playing those wonderful high A horn parts, but I also recall realizing that I was in those four measures part of a ten-piece wind choir. It is tiresome to hear in recording after recording very little or no woodwinds, that is until the first flute pops out on its high A in m.28. Similarly, it is tiresome in recording after recording to hear only first violins—no seconds and violas—in mm.33–36.

Articulation problems, similar to those in the Seventh's first movement, arise in much of the Finale. Beethoven is absolutely consequent and logical about differentiating between ♩. ♫ . ♪ and ♩ 𝄾 ♫ 𝄾 ♪. Yet these phrasing distinctions are for the most part ignored, conductors making arbitrary choices which are often the exact opposite of what Beethoven notated. Measures 53–62 should not be played ♩ 𝄾 ♫ 𝄾 ♪ (especially with a reverse bowing), as one is forced to hear on the vast majority of recordings. Beethoven's intention here is to play on the contrast between the short, incisive staccato winds and timpani on the one hand and the more sustained strings on the other.

Two notational errors are embedded in the conventionally used parts, one of them is wrong in the score as well. In m.63 the first violins' quarter-note F♯ should obviously be an eighth-note (as it is in Beethoven's autograph). All the winds and remaining strings have an eighth-note and it is somehow ludicrous and bizarre for conductors to express their fidelity to the score in the one place where there is an obvious error, having in the meantime paid very little attention to a thousand other Beethoven markings. A glance at m.284, the corresponding place in the recapitulation, where the notation *is* correct, might have been useful. It is truly astonishing to me that intelligent world-famous conductors—like Karajan, Jochum, Abbado, Kubelik, Casals, Reiner, Dorati, and Gardiner—insist on or allow the long quarter-note F#.[21] (One might also then ask the question why not make the D in m.284 long as well?)

The other error is in the second horn part in mm.288–89 in which

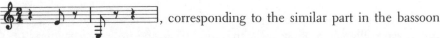, corresponding to the similar part in the bassoon

in mm.68–69, is missing. But it is an error which was discovered decades ago[22] and is nowadays almost always played correctly. Some of the conductors of today who don't seem to know about the missing second horn part are Maazel, Harnoncourt, and the Collegium Aureum.

Caution must be used in mm.64–74 and all similar places to not let Beetho-

it does in a number of recordings, such as Cantelli's, and the Collegium Aureum's.

A fierce balance problem arises in mm.105–22 (and the corresponding place, mm.329–44), once again involving timpani parts. As Ex. 19 shows, the bass and timpani parts criss-cross continually. The timpani's two notes, A and E, not only

Ex. 19

Vc., Cb., Bns. = stems down; Timp. = stems up

21. In Fricsay's recording with the Berlin Philharmonic, a performance ruined to a large extent by a constantly excessively loud and booming timpani, one cannot hear the violins' F# (or for that matter any of the orchestra) at all, as the timpanist releases a *ffff* cannon shot here, the likes of which I have never heard on recordings.

    A close contender for the recording most ruined by a timpani player is Barenboim's with the same orchestra—probably the same timpanist.

22. These horn notes are missing from some of the earlier conductors' recordings, for example Stokowski's 1928 Philadelphia, Mengelberg's Concertgebouw, Ansermet's Suisse Romande, Walter's Columbia Symphony, Furtwängler's Berlin Philharmonic, and Cantelli's NBC Symphony performances. In Klemperer's 1950s Philharmonia recording the maestro, having by then learned of the mistake in the horn part, attempts to correct it, but fixes it incorrectly: the two missing horn notes now coming one bar too late which, when dutifully repeated two more times, causes the elimination of the two-horn octave in mm.293–94.

do not correspond at any time to the bass notes but serve different harmonic *functions* in the entire chord progression: in mm.105–106, 108, 110 the timpani's A's are the sixth in a C diminished chord; in mm.107, 109, 111 the E's are the third in a C♯ minor chord; in m.112 the A is the third in an F♯ minor chord; in m.113 the E is the third in a C♯ minor 6/4 chord, and so on. Clearly, if the timpani plays its fullest *ff*, followed by seven cannonading *sf*'s, it will either overpower or at least obscure the all-important bass line, thus also obscuring Beethoven's remarkable harmonic progression here. How tremendous this passage can sound when the timpanist blends into the harmonic fabric, rather than performing a 'timpani concerto,' can be heard above all on Dorati's excellent London Symphony recording, but also on Thomas's, Steinberg's, and Toscanini's (New York Philharmonic) recordings. How horrendous this passage can sound can be sampled on the recordings of Barenboim, Walter, and Mengelberg.

It is a serious misrepresentation of Beethoven's score to slow the tempo at m.130 or to stretch the held notes here with fermatas, as so many conductors do. Can they not realize that, after some 280 measures(!) (counting the repeats) of more or less relentless, fast rhythmic activity, mostly sixteenth-notes, Beethoven has in mm.130–33 already slowed down the motion of the music by a factor of twelve, or at least six? Additional slowing down, stretching and distorting the tempo, is totally unnecessary and unwanted. Beethoven could certainly have put a fermata over m.131 and m.133 if that was what he wished. Big offenders in this regard are Furtwängler, Weingartner, Klemperer, Reiner, Mengelberg, Barenboim, Thomas, Colin Davis, Ashkenazy, Mehta, and Harnoncourt, with the worst interpretations coming from Stokowski (huge fermatas *and* vulgar slides) and Abbado (who adds a whole measure at m.131 and m.139), drastically recomposing Beethoven's closely structured music. Strauss, who often spoke out against this particular type of tempo distortion, is guilty of it himself in his 1926 Berlin recording. Anyone interested in savoring how this passage *can* sound when played in tempo should hear any of the following: the two Kleibers (father and son), Kubelik, Casals, Masur, Böhm, Dohnanyi, Solti, Steinberg, and, above all, Toscanini and Dorati. Karajan's 'interpretation' is perhaps the most curious and perverse of all in that he speeds ahead in m.127 beyond his already fast basic tempo of ♩ = 76, but then as irrationally suddenly slows down at m.138.

The development section of the Finale is one of Beethoven's most awesome creations, but—again, alas—rarely given its full due in performance. The reader may recall that I referred earlier to what I called Beethoven's double syncopations at the beginning of the movement. They return now, not only in the 'false' recapitulations in mm.127, 134 and 147, but they play an important developmental role beginning at m.163, where Beethoven once again, as is his wont, turns the original passage literally upside down. The bass line of mm.5–12 is now in effect put high up (four octaves higher) in the first violins, while the erstwhile high woodwinds' *sf*'s are now dropped down to the cellos and basses (see Ex. 20). Even more ingenious—and startling to discover—is that the lower

Ex. 20

part is a rhythmically exact retrograde of the upper part (♪♩♩ is the exact reverse of ♩ ♩♪). Incidentally, the *sf*'s in the cellos and basses in mm.163–64 must be continued all the way through m.195. They were inadvertently left out by Beethoven or assumed in context to be continued.

Two of the four elementary thematic cells I pointed out at the beginning of the discussion of the Finale | ⅞ 𝄽♩♩𝄽 ⅞ | and |♩ ⅞ ♩ ♩♩ ⅞ ♩| are here

given a developmental workout: they are both of equal importance, thematically interlocked as one can see not only in the exposition but here in the development section. Yet conductor after conductor demotes the ♩♩♩ figure to a secondary status in mm.167–74 and, even beyond that, in the first violins in mm.175–96. This is again a serious misinterpretation for the following obvious —one would think—reasons. That three-note unsyncopated figure, along with the second violin and viola figurations, is the only stabilizing metric element—the rhythmic anchor as it were—against and from which the off-beat *sf* syncopations can rebound. Second, it is also an important harmonic determinant (again in conjunction with the second violins and violas) in mm.163–70, clearly defining the four tonalities in progression here: C – F – B♭ – C. This harmonic function becomes even more pronounced some measures later where the reiterated D's of the first violins (along with the trumpets) are the single stabilizing element against the constantly and rapidly shifting harmonic stations. The figure's importance increases even more when it moves to the pitch E, the dominant of the tonic key, presaging the arrival of the recapitulation (which is, however, avoided by a deceptive cadence). Beyond that, as Ex. 21 shows, the

Ex. 21

repeated insistent E's remain a constant, while the intervallic gap between them and the gradually rising 'bass line' is diminished to a minor second, converging finally in the semitone (D#/E) clashes in mm.194–98. The three-note figure's primacy is further confirmed in a quasi-recapitulation (like a faint echo of the movement's opening) in mm.200–203, in the remote evasive key of F major. The real recapitulation finally arrives in m.222 after many false foretokens, detours, and delays. All that I have discussed regarding the exposition should obviously apply here as well.

When a recapitulation has run its course, most composers would have gone directly to a coda of some kind. But Beethoven, as we have already seen in the first movement of the Fifth Symphony, is always intent on extending and developing his materials further, what we have called a *Schlußdurchführing*. Accordingly in m.351 another deceptive return to the opening of the movement is quickly diverted into a grand elaboration of what constitutes the second subject of the movement, originally heard in m.37–52. Whereas in its initial appearance Beethoven holds to the tonic key and its relative minor (F# minor), here Beethoven moves to B minor and from that somewhat remote tonal position unleashes an extraordinary spinning out of the second subject's elements, including the five-note figure cited at the beginning as one of the four elementary motivic cells (p.269, illus. c). These whirling fragments are set against a descending bass line—I call it 'the grand descent'—of the most extraordinary boldness and originality. Nothing like it was heard in music again until half a century later in some of Wagner's late operas, where similar chromatic progressions occasionally occur.

Beethoven's 'grand descent' starts in the tonic key of A (m.374)—see Ex. 22— and moves through some fifty (!) harmonic positions to the dominant E, but

Ex. 21

Bbm E° Bbm Gb Bbm E° Bbm F   Am   F   Am   F   Am D#° Am   F

Am D#°   Am D#°   Am D#°

then oscillates back and forth between E and D# for another *twenty* bars, the D#'s grinding hard against the prevailing tonic/dominant harmonies. The entire passage is marked with reminding *f*'s and certainly does not display even a hint of any diminuendo. Yet great numbers of conductors make varying degrees of decrescendo around m.376 in order to build again, some twenty bars later, to a huge *ff* climax (at m.415). This is not only one of the most obvious self-indulgent, audience-pandering ploys, but constitutes a particularly gross distortion of one of Beethoven's grandest inspirations.

May it also be said that m.415 is *not* the 'arrival point', the 'climactic release' that so many conductors make of this, because they happen to see a *ff* there, forgetting that the music has been—or is supposed to have been—*ff* all along. The real climactic peak is in m.437. But unfortunately with too many conductors, who, seeing m.415 as the peak and then have nothing more to give, the glorious climax at m.437 becomes an anti-climax. The only appropriate way of rendering this magnificent passage is to maintain, especially in the cellos and basses, the utmost intensity throughout, never letting up for even a split-second. In that way it becomes an overwhelming experience. A number of recordings achieve this, most notably and magnificently Toscanini's, Dorati's, Bernstein's, Masur's, Jochum's, Abbado's, Gardiner's, and Carlos Keiber's.

A final caution about the timpani part: at mm.415–18 the cellos and basses keep grinding away on their low E/D# pedal point (and again five bars later, mm.423–26). These last measures of the 'grand descent' must not be drowned out by the timpani, something that happens, unfortunately, on at least two-thirds of the recordings sampled.

There remain only some lesser 'housekeeping' details to be mentioned. The descending violin sixteenth-note passages, mm.427–36, must be played in a continuous unrelenting *ff*, not with accents or *ffmp*'s on the head note of every two-bar phrase, as one usually hears here. Also, as tumultuous as these climactic moments of the Finale are—Beethoven has demanded a continuous *ff* from the orchestra ever since m.329, some 108 measures in all, one and a half *minutes* in duration—some energy and projection must be kept in reserve to achieve the triple forte in m.437. Beethoven used this dynamic very sparingly in his symphonies; thus, when it is used, it is meant to produce a very special, dramatic, all-shattering, towering climax. Many conductors and orchestras give their all much too early and have nothing left for the *fff*. In addition the second trumpet must be urged to play a very strong low D in mm.439–40 to form a balanced triad with the other trumpet and the horns ♪♩. Some conductors have at-

tempted to 'improve' and update Beethoven's orchestration by putting the second trumpet's D, an octave higher; or, worse, the second trumpet's G, two bars
earlier, an octave *lower*. Beethoven could have made the former choice himself,
but didn't; the latter he could not have made, but he knew that very well and
arranged his voice leading accordingly, taking that 'limitation' into full consideration—with the result we now see in his score. Indeed, it is that second trumpet's fantastic leap—down an octave and a fourth—which when well projected
adds so immeasurably to the overpowering effect of this climax.

In mm.441–44, I have never heard the winds of the orchestra play those four
half-notes fully sustained. One invariably hears one or the other of the following

interpretations: ♩ ' |♩ ' |♩ ' |♩ ' | or

|♩·· ♪ |♩·· ♪ |♩·· ♪ |♩·· ♪ | or

♩· ♪ |♩· ♪ |♩· ♪ |♩· ♪ |, as if *sf* means shortening the duration
of the applicable note.[23]

We come now to the final fifteen measures of the movement. As in mm.25–
28 mentioned earlier, here too, in mm.462–68, I am so tired of hearing in the
vast majority of performances only the two horns or even worse only the first
trumpet; or alternatively only the timpani. Once again there is an ensemble of
ten winds (twelve counting the trumpets) that should be heard as a unit, as a
balanced ensemble. Even more important is that the violins should not be covered here, for they have, after all, the most important thematic/motivic material
of the entire movement, that sixteenth-note cell (*c*) that I cited at the beginning
of the movement. On all but half a dozen recordings the violins are quite inaudible—and one might as well just forget about the descending fourths and fifths

---

23. Only Celibidache plays these durations correctly. One of the most grievous sins committed here
by any conductor is the recomposing of the horn parts by Mengelberg in mm.441–44 (and 457–60).

Instead of Beethoven's wonderful and unusual

he demands the very ordinary cliché, used in many 'horn fifths' situations

 . First of all, Beethoven could have written those

'horn fifths'—they were available on the A-horns of the day—and the fact that he chose not to do
so ought to be reason enough not to tamper with or rewrite the passage. Second, Mengelberg did
not seem to know that according to the basic rules of classic voice leading and counterpoint in a
first inversion harmony, the third, already strongly represented in the bass, should not be doubled
in the upper structure of the chord. If Beethoven had not learned that rule in his youth, he is
certainly likely to have gotten it from Albrechtsberger and Salieri. It was incumbent upon Mengelberg not to break that rule, especially in a work by the master, Beethoven.

of the cellos and basses. These players might as well have gone home before the end of the recording session.

There are two perversions—'cheap tricks' is perhaps a more appropriate term—that many conductors indulge in towards the end of the Seventh's Finale. One is an acceleration of the tempo, always a sure-fire way to bring an audience to its feet shouting bravos; the other is to let the timpani take over completely with an enormous final crescendo, especially in the last five bars. Previn's timpanist, for example, 'kills' the rest of the orchestra—a technical recording triumph to the hi-finatics, but artistically an offense and a gross corruption. The last-minute speeding up—some conductors start gunning the tempo as early as m.427 (Fricsay, Dohnanyi)—is, as I say, the ultimate distortion and cheapening of Beethoven's monumental Finale ending. The conductors who seemed the least able to resist this childish temptation are Ansermet and Harnoncourt (up to $\textit{d}$ = 80); Furtwängler (up to $\textit{d}$ = 84); Dohnanyi (up to $\textit{d}$ = 86); even Kletzki, generally a tasteful, disciplined conductor (up to $\textit{d}$ = 88); Stokowski (up to $\textit{d}$ = 92); and Barenboim (up to $\textit{d}$ = 96).

Any reader who would like to hear how fantastic the Finale's climactic ending can sound, without any of the above exhibitionistic shenanigans, need only listen to the recordings of Carlos Kleiber, Gardiner, Masur, Dorati (London Symphony), Reiner, and Toscanini (New York Philharmonic). The last-named's recording of the Seventh, except for a few minor foibles here and there, is a remarkable document, especially for its time (1936); and its last movement, especially its ending, is an absolute triumph of a performance.

Brahms: First Symphony

If Beethoven and the best composers of his time were remarkably precise and careful in the notation of their musical creations, Brahms was even more explicit, more detailed and exacting, if only for the reason that musical notation in general by the second half of the 19th century had become more explicit, more refined and sophisticated. This was inevitable as the increasingly complex demands of music required additional notational devices and terminology. Also, composers by and large had for some time felt the need to protect themselves against the vagaries and indulgences of their interpreters, and had invented notational techniques with which they expected to achieve that goal.

Brahms was not only a meticulous worker, constantly self-critical and self-examining, but also at heart a strict classicist. For all the new complexity and modernity of his musical language, especially in his symphonies, Brahms's music develops from the classical lineage and indeed more often than not adheres to a basically classical conception—precise, succinct, clear in form and continuity. The fact that we don't often hear Brahms played that way is unfortunate and is, of course, not Brahms's fault. The opinion held in many quarters that Brahms's music is heavy and turgid, rather square, and even 'academic,' exists primarily because most *performances* of his music are 'heavy' and 'turgid,' emotionally overladen, indulgent in Romantic exaggerations and distortions which do severe damage to the music.

The irony in all this is that it took musicians, orchestras, conductors many decades to learn how to play Brahms's symphonies with technical ease and control—Brahms's music is still, even today, among the most difficult of the entire 19th-century repertory, especially in regard to harmonic understanding, intonation, and certain particularly Brahmsian rhythmic/metric problems. But, once fully accepted into the repertory, his orchestral music was then artistically suffocated by an excess of misinterpretations and blatant disregard of his scores and their wonderfully explicit notations. Very few early 20th-century conductors rec-

279

ognized or appreciated the classical clarity and balance in Brahms's music, its absolute integrity and coherence of musical construction. Two rare exceptions were Toscanini and Weingartner, who intuitively cut through all the previously acquired interpretational encrustations to the real heart and essence of the music as envisioned in *the score*. Because Toscanini was more celebrated for his volatile Beethoven interpretations and his devotion to both Wagner and Verdi, his service to Brahms went virtually unrecognized and is, I think, even today generally little appreciated. Later remarkable Brahms conductors, more or less in the Toscanini lineage—Reiner, Haitink, Suitner, Skrowaczewski, Carlos Kleiber—are fine representatives of how an inspired, imaginative 'interpretation' can be developed from an exacting reading of the score. One would certainly not consider Stokowski a conductor in the Toscanini mold, given to textual fidelity and an innate respect for the composer's notations. But among the hundreds of recordings Stokowski made in a long career, many of which were wildly revisionist and disrespectful of the composer's intentions, his very early (1927) recording of the Brahms First, apart from one or two eccentric aberrations, stands out as a paragon of musical sensibility and taste. Stokowski seems to have had a special affinity and love for this work, and his intelligent, respectful, loving approach, expressed through the sumptuous sounds only he could elicit from the Philadelphia Orchestra, makes his recording a very special and much to be admired document.

How explicit Brahms could be in his notation—a few exceptional anomalies and ambiguities notwithstanding—we will learn fully to appreciate in the ensuing perusal of the First Symphony and an analysis of its many recordings. Once again, this traversal of the work will emphasize those aspects that are most pertinent to the conductor and to achieving a faithful, relevant, and inspired performance-realization.

The symphony opens with one of the grandest, most profoundly moving and overwhelming passages in all the symphonic literature. But like almost all great moments in music, this opening contains hidden performance problems and interpretational questions. It also contains what many consider a terminological ambiguity: the sole tempo indication of *un poco sostenuto*. It is indeed one of the rarer tempo designations, having been used (to my knowledge) only a handful of times previously, most notably in the opening of Beethoven's Seventh Symphony (but there backed up by a metronome marking). The term as such has, of course, no explicit tempo denotation; it simply means "somewhat sustained" (literally, a little sustained) but offers no indication as to *which* tempo is to be played in a sustained manner. That Brahms was averse to using metronome indications is, of course, well known, and in this case that does leave the tempo question quite open to speculation.

Many conductors have simply assumed that *sostenuto* here means something slow, but again without any specific agreement on *how* slow. And indeed recorded performances show that the tempo variances can run all the way from $\eighth = 70$ (Abbado) to $\eighth = 72$ (Klemperer, Furtwängler) to $\eighth = 100$ (Szell,

Toscanini), and lately, Norrington $\flat$ = 114 ($\downarrow$. = 38). It seems to have occurred to very few conductors that Brahms may have intended a very specific
metrical relationship—what are nowadays called "metric modulations"—between the introduction and the body of the *Allegro*, namely, the eighth of the
opening 6/8 equaling the dotted quarter in the *Allegro* (i.e.$\leftarrow\flat$ = $\downarrow$.$\rightarrow$). What
lends credence to such a notion is the fact that, as a result of Brahms's studies
of Haydn, Mozart, and Beethoven, he was all his life an ardent devotee of classical symmetry and purity of form. Precise proportional tempo relationships between slower introductions, and the main *allegros* abound in the classical literature, e.g. Haydn's Symphonies No.101 and No.103, Mozart's *Linz* Symphony
(No.36), Beethoven's First and Fourth), to provide but a few of many, many
examples. We also know that the introduction of Brahms's First was an afterthought, after he had already composed most of the *Allegro*, undoubtedly then
instinctively relating the two tempos. For all these reasons it is most likely—for
me it is a given—that Brahms intended the tempo of the opening to correspond
in some direct relationship to the *Allegro* at m.38.[1]

This concept still does not, of course, tell us *which* tempo is to be taken, but
I think it helps us to establish at least a range within which the two tempos
can—in Bruno Walter's phrase—"allow for technical exactness," as well as permitting "the musical meaning and the emotional significance" of the music to
project "to best effect ." Too fast an *allegro* and the music is drained of its
meaning and may even lead to 'technical *in*exactness' and become more or less
unplayable. By the same token, too slow and ponderous an opening 6/8 can
produce a similarly negative effect, draining it of its "emotional significance"
and perhaps even making the long line of the first eight measures virtually 'un-
sustainable.' A tempo of around 92, give or take a few points, seems to me to
satisfy most happily both sections: a bright, lively, energetic *allegro* and a not-
too-ponderous, overly notey opening 6/8. As for the latter, let us remember that
Brahms includes the word "poco" in *un poco sostenuto*. Imagine that he had
written *molto sostenuto*; how different in feeling the opening would be. But the
point is that he wrote *poco* and as performers we ought to respect that. (But see
pp. 316–17 for a quite different alternative to this tempo question.)

But why did Brahms not give a proper tempo designation, such as *andante*
or *allegro?* We shall probably never know the answer to that question, given that
the autograph to the symphony's first movement appears to be irrevocably lost.
It might have contained some indication, some clue, some first and second
thoughts, some notational alterations and changes—these are often more revealing in many of Brahms's autographs than the final editorial determinations—as
to the composer's decisional process in withholding a specific tempo marking. I
can only conjecture that Brahms, having written the main *Allegro* and having

1. It is also a given for several other conductors, most notably Weingartner, Toscanini, Karajan,
Szell, as well as for David Epstein, conductor-composer theorist, who in his invaluable "Brahms and
the Mechanisms of Motion" contained in *Brahms Studies* (Oxford, 1986) has shown how not only
the introduction and body of the first movement but all four movements and their subdivisions
can—and according to him *should*—relate metronomically (pp. 212–13).

arrived at (or already composed) the *meno allegro* coda, felt that the movement needed an introduction, as a counterbalance to the coda—a formal framing, as it were, of the body of the movement—but in his excitement and, for all we know some haste, forgot to specify a tempo marking on the assumption that it would be clear to everyone that his retroactive *poco sostenuto* would imply a slightly more sustained version of the main *allegro*. If we assume that as a possible explanation and hypothesis, it would mean that the tempo of the introduction's eighth-notes were meant to be only slightly slower than those of the *Allegro*, meaning in turn that the introductory 6/8 would be felt in 'two', not in 'six.'

But what about the term *sostenuto* itself? What does Brahms mean? Does it connote only the intended feeling and character of the music, or does it have a tempo implication as well? It is my conviction that it refers to both tempo and character, with the emphasis, however, more on feeling and character, because, in my judgment, as I have suggested above, Brahms assumed the tempo of the introduction to be determined in relation to the *Allegro*. Moreover the *sostenuto* character of the music is re-emphasized and confirmed quite specifically in the first measure of the score, most particularly in the bass and contrabassoon parts (adding the word *pesante* (heavy) to the bass part). There Brahms adds a tie

underneath each measure's six eighth-notes, , a notational

character that in itself signifies unequivocally the full sustaining of the notes. (Hear how magnificently Stokowski [in his 1927 recording] elicits from his Philadelphians, especially the basses, an astonishing singing *sostenuto*, at the same time avoiding the usual overplaying of the timpani part.) Further confirmation can be found in the word "legato" in the woodwinds and strings, an admonition not really necessary in view of Brahms's long slurs in all these parts. But obviously he wanted to achieve the utmost in a sustained *legato* line, by the way not an entirely easy thing to do in the winds since, unlike the strings, they have to breathe, thus automatically interrupting the sustained phrase lines here and there.

In this context it is clear to me that the timpani part should also be played in a sustained way, not in the horrific pounding manner one almost always hears. We should note in this connection that the prevailing dynamic, inclusive of the timpani, is *f*—not *ff*. This is of considerable importance since the recapitulation at m.25 of this opening phrase is marked *ff*. Alas, most recordings do not achieve the intended difference in intensity, because the opening eight measures of the symphony are in most cases already played so loudly that no contrasting *ff* intensification can be attained. Add to this fact that the louder one plays the opening, the more bow changes in the strings and disruptive breathing in the winds become necessary. Note also that Brahms's *ff* re-rendering of the opening at m.25 is only half the length of the original phrase—surely not an accident. Its shortened form allows the passage to be played at full intensity, 'in one breath,' as it were, perhaps not literally but figuratively speaking certainly. Conductors who managed to keep the timpani from pounding *ff* are very few but all the more worth mentioning: Stokowski (1927), Horenstein, Suitner, Furtwängler, Celibidache, Skrowaczewski, Leinsdorf, and Toscanini.

Returning to the opening measures, an extraordinary example of achieving that grand *sostenuto/legato* line I speak of, can be heard on Furtwängler's Vienna Philharmonic recording. Furtwängler, the master of the 'grand line,' could within any given tempo produce an almost incredible stretch, the ultimate *sostenuto*, and at his best, without any sense of dragging the tempo, even in his famous 'slow tempos.' In his just mentioned recording, the first eight measures of the symphony are played with a depth of sound and an inexorable beat that make this passage a truly overwhelming and sublime listening experience, a striking example of how a remarkable conductor can on occasion with certain special gifts be utterly convincing, even in what by some other criteria might be considered a 'wrong tempo.'

If inexorableness of tempo and beat is a performance requisite in these opening measures, Ozawa in his recording certainly misses the point completely. Starting with an excellent tempo of ♪ = 90, he seems unable to hold it, accelerating considerably by the fourth measure and rushes even further into m.8, only to make a big compensatory ritard in the 9/8 of m.9.

One of the most fascinating and original aspects of Brahms's opening idea, so rich in interpretational options and—therefore—obligations, is the horizontal layering of two magnificently spun-out lines over the sustained stationary pedal point. Brahms, ever the great contrapuntist, designs two majestic lines in contrary motion, the one—in the winds and violas—descending (primarily) and harmonized in thirds and sixths, the other—in the violins and cellos—rising and in lean three-octave unisons. The descending wind line is also initially cast in three octaves but is expanded to four octaves in the third measure (Ex.1). Already this contrapuntal strand takes considerable care on the part of the con-

Ex. 1

ductor and the wind players to balance, and to balance consistently throughout its eight measures. The flutes tend to disappear during the second measure and pop out again, as it were, in the middle of the third measure as they regain a more favorably projecting register. In the meantime the bassoons are almost never heard in most performances, especially in the third through fifth measures where their line dips down into a less projecting range. (The same applies to the clarinets in the fourth through sixth measures.) All of this can be easily balanced by the players adjusting and compensating for their less projecting registers: the flutes, for example, maintaining their *f* level by a subtly compensating crescendo in m.2 (and again in the downward octave jump in m.6), the bassoons compensating in mm.3–5. All this careful balancing comes to naught, however, if the timpani is allowed to pound away with an all-obliterating *ff* or *fff*. It also comes to naught when the third and fourth horns blast in with an over-blown *f* in m.7, totally unaware that here they are part of the woodwind section. Sadly, such dynamic balancing is almost never undertaken—in my sampling of fifty-odd recordings of the Brahms First only a few conductors seemed to have been aware of this problem and at least *tried* to deal with it (Stokowski, Toscanini, Walter, Boult, Skrowaczewski, Haitink, and Suitner).

But the internal balancing of the woodwinds, two horns, and violas is just one-third of the problem in achieving an over-all orchestral balance in mm. 1–8. There is the matter of the three-octave string line in ascending motion and, of course, as the third element, the C pedal point. Generally the three string lines (Ex.2) balance well among themselves, but in many orchestras

Ex. 2

and performances—especially with conductors who continually emphasize, in fact, conduct only the strings at the expense of the winds—the strings overpower and overbalance the wind/viola line (except usually for the horns). When that happens—and it happens all too often—Brahms's wonderful contrary-motion counterpoint goes out the window. The loss is considerable, because now both the exciting *harmonic* tensions Brahms has built into his two—actually three— lines (see Ex.3a) and the extraordinary sensation of the two contrary-motion lines are all essentially precluded. The rich and wholly new polyphony achieved here by Brahms can best be appreciated in condensed reduction (Ex.3a). Note

Ex. 3a

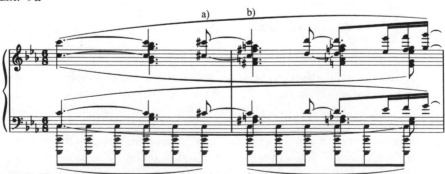

Ex. 3b

particularly the remarkably powerful harmonic tension (dissonance) points, marked a) through e) in Exx.3a and 3b, harmonic tensions that are woefully underplayed or ignored in a vast majority of performances.[2] Only Stokowski,

2. Over a lifetime of listening to countless performances and recordings of the great masterpieces of the Baroque, classical, and Romantic literature (from Bach and Rameau to Brahms and Tchaikovsky) I have noticed the distressing habit of an overwhelming majority of conductors, performers, soloists—whoever—of suppressing or ignoring the dissonances and harmonic/chromatic tensions in the music. It is as if they are afraid of these dissonances—assuming they are even aware of them—even though such dissonances are almost always resolved, and are afraid of offending the ears of

among the earlier conductors, seemed to understand (and hear) the almost radical harmonic invention contained in Brahms's masterful melodic-contrapuntal construction, while among present-day conductors only a few (Celibidache, Sanderling, Suitner, Norrington) seem to be aware of the importance of these harmonic tension points.

A further example of Brahms's masterful contrapuntal construction, combined with the utmost economy of means, can be seen in the very first four bars. In keeping with the contrary motion of the two lines, the third and fourth bars of the violin line are a pitch retrograde of the first *two* bars of the flute (G-Ab-A-Bb in the violins, Bb-A-Ab-G in the flute). The wonder of such melodic manipulations is that Brahms does this so naturally, so effortlessly—and I would say, so unostentatiously—that one might hear this passage many times, reveling in its wonderfully simple tunefulness and emotional outpouring, without realizing or hearing that it is also a remarkable example of the musical mind and intellect at work.

It may be clear by now that it is not entirely easy—let alone automatic—to realize the full intentions and implications of Brahms's majestic and wholly original First Symphony 'prologue.' Indeed, it probably can only be achieved by taking the music apart, so to speak, dividing it into its separate components. For, once the musicians hear the three strands of music individually, they will quickly hear and understand how the dynamic compensations and internal as well as over-all balances need to be adjusted. Thus, having the woodwinds, violas, trumpets, and third and fourth horns play mm.1 to 9 alone, correcting all those balances and making the necessary register adjustments, then having the violins and cellos play their lines alone, followed by putting both strands together again, will in all likelihood produce a well-balanced polyphonic composite. If in the meantime the timpani player will have listened to the foregoing 'sectional' rehearsing, he or she will, it is hopeed, also have understood how and why Brahms's wonderful texture ought not to be drowned out by a barrage of pounding timpani fusillades; and wonder of wonders, one may then perhaps even realize the possibility of actually hearing the contrabassoon.[3]

For all the reasons given earlier in this study, there should not be any ritardando or broadening in m.8—as so many conductors somehow feel the need to do. Brahms's crescendo is sufficient to achieve the desired result of a phrase climax on the downbeat of m.9.

In m.9 we encounter for the first time in this study one of the two absolutely most original ideas Brahms contributed to the development of music: the shift-

---

their audiences. This is as true of conductors and orchestras playing Bach or Mozart or Beethoven, music laden with powerful, pungent dissonances, as it is of pianists playing Chopin, for example, whose richly chromatic writing is almost always suppressed and concealed in favor of the simpler melodic and harmonic elements. No wonder such conductors cannot deal with 20th-century or atonal music, when they can't even handle 'dissonances' in earlier music.

3. As in the case of the Beethoven Fifth recordings, the contrabassoon is, with but a few rare moments, totally inaudible on all Brahms First recordings.

ing or dislocation of phrases rhythmically from their expected metric placement
to other positions, very often displaced by one beat.[4] The other major innova-
tional concept Brahms developed—it also became a virtual obsession with
him—was the rhythmic juxtaposition of triple and duple rhythms, either in horizontal succession or vertical simultaneity.[5]

In mm.9–18 (Ex.4a), Brahms moves his motivic material, which with any
other composer (except perhaps Beethoven) would have been written simply as
in Ex.4b, one eighth-beat later in the measure. This and all similar passages—

Ex. 4a

Ex. 4b

not only in m.13 but in the many phrases shifted a half bar (mm.51–52,53–56,
59–63,63–68,145–52,291–320)—are usually played as in Ex.4c (♩ denoting a
stressed beat, ⌣ an unstressed beat): in other words, as if Brahms had written
the passage as in Ex.4b. But the fact is that Brahms wrote it as in Ex.4a, by
which he meant to achieve a quite different effect than that which results from
playing it in the ordinary version (Ex.4b). We must *hear* the rhythmic dislocation

Ex. 4c

in performance, a kind of 'syncopation' of the phrase, which we will definitely
not hear if we simply turn the second eighths of m.9 and m.10 into downbeats,
stressed beats, and thereby 'underweight' the first and fourth eighths in mm.9

4. In this respect Brahms was a forerunner of Charles Ives, whose fondness for dis- and mis-placing
themes, melodies, rhythmic figures, whole phrases, is by now legendary. By the same token, Brahms
was not, of course, the first to explore metric/rhythmic beat-shifting. See, for example, the opening
measures of the Finale of Beethoven's Seventh Symphony. Brahms, however, took hold of this idea
and, in an almost obsessive way, made it into one of his principal *modi operandi* in his symphonic
and chamber music.
5. For much more on this, see also the succeeding discussion of Brahms's Fourth Symphony.

and 10, which logically are stressed beats inherent in a 6/8 meter. It is inconceivable that Brahms, at once the most rigorous classicist and the most radical rhythmic and metric innovator[6] of the entire 19th century—except perhaps for Strauss in his *Till Eulenspiegel* (see discussion of same herein, p.430)—would have notated such passages as he did if he meant them to sound simply as in Ex.4b. Rather, the classicist in him would have wanted to preserve the integrity and symmetry of the basic meter (a pendulant 6/8 in this case); while the rhythmic innovator in him wanted to introduce a conflicting rhythmic force, a counterweight that would stand in constant contention with the underlying rhythmic/metric pulse, but would nonetheless remain in equipoise with it, neither overwhelming the basic pulse nor being subjugated by it.[7] In that conception a more precisely defined notation would be the obverse of Ex.4c, namely Ex.4d.

Ex. 4d

6. There is a passage in the Brahms *Fourth Symphony* (first movement, m.130–32) that is as modern and complex as anything in Stravinsky's *Sacre du Printemps*, but which unfortunately has to my knowledge been played correctly (or even very close to correctly) on only one recording: Skrowaczewski's with the Halle Orchestra. I *have* heard it played correctly on a tape of a live performance by the Harvard-Radcliffe Orchestra, conducted by James Yannatos, but never otherwise.

7. It is analogous to the situation of a hemiola, the overlay of three binary entities (2/4,2/8,2/2) on two ternary ones (3/4,3/8,3/2), which I always describe as a contest or a battle between 'two' and 'three', but a battle which neither side wins. Both rhythmic feelings, the 'two' and the 'three,' must be equally represented, equally felt and heard, as, for example Spanish and Mexican musicians do as second nature in their native music. Unfortunately in most performances of European music containing hemiolas—from Beethoven to Dvořák and Brahms, and including the Spaniards and the 'French Spaniards,' Chabrier and Ravel, *all* of whose works are rich in hemiolas—the 'two's' always win out, and the underlying pulse and meter are constantly sacrificed. Take, for example, Dvořák's famous *Slavonic Dance No.1* in C major. The first eight measures are almost always played, even by Czech, Slowak, and Bohemian-tradition conductors as if Dvořák had written merely a series of twelve 2/4 bars, when, of course, he wrote eight 3/4 measures with an overlay of 2/4 phrasing/

articulation patterns: 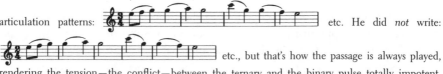 etc. He did *not* write:

etc., but that's how the passage is always played, rendering the tension—the conflict—between the ternary and the binary pulse totally impotent. The correct rendition—and this would apply to many *hundreds* of similar passages in the classical and Romantic orchestral and chamber literature—is to make clear *both* rhythmic/metric patterns, the 'three' and the 'two.' By *feeling* both, adhering to Dvořák's phrasing (and bowing) and at the same time not losing the underlying stress (the weight) in the second and fourth bars of the 3/4 meter, the true essence of the music is imparted.

One more, slightly different but related, example: and let me stay with Dvořák. In his *New World Symphony*, there are passages—very often given to the horns—which involve a 3-against-2 or 2-against-3 (i.e hemiola) situation. In the third movement (mm.253–56) and several very similar

I see in this an exact parallel in the realm of rhythm and meter to Brahms's profound commitment to counterpoint and polyphony. His metric/rhythmic displacements and unexpected syncopations produce a kind of 'rhythmic/metric/agogic counterpoint.' It is the vertical analogue to the horizontal aspect of linear counterpoint. Just as two or more contrapuntal lines inherently produce to some degree or other a conflict and tension between those lines, remaining nonetheless in a balanced equilibrium, so Brahms's displaced rhythms and phrases produce a conflict and tension between rhythm and meter, remaining nonetheless also in a balanced equilibrium—or so they should.

To achieve what I believe is Brahms's true purpose, to produce that slightly uneasy feeling and rhythmic tension in m.9 and m.10—coming after, let us remind ourselves, fifty-one steady, regular, conventionally stressed beats in mm.1–8—all the conductor has to tell the musicians is to feel the fourth and first eighths in those measures as stressed or strong beats, and to underplay (or at least not unduly stress) the second and fifth eighths. The second and fifth eighths must be felt and heard as gentle, subtle syncopations coming off the first

passages we see [music]. What we usually hear—indeed almost always—is [music], a notation Dvořák *could* have written but didn't. Simply by being aware of the fact that the underlying metric infrastructure is a triple meter (3/4) and giving the appropriate weight (stress, not accents) to the second and fourth measures, the true musical/rhythmic essence of the passage will be honored, rather than a binary falsification.

That horn motive first appears in various notational guises in the symphony's first movement introduction and *Allegro molto*. In the latter we see—again in the horns—

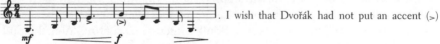

. I wish that Dvořák had not put an accent (>)

on the fourth note, because it leads musicians even more to shaping the passage erroneously, as

follows: . Again, preserving—feeling—the integrity

of the underlying 2/4 will guarantee the right interpretation in which the two competing rhythmic feelings will both be honored and heard.

Staying with this theme one more time, one of its last appearances—again in the horn(s)—occurs in m 267 of the last movement.

. What one invariably hears is

as if Dvořák had written [music] etc. Gone is the pull of the syncopated-against-the-underlying-pulse feeling that Dvořák so ingeniously conceived. (Note too that in such misinterpretations, the eigth-note rests in mm.268–69 are also thrown to the winds.)

and fourth stronger beats, instead of the usual bland rendition, which, if no-
tated, would look as follows:  and

 Just thinking and feeling the

'weight' of the first and fourth beats, even when tied over (as in the winds),
will produce the desired effect of both rhythmic essences being fully, correctly
represented and respected.

Some 'intellectuals' and 'theorists,' intent on modernizing and updating
Brahms, will want to argue that what Brahms really had in mind was an asymmetrical contouring of the phrase as perhaps in Ex.5. They will also argue that

Ex. 5

these asymmetrical patterns are in retrospect all the more 'exciting' and 'logical'
when Brahms eventually 'straightens out' the rhythms in conformity with the basic
meter as in mm.19–37, letting the rhythms, so to speak, back into the metric fold.
But this is imposing a completely alien viewpoint and aesthetic on Brahms's art.
For all the daring and relative complexity in his music, he never broke out of the
bounds of his classical orientation. Indeed, a metric rethinking of the passage as
shown above (Ex.5) is much less interesting, much less 'exciting' than Brahms's
highly original—and much more subtle and less obvious—conception.

In my sampling of nearly fifty recordings of the Brahms First there were only
a few conductors who seemed to be aware of the rhythmic/metric issues just
raised: Stokowski, Walter, Klemperer, Boult, Horenstein, Haitink, Abbado, Suitner, and Skrowaczewski.[8]

It is only in recent times that Brahms's remarkable rhythmic/metric innovations have begun to be assessed and understood, however, alas, not by any of
the major Brahms conductors but by a few musicologist/theorists, most notably
Walter Frisch and David Epstein.[9] But even here, while both writers have singled out and analyzed from a conceptual point of view some of Brahms's most
radical metrical displacements, both in his chamber music and his symphonies
and songs, neither has dealt with the question of *how to perform*—how to properly realize—these metrical/rhythmic dislocations.

8. Giulini and Rowicki really distort this passage beyond all recognition by delaying the second-beat
A♭ (in m.9), Giulini by *a whole eighth-note beat*.
9. See Frisch's "The Shifting Bar Line: Metrical Displacement in Brahms" and Epstein's "Brahms
and the Mechanisms of Motion: The Composition of Performance", both in the aforementioned
Brahms Studies (Oxford, 1990).

I am quite convinced that Brahms's metrical/rhythmic displacements are not to be merely admired intellectually, analyzed conceptually, but are in fact to be made audible in performance. I must digress temporarily to deal with this point, for to my knowledge it has never been examined in writing, let alone explored in performance.

First to really understand fully what Brahms's rhythmic innovations in the realm of 'metrical displacement' comprise, first we must remind ourselves of two essential points: one, that in the classical European tradition there exists in any metric design (4/4 or 3/4 or even 2/4) a hierarchy of weights and stresses. (It is incorrect to call these 'accents'; perhaps 'emphases' or 'stresses' are the better and more neutral terms.) In that hierarchy, beat 'one' in a 4/4 measure has the strongest weight and emphasis, followed in declining order by beats 'three,' 'two,' and 'four.'[10] That was the basis of all classical metric conceptions, a tradition which Brahms inherited and strongly adhered to, and which he saw as the un-shakable foundation of all musical structuring.

Also, we should realize that it makes absolutely no sense for Brahms to have written his many phrase displacements if he had not meant something more special, something different, than merely having the phrase appear one beat earlier (or later, or whatever the degree of displacement.) And he was surely not interested in merely composing a series of uneven, irregular, asymmetrical subphrases, delineated in diverse metric durations. This is proven by at least two facts: *First* that he did *not* write a series of irregular meters, say, a sequence of 4/4 - 5/4 - 4/4 - 4/4, as he might have done (but didn't) in the first four bars of the slow movement of his Second Symphony (see Ex.6). The 5/4 meter as a rhyth-

Ex. 6

10. We saw in the chapter on Beethoven's Fifth Symphony that this applies as well to four-measure groupings—what Schenker called "Viertaktigkeit" and David Epstein has called the "hyper-measure"—substituting only the word 'measure' for the word "beat."

mic notation was certainly available to him by 1877; it was both a technical/
notational and conceptual option for him. But changing meters was not what
interested Brahms, for that would only produce a horizontal linear effect,
whereas what Brahms was fascinated with and constantly striving for was the
vertical (therefore polyphonic) interplay of lines and rhythms. He could *not*
have found in 1877 a metric design to accommodate a passage like the one in
the development section of the Second Symphony's first movement (mm.135–
152, see Ex.7a). That is to say, if he had wanted the three strands of polyphony
to be felt and heard in a metric design corresponding to the notated phrase slurs
and ties—and as they are always played—then he would have had to have written the passage in polymetric notation (as, for example, in Ex.7b). But this was
not a technical/notational option available to him.

Ex. 7a   (mm.136–40)

Ex. 7b

Second, if we take into account the harmonic structuring in this passage, and try somehow to justify an irregular metric sequence (as, for instance, in Ex.7b) in relation to its harmonic sequence (i.e. tonic and other degree positions), we still will find little congruence of meter and harmony, for these elements are operating on their own separate time tracks. The significant harmonic stations, i.e. the main audible harmonic stress points, as composed by Brahms, are also out of phase with the metric structuring. Furthermore, to unanchor the passage even more, to make it even more harmonically ambivalent, Brahms introduces the A# in the strings' melodic lines, where most ordinary composers would have used an A♮, the A# in this context materially undermining the prevalent E major tonality.

In the end, we have to rule out all of these other possibilities, because all the internal evidence offered by the many passages of metrical displacement scattered throughout Brahms's works demonstrates that what Brahms was after was to create a tension, a tug of war, as it were, between the actual heard rhythm and phrase, and the underlying metric pulse. In addition, the harmonic design might often also be in a state of tension—and contention—with both the rhythmic and metric design, creating, in other words, three lines of polyphonic tension.

When these passages are played in such a way as to preserve the integrity and feeling of the underlying metric structuring, that is, 4/4 in the slow movement of the Second Symphony, 3/4 in the first movement, the music is transformed into something totally different: imaginative, original, and exciting, not ordinary, routine, and prosaic, as one usually hears it.

I cite one more relatively simple but nonetheless very striking Brahmsian example of rhythmic shifting, because it is so prototypical (countless similar examples can be found in many other works of Brahms). It comes from Brahms's Horn Trio, Op. 40, in the very opening of the work, where the chordal piano accompaniment to, first, the violin, and then, the horn, is for 16 measures consistently set on the second beat (in a 2/4 meter). The confusing and unsettled feeling this theme statement generates—all the more unsettling and confusing when it is played incorrectly—results from the fact that the single notes of the violin (and horn) on the downbeats are opposed by the weightier six- or seven-part quarter-note chords Brahms has placed on the second, presumably weaker, beats. But if Brahms wanted the music to sound as it is almost always rendered (as in Ex.8a), then surely Brahms would have written it as in Ex. 8b. But he

Ex. 8a

(piano part simplified, reduced to one note)

Ex. 8b

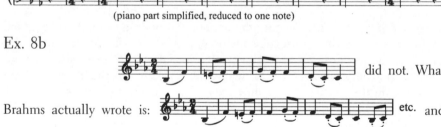

did not. What Brahms actually wrote is: etc. and

therefore by no logic known to me should this theme be 'interpreted' in such a simplistic way as in Ex.8a or Ex.8b. It is relatively easy to achieve the correct subtle balance of weights and stresses by having the violinist—and later the hornist—gently emphasize the downbeat feeling (and it is enough to think it and feel it), while at the same time having the pianist play the second beat chords with a light 'lifted' touch, feeling the passage as a constant unresolved chain of anacruses. These piano chords should not 'sit'—and certainly not heavily—but should 'float.' With just that much subtle adjusting and reorientation on the part of the three players, a whole new wondrous rhythmic/expressive world opens up, one that, once heard, is truly unforgettable. In such a rendering without any tremendous effort or rewriting, *both* Brahms's 2/4 pulse *and* his rhythmic dislocations are honored and made movingly audible.

I go so far as to suggest that the terms 'metrical ambiguity' and 'harmonic ambiguity,' as used by Frisch, Epstein, and other writers, are incorrect, that is, they do not apply to Brahms's music. 'Ambivalent,' yes, but 'ambiguous,' no. There is to my eyes and ears nothing ambiguous about Brahms's rhythmic displacements, nor is the term 'metrical displacement' entirely correct or applicable. It is not that the metric design is displaced or dislocated; it is the *rhythmic* shapes, phrases, and patterns which are dislocated *over* an inviolate and immutable regular metric sequence. It is the tension(s) between and among those contending forces that Brahms wishes to express, and this does, of course, result in a degree of ambivalence. But ambivalence suggests a condition in which one or more options, one or more understandings, one or more feelings, co-exist co-equally, and which therefore must be interpreted and realized co-equally. This in turn means that two opposing rhythmic/metric forces are to be heard and made audible in intimate, inseparable equipoise. It is in this way that Brahms so ingeniously and imaginatively managed to cling to earlier classical traditions while at the same time dramatically renewing and re-interpreting those traditions in wholly new expressive ways.

I am not generally given to hermeneutic interpretations, to superimposing onto music—least of all Brahms's music—extraneous, extra-musical scenarios. But it is fascinating to speculate that the ambivalence one finds in so much of his music is in itself a reflection of the varied and myriad ambivalences which we know by all accounts were an intrinsic aspect of his personality, manifestations of which ran as a constant thread throughout his entire life. The stories documenting Brahms's ambivalence of feelings, of commitments, in making decisions, are legion.[11] I cite only two which are especially fascinating and, I think, especially relevant to matters musical. It is well known by now that Brahms late in life agreed to make a cylinder recording for the Edison Company. An eyewitness account of the event reports that at the appointed time for the recording, Brahms, in a very agitated mood, at first refused to participate, but a short time later, suddenly sat down at the piano, impatiently wanting to begin recording, when all the technical preparations necessary for making the recording had not yet been completed.[12]

A second striking example of Brahmsian ambivalence is his obsessive desire on the one hand to achieve positions of prominence in Hamburg and Vienna, but on the other hand, when accorded such positions (director of the Wiener Singakademie or the venerable Gesellschaft der Musikfreunde, for example), he resigned soon after accepting the appointments. Similar examples, both in his personal and professional life, abound, especially in his attitude towards marriage and female companionship.

---

11. See any major Brahms biography (Geiringer, Kalbeck, Schauffler, Specht) but especially Karl Geiringer's article "Brahms the Ambivalent" in *Brahms Newsletter*, Vol. I, No. 2, (Autumn 1983).

12. I am indebted for this account to George Bozarth, who in turn drew upon the memoirs (*Klänge um Brahms*) by Richard Fellinger, the son of Brahms's Viennese friends at whose home the Edison recording was made in 1889. (See "Brahms on Record," in *Brahms Newsletter*, Vol. VI, (Spring 1987), p.5.)

Since great creative artists reflect in their art their personalities and character traits intimately—the greater the artist the more directly and intimately are these reflected—is it not possible to think that Brahms's penchant for structural elision, evasive modulations, harmonic/metric incongruence, disguised cadences—his entire arsenal of rhythmic/metric permutations and shiftings—are all in subtly varied ways direct reflections of his mental and emotional attitudes? The miracle of Brahms's art is that, although he was in much of his life unable to reconcile the contradictions in his character and attitudes in his personal life, in his music he was able to achieve a remarkable symbiosis of opposites and conflicts, of old and new, of traditional and innovative—the "classical Romanticist," as Wagner called him.

Brahms's metric/rhythmic ambivalence is just one aspect of his rhythmic innovations. While this concept of rhythmic-to-metric counterpoint is based on a regular, steady underlying pulse, Brahms achieves rhythmic and continuity flexibility—speed modulations, as it were—by frequently (throughout his work) changing the amount of rhythmic activity within the bar and/or within the beat. At such times the music completely changes in character—without the basic pulse itself being altered. (For a striking example, see Ex.12 in the discussion of the Brahms Fourth and the surrounding commentary. See also footnote 46 below, p. 356).

By these means Brahms achieves fluctuations and modifications of motion *without* resorting to ritards and accelerandos. The motion (speed) of the music remains under his control by virtue of the steadiness of the underlying pulse, beat, and meter. But this means in turn that we as interpreters may *not* willy-nilly change the basic tempo, and, as I've said before, *not* slow down the basic tempo when Brahms has already decreased the amount of rhythmic activity. This is especially mandatory when the underlying pulse happens to be silent; then the audible surface of the music has changed (slowed) while the underlying pulse and feeling of the music remains constant, and remains under Brahms's control—or so it should, and would if conductors could learn this basic lesson. This also explains why Brahms has so few ritardondos and accelerandos in his music, especially his symphonies.

To return to the Brahms First musical examples cited earlier, it is quite possible—indeed, readily so with a little extra rehearsal effort—to do justice to and render audible both the rhythmic dislocations and their underlying metric pulse. Anyone who has not heard the remarkable expressive effect that results from a balanced, equipoised realization of these opposing, even colliding, rhythmic/metric and harmonic forces probably cannot imagine how exciting, how powerful—and how right—such realizations sound. Unfortunately the reader will find it difficult to hear such on recordings, since there are very few conductors recording Brahms symphonies who have understood this fundamental premise of Brahms's art, and therefore very few recordings on which its rightness can be assessed or, for that matter, argued—and even among those few there are none which are entirely consistent in rendering this aspect of Brahms's art unerringly.

The next interpretation/performance problem in the symphony's first movement occurs in m.15, where most conductors allow (or elicit) a premature and

exaggerated crescendo, reaching the *f* of m.19 in m.17, thereby destroying
Brahms's intended asymmetrical arching shape (Fig. 1).

Care must be taken that the crescendo/diminuendo wedges (< >) in

Fig. 1

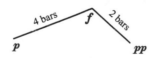

mm.15–16 not be exaggerated *and* that the diminuendo half of this expressive
nuancing returns the dynamic to *p*. What usually happens instead might be
rendered in notation as (Ex.9). This is, of course, once again part of a

Ex. 9

much larger widespread and long-standing 'interpretational' problem, namely,
the world-wide epidemic of making premature and/or exaggerated crescendos,
basically rendering all real climaxes anti-climaxes, undercutting their real intended impact. It seems to be typical of human nature—musicians and conductors included—to be generally unable to resist these temptations to overstate.
And yet, I know from many personal experiences that if a conductor insists
firmly on controlling the pacing and curvature of crescendos (the same can be
said about diminuendos)—also controlling his own baton/technical pacing—
such nuances can be rendered accurately, feelingly and, in the end, much more
effectively than is typically the case.

As a 20th-century composer composing in an atonal, highly chromatic language, and being deeply aware of the innovations of my creative precursors,
among them Brahms and Schönberg (the latter being as much influenced by
the former as he was by Wagner), I am fascinated by the wide-ranging chromaticism of the melodic line between m.9 and m.19—only one of hundreds of such
passages in Brahms's *oeuvre*. As shown in Ex.10, all twelve pitches of the chromatic scale are touched, not in absolute succession, of course, as in a twelve-
tone or serial work, but nonetheless clearly heading in such a direction: eleven
pitches in all, the only pitch not represented being A.

Ex. 10

Similarly, harmonically the passage moves around chromatically in a way that
one would not find, for example, in Beethoven or Schubert or even Schumann,
but which again presages what early 20th-century composers like Reger and
Schönberg—not to mention Brahms's arch rival, Wagner—did as a matter of

Fig. 2

m.9/10	m.11	m.12	m.13/14
G^{-9}	D^{b7} G^b	D^0 B^{b7} D^0 Cm	G D^b ^{0} (implied C^{-9})

m.15	m.16	m.17	m.18	m.19
G^{b7} C^b	G^0 E^{b7} Fm	A^0 F^7 Gm	B^0 A^b D^0 G^7 A^b	D^b

course (Fig. 2). Nearly twenty different chords—harmonic stations—are passed through in the short span of only ten measures. Particularly striking is the abrupt dramatic move from G major to D♭ diminished (with the implied C dominant minor ninth), the modernity of which (for its time) has been lost to most contemporary ears with over-familiarity.

Almost everyone gets the wonderful magical moment at m.21 right—the *pp* filled with a sense of mystery—but trouble starts again at m.23, where too many conductors accelerate the tempo (also indulging in a premature crescendo), only to slow up dramatically in m.24, before the *ff* of m.25. The worst of these unwanted and unnecessary ritardandos were perpetrated by Muti, Giulini, and Bernstein. Oddly enough, going quite against the normal trend, Klemperer and Stokowski actually accelerated the tempo in m.24, arriving at a suddenly faster tempo in m.25. How effective the passage can be without ritardando or accelerando can be heard on the recordings of Furtwängler (surprise!), Toscanini, Horenstein, Skrowaczewski, and Walter.

At mm.25–28, a shortened and slightly re-orchestrated recapitulation of the symphony's opening, now situated on the dominant of the base key, we encounter another one of Brahms's most daring harmonic utterances. I mean the clash of the pedal point G with the C♯ minor harmony at the beginning of m.26, more powerful and terrifying than the parallel place in m.2—more powerful because this time the trumpets are brought in to sustain the G in the upper range against the alien C♯ minor (Ex.11). Unfortunately, very few conduc-

Ex. 11

tors and orchestras are even aware of this moment, let alone exploit it. It is important not to let the trumpets diminuendo on their sustained G's. How extraordinary and overwhelming in its expressive power this passage can sound can at least be savored in Järvi's recording with the London Symphony, Suitner's with the Berlin Staatskapelle, Lehel's with the Budapest Philharmonic, and Furtwängler's with the Vienna Philharmonic. Let us also note in passing Brahms's unusual crossing of lines at m.27 in the violas, second violins, and cellos, the latter two suddenly dropping *down* an octave, the violas contrarily jumping *up* an octave. Norrington ruins this magnificent four-bar passage entirely by adding—God knows whence he got this idea—a big diminuendo in the strings

 in m.26 (thereby totally vitiating

the C♯ minor/G♮ harmonic clash), followed in quick succession by a swooping crescendo, a *p subito* and *another* crescendo, all of these not even remotely indicated (or intended) by Brahms.

The rest of Brahms's introduction presents no particular performance problems, and is generally played well, *comme il faut*, except perhaps for the final two bars (m.36–37), where many conductors make a ritardando—often an excessive one. It is better to keep these bars in tempo, particularly if one is intent on showing a tempo relation between the introduction and the main *Allegro* part of the movement (as discussed and suggested earlier). Many conductors also make such an excessive diminuendo in m.36–37 that the resolution to G major in the final measure of the introduction is virtually inaudible. Brahms's dynamic is *pp* in m.34, and, of course, he could easily have asked for a further diminuendo and marked the final notes *ppp*. But he did not, and thus it is incumbent on conductors and performers not to fade away to virtual silence for some obvious 'audience effect.' For as they do so, they destroy the link to the main *Allegro* which in turn vitiates the whole point of the introduction: it doesn't 'introduce' the main body of the movement, but ends up being an independent closed movement, disconnected from the whole.

After the gentle resignation of the final introductory bars, the ensuing *ff allegro* (m.38–41) bursts upon our ears with a power and explosiveness that must have been a real shocker in Brahms's own time.[13] One can readily imagine why Brahms on second thought felt this burst of *ff* was too abrupt and abrasive a start for his symphony and thus added the introduction. It is a dramatic call to action, and what remarkable actions and activities it now summons forth!

If we didn't know the introduction was composed after the *Allegro* had been finished, we would assume the rising chromatic motive at m.38

 to be a variant of the movement's first two bars in

the violins and cellos (mm.1–2). The reverse is, of course, the case.

13. As explosive and powerful as this passage is, care should be taken that the two horns not overpower the woodwinds, but rather blend with them. Far too many performances (and recordings) are marred here by overly rambunctious modern hornists who, seeing only the *ff* in their parts and paying no mind to the fact that they are playing with woodwinds, enter the fray with a (misplaced and mindless) vengeance.

Performance problems abound here, mostly unsolved or misinterpreted, in the great majority of sampled recordings. Take, for instance, the progression from *f* through *piu f* to *ff*—so typically Beethovenian—in the strings in mm. 42–51. In nine out of ten performance/recordings the strings—either permitted or urged on by their conductors—plow into the rising figure at full tilt and

 with an aggressive, hard sound—

simply continuing the previous *ff*—that, of course, permits no dynamic augmentation at m.46 and m.51. Therefore there is no growth, no intensification, parallel to the harmonic expansion, in the entire passage. It is just boringly loud, stuck on its *ff* plateau. Notably, Stokowski, Weingartner, Toscanini, van Beinum, Kondrashin, Abravanel, Jochum, Järvi, and above all, Suitner, are among the few conductors who got this passage and its subtle dynamic distinctions right.

Another, even more challenging, performance problem is the necessity of observing and dealing practically with Brahms's amazingly varied phrase-ending release notes—all of this unobserved or consciously ignored by almost all conductors. These releases come in all 'shapes and sizes', ♪ (as in m.51), ♩ (as in mm.44,52, and 171), ♩ (as in m.74 in the woodwinds or m.195), and even ♪ (m.263 in the strings) and ♪ (m.337 in the strings). Not only does Brahms constantly make these articulation distinctions throughout the movement with remarkable consistency, but he frequently differentiates with different articulations between, say, strings and woodwinds (as in mm.63, 74, and most important in mm.90 and 91). These are fascinating examples of how meticulously and precisely Brahms marked his scores, and conductors who simply willfully change or ignore these markings are conductors without any artistic integrity or musical imagination.

Even Brahms's meticulously marked scores are, of course, not without notational ambiguities and discrepancies which can confound even the most intelligent and exacting minds. For example, what does Brahms mean by the marking ? Is the B♭ to be re-articulated, or is it to be slurred into but then played staccato, i.e. short? Without being able to consult Brahms in person and without access to the autograph score we cannot be sure. But my sense is that the two notes should be slurred, primarily because that is how they appear in their original formulation in m.11. There is world-wide confusion about this phrase marking among not only composers but music editors, engravers, and publishers, with the result that no one seems to know with certainty how to interpret such notations, particularly when they can yield different interpretations between string and wind players. For the former they can have bowing as well as merely articulative or phrase implications. For the latter they can have particular tonguing implications.

I have an easy solution and suggestion to make on this point, one that I have used in my own scores for many years. It is to put the dot *between* the end of the slur and the notehead if the new note is to be separately articulated (tongued in the case of the winds) 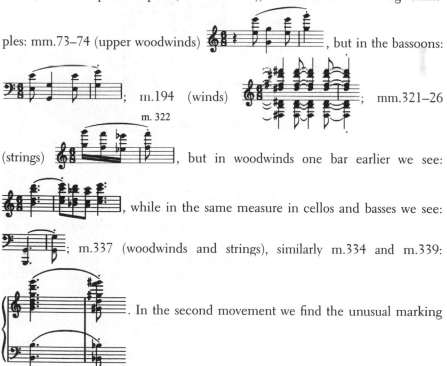, and to put the dot *above* the slur, away from the notehead, if the new note is to be slurred into .

The usually meticulously precise Brahms is inconsistent in this respect, as are most composers past and present, everybody assuming that somehow performers can deduce the right articulation from the musical context. While this is often the case, there are far too many instances in the literature where it is quite impossible to tell which articulation the composer really intended, where both basic ways—articulated or non-articulated (slurred)—are possibilities.

Not only is Brahms's autograph score inconsistent in these respects, but so are all printed scores of the symphony, more or less following Brahms's markings, and in some cases, to make matters worse, adding still different articulation variations. Staying for the moment only with Brahms's First Symphony, let me point out the following examples in which the phrasing/articulation—and in the case of the strings therefore also the bowing—are not unequivocally clear, and indeed are either interpreted variously or in opposition to Brahms's apparent intentions.

In the first movement, apart from the passage already referred to (mm. 54–66, and its parallel place, mm.355–67), consider the following exam-

ples: mm.73–74 (upper woodwinds) , but in the bassoons:

; m.194 (winds) ; mm.321–26

(strings) , but in woodwinds one bar earlier we see:

, while in the same measure in cellos and basses we see:

; m.337 (woodwinds and strings), similarly m.334 and m.339:

. In the second movement we find the unusual marking

(in the first violins) in mm.9–11 , and in mm.115–16

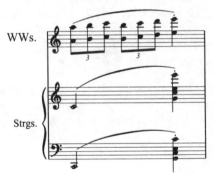

, also identically in the wood-

winds in mm.121–22. Just how confused performers are, is shown by the fact that in the majority of performances and recordings the violins re-articulate the B and A♯ in mm.115–16 (but not on Weingartner's and Celibidache's), whereas the woodwinds usually slur into those notes in mm.121–22.

In the fourth movement we see in mm.129–30 in the strings

and in mm.176, 178, 179 (and its parallel place, mm.360, 362, 363).

WWs.

Strgs.

From all this 'evidence' one can see why interpretations can—and do—vary considerably, although certain 'traditions' have evolved, with winds interpreting these notations mostly one way (slurred and short), strings variously, depending on dynamics, durational values, a concertmaster's decision, while most conductors stand by, either unable to referee the question or, in most cases, even unaware of a problem.

I do not pretend to know unequivocally what in each instance is the correct interpretation. But at least we should be aware that there is a problem here which needs to be given very serious thought. What makes no sense to me is, for example, to have the winds tongue the eighth-notes in mm.54–66 (first movement), but to slur them, given the same notation, in mm.121–22 of the second movement.

In any case, the phrasing in mm.54–66 that has been almost universally adapted is the separated, newly articulated one. In my sampling of recordings only Stokowski and Rowicki slurred the two notes, which is in my view the correct realization.

It is really bad when the violins, for example, play in mm.53–55 as follows  or, worse yet,

My suggestion for a bowing which retains the slurring I believe Brahms intended, as well as helping to preserve the integrity of the basic binary meter is:

Another not easily solvable problem is caused by the discrepancy in notation between m.51–52 in the exposition and m.352–53 in the recapitulation. In the latter instance Brahms writes ⟩ under each pair of notes, but not so in the exposition.[14] Which is right? Or are both right? It is a crucial question, not only per se, but it impinges on the larger issue raised before of how to deal with Brahms's penchant for off-beat or syncopated phrasings. For if a diminuendo is made in m.51 off the second beat, then it will be extremely difficult to maintain a sense of the underlying 6/8 meter in the next and succeeding measures.

The term "pesante" in m.60 and m.64 (meaning, of course, heavy, weighty) implies as well that the dotted quarters should be played in a very sustained manner.

In m.63 Brahms's *più f* should undoubtedly read *sempre ff*, a *più f* necessarily driving the dynamic up to *fff* instead of Brahms's *ff* in m.68.

Before we move on to the next section of the *Allegro's* exposition, let us note—and appreciate—the remarkable harmonic progression (mm.59–69), ingeniously returning us to the tonic key, surely startling in its time but rendered almost routine for modern ears by constant re-hearing (see Ex.12).

Ex. 12

14. These two measures are, of course, a variation of mm.9–10, just as the entire section from m.9 to about m.16 is brought back and reconstituted in m.51 to about m.67.

In the entire section between m.70 and m.114 the articulation questions I raised before are of paramount importance and must be rigorously observed, as opposed to the indiscriminate changing and ignoring of Brahms's distinctions. Particularly important is the retention of Brahms's clear differentiation between woodwinds and strings in mm.90–98. That these are not some accidental differences—much less something that can arbitrarily be changed at a given maestro's discretion—should be clear from the fact that Brahms's indications here are absolutely consistent and logical. The strings and brass in combination alternate with the woodwinds, so that when the former have a full quarter-note, the latter have sharply articulated short eighth-notes, and vice versa. The sustaining of the quarter-notes is especially crucial when the other instrumental choir has a dissonance against the basic chord, e.g. the woodwinds' Ab against the G chord (m.90), the C against Bb in m.92, the Eb against D major in m.94, etc. The only conductors, as far as I can tell, who are aware of these notational and harmonic distinctions are Stokowski (1927) and Celibidache.

In many editions the accent (>) for the strings' Cb in m.99 is missing. It is meant to be a slightly milder version of the *sf* in m.97.

Great care must be taken that the violins' pizzicatos in mm.105–109 be well heard, especially the lower-pitched notes. The full realization of this, once again, wondrous harmonic progression (Ex.13) depends on the correct

Ex. 13

balancing of both the pizzicatos and the wind chords which, following the first violins' rising intervals, also move upwards in fourths and fifths:

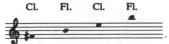

The pizzicatos of the second violins in m.114 and m.116 against the third horn's B♭'s are equally important, as are the ensuing A's in the first violins and cellos in mm.117–118. The horns' unusual fifths in m.114 and m.116 also should not be played too softly, too blandly, but instead perhaps even with a gently swinging accent, for they are the notes from which the woodwinds 'bounce off.'

In order for one of the first movement's main themes to be properly heard, appearing now in the cellos at m.121, the woodwinds and horns must not play too loud, or crescendo too much. The winds should not reach their f until m.128, when—interestingly—the strings have already begun their diminuendo. This is what the score prescribes; no recording or performance that I know of has ever observed this very special and remarkable nuance. Indeed, most conductors make the violins crescendo into m.129, then call for a big diminuendo, in the meantime often losing the F eighth-note, and thus the proper resolution to the temporary tonic of B♭.

The third episode of the exposition, mm.130–56, with its elegant oboe, clarinet, and third horn solos, looks simple enough on the face of it. But Brahms is never that simple; his harmonic and timbral (instrumental) interconnections are so imaginatively devised and, also, constantly regrouped, rearranged, recast,[15] as to provide ever-new, often subtle performance challenges. For instance, it would seem obvious that the oboe and clarinet exchanges (mm.137–41) ought to match in phrasing and dynamics. Yet, judging by the recordings, that is in fact rarely the case. Indeed, it distresses me that in these exchanges, but even more so in the clarinet and third horn exchanges in mm.148–52, the phrase emphasis seems always to be on the anacrusis notes, as if Brahms had written

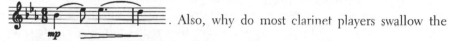 . Also, why do most clarinet players swallow the

final note D, sometimes to the point of inaudibility? (Oboists seem to follow through much more consistently.)

I think these divagations, giving the players for the moment the benefit of the doubt, are caused by the fact that, starting subtly in m.137 (2.oboe, 2.bassoon) and more obviously a few bars later (mm.142–48), Brahms shifts the accompaniment phrasings to the second beat of the bar. But that should not be a reason for the winds to change their agogic emphasis, especially since Brahms provides a substantiating crescendo (⸺) for the two anacrusis notes, clearly indicating a stress on the ensuing downbeat. In fact, it is, as I have suggested earlier, precisely the conflict between two types of rhythmic phrasings, in this case one beat apart, that fascinated Brahms so and that he worked into his music at every possible opportunity.

An additional problem that often arises in this episode is an immoderately

15. It is Brahms's extraordinary talent for permutation and variation of his musical materials (themes, motives, harmonies, instrumentations, etc.) that so impressed and influenced Schönberg, moving him in turn in the direction of two major compositional principles: constant variation and non-repetition, and thus paving the way for two of the essential traits of the Second Viennese School. See Schönberg's essay "Brahms the Progressive" in his *Style and Idea* (New York, 1950).

loud pair of horns (mm.137–42), clearly overbalancing the solo woodwinds, as one can hear on any number of recordings, notably Dohnanyi's, Rowicki's, and Boult's.

Why would any conductor change Brahms's notation—but not just the notation, the very conception, the very sound of the music—as so many conductors have done in mm.145–52? I find this a particularly grievous distortion of Brahms's intentions, especially since, as I have already shown, he was so explicit in his delineation of all manner of rhythmic durations. To change

willfulness. The conductors who indulged themselves in this particular whim are Szell, Horenstein, Abravanel, Klemperer, Munch, Boult, Rowicki, and Stokowski. I suppose that some of them thought that Brahms must have made a mistake, meaning to observe the phrasing as in a similar but not identical passage in mm.287–93. If they thought that far, they didn't think far enough, for the real parallel to mm.145–52, confirming that Brahms knew exactly what he was doing, occurs in the recapitulation in mm.418–25. To not observe Brahms's durations is obviously wrong, but easy to correct. More difficult, although certainly not impossible, is preserving the integrity of the duple meter (as extensively discussed earlier). It will suffice if each string player (in m.145–52) *feels* the downbeats in these measures, and the correct balance between the actually articulated notes and the underlying pulse will be achieved.

After this quiescent episode, Brahms returns to the *Sturm und Drang* mood of the earlier parts of the exposition. In four brief harmonically and rhythmically striking measures (mm.157–60)—a kind of stretched-out anacrusis gesture—the music rouses itself to full force in m.161. But unfortunately these four bars are rarely rendered to their fullest expressive potential. On the harmonic side, Brahms's startling, unexpected, and abrupt dominant minor ninth chords in

mm.157–58 are rendered *unstartling* and anonymous in most

performances by the suppression or underplaying of the major third, A. How something so elementary and obvious can be so consistently misrepresented defies explanation. It is not that nobody plays the A, but somehow in the pizzicato doublestops of the violins and triplestops of the cellos, the A gets lost. The point is that almost no conductor seems to hear that this important pitch—the one that not only makes it a real chord but needs to be fully sounded to balance against the 'dissonant' minor ninth (G♭)—is underrepresented. The problem is usually exacerbated by the fact that the violas play their arco notes anything but *p* (Brahms's dynamic), while the pizzicatos in the remaining strings are barely grazed, the two approaches obviously resulting in a serious imbalance. Matters are even worse in the parallel place in the recapitulation, where the all-important F♯ (in a D dominant minor-ninth chord) appears only in the first violins.

Rhythmically the passage fares not much better.[16] The cross-rhythms in mm.159–60 (and mm.433–34)  are hardly ever balanced properly, and their 6/8 feeling is usually distorted into a 3/4, as if Brahms had written [music] .

There are very few problems after that, except to note that the thematic material beginning with [music] (m.161) has staccato dots while the lower line in violas, cellos, basses, and bassoons has none. When the lines are reversed nine bars later, the same respective articulations are retained. This simply means that at m.161 the lower line (and at m.169 the upper line) should be played in a slightly more *tenuto* manner. Nonetheless this difference in marking has confounded many conductors who have edited the violins' parts, for example, variously as  or

[music] . The only conductors who have made Brahms's articulation distinctions clear are Stokowski, Celibidache, Dohnanyi, Jochum, and Abbado.

But a much worse sin is generally committed by conductors and their strings in m.180 and m.184—I should add by conductors who either (a) can't read a score correctly; (b) do not trust the score or Brahms; (c) assume that Brahms

16. Unbelievably, Horenstein turns m.157 and m.158 into 9/8 measures, adding a whole extra beat of silence at the end of the bar!

must have made a mistake; or (d) who do not hear that what is being played doesn't correspond to what is in the score. I'm referring to the unusual but very ingenious differentiation in articulation and duration between the string (and brass) parts and the woodwind parts. The last have staccato quarter-notes (♩̇), the others have dotted quarter-notes (♩˙): a significant difference. Despite this clear distinction, on only four recordings—those of Sanderling, Jochum, Abbado, and Klemperer—do the strings hold their notes at full value. Really curious is the quite illogical interpretation by a number of conductors (Karajan, Furtwängler, Szell, Jochum, Horenstein, Giulini, Leinsdorf, Janowski, van Beinum) in which m.180 is played *tenuto* but m.184 is not, and the even more illogical one by Wand, namely, m.180 short, m.184 long!

Following this thunderous peroration (in E♭ minor) and a quick move to B major,[17] we find ourselves quite suddenly in the development section—starting actually in m.189. Two warnings: (1) it is not at all necessary or desirable to slow down in and around m.196—not even for the ten-bar 'descent' in m.215— and (2) a half-hearted *p* and *pp* in m.197 (and 205) will not do, for it will not achieve the air of mystery mixed with a suppressed tension that is so unique to this passage. How extraordinary this passage can sound when played with a real *pp sempre* (Brahms's precise marking) and a very quiet, introspective *p* in the winds can be heard to best effect on only nine recordings: Toscanini's, Karajan's, Haitink's, Jochum's, Böhm's, Chailly's, Herbig's, and above all, Suitner's and Furtwängler's. Note the way the violas surreptitiously shadow the bassoon (mm.197–204), later the flute and oboe.

The gossamer texture of these measures must be maintained at all cost when the music modulates back to C, although this time to C major (only briefly). A recurrence of the third principal theme of the exposition brings with it, in keeping with Brahms's penchant for additive and variational procedures, a series of clearly enunciated and obvious references to Beethoven, particularly the famous opening four-note motto of the Fifth Symphony. For some forty-odd measures Brahms hammers home Beethoven's motive, and yet, as clear and obvious as this is, it is astonishing how few conductors cause(d) or allow(ed) these refer-

---

17. In earlier times, no one ever made the repeat via the first ending. Of late, however, under the general pressure of the trend towards 'historically informed authenticism,' a number of conductors have recorded the first movement with the repeat. For me there is, I must confess, a certain awkwardness in the way Brahms returns to m.38, an abruptness in the harmonic progression which I can't quite analyze. On the face of it the E♭ minor of m.185 should elide quite nicely with the E♭ diminished chord of m.189 (i.e. m.38). However, in practice the effect seems constrained, lacking Brahms's usually infallible harmonic sense. If one were ever to consider a revision of this transition passage—a fairly outrageous thought to begin with—I would offer two suggestions: (1) I believe one problem is the D♭ at the end of m.189, which works magnificently when the music moves to B major but seems less felicitous in the move back to C minor. I would suggest replacing that D♭ with a D♮ in m.189 (the first time only, of course); (2) It helps to soften the abruptness of the return to m.38 if the brass, bassoons, and timpani on the downbeat of that measure are eliminated.

ences to Beethoven, the "giant" whose "tread" Brahms constantly heard behind him and feared when writing his C minor Symphony, to become audible. Occasionally, on a few recordings, as if by accident, they will surface here and there, perhaps in the horns or trumpets. But the whole chain of repeated Beethoven quotations is clearly projected in only a few recordings: those of Toscanini, Böhm, Szell, Abravanel, Leinsdorf, Dohnanyi, Janowski, Jochum, Tennstedt, and, best of all, Skrowaczewski and Chailly.

An idea of the compactness of structure and economy of means Brahms commands, almost at the level of Beethoven's Fifth, can be gained not only by the way all the primary thematic material of the exposition is reworked— expanded, contracted, varied, inverted, reorchestrated—but also in the way Brahms exploits even secondary material for further recycling. Note, for example, how the innocent looking chromatic bass line appearing in the cellos and basses first heard in mm.265–66, then again, twice, in mm.269–72, suddenly emerges as primary melody in the violins four octaves higher in m.273. But then we see that this passage is, in addition, a veiled reference to the very opening of the symphony: over a G pedal point in gently reiterated eighth-notes, two lines are spun out contrapuntally and in contrary motion. Again, the one in the woodwinds and violas, descending and in harmony (at least in thirds), the other in violins in octave unisons rising and falling, wending its way gradually downward in a long (21–bar) diminuendo. Despite the clearly differentiated rhythms and phrasings and the unequivocally specified *continuous* diminuendo, conductors are constantly changing the rhythms, punching holes and breaks into Brahms's sustained lines where there shouldn't be any, and making crescendos, as in mm.274, 278, and 282, where there clearly aren't (and shouldn't be) any. This is mere self-indulgence and a "know-better" contempt for the score.

With this beautiful long dynamic and registral descent, we reach one of the most moving moments of the entire symphony: when, led by a darkly mysterious combination of contrabassoon, low cellos, and basses in m.293, the long ascent to the *ff* full orchestra pinnacle at m.321 begins. There are two performance problems here, however, which are roundly ignored by almost all performers of this work. One concerns dynamics, more specifically the temptation—almost always yielded to—to crescendo too much too early. Instead of climaxing with a *ff* at m.321, most performances reach that dynamic level much earlier—Ormandy, for example, at m.303, nearly twenty bars too early—thus turning the intended climax into a big anti-climax. Brahms gives us excellent clues for pacing the 37-bar crescendo in the clarinets and oboes. But the real problem is usually in the strings, particularly in the first violins, who (a) in general love to play loudly, forcefully, when on the G string, (b) who when seeing the crescendo wedges in mm.295,297,299 tend to make overbearing crescendos and (c) in the alternate measures never return to a *pp*. The same tends to happen with the cellos and basses. To keep the crescendo in check so that it is truly very gradual, it is well to add the following dynamics: *p* at m.302, *mp* at m.305, *mf* at m.309, and *f* at m.313. I would also suggest that a distinction be made in the clarinets and oboes between what is 'soloistic' and merely harmonic/accompanimental, at least in the first two entrances, i.e. clarinets' *pp* on the second beat m.298, the oboes' *p* on the second beat m.302. These instruments should track (except for their four-note 'solos') the dynamics of the two bassoons. Care should also be taken that the brass's and timpani's Beethoven reiterations be well heard and follow the same gently rising cre-scendo line.

The other problem, judging by numerous performances and recordings, is apparently even harder to resolve: namely, to prevent the beat from being turned around. This happens quite naturally unless guarded against, because there is no articulated downbeat in every alternate pair of measures in the main leading voices: for example, m.295 and m.297 in the first violins, m.294 and m.296 in the cellos, basses, and contrabassoon. This ambivalence of pulse is exacerbated by the fact that the two leading lines are in canonic imitation, one bar apart, with the result that when, for example, the bass line *does* have an articulated downbeat, the other line doesn't, and vice versa. Now surely, Brahms—ever the explorer of new rhythmic ideas—wanted here to create a degree of unsettledness but, I am convinced, not to the point of being completely disoriented metrically. It is inconceivable that Brahms would have wanted the violins, for example, to

sound  , which, incidentally,

when played that way, results just before the climax (m.321) in a disruptive bar of either 9/8 or 3/8.

The reason that the phrasing in the bass line starts on the second beat is that it is merely a transposition (a tritone down) of the top notes at the very begin-ning of the *Allegro*, mm.38–39 (Exx.14a,b). But this does not necessarily mean

Ex. 14a   m. 38

Ex. 14b   m. 293

that the *beat* should be allowed to turn around, in effect moving all the bar lines one dotted-quarter beat later. The entire passage can—and in my view—certainly should be played to preserve both sensations: the rhythmic unsettled-ness mentioned above and the regular pulse of the underlying meter, in a word 'to have our cake and eat it too.' For it is once again, as so often in Brahms, the conflict, the tension, between two opposing rhythmic/metric forces that he wants to create, and that we, as performers, must render appropriately. All that needs to be suggested to the players is to keep the 6/8 pulse in mind, to feel it, especially on the unarticulated downbeats—this might be rendered notationally as  —and the balance between the syncopated melodic line and the underlying beat will be preserved.

But we lose even more when we allow the beat to be turned around; we lose all those remarkable and for their time daring upward leaps starting in the bass instruments in mm.296–97, and in mm.307–308 in the first violins. As the crescendo mounts, these upward thrusts—like the slow initial tremors of a volcanic eruption—must become increasingly pronounced, canonically one measure apart, at the same time increasingly emphasizing the 'ones,' the downbeats, of the measures. When all of this is done properly, the climactic release at m.321 is overwhelming, because it evolves absolutely organically and is not felt as a strange unexpected rupture (or, as mentioned earlier, an anti-climax).

The powerful dissonantal clashes at the climax in most cases need to be worked out. I am speaking of the A♭'s against the G's in m.321, the clash of the violins' D against the woodwinds' C in m.322 (repeated one tone lower in m.324). But it is perhaps even harder to make the violas and cellos heard in m.322 and m.324, as they answer the violins, again canonically. The woodwinds must be cautioned to hold all their dotted quarter-notes full length to differentiate their rhythmic contour from that of the brass and timpani.

But the biggest problem, because least noticed and appreciated, is the careless habit of the strings playing instead of . One can hear this on virtually all recordings, most notably Chailly's, Barenboim's, Muti's, Herbig's, and even Furtwängler's and Toscanini's.

Brahms writes a reinforcing *ff* for the strings in m.329. All string sections, of course, play this passage *ff*, but are more often than not acoustically drowned out by timpani and winds. On only a few recordings (Walter, Toscanini, Järvi, Rowicki, Tennstedt) can the strings be heard at all. (Can't these world-famous Grammy-winning record producers hear that the strings are totally covered?)

With the magnificently ponderous bass line in m.339, we have arrived at the recapitulation. But note how ingeniously Brahms modifies its first four bars, also reversing in effect the orchestration: what was in the winds in the exposition is now in the strings, and vice versa.

As so many composers, starting with Haydn and Mozart, have done, so too does Brahms foreshorten the recapitulation. The whole section of the exposition from m.69 to m.96 is not represented in the reprise. This was partially necessitated by Brahms's need to return to C minor (whereas the parallel sections in the exposition were basically in E♭ minor). Although the material in the recapitulation is essentially the same as in the exposition, Brahms introduces many variations, some very subtle, and mostly in terms of reorchestration. Obviously, all comments and suggestions mentioned in reference to the exposition apply to the recapitulation as well.

Just as the exposition comes to a close at m.188, so now the parallel place in the recapitulation (m.461) must find a new way to continue. And how brilliant is Brahms's solution! Whereas at that juncture in the exposition Brahms moves from Eb minor to Cb major (actually enharmonically B major), in the recapitulation he moves from C minor to a C major dominant seventh in a third inver-

sion, intensely orchestrated in horns and bassoons .

Unfortunately very few conductors take the trouble to balance this chord in its six reiterations, the problem being that the note which makes it a major chord, E♮, lies rather low in only the second horn and inherently projects much less than the G in the third horn a tenth above. But, of course, the chord *can* be balanced; it just takes a little rehearsing and a pair of caring ears. Of the many recordings I heard, there were only nine where these chords were properly and excitingly balanced, and/or not drowned out by the timpani: those of Stokowski, Jochum, Skrowaczewski, Wand, Tennstedt, Abbado, Ormandy, Suitner, and Järvi. On too many recordings the horns and bassoons were allowed to play instead of . The full sustaining (and correct balancing) of these chords is what makes this passage so thrilling to hear.

In m.460 and m.464 Norrington has the bizarre idea of adding *subito p*'s (followed by immediate crescendos, of course), perversely impeding the flow and drama of this climactic passage.

In m.466 Brahms builds a kind of *stretto*, using his basic chromatic main motive which we first hear at the very opening of the *Allegro* (mm.38–40), now again transposed down a tritone and set against a contrary-motion bass line (Ex.15). The prominence Brahms gives this chromatic motive F#-G-Ab at this

Ex. 15

point, seems to me almost preordained, for he has previously used it numerous times in the movement, always at critical formal junctures. Fig. 3 displays a selection of these as they occur and re-occur variously in the movement, demonstrating Brahms's Beethovenian compactness of architectural design and economy of means.

Fig. 3

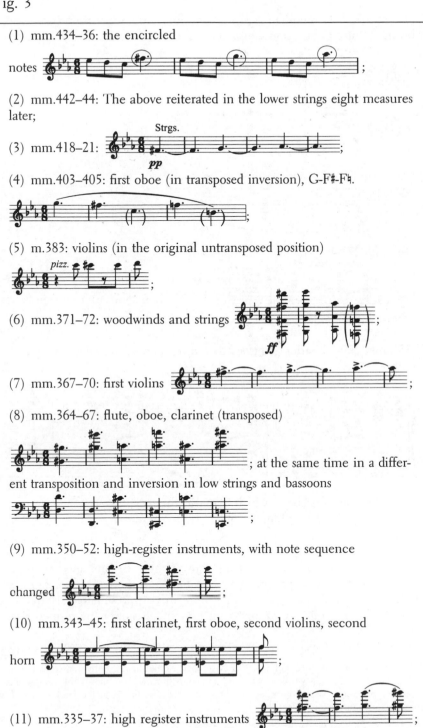

(1) mm.434–36: the encircled notes

(2) mm.442–44: The above reiterated in the lower strings eight measures later;

(3) mm.418–21:

(4) mm.403–405: first oboe (in transposed inversion), G-F#-F♮.

(5) m.383: violins (in the original untransposed position)

(6) mm.371–72: woodwinds and strings

(7) mm.367–70: first violins

(8) mm.364–67: flute, oboe, clarinet (transposed)

; at the same time in a different transposition and inversion in low strings and bassoons

(9) mm.350–52: high-register instruments, with note sequence changed

(10) mm.343–45: first clarinet, first oboe, second violins, second horn

(11) mm.335–37: high register instruments

(12) mm.339–41: simultaneously, original untransposed version (violins, horns, 2.flute, 1.oboe) and transposed retrograde as well as inversion

(13) mm.293–95: contrabassoon, cellos, basses

(14) mm.273–77: violins, a) transposed b) retrograded

; then spun out in various other transpositions;

(15) mm.229–31: strings transposed, and in canonic imitation (encircled notes)

All the examples above, which figure in the recapitulation, appear, of course, in still different transpositions in the exposition.

At m.474 and the remainder of the movement,[18] Brahms presents us with one of the most serious interpretational problems in the entire symphony. It is not clear how the performers are to get from m.474, presumably at full *allegro* tempo, to the *meno allegro* at m.495, or how much slower the *meno allegro* is to be.

Any move to a slower tempo, whether via a ritardando or not, involves *above*

18. Here many a performance and recording have been ruined by an overly boisterous timpanist. The brutality and insensitivity with which some timpanists literally slam into m.474 is staggering; in the process, of course, 'wiping out' the entire rest of the orchestra. (Hear the Janowski and Paita recordings for confirmation.)

all a firm decision as to how slow that slower tempo is to be. There are very few places in the symphonic literature where that decision is so difficult to make as in the Brahms First. *Meno allegro* is such a relative term and with no metronome indication as a clue, it is anyone's guess what Brahms may have had in mind. Ironically, matters are also made difficult by the knowledge that the movement's *un poco sostenuto* introduction, to which the *meno allegro* coda is obviously closely related, was composed *after* the main body of the movement had been completed. That information now leads to the tempting thought that the introduction and coda ought to be identical (or at least 'closely related') in tempo. Such thinking in turn leads to the possibility that both the introduction and coda should be in a *moderato* or *allegro moderato* tempo, conducted in 'two,' only moderately slower than the full *allegro*.

It is not difficult to find a rationale for such an approach. After all, at m.495 Brahms simply says *meno allegro*, i.e. less lively. He does not say *andante* or *adagio*; and, as I have already pointed out, his *un poco sostenuto* of the introduction may signify a similarly moderate divergence from the main *allegro* tempo.

Given the ambiguousness of Brahms's tempo indications here, conductors have resorted to all manner of 'solutions,' purely intuitively in most cases, probably without any particular intellectual or analytical rationale. A relatively few (Rowicki, van Beinum, Leinsdorf, Abravanel) have taken Brahms's *meno allegro* at face value, conducting it in 'two' in what might be called an *andante con moto* or *allegro molto moderato* (the dotted quarter in the metronome's upper 50s, lower 60s).[19]

It is also significant that Brahms wrote *meno allegro*, and not, for example, *più andante* or *più adagio*. In other words, he meant the tempo to stay in the *allegro* realm, with the implication that his tempo indication refers to the dotted quarter ($\downarrow\cdot$), not the eighth-note.

Another fact that lends credence to interpreting Brahms's *meno allegro* in a relatively 'lively' tempo is the fact that he does not indicate any ritardando prior to m.495. Unless we simply want to assume—without any justification or proof—that Brahms just forgot the ritardando, his *meno allegro* has to take on a quite different meaning than it has heretofore usually been accorded. Most conductors have taken a tempo of the dotted quarter between 40 and 50, with Furtwängler, Klemperer, Bernstein, Chailly, Skrowaczewski, and Horenstein even below that (at $\downarrow\cdot$ = 34, 36 and 38). The problem for all of them has been how to get from the full driving *allegro*, say, at m.474—usually around $\downarrow\cdot$ = 92—to their much, much slower *meno allegros*. And again, every possible logical and illogical option has been attempted by someone at one time or another, from an immediate sudden pulling back of the tempo at m.475 to a judicious almost imperceptible slowing over twenty bars, and many gradations in between (see below for more details).

19. If indeed some of those conductors ever thought about also taking the introduction in a similar *moderato* tempo, as a counterpart to the coda, they probably were dissuaded from doing so by the long-standing entrenched tradition of doing the opening in the familiar ponderously slow tempo, conducted in 'six.' To my knowledge only Norrington has dared to pace the opening in a fairly lively 'two.'

But what about those wonderful duples in the cellos in mm.294–93? What is their significance? In my view they are Brahms's ingenious way of gently relaxing and calming the *feeling* of the music, without appreciably disturbing the *tempo*.

David Epstein, in his his aforementioned discussion of tempo relationships between movements and within movements in Brahms's First Symphony, proposes a very interesting answer to the tempo question in the *meno allegro* by suggesting that the cellos' duple quarter-notes should equal the eighth-notes of the slower 6/8 (←♩ [or ♪.] = ♪→). In his suggested tempo range of ♩. = 96–104, the eighth-notes in m.495 would then be in the range of 192–208. Be it noted that in Epstein's suggestion there is no possible consideration of the normally adopted ritardando between m.475 and m.495.

I realize that most conductors will heatedly argue against such a 'ridiculous notion.' And the established tradition will automatically be invoked, supported by generations of world-famous maestri, to the effect that the *meno allegro* has "always been done slow, and so it should be".[20] But I invite anyone to sing through the last nine bars or so of the *Allegro* at a tempo, say, of ♩. = 92, noting the calming effect of the cellos' duples (but without slowing down), and at the *meno allegro* dropping down to, say, = ♩. 60. (Norrington takes ♩. = 62.) Any reasonable, objective, unprejudiced mind will have to admit that such a conception works very well, and is, at least, as reasonable an option as any other that has been offered through the years. It cannot just be peremptorily ruled out as a possibility, rejected out of hand.

This point of view raises, in retrospect, as I implied earlier, the fascinating possibility that Brahms meant *both* the opening of the symphony and the coda (m.495) to be in a moderately moving tempo, that is, felt and conducted in 'two,' not 'six' (or optionally, a subdivided 'two'). Again I invite the incredulous objector, horrified at the very thought, to sing through the entire introductory section at a tempo of about ♩. = 50–60. Any remotely objective person will have to admit that it *is* a possibility. Brahms's long lines flow beautifully, the integrity of the 6/8 meter and its 'two' feeling is more readily maintained, and even the lovely sequence of oboe, flute, cello (mm.29–37) gains a gently undulating, lyric quality. Admittedly, at this tempo, the introduction loses its painfully anguished, heavily tragic mood. But then whoever said—or presumes to know with certainty—that Brahms intended the movement to have this tragic ponderous quality? In all likelihood the various hermeneutic interpretations of the Brahms First's first movement as "despairing," "gloomy," "elemental," "tragic"

20. In point of fact, I doubt that anyone—certainly not anyone living today—knows when that tradition started, who started it, and even whether it was (as is so easily assumed) a tradition created by some interpreter in the early years of this symphony's existence. Did Otto Dessoff establish the tradition when he first premiered the work in Karlsruhe in 1876? Or was it Brahms himself when he conducted the symphony in later performances? Or was it Bülow? We don't know. All we do know is that by the time the first recordings were made (by Stokowski and Klemperer) in the 1920's the 'tradition' was graven in stone, and has been—I think, thoughtlessly—perpetuated ever since.

(see Tovey, for example) all came to mind as a result of performances which, erroneously, interpreted the work in that fashion. Or perhaps it was the other way around.

Taking the introduction in a slow 'two,' of course, destroys the idea of the previously suggested 3:1 metric-unit relationship (← ♪ = ♩· →) between it and the *Allegro*. At ♩· = 96–104 in the *Allegro* and ♩· = 48–52 in the *meno allegro*, the relationship would obviously be 2:1 (♩· = ♩·) which, again, is another reasonable interpretive possibility.

I do not claim definitive knowledge in regard to these tempo matters (I wish that others would also not claim such authority); I am simply suggesting some alternatives on the basis of what Brahms's score actually says, alternatives that, it seems, very few conductors have ever considered before, suggestions that have enough validity to be taken seriously, to be at least discussed and argued.

What conductors *have* done in the final 38 bars of the movement (starting at m.475) is to introduce arbitrarily various types and degrees of ritardandi, depending on how slow (again arbitrarily determined) the *meno allegro* tempo was to be. Some conductors—Klemperer (the Berlin State Opera recording of 1928), Walter, Kondrashin, Skrowaczewski, Bernstein, Boult—have started the ritardando immediately at m.475—sometimes in a gigantic jamming on of the tempo brakes—so that by m.478 (just three bars later) they are already in a substantially slower tempo. Then they have ritarded even more, arriving finally at a lugubriously slow *meno allegro*. Still others (Levine, Stokowski, van Beinum) have waited until m.478 to commence their ritard; still others (Szell, Dohnanyi) have waited even longer, ritarding only in the final eight or nine measures (ca.m.486). Some few conductors (Rowicki, Abravanel, Haitink, Toscanini, Jochum) have very gracefully and tastefully calibrated a subtle, almost imperceptible ritardando over the twenty bars (mm.475–95). But others have come up with really bizarre 'solutions,' like Järvi, who slows up at m.474, springs back into tempo in m.475, starting another ritard around m.481 and arriving finally at an *adagio* version of the *meno allegro* of ♩· = 44; or Furtwängler and Janowski, the opposite of Järvi, who *accelerate into* m.474, then pull back enormously at m.475. Still other conductors, like Tennstedt, Abbado, and Chailly pull back suddenly at m.474 *(sic)*, which makes no sense at all. But perhaps the most blatant aberration comes from Celibidache, who makes a sizable fermata (⌢) over the quarter rest of m.475. Herbig's approach is also very strange. After slowing down slightly around m.476, he makes a huge ritard starting at m.491, only to go *quasi a tempo* (♩· = 52, ♩· = 56) at the *meno allegro*. To make the ritard at m.491 is (1) to ignore the fact that there is none indicated by Brahms, (2) is to ignore completely the duplets (and their meaning) in the cello in mm.492–93, and (3) is to ignore the great probability that Brahms wanted the tempo of the last few measures before the *meno allegro* to elide almost imperceptibly into that coda.

Beyond that, minor bad habits and faults have crept into renditions of this section. For example, virtually all violin sections make diminuendos in m.481

and m.484, when it says quite explicitly *poco a poco cresc.*, and when the resolving notes G and A♮ (respectively) should be fully intoned.

I wish that more conductors and orchestra musicians would appreciate Brahms's penchant (in all his orchestral works) for an organ-like orchestration, as, for example, in mm.477–79, where horns represent an eight-foot stop, the clarinets a four-foot stop, and the flutes a two-foot stop.

During the *meno allegro* itself several problems can arise. First, there is the problem of a good balance between horns and timpani, particularly the low C of the second horn and also the *sostenuto* character of the timpani. Second, I find that often the various chromatic phrases, echoes of the very opening of the symphony, do not start *p* (see mm.495,497,499,501). Last, let me plead that the contrabassoon's low C in m.508 be well heard. (It is quite inaudible on the vast majority of recordings.)

I have no strong particular opinion as to how long the winds should hold the final C major chord. It seems to me that it works well at least two different ways, for example, the winds releasing with the strings' final pizzicato, or alternatively holding the chord a little beyond the pizzicato. Most conductors, myself included, opt for the latter interpretation, but I recall being severely chastised on one occasion by a very respected musician colleague, who claimed to *know* that the winds' chord "should never sustain beyond the 'pluck' of the strings' (and timpani's) pizzicato." He may have asserted that in response to the many conductors who hold the last chord inordinately long, e.g. Stokowski, Ormandy, Abbado. (All that has been said on this point applies equally to the final measure of the third movement.)

The second movement's major interpretational problem is one of tempo. Brahms gives us *andante sostenuto*. This seems relatively explicit to me, particularly that word 'andante,' and I therefore cannot understand why the vast majority of conductors, past and present, insist on playing this movement as an *adagio*, indeed an *adagio molto* or *adagississimo*. Some conductors (Kertesz, Rowicki, Klemperer, for example) play this movement *so* slow that the triplet eighths in m.6 feel like individual quarter-notes. In such instances the triplets lose all feeling of triplets, have no flow or line; they just sit there: stolid, stiff and heavy, unmoving—indeed unmovable. The tempo such conductors take is not on any metronome, not even on modern metronomes which usually go down only to 32.

I am, once again, not sure who started this 'tradition' of such extremely slow tempos in this movement, nor why. I could guess that, once the late 19th-century hermeneutic interpreters with their extra-musical anecdotal fantasies got hold of Brahms's First, his symphony was seen as the great tragic, anguished statement of a genius struggling not only with fate (as Beethoven was imagined to have similarly struggled in his Fifth Symphony), but with the spell of the universal canonization of the very same Beethoven. The fact that both Beethoven's Fifth and Brahms's First were in C minor became a convenient polemical premise by which Brahms could be elevated to the true heir of the throne recently vacated by Beethoven, at the same time investing him with the mantle of the great tragic, suffering, struggling Romantic artist. And such tragedy, such suffering, could, of course, only be expressed in the slowest and most anguished of tempos.

But the fact is, all anecdotes and fanciful figments aside, the second movement of Brahms's allegedly 'tragic' symphony is in E major, an untragic, warm, luminous, almost sunny key; and it is an aria, a song, sung by some of the brightest singers of the orchestra: the oboe, the clarinet, the violin, and the horn.

It will come as a shock—an act of heresy—for those who are used only to the thickly, massive, creeping, lugubrious renderings of this movement that a tempo of, say, ♩ = 52–60, the lower end of the *andante* scale, works beautifully. The music then easily achieves its essential, almost Schubertian song-like quality,[21] its long arching lines, without any loss of passion or drama, allowing even for the appearance of the occasional 'darkening cloud' along the way (for instance, mm.3,16,49,53,70).

Another major reason for not adopting the slowest possible tempo in the second movement[22] is its remarkable—better said, extraordinary—phrase construction, no aspect of which is realizable or audible when the tempo is so slow as to pull all phrases, even individual measures, into small clumps of isolated disconnected sounds. No conductor to my knowledge—even the 'greatest' Brahmsians—has encouraged an orchestra to feel and hear—and communicate—the unusual periodizations in this movement. Without an understanding of these heterogeneous arching lines, performers are reduced to merely sloshing through the notes, bar by bar, without any sense of the structural continuity, of where they are in the over-all form of the piece.

Consider this remarkable, unorthodox, and highly original structural plan (schematically represented in Fig. 4), clearly discernible from even a casual reading of the score.

The bad habits—bad traditions—and misbegotten interpretations, as well as plain ignoring or rejection of Brahms's score, are legion in this movement, and a recital of these does not give me much pleasure. But they are so insidious, so ingrained in almost all performances, that they must, I feel, be addressed and exposed.

The tempo question already referred to is, of course, the most serious problem and in my view, a real obstacle to a faithful performance (see Fig. 5 for a table of various conductors' tempo choices). For if Brahms's *andante* is observed, all

21. The composer Douglas Townsend has made the case very well in his writings—including his superb liner notes for Rowicki's recordings of the four Brahms symphonies—for the strong impact of Schubert's influence on Brahms. Townsend's reflections on the subject of Brahms's early influences are worth citing here: "Brahms's musical genealogy might read something like this: great-grandfather: Bach; paternal and maternal grandfathers: Mozart and Beethoven; uncle: Schubert; cousin: Mendelssohn; and father: Schumann." Townsend goes on to say: "Schubert's influence can be observed in many of Brahms' own melodies, which, however Brahmsian, have as their point of origin early nineteenth-century Vienna as it is expressed in the music of Schubert."

22. It occurs to me that Mahler in similar circumstances would have written 'langsam aber nicht schleppend' (slow but not dragging).

Fig. 4

Number of meas.	5	3	5	3	1	10
Meas. numbers	mm.1–5 (or 4+1)	mm.6–8	mm.9–13 (or 4+1)	mm.14–16	m.17	mm.18–27 (6+2+2)
						3 mm.1–3 (repeated)

kinds of performance aspects will fall automatically, nicely into place. Assuming, then, that a reasonable and moderate tempo is taken, there are still a variety of musical misdemeanors which have become part of the 'tradition' of performing this movement. For example, most orchestras and conductors make a tremendous crescendo in the very first measure, undoubtedly wanting to show how emotional, how profoundly expressive, they can be, as opposed to the allegedly 'cool,' overly 'intellectual' 'remote' Brahms,[23] who didn't even have the

Fig. 5

♩=28	Klemperer (1928), Giulini	♩=42	Dohnanyi, Leinsdorf,
♩=30	Bernstein, Ozawa, Tennstedt, Abbado		Kondrashin, Järvi, Walter
		♩=44	Wand
♩=34	Munch, Janowski, Böhm, Sanderling	♩=46	Karajan (Vienna)
		♩=48	Boult, Suitner (who slows
♩=36	Weingartner, van Beinum, Stokowski, Ormandy, Klemperer (1955–57), Steinberg, Jochum		to ♩=42 by m.5)
		♩=52	Norrington
♩=38	Horenstein, Furtwängler (Vienna), Rowicki (who slows to ♩=32 by m.5), Celibidache	♩=60	Toscanini, Muti (who slows to ♩=42 by m.3)
♩=40	Skrowaczewski, Levine, Szell, Abravanel, Kertesz, Chailly, Haitink, Paita, Herbig, Lehel		

23. These were indeed some of the early and immediate reactions to Brahms's First Symphony, especially, of course, by the Wagnerites of the day. The general consensus in most circles was that Brahms's new symphony, for all its skill—or perhaps because of it—was "too intellectual," "remote" (*fremdartig*), "revolting" (*abstossend*), "aloof" and other similar epithets. Early reactions to Brahms's symphonies in the United States were no different. In Boston, the critic Philip Hale, evidently speaking for many American music lovers, suggested that the doors in Symphony Hall be equipped with signs reading "Exit in case of Brahms." Harry Ellis Dickson in his memoir recalls that at the first performance of the Brahms Fourth Symphony by the Boston Symphony the work "was removed from the Saturday program, and a symphony by Schumann was substituted. [Conductor] Gericke announced to the press that the Brahms symphony was 'incomprehensible.'" (Harry Ellis Dickson, *Beating Time—A Musician's Memoir* (Boston, 1995), p. 74.)

imagination or decency to put a ‹ › in the first two bars! What such 'interpreters' forget is that Brahms was quite capable of writing such 'hairpin' crescendo-diminuendos, when and if he wanted them. Indeed, by my count, in this second movement alone there are no less than twenty instances of this particular dynamic nuance.

Someone will now undoubtedly argue the case that Brahms simply *forgot* to put a ‹ › in the first two bars. Any objective, rational mind will have to concede that that is a possibility; Brahms *could* have forgotten. But what I don't understand is how we go so quickly from this *possibility*—to my mind very remote, given Brahms's notorious fastidiousness in details of notation—to the assertion (and *absolute conviction*) that Brahms obviously *must* have forgotten this expressive nuance.[24] Why must the possible and probable become automatically the absolute, unarguable, untouchable, unalterable *sine qua non* of performing traditions? Why not explore with a little imagination, how what is presumed to be a mistake or an omission or a miscalculation could be in fact absolutely feasible, and—perish the thought—the best solution after all?

The opening of Brahms's second movement is an appropriate case in point. Its little six-note melody (mm.1–2) has a simplicity, an unadorned beauty and artless charm, that is completely destroyed when gussied up with extraneous dynamic swells and heavings, especially offensive when additionally the tempo is twice as slow as it should be. For, the slower the tempo the more painful these interpretational exaggerations become. My motto, and my admonition to orchestral players who have been taught to play this opening phrase in such a vulgarized way, is: let Brahms's beautiful notes do the talking. In their wonderful purity they don't need any additional shoring up, improving, embellishing, supplementing, interpolating; they communicate quite well enough without our pitiful 'interpreter' overlays.

The same problem is at issue in m.3, where, again, conductors and violinists want to add a presumably missing ‹ › to the phrase, justifying their decisions, I assume, by the fact that Brahms adds just such a nuance the next two times this little phrase appears (violins m.15, violas m.16). There is, of course, no substantiable logic behind such an assumption; and furthermore it precludes, again, the more interesting option of performing mm.3–4 as written (*come è scritto*, as Toscanini would say), that is, to seek out the meaning behind Brahms's avoidance of a crescendo-diminuendo, rather than blithely assuming that its absence is a 'mistake.' In fact, the meaning of this nuance-less *pp* is, it seems to me, quite clear. By whatever metaphor one may wish to describe this phrase—m.3 always evokes for me the image of a small passing cloud that for a moment partially darkens the sky—it is first and foremost a sudden softening, darkening, distancing of the music from its two-bar antecedent phrase. And just

---

24. It seems to me that such assertions and undocumented (indeed undocumentable) assumptions are at the heart of the matter of what ails most performing of the Romantic and classical repertoire. A possibility, a probability, is quickly turned into a certainty and a sacralized tradition by those who would righteously presume to know better than the composers themselves what these composers intended their scores to reveal.

as that first phrase has no ＜ ＞, so its 'echo' also doesn't—and, of course, shouldn't.

As an ex-horn-player I cannot resist commenting on the anomalous—actually incorrect, but only slightly so—marking, *gestopft* (stopped), in the horns in m.3. On the natural horn, for which Brahms always wrote despite the fact that the valved horn had come into common use as early as the 1830s—another example of his adherence to certain conservative or classic viewpoints—the written F♯ (sounding A♯ in actual pitch on the horn crooked in E) could only be produced by partially closing the bell of the horn with the hand—emphasis on "partially." For if a horn player were to completely close the bell, i.e. the true *gestopft*, the pitch that would result would be a written A♭,[25] in this case a sounding C♮, which would, of course, not fit into Brahms's F♯ dominant chord. Brahms was half right, half wrong. He knew that some degree of 'stopping' was necessary to get the A♯, but was wrong in calling it *gestopft*, which is a very specific muting technique resulting in a particular sound quite different from what Brahms had intended.[26] Partially closing the bell gets exactly the sound Brahms needed here: softer, darker, more distant. Horn players have been confused by this marking for many, many decades, but surprisingly few have given much thought to its meaning or what Brahms intended to indicate by it. Most horn players have simply played it—and continue to this day to play it—in the conventional hand-stopped manner,[27] producing quite the wrong sound: a nasal, buzzy, piercing, edgy sound which is inappropriate here. What horn players should do to render Brahms's intended effect correctly is to 'half-mute' the note, fingering a sounding B, closing the hand into the bell just enough to produce an A♯, a move that will also automatically darken and distance the sound—exactly what Brahms wanted. The other alternative—not as good—is to play m.3 open but suddenly softer(*pp*), and to darken, perhaps even slightly muffle the sound with the embouchure.

I indicated in Fig. 4 that the first five-bar phrase could also be thought of as 'four-plus-one.' The 'one,' actually 'one-and-a-third' bar is, in fact, slightly separable from the first four measures and can be thought of as either an extension of those first four bars, or a long anacrusis gesture, an extended upbeat in effect to m.6. Even more interesting is the fact that these four beats of music

 have already made an earlier ap-

pearance in this symphony, in fact several times. Incredibly, these notes are, but for the transposition a third lower and a slower tempo, exactly the same notes as in the opening of the *Allegro* (m.38–40) in the first movement. This

25. For a full explanation of this acoustic phenomenon and its technical realization, see this author's *Horn Technique* (London: Oxford University Press, London, 1962,1992); pp.60–69.
26. Brahms was also wrong in not putting the word *offen* (open) in m.4, the term used to cancel any previous stopping.
27. I remember as a young horn student not only being taught by my teacher, a member of the New York Philharmonic horn section, to play this note hand stopped, but hearing it played that way in all performances in those days.

was one of Brahms's delights in his major works: to bring back—to slip in—in
one movement material that had appeared in a previous movement, or, its opposite, to develop a whole new theme, even a whole new movement, out of some
incidental material that had occurred earlier.[28]

The serenity and gentle lyricism of the opening four measures are quite suddenly transformed via the fifth bar into a most extraordinary passage (mm.6–8)
of quintessential Brahmsian polyphony, a passage which, however, loses most of
its extraordinary quality when played too slowly. The three individual rhythms
used to generate this rich and complex contrapuntal fabric—eighths, triplet
eighths, and dotted eighths-sixteenths—need a certain momentum, a tempo
flow, to unveil their particular rhythmic qualities and characteristics. At too slow
a tempo they simply disintegrate into some unfeelable mass, so that the tension
and conflict at close quarters between and among these rhythmic configurations
are lost. An idea of what I mean by conflict and complex counterpoint can be
gained with even a cursory glance at the musical example (Ex. 16a), which
displays a linear representation of the composite rhythm produced by the three
aforementioned rhythmic units seen in m.8. Separated out, the three truly *contrapuntal* rhythmic lines in m.8 look as in Ex. 16b. This is obviously an ex-

Ex. 16a

Ex. 16b

tremely complex rhythmic design, and as specifically expressed by Brahms, particularly in mm.7–8, needs to be played with the utmost rhythmic accuracy.
Failing that, even slightly, Brahms's rhythmic polyphony turns into rhythmic
mush and rhythmic anonymity.

Some might argue that the distinctions between the three rhythmic lines will
be more clearly audible the slower the basic tempo. Possibly; at least on one
level of perception. But what might perhaps be gained in rhythmic clarity and
discernibility in a very slow tempo, will be vitiated by the loss of momentum
and tension. The three rhythms, when stretched to excess, lose their strength,
like an old, worn-out rubber band. Surely, Brahms did not create such complex
contrapuntal passages in order to create something easy to listen to, something
conflict-free. Furthermore, this passage works musically only within the frame

28. We shall see a brilliant example of the latter in the discussion of the Brahms Fourth. These
little motivic or thematic cameo appearances or re-appearances in Brahms's works remind me of
Alfred Hitchcock's penchant for appearing in his own films in some virtually unnoticed, tiny,
'cameo,' capacity.

of reference of a quarter-note beat. Conductors who do this passage so slow (van
Beinum, Tennstedt, Bernstein, Giulini) that they are almost forced to subdivide,
in other words conduct in eighths, produce only a chunky, ponderous, muscle-
bound feeling, which is quite antithetical to Brahms's conception. How powerfully expressive and exciting the three bars in question can sound can be heard
on Haitink's altogether splendid recording.

From the high levels of intensity of mm.6–8, Brahms now returns to the
tranquillity of the opening in three exquisitely harmonized and nuanced measures. Unfortunately, the ninth chord in m.10 and the thirteenth chord in m.11
are almost never fully realized. For some strange reason—is it a latent fear of
dissonance, of chromaticism?—the A in the second violins in m.10, and the D♮
in the violas in m.11 are almost always under-played. Many conductors really
do not hear very well harmonically and seem to have no interest—nor the necessary ear—to bring out the pungent harmonies all great composers blend into
their music. This incapacity severely undermines the full effect—and effectiveness—of the music of the great harmonic masters, especially Brahms. I could
find only two recordings in which these harmonic shadings were properly realized (Sawallisch and Celibidache).

Measure 12 is rarely performed correctly, mainly because no one seems to
know that *rf* is not the same as *sf* or *sfz*. The marking *rf* is simply an abbreviation
of the Italian term *rinforzando* (meaning reinforcing). Thus the downbeat of
m.12 is not some huge *f* accent, but a gently expressive 'reinforcement.' Coming
from the *pp* in m.11 the actual dynamic level at m.12 ought not to be more
than *mp*. Also, no one ever seems to pay any attention to Brahms's meticulously
precise placement of the dynamic nuances in m.12. The winds crescendo goes
to the second beat (not the first), one beat later than the strings'. It is almost

always erroneously played as .

It is important that the downbeat of m.13 in the winds be *p* (something much
harder to achieve if m.12 is played *f*), for it might then remind conductors and
string players that the prevailing dynamic here is, in fact, *p*. Recording after
recording offers this beautiful, elegant, gentle phrase in a full, thick, fat *f*, further
vulgarized with gratuitous saccharine *portamenti*, (as represented in Ex.17), in

Ex. 17

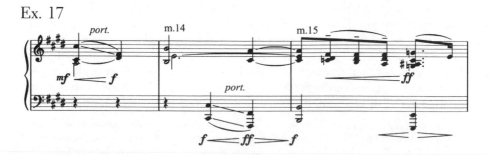

my view all a grievous trivialization of Brahms's intentions. There is a heavenly repose in this passage (mm.13–17) when it is played softly in a contemplative, inner-directed way, as Brahms wrote it. Among the very few conductors who caught this special mood, I would like to cite particularly Suitner, Haitink, and Sawallisch.

Speaking of harmonic neglects, Brahms's C♮ in the second cellos in m.16 (and the parallel place, m.70) is rarely fully projected. Most composers would have written the much less interesting C♯ here. But Brahms, always intent on varying his materials, constantly reinventing, having already used the plain diminished chord (in m.15), darkens the harmony with the C (making it a 'German' sixth chord). But the beauty of this subtle harmonic touch seems to escape most performers, although not Stokowski and Abbado.

The quiescent mood continues with a poignant oboe solo accompanied by soft winds and violas—only here again, too often the mood is destroyed by the 'mezzo-fortissization' that seems to plague so many orchestras. This is one of Brahms's most poignant melodies; it has a feeling of intimacy, even of fragility and vulnerability. But when the oboist swaggeringly trumpets out his solo, the other instruments automatically come up in dynamic level; and unless the conductor prevents this dynamic distortion, the whole transcendent beauty of this passage is gone.

What also gives this passage its poignancy is the bittersweet dissonance in the clashing of the A♯'s and B's in m.19 and m.21: ⧉. The absence of this expressive dissonance in most performances can be ascribed to two factors: (1) the aforementioned tendency of most conductors (and most musicians, for that matter) to avoid any 'alien' dissonance in tonal music, and (2) once again, the misreading or ignoring of Brahms's carefully placed dynamic markings. The 'hairpin' nuances in m.18–21 are usually played ♩ ♩ ♩ ♩ ♩ 𝄽 instead of Brahms's ♩ ♩ ♩ ♩ ♩ 𝄽. The result is that precisely where the A♯'s clash with the E minor chord, the players have backed off into a tensionless *p* or *pp*, rather than the expressive *mp* Brahms's notation implies. On only five recordings—those of Abravanel, Janowski, Järvi, Stokowski, and Suitner—is this very Brahmsian touch exploited.

In the middle of this phrase Brahms devises one of the most extraordinary, brilliant musical/structural ideas any composer had conceived up to that time. It consists of the encounter between two totally unrelated musical ideas: the movement's main theme (Theme I) as first stated in mm.1–4, and the second subject (Theme II) as stated by the oboe. I have already alluded to this in Fig. 4, but perhaps the radicality of Brahms's idea can be appreciated even more in the following graphic representation of Fig. 6.

Fig. 6

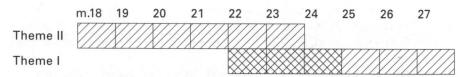

What Brahms has done in effect is to slide Theme I in under the peak measure (m.22) of Theme II,[29] then abandon Theme II two bars later while letting Theme I continue (m.24)—again, be it noted, without any dynamic nuances. Then, in m.25, Brahms picks up the interrupted oboe melody, fashions a variant of the oboe's last three notes as a link-measure to allow the melody to be finished out in mm.26–27 in a final resolution to the tonic key of E, thereby also rounding off the first major episode of the movement. By rights the resumption of the interrupted (oboe) melody in m.25 should have been given to the oboe. But Brahms decided to turn it over to the strings, the 'intruders' who surreptitiously crept in three bars earlier. The whole passage (mm.18–27) is at once, in its quiet and subtle way, one of the most radical and sophisticated polyphonic conceptions, a worthy extension of the lessons Brahms learned from his studies of Bach's polyphonic masterpieces.[30] Unfortunately very few conductors, let alone orchestral musicians, have appreciated the originality and technical sophistication of Brahms's invention, and the whole marvelous thematic interplay here is rarely fully realized in performance and recordings.

Before leaving this section, I must reiterate that the *rf* in m.25 is, once again, not some super explosive *sfz*, but a gentle expressive accent in the context

29. What happens in mm.22–23 can be likened in the visual realm to our suddenly seeing a photographic double image, one picture superimposed upon another, but slightly askew.

30. This is, of course, as I have previously suggested, not mere imitation of straight Bachian fugal and canonic writing. As Douglas Townsend puts it in the aforementioned liner notes: "when we say that musically speaking [Brahms'] great-grandfather was Bach, it is not because his music is so polyphonic, but rather reflects the *knowledge* of counterpoint, even when it is not contrapuntal." I would add that it not only *reflects* that knowledge, but in highly imaginative ways extends and augments that knowledge, investing it with wholly new concepts, reinventing and renewing polyphony in relation to the newer symphonic forms. Townsend continues: "[Brahms's] knowledge of counterpoint and fugue are manifested in most of his work by the manner in which the *texture* of his compositions is constantly varied from the polyphonic to the homophonic and back again."

of the essentially calm, quiet, serene main-theme. I must also remind us that the $<\ >$ in m.26 peaks on the fourth eighth-note of the measure, not the third beat—which is the way, alas, it is universally, but erroneously, played.

The next episode (mm.28–38) is one of the most gloriously singing in the entire symphony. It is so beloved by conductors, as well as string players, that it is generally treated with great respect and love and thus relatively well performed. But what sometimes happens to this passage is (a) too much crescendo in m.29 (the *p*'s in the accompaniment in m.30 provide a clue that the dynamic in the violins and violas ought not to exceed *mp*—or at most *mf*); (b) rushing the tempo out of sheer uncontrolled excitement in mm.29–30 and mm.31–32, also—and even more likely—crescendoing in those two measures (Brahms delays the crescendo until m.34); and (c) an erroneous, often very vulgar ritard in m.37, followed by an equally mindless accelerando in m.38. This last aberration is particularly disturbing because it destroys the whole momentum and pulse of the music, which it is most important to preserve in order to properly set up, as it were, the next 'kinder and gentler' oboe and clarinet episode (mm.39–*ca*.48). The sad fact is that the inordinate slowing up of the tempo in m.37 and speeding up in m.38 are often caused by the string players, especially the violinists, who want to use huge full bows (four of them) in m.37, but who quickly run out of bow in m.38. It must be thousands of times in the history of the piece that the strings have already reached the tip of the bow by the second or third eighth of m.38, with the result that there is a very quick one-beat diminuendo; for the rest of the measure the strings then hang on weakly, waiting for the next downbeat, instead of maintaining a long *three-beat* diminuendo. This can easily be done with a little thought and care by balancing bow pressure against bow speed—what string players call "saving the bow." But too many conductors have caved in to the string players, accommodating to their bad habit by accelerating the tempo in m.38, when in fact by all musical logic and feeling, after the almost ecstatic passion and grandeur of the entire previous phrase, its resolution, its resting point in m.38, should also be grand and sustained. Among the few conductors who avoided this particular distortion, I single out especially Leinsdorf and Suitner.

Up to this point in the second movement, the only conductors, among the fifty-odd recordings sampled, who offer truly satisfactory, respectful yet inspired performances are Weingartner (in his late-1930s London Symphony performance) and Suitner in his recent recording with the Berlin Staatskapelle. Except for a rather slowish over-all tempo (Weingartner: $\quarternote = 40$, Suitner: $\quarternote = 48$) and a weak unappreciated C♯ in m.16, all the points covered thus far are beautifully handled with warmth, taste, and intelligence.

I must confess that when I first began conducting this symphony, I did not understand the dynamics in the accompaniment in mm.34–37, especially the *p* in m.35. Why should the accompaniment drop out, so to speak, in mm.35–37, rather than fully supporting the upper strings? I have since realized that

Brahms's dynamics here are not simply 'wrong'—as one might too hastily assume—but that, in fact, Brahms is after another daring and unorthodox effect, namely, that of the supportive harmonic accompaniment functioning on its own separate dynamic track: a kind of polyphony of dynamics and of texture. In effect, the accompaniment recedes, almost disappearing like some underground river, only to re-emerge in m.37 in complete support of the melodic line. This is then an idea to be explored fully and exploited rather than changed or rejected out of hand.

   We should note in passing how ingeniously Brahms maximizes the use of his musical materials. By criss-crossing his two top melodic/contrapuntal lines (Ex.18a), he is able to stretch m.34 out to three measures (as shown in Ex.18b). It is this three-stage bar-by-bar descent—descent both registrally and dynamically—that undoubtedly prompted Brahms to conceive the aforementioned unusual disposition of the accompaniment here.

Ex. 18a

Ex. 18b

   The oboe's long sinuous, exquisitely spun-out line rises almost imperceptibly out of the receding diminuendo of the strings in m.38. When this movement is played at the right flowing (*andante*) tempo, this middle episode has a refreshing, light, airy character that provides a wonderful contrast to the three previous sections.[31] The music seems to be floating on buoyant, feathery-light springs, a feeling which is best achieved in the strings by a subtle relaxed subdivision of the conductor's beat. In many recordings the strings' syncopated accompaniment is anything but feathery-light; instead it is heavy, stiff, chunky, and, surprisingly often, untogether. Care must be taken that the sixteenth-notes not be cut too short, a bad bowing habit that here destroys the grand line and

31. Let us also note that the pitches in the first violins here (m.39–40) are, with but one exception

(the B♯)

identical to Brahms's principal opening theme. With the entrance of the clarinet five bars later, Brahms inverts the accompaniment C B♭ G A♭ B♭♭.

dissects the music into countless tiny little fragments. String players must be cautioned to play the sixteenth-notes full length and think across the interceding rest, so that a long uninterrupted chain of floating syncopations results. It is surprising—and disappointing—to hear in both Karajan's and Kertesz's recordings, both with the Vienna Philharmonic, the strings play the accompaniment in mm.39–49 not only very stiffly, unfloatingly, but in such a manner that the syncopations seem turned around, as if Brahms had written

instead of

. (This is probably the result of misguidedly

bowing the passage .)

Before the clarinet's response to the oboe is completed, cellos and basses sneak in with the oboe's sinuous m.39 theme, ominously darkening the mood of the music. In m.48 care must be taken that the clarinet's final measure is not drowned out by the entering flute and bassoon, as happens unfortunately on dozens of recordings.

The upper strings' syncopations now are used by Brahms to considerably agitate the mood (mm.49,51–52). Surprisingly, these are again the melodic notes of m.3, rhythmically and transpositionally varied.

becomes, a third lower and enharmonically

re-spelled, . As in a seething sea,

waves of churning syncopated rhythms roll forward, crashing ultimately (m.53) onto the barrier of a five-octave-deep G#, bringing the music back to the temporary base key of C# minor. The *sfp* here must be sharply articulated to fully represent the sudden arrestation of the music's rising momentum. It is too often played as a mediocre *sfmf*, which hardly does justice to what is thus far in this movement its most dramatic, almost shocking, moment. But the tempestuous surge of the music cannot be held back for long. It breaks forth into a passionate, richly textured outpouring which comprises the climax of the whole C# minor middle section (mm.39–62) of the movement.

Its tail end (mm.61–62) presents a real performance problem, though. It is my view that the sixteenth note groups, divided among various woodwind and string choirs, ought to be played so as to create one beautiful *Klangfarben* sequence.[32] I can well imagine Brahms playing these two bars at the piano (for Clara Schumann?) in one unbroken descending line, as represented in Ex. 19.

---

32. There is a very similar passage in the slow movement of the Fourth Symphony (mm.57–59), where the same performance problem of connecting three disparate instrumental groups into one single line exists.

Ex. 19

This is much harder to achieve, of course, among different instrumental sections than it is on one piano played by one person, whose one ear and one mind are controlling ten fingers. Most of the time, out of sheer inattention or laziness, the first four successive groupings (mm.61–62) are played in a disjointed manner, mainly because of that universal bad habit—a veritable plague—of cutting off last notes before a rest. This passage can be done correctly if all the players involved hold their last sixteenth full length—not longer, not shorter—and realize that they have to hand their last note over to another player or group of players, just as incoming players have to realize that they are taking over from another set of players. It may take a bit of rehearsing to achieve this continuity of line, but it is well worth the effort. Furtwängler's Vienna Philharmonic recording offers a shining example of how beautiful this passage can sound when played in the conjoined manner I am suggesting.

I have several times referred to Brahms's fondness for playing around with the beat and the meter. I have also not concealed my strong conviction that despite—indeed because of—these rhythmic shiftings, Brahms meant the underlying pulse and meter also to be felt, to be implied. These rhythmic shiftings are like enlarged syncopations: they are not just one note being syncopated against its adjacent beat, but a whole phrase syncopated against a whole measure, or several. Measures 63–65 are a case in point. Whereas we almost always hear in performances and recordings these measures played as if Brahms had written as in Ex. 20a (in piano reduction), they should, in fact, sound as *actually* set by Brahms, namely Ex. 20b. Anything less than this vitiates the whole point and

Ex. 20a

Ex. 20b

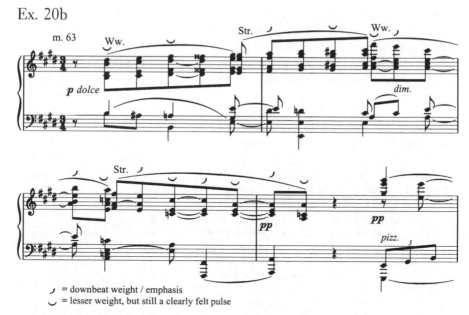

⌐ = downbeat weight / emphasis
⌣ = lesser weight, but still a clearly felt pulse

meaning of Brahmsian syncopations, which should have an against-the-beat lilt to them. And it is very disturbing (as well as wrong) when we are obliged to hear an inadvertent 3/8 (as bracketed in Ex. 20a) in m.66.

Furtwängler, who had just negotiated mm.61–62 so beautifully, now loses himself in such a slow tempo in m.63—♩ = 46 (even slower in his Berlin Philharmonic recording), the eighths almost as slow as his quarters were at the beginning of the movement—that the music, not to mention the form, loses all shape and coherence. Toscanini, who also negotiated m.61–62 relatively well, takes the opposite approach to Furtwängler's in m.63, lunging precipitously into a tempo of ♩ = 58, completely out of context with his basic tempo for the movement and the kind of tempo waywardness that Toscanini almost never permitted himself.

We have arrived at the recapitulation, but not the ordinary reprise of classical vintage. It is the new genre of recapitulation first proposed by Beethoven in his Fifth Symphony, based on the principle of non-repetition and perpetual variation. The recapitulation at mm.66–67 is a glorious revisiting of the main (and opening) thematic material of the movement. Measures 1–27 are reanimated, vertically and horizontally expanded (extra bars are interpolated three times: mm.70,80,85–87), orchestrationally refurbished, and contrapuntally enriched, to the point that the half-attentive listener (and musician) may not even realize he is hearing a recapitulation.

Though the orchestra used now is much larger than in the exposition,[33] the dynamic levels are—and should be, but almost never are—the same. There is nothing more exquisite in music than a full symphony orchestra playing *pp*, and

---

33. It is worth noting that trumpets and timpani make their first appearance in the movement at this point, having been saved until now for just this purposeful entrance.

Brahms definitely calls for it here (mm.67–71), the little crescendos never exceeding *mp*. But sadly, this is universally ignored; an obese *mf* is usually substituted (only Skrowaczewski managed to achieve the almost Debussyan transparency and tenderness of these measures). Even worse is the predilection on the part of so many conductors to forever conduct only the strings, an approach that is severely damaging to the passage in question, for the primary (thematic) material is in the upper woodwinds,[34] *not in the strings*. The upper strings have only a simple counterline which certainly should not cover or obscure the delicately orchestrated woodwinds. But too many conductors exhort the violins and violas to a passionate outburst, starting not *pp* but *mf*, followed by an enormous crescendo. Under these exhortations, the string players have no choice but to change Brahms's bowing in mm.67–68 to

 or, worse yet, to

Between the original m.4 and m.5, as recapitulated here, Brahms interpolates mm.16–17 (reorchestrated and revoiced, of course), the model for this being the exposition's mm.15–17. The original m.5 then returns in m.72, embellished with string pizzicatos. Here great care must be taken by the first flute and first clarinet to play the softest *pp* possible in m.71 and not to make too big a crescendo. On the other hand, the second flute, first oboe, and second clarinet should know that they have the leading melodic notes. (I know of no recording where this bit of sophisticated voice leading and balancing was properly handled.) In mm.73–75 (the original mm.6–8), Brahms modulates to a new tonal region. Again, the sweet 'dissonances,' first encountered in mm.9–11, must be brought out: the ninth chords in mm.76, 77, and the A's in the 13th chord of m.78. The voicing is quite close, almost as in jazz block-chord writing (Ex. 21).

Ex. 21

Notice particularly the close voicing of the second clarinet and first horn. There are also the delicate melodic and rhythmic clashes of the resolving eighth-note F♯'s against the triplet E♯'s in mm.76–78, which require very precise timing to bring off correctly. In m.79 the contrasting dynamics (the woodwinds peaking on the second beat, the strings on the third beat) are almost never observed, nor is the need for an enormous diminuendo from the *ff*'s of that measure to the

---

34. Brahms undoubtedly would have wanted to take the first flute up to the highest C♯ in m.67, but fearing—rightly so—that it would be too loud (a high C♯ on the flute cannot be played *p*), he took it down the octave. Jochum and Ormandy are the only conductors I know of who take the C♯ up an octave, not to the best effect, I'm afraid.

soft, delicate, transparent *p* textures of m.80. If anyone were ever to observe Brahms's diminuendo in the strings in m.80, the next bar would commence *pp*, and would thus be the appropriate recapitulatory parallel to mm.13–17.

Brahms now interpolates into the recapitulation two previously heard phrases —small recapitulations within the larger recapitulation, as it were. The first is the woodwind phrase of mm.85–86, a slightly expanded reworking of m.80; the second an altered 'repeat' of mm.81–84, adjusted by Brahms to prepare the way for the return of the second subject (the oboe melody of m.18). Again much depends on the respect the conductor and performers have (or do not have) for Brahms's dynamic indications. As I have emphasized several times, not much in the way of color and timbral richness can be achieved if dynamic variety, especially at the softer levels, is suppressed; that is, if the ubiquitous *mezzofortissimo* is put into operation. Brahms's mixing of colors, of lines, of particular sonoric effects, of rhythms, here is so ingenious, so delicately calibrated, that every detail of his scrupulous notation must be observed and rendered accurately.

To begin with (see Ex.22), the three 'solo' instruments—oboe, horn, violin—

Ex. 22

have to be well balanced among themselves in mm.91–96. (I have lost track of how many recordings fail to achieve this; usually one hears primarily the horn and violin.) Next, the crescendo swells in m.91 must be handled very discreetly lest they overpower the soloists, who in self-defense will then, of course, resort to a *mf* dynamic or more. Through this relatively dense texture the harp-like triplet pizzicatos of the cellos must be able to project without forcing.

Next, the A♯'s previously referred to in connection with m.19 and m.21, must be well balanced against their neighboring B's. Almost always ignored is the articulation Brahms gives most of the accompanying instruments in m.92 and m.94: ♪ ♪ not ♪ ♪. Note that the trumpets and second horn maintain this notation for the entire first four measures (mm.91–94), Brahms's way of showing that these notes should be played with something approximating the sound of timpani notes when played with soft sticks, a sound like ♪ ♪, as seen in notation [35] The combination of *cresc.* and ⎯⎯⎯ ⎯⎯⎯ in mm.93–94 has confused musicians and conductors for years. But it is really very simple: in those two

35. Here, incidentally, on this small point the question of tempo arises again. For if the potential duration of a timpani note played *p* in average acoustics is, say, a half to three-quarters of a second, then in an *andante* the duration of the timpani notes will correspond well to Brahms's notation; if on the other hand a tempo of ♩ = 30 is taken, the timpani notes will fill only a fourth of that duration, sounding like ♪.

bars, one is to make an overall crescendo from *p* to *f*, at the same time incorporating an additional *cresc.– dim.* (≤ ≥), which peaks temporarily on the downbeat of m.94, but only in those instruments (upper woodwinds, trumpets, and timpani) that have this dynamic overlay.

Measure 95 is difficult to balance, especially in live (non-recording) performance, for several simultaneous demands of the music must be fulfilled: the three solo instruments must not be smothered by the *f* of the other instruments; at the same time the main opening theme, entering in low strings, bassoons, and contrabassoon must also be fully (i.e. deeply, richly) represented, all of this, one hopes, not obscuring the pizzicatos in the cellos.

Many conductors make a little break, a tiny *caesura*, between m.96 and m.97. This is wrong, for the three solo instruments' melody (E-B-G♯) should carry directly into the tune's continuation in the flute and clarinet. The interruption comes in the *next* measure, there actually composed into the music by Brahms. Other tempo distortions at the hands of many conductors abound in this recapitulation (m.67), but none worse than those of Bernstein, and to a lesser extent, Rowicki. After turning the beat completely around, for example, in mm.63–64 Bernstein now makes a *huge* ritard in m.65, followed by an enormous fermata on the second beat of m.66. Measure 67 is then taken up *adagississimo*, but then—ludicrously—m.71 is suddenly much faster, but only for two bars. For at m.73 Bernstein slows up dramatically again. Similarly, some time later, at m.89 Bernstein imposes the hugest ritard of all, but with the entry of the horn solo (m.100) rushes suddenly forward again. However, next one can hear the horn soloist pull the tempo back to where in the larger context it actually should have been all along. There is in all of this just too much of an "oy-vay" *Weltschmerz* to be bearable. I guess I will never fully understand how a man of Bernstein's basic talent and intelligence could allow himself such tempo excesses, wreaking havoc with Brahms's classic form, and in effect recomposing and restructuring the music to his own whims and ego-driven fantasies.

Not quite in Bernstein's league, Rowicki nonetheless competes valiantly in the 'tempo distortion' game. Having embarked on the horn solo in m.100 at his basic tempo for the *Andante* movement, a sluggish ♩ = 38, he suddenly jumps the tempo to ♩ = 46 in m.104, but two bars later is back down to ♩ = *ca.*40. What *was* he thinking of?

Astonishing mixtures of sonorities, of rhythms, also abound in the second subject's recapitulatory extension, set forth by a solo horn (mm.100–104). Let us assume that the horn, a well projecting instrument after all, will not feel the need to play loudly, will in fact start the solo in *p*. This is not only what Brahms's score calls for—reason enough to respect it—but beyond that, the mixture of soft flutes and clarinets (the latter continuing the previous pizzicato of the cellos), soft timpani, and strings, all set in waltz-like triplets over a sustained pedal point in cellos, basses, and one low horn, will be simply blotted out when the horn is too loud. Then there is the solo violin to be reckoned with. Many a concertmaster has had to play this solo *f*, in order merely to be heard. But all such forcing of the sounds, whether in the horn or the violin, or other instruments competing to be heard, ruins the delicate texture and airy dance-like lilt of the music. One of the most elegant and lyrical renditions of this passage (mm.100–104) can be heard on Chailly's Concertgebouw recording, with Toscanini's, Levine's, Suitner's, Sawallisch's, and Szell's, as close runners-up. The last named features a particularly elegant and tasteful violin obbligato by Rafael Druian.

There is considerable confusion as to how the sextuplets in the solo violin in mm.103–104 should be played: in three groupings of two or two groupings of three. I lean towards the latter choice for three reasons: (1) the violin solo is first and foremost an accompanying ornamentation of the horn solo and as such should preserve the duple division of the beats; (2) mm.103–104 being a variant of mm.101–102, it seems logical that Brahms was thinking to extend the earlier sixteenths to sixteenth triplets, preserving the duple division of the beat for that reason; (3) I suggest that Brahms was setting the two solo partners, horn and violin, *against* the underlying triplet accompaniment—again his fascination with two over three, two *against* three. Here it is very important also to bring out the somewhat under-orchestrated cello D (the seventh of E major) in m.103, and the 'dark' C♮ in m.104.

It is astonishing with what extraordinary economy Brahms consistently works. As in the first movement of Beethoven's Fifth, every tiny scrap of material is of significance, regardless of how *in*significant it may look or sound at first hearing, and is used and re-used in the most imaginative and original ways. I don't think many conductors or orchestra musicians have realized that in mm.105–108 (and in a varied form in mm.109–11) the three-note melody of the movement's opening theme is used as the bass line (Ex.23). Furthermore, the melodic line, split between the flute and the violins, is taken from mm.21–22 of the second subject oboe theme. Care must be taken that the over-all line, from the pick-up eighths in the woodwinds in m.104 through m.111, not be broken. The sonoric exchanges between woodwinds and strings should be carried out very smoothly. The dynamic in the strings in m.106 should probably be *mp*.

Ex. 23

The whole movement is constructed with such a wondrous over-all line, with themes and motives merging almost imperceptibly into one another, eliding and overlapping, that it is sometimes difficult to tell where one section ends and another begins. The coda is a case in point. I suppose one might place it in m.114, but here too one phrase (in the winds) elides and overlaps with another incoming one (in the strings), making any clear structural delineation ambiguous. I have often felt, despite what my mind was telling me, that the coda starts in m.112, when the clarinet and solo violin phrase has come to a resting place, resolving on the tonic key.

More thematic recycling now: what is in the winds in mm.114–16 re-appears in the strings in mm.120–22, and vice versa. At the same time the chromatic transitional phrase that we first encountered in m.5 is used twice more, once leading to a false cadence, the second time to the final resolution in E major. Again, there are many special—I would say unique—Brahmsian touches, that should be but rarely are observed. In m.118 the trumpets' and second horn's E's are not audible in most performances and recordings. As the sevenths in a second inversion F♯ dominant chord, these E's give an unusual color to the harmony, also, of course, presaging the tonic key to which all will soon be resolved. Also, nobody seems to want to observe and respect Brahms's unusual dynamics in mm.118–19: the flute and clarinet crescendo into the downbeat of m.119, while all the other instruments have a long two-bar diminuendo. The $<$ $>$ nuance in mm.122–23 is also very special and rarely performed as written. The crescendo peaks—it is a mild crescendo in any case, only to p or at most mp—in the middle of the first beat (m.123). This makes the crescendo shorter (three eighths) than the ensuing diminuendo (five eighths). When the whole orchestra observes this dynamic nuance faithfully, it is a ravishingly beautiful effect, and in the most exquisite way prepares the closing five bars, contrabassoon and timpani (the former hardly ever heard in performance, alas) having the final all-resolving word.[36]

The third movement has the somewhat unusual and slightly ambiguous marking of *un poco allegretto e grazioso*—"slightly ambiguous" because there has been from time to time in some circles an uncertainty about the term *allegretto* (is it a somewhat slower *allegro* or just a lighter one?) and because of the word *poco* (does it modify the *allegretto* in a faster or slower direction?). It is really not clear, and in the absence of any metronome marking it is anyone's guess what the ideal tempo might be. And indeed, the music works well in several tempos within a certain range. In the many recordings I have sampled the slowest was Bernstein's (♩ = 74), the fastest were Horenstein's ♩ = 94 (surprisingly,

---

36. Brahms uses the same effect of a single instrument holding through two separate concluding chords as Wagner had done earlier in *Tristan und Isolde* at the very end of the opera. The idea is thus not original with Brahms, who, despite the feuding between the Wagner and Brahms disciples, knew and respected Wagner's late operas well, and was certainly aware of the *Tristan* ending.

Abbado has the trumpets also hold through with the solo violin. I assume he garnered this idea from Brahms's autograph facsimile, which indeed has the trumpets tied across the last two bars. But this is clearly an oversight on Brahms's part. Brahms originally had *all* the winds tied across the last two bars, but then crossed those ties out. I am certain that Brahms meant to eliminate the trumpets' ties as well, but inadvertently failed to do so.

a conductor generally given to slowish tempos) and Klemperer's (in his 1928 Berlin recording, ♩ = 94); and they all seemed to be possible interpretations. (Talk about *elastischer Takt!*) My own preference is for ♩ = *ca.* 80, a tempo that gives the music a *grazioso* lilt, is unhurried, and yet enables the music to be heard in its phrase lengths *as* phrase entities, not just an arbitrary succession of 2/4 bars.

These phrase lengths are a matter of some import and should be felt, heard, and understood by the conductor and the musicians. They are rather unusual and unorthodox. The two first phrases (mm.1–5, mm.6–10) are, for example, five-bar phrases. But when this music is recapitulated in mm.19–25 and mm.26–32, Brahms has extended them to seven-bar phrases. In truth, the former are four-bar phrases extended by one bar, the latter four-bar phrases extended by three bars. In many recordings there is absolutely no sense or awareness of this, nor of the idea that the strings' entrances in m.4 and m.9 are nothing more than a color added to the prevailing winds. They should blend with winds, not take over from them.

The opening principal ten-bar theme is also a brilliant example of Brahms's inspired ability to construct musical ideas out of the simplest materials and by the simplest, most economical means, for the second five bars (mm.6–10) are a melodic inversion of the first five bars.

Measures 11–18 are almost always played too loud, both in the woodwinds and in the strings (the ones notated *p*, the others *pp*)—only four recorded performances manage to achieve the textural contrast at m.11, those of Toscanini, Stokowski, Kondrashin, and Skrowaczewski. Here too, the woodwinds in some orchestras play triplets (instead of dotted eighths-sixteenths); and the second flute and second clarinet in mm.16–18 often are unaware that they have the main voice here (for example, the Chicago Symphony players in Levine's recording). The wonderful softly pulsating pizzicatos in the basses are also often underplayed or acoustically blurred. When mm.11–18 are played too loud, the contrasting dynamic Brahms calls for in m.19 cannot be realized. Indeed, in some recordings the intended effect is quite reversed: m.19 being *softer* than mm.11–18. In any case, the first violins should now sing out, playing the clarinet's melody at a slightly fuller dynamic level. Occasionally clarinetists may have to be told that the triplet passage (mm.19–22) is not a 'solo,' that it is in fact a discreet accompaniment to or embellishment of the violins' tune.

What has been said about mm.11–18 applies, of course, to mm.33–44, a subtle re-working of the earlier passage. Something quite new and different has also been added, an echo phrase in mm.39–40 It is amazing and shocking how many performances and recordings blithely ignore this exquisite effect, riding roughshod over these two bars as if there were no *pp*'s and *ppp*'s there.

Indeed, it is depressing to realize how many recorded performances by world-famous maestri completely ignore Brahms's wonderfully subtle dynamic and textural contrasts in the *entire* third movement exposition (mm.1–44). The list of offending conductors is a long one and includes, most notoriously, Böhm, Horenstein, Boult, Dohnanyi, Klemperer, Rowicki, Ozawa, Wand, with perhaps the worst offenders Munch (with the Orchestre de Paris) and Janowski, both of whom charge in with a healthy *mf* at m.11, play m.19 *softer*, m.33 loud again, and barely give any notice to Brahms's echo phrase at m.39. The conductors who seem to understand and respect Brahms's intentions here are, again, Toscanini, Stokowski, Kondrashin, Skrowaczewski, and, in a half-hearted sort of way, Muti with his Philadelphians.

Many conductors get overly excited in the next episode (starting at m.45) and push the tempo, having then, of course, to relax it again shortly before m.62. Actually, a little *rubato* in this very Hungarian-gypsy-like music is quite appropriate—in the sense of Beethoven's "Tempo des Gefühls" (tempo of feeling), but this should not be exaggerated. Be it noted, nonetheless, that for all the rhythmic drive and *Schwung* of this passage (mm.50–53), it is a variant of the clarinet's opening theme, as are the rolling accompanimental figures in the strings in mm.59–61. Finally, I find it awful when the woodwinds clip off their last eighths in m.58 and m.60, as one can hear, alas, in far too many recordings.

On the return of the opening theme—now the third appearance—Brahms extends it from its previous seven bars to nine bars. This elongation and its concomitant diminuendo have led many conductors to slow up the tempo in mm.67–70, unfortunately arriving then at a slower tempo at the 6/8, which is in effect the 'Trio' of this dance-like movement. (In earlier days the third movement would have been a Menuet or a Scherzo, or in later days a Ländler, as in Mahler). In any case, the basic tempo of the outgoing 2/4 and incoming 6/8 should be the same (←♩ = ♩.→), another manifestation of Brahms's consistent fascination with relationships of two to three.

It is quite unmusical when the winds drop their dotted quarters rhythmically and/or dynamically (mm.71–72, mm.75–76, etc.), for it dissects what is clearly intended to be a four-bar phrase into three little one- and two-bar phrases. Note also Brahms's careful notation of the winds' dotted quarters, but the strings' plain quarters.

I do not know what to make of the curious articulation—unslurred—in the violins in mm.85–86, except that I think it is a mistake, in both Brahms's manuscript and the printed editions. One can be sure that, since the horns are slurred, the violins were likewise meant to be *legato*. This scriptural lacuna has led conductors to some of the strangest interpretations, more often than not heavy *détaché* strokes (which to my ears sound completely out of context). But the most bizarre solution was adopted by Klemperer in his 1928 recording in which he had the violins play the two bars with a *secco* Stravinskyan *staccato*, a decision about which he obviously had second thoughts, for in his later Philharmonia recording he opts for ordinary *tenuto* separate bows. I find it ironic—even a bit perverse—that so many conductors, past and present, who have had no compunctions about ignoring or changing Brahms's score at will in hundreds of places, here in this one place where logic would dictate that there really is a textual error, they all adhere stubbornly to the misprint. Only five conductors (among the recorded samples) have had the good sense to slur these two bars: Furtwängler, Toscanini, Munch, Norrington, and Rowicki.

Apart from an almost universal neglect of the *sf*'s in the winds in m.90 and m.94, there is the tendency here on the part of many conductors to push the tempo enormously, either for the effect of a 'cheap thrill' for the audience or because of an utter lack of tempo control. How powerful the climactic passage at mm.96–107 can sound *without rushing* may be heard to excellent effect on the recordings of Suitner and Klemperer (Philharmonia). What is also almost always neglected is Brahms's important and careful notational differentiation in the bass part between the quarter-notes of mm.93–95 and the dotted quarters of mm.96–101. The latter must be played *molto sostenuto, espressivo,* powerfully singing with great intensity. No recording really achieved this but a few, Abravanel's, Klemperer's (Philharmonia), Suitner's, and Kondrashin's, come close.

In my view mm.106–107 are not a mere succession of 3/8-3/8-5/8 entities, as they are almost always played and heard, but a series of strongly syncopated accents set against (but not obscuring) the underlying 6/8 pulse. In order to achieve the twin result of hearing both the asymmetrical cross-accents and the basic pulse, all the conductor has to tell the musicians is to feel and think the beats (the first and fourth eighths in both measures) underneath and inside the stated notated rhythms. How exciting this can sound when performed correctly can be savored on Stokowski's and Dohnanyi's recordings.

In the first ending Brahms seems to have omitted a crescendo in the brass and strings. One needn't worry about the brass, however, since the eighth-note figure, rising over an octave, will almost automatically be accompanied by a crescendo.

One of the most annoying liberties—because so naive, so self-indulgent and thoughtless—taken in Brahms's First Symphony is the ritardando almost everyone makes in the second ending (mm.108–114). It is 'self-indulgent' in that it is done simply because 'it feels good' or 'I just like it.' It is 'thoughtless' and 'naive' because (a) Brahms would certainly have written a *poco rit.* if he had wanted it, and (b) because it must be quite clear from an intelligent reading of the score that Brahms intended for the 'threes,' now written as eighth-note triplets (in m.110 and mm.115–19) but equivalent to the eighths of the previous 6/8, to be identical (Ex. 24). But not only that: Brahms obviously wanted a smooth and

Ex. 24

imperceptible return to the 2/4 music, the reverse of what he had achieved when the 2/4 earlier on was imperceptibly transformed into a 6/8. Again, the old Brahmsian fascination with 'three' versus 'two.' How ingenious and wonderful Brahms's overlay and extension of the 6/8 feeling over the return of the main theme (mm.115–119)—and then beyond that in the clarinets and bassoons (mm.120–124)! I can find only one recording in which the transmutation from the Trio back to the 'Menuet,' that is, without any tempo disturbance, is handled correctly: Leinsdorf's with the Boston Symphony. Others, who make a modest, tasteful ritard in mm.109–114 and thus do not substantially disturb the flow of the music, are Skrowaczewski, Abravanel, Janowski, Wand, Klemperer (both recordings), Rowicki, and Ozawa. Some conductors, Toscanini, Norrington, and Karajan, are of the category who, having gotten a little too excited tempo-wise in mm.96–108, are forced to ritard back into their more moderate original Tempo I—a mild and pardonable sin.

But what some other conductors generally perpetrate here borders on the bizarre and the incredible. There are those who make a sudden, abrupt *meno mosso* at m.109—among them van Beinum, Dohnanyi, Giulini, Szell, Suitner, Kondrashin, Järvi, Walter, Celibidache, Bernstein, Paita—who then have to jump about 20 metronome points at m.115 to get back to some kind of normal recapitulation *Tempo I*. A few others, such as Böhm and Furtwängler, start the 2/4 in tempo and *then* make a huge ritard. All these conductors think, I suppose, that they are, like old Bülow, delineating the form for the audience, when in fact they are tearing it apart. Two of the strangest aberrations visited upon these six transitional measures (mm.109–14) are by (1) Stokowski, who, in a mostly superior recording, especially considering its early date (actually the first recording of the Brahms First), makes a precipitous *meno mosso* (♩ = 52) at m.109, but two bars later jumps twenty points (*sic*) forward (♩ = 72) and then accelerates further (*sic*) into a ♩ = 78 *Tempo I*; and (2) by Horenstein, who at first seems to want to maintain the tempo at the 2/4, but then suddenly changes his mind four bars later and crams a huge ritard into the final two bars of the phrase.

Once again, the clarinet must lead in mm.115–19, as at the beginning of the movement. Violins and violas add a wholly new color in mm.118–19. In m.120 Brahms writes the relatively rare marking *molto dolce*, meaning in this context *pp* and 'very sensitive' and 'gentle.'[37] It requires a very quiet, warm, and sensitive sound in *pp*. This also explains Brahms's bowing: all four bars (mm.120–123) on one bow. Yet recording after recording ignores both the *molto dolce* and the indicated bowing, in innumerable recordings the violins blithely sawing away at a healthy *mf*—with one bow per bar. What then results is that the lovely rolling triplets of the clarinets and bassoons are barely audible or even downright inaudible (as they are on 50 percent of the recordings sampled). Let us note, too, how subtly Brahms recasts the second part of the main theme of the exposition—originally a five-bar, later a seven-bar phrase—as a *six*-bar phrase. Brahms does this so ingeniously, so naturally, that the alteration can easily go by unnoticed.

Especially ingenious is the way Brahms foreshortens the entire recapitulation. The exposition, originally seventy bars in length, is now reduced to not quite forty bars, reshaped into a new sequence by some typically skillful (and painless) Brahmsian surgery (in effect cutting mm.19–58 from the exposition). Thereafter, the 'Trio' is briefly recapitulated and renotated in 2/4 (instead of the original 6/8), now serving as the coda of the entire movement.

This coda has been so variously interpreted and misinterpreted as to be almost legendary as a 'problem piece.' Actually it is quite simple and clear. Brahms writes *poco a poco più tranquillo*, starting in m.152, meaning in plain language 'gradually quieter,' but quieter implying also 'quieter in tempo.' This translates, in other words, into a very gradual relaxing of the tempo till the end.[38] Once again, however, it is beyond comprehension why the vast majority of conductors can't read or understand Brahms's marking, or why they simply reject it as 'wrong.' Some start the ritardando six or eight bars earlier than indicated, some even before that (as early as m.142!). The problem with these very premature

---

37. *Dolce*, of course, means sweet in everyday Italian, but in music in the 19th century, particularly with Beethoven and composers after him, it took on a range of related and overlapping meanings, most often "gentle," "delicate," "soft," "quiet," and perhaps a combination of these. *Dolce* was also used by Beethoven—and Brahms, in emulation of Beethoven—to mean simply *p*. There are hundreds of instances in Beethoven scores where *dolce* substitutes for *p*. Here in Brahms's m.120 it means *molto p* or *pp*.

38. I am at a loss to explain the double bar at m.154. Could it be that it represents the point where Brahms intended the *più tranquillo* to start, and the words *poco a poco* accidentally were written in two bars earlier? There is evidence to that effect in the autograph manuscript, in which there appears to be a slight difference in the hand writing between *poco a poco* and *più tranquillo*, suggesting that *più tranquillo* was Brahms's initial impulse—it would also explain the double-bar at m.154—but that at some later point he thought of the even better idea of a continuous gradual relaxing of the tempo to the final bar of the movement. Finding no room at m.154 to write in the afterthought *poco a poco*, he wrote it in two bars earlier, which would leave open only the question of where the slackening of tempo should start: m.152 or m.154; not *whether*. But this seems to be a moot point, since most conductors ignore all of this anyway.

A similar tempo modification in another celebrated masterpiece, Debussy's *L'Après-midi d'un faune*, is likewise ignored by the majority of conductors. There Debussy places very clearly five bars from the end *Très lent et très retenu jusqu'à la fin* (very slow and very held back until the end). But almost no one seems to take notice of this marking, the last three bars being generally played almost twice as *fast* as the previous two. Similarly, Strauss's long ritardando (*poco a poco più calando sin al fin*, stretching across seventeen bars) at the end of *Death and Transfiguration* is consistently ignored.

ritardandos is that, if they are carried out progressively, they lead to a virtual standstill of the music by the last three to five measures, as in the case of Furtwängler, Chailly, Abbado, and Sanderling. In the last-named's recording, for example, the tempo in the last three bars is ♩. = *ca.*21, which with all due respect cannot be a possible tempo in an *allegretto* movement—unless one simply wants to abandon any notion of formal balance, of proportions, and of boundaries of taste and logic.

Beyond such considerations, too slow a tempo in mm.154–64 makes the coda sound maudlin and overly sentimental—like something on an 'easy listening' radio station—losing that tensile strength that is always, even at its most lyrical and romantic, an essential feature of Brahms's music. A lithe, flowing, very gradually ritarding tempo, ending in the last few measures of the movement, say, around ♩. = 63 is not only intrinsically appropriate but, in my view, provides the right parallel to Brahms's wonderfully delicate, transparent, gossamer instrumentation of multiple divided strings and winds. Last, I would offer the thought that a moderate tempo makes audible—makes intelligible—the three-against-two polyphony in mm.156–57 and mm.160–61, which when played at too slow a tempo loses all motion, all flow: the dialogue between the two rhythms becomes so distended as to be almost meaningless.

The antipodes of interpretations on recordings are represented on the one hand—the excessively slow tempo and premature ritard—by Furtwängler, Levine, Muti, Kondrashin, Järvi, Stokowski, Abravanel, Rowicki, Bernstein, Ozawa; on the other hand—the moderate, tasteful, almost imperceptible slackening of tempo—Skrowaczewski, Leinsdorf, Kertesz, Toscanini, and Walter. Somewhere on a middle ground are a host of others, divided into two basic categories: those, van Beinum, Haitink, Szell—is this a peculiarly Dutch tradition?—and Wand and Jochum, who make no ritard at all until the very end (perhaps the last three to five bars); and those, Dohnanyi, Giulini, Abbado, Tennstedt, Chailly, Munch, Janowski, Karajan, Böhm, and Suitner, who provide a kind of 'roller coaster' interpretation, ritarding at first but then rushing forward with the crescendos in m.156 and m.160, only to ritard a lot in the final three bars. An especially willful interpretation is that of Bernstein, who, after giving us the 'roller coaster' treatment, incredibly suddenly goes much *faster* (*sic*) for the last three bars. With such a divergence of tempo interpretations, one begins to wish that Brahms *had* used metronome markings, to at least provide a general clue as to his tempo feelings. But probably that wouldn't have helped with a Bernstein.

The Finale movement of the Brahms First is fraught with major performance problems, mostly in the realm of tempo questions which have puzzled interpreters for over a century and which, to my knowledge, have never been satisfactorily resolved. And perhaps they will never be, at least not unequivocally, definitively. I am referring to the tempos of the opening *Adagio* introduction, the *Più andante* of m.30, and the main body of the movement, the complexly named *Allegro non troppo, ma con brio* at m.62, and the possible tempo relationships between and among these three major structural junctures.

As in the first movement, Brahms felt the need for a slower introduction to the movement proper, only this time in two phases (mm.1–29 and mm.30–61), both of these further partitioned by various thematic interventions and tempo modifications. And as in the first movement, this twin introduction supplies all the relevant thematic/motivic material that will constitute the main argument of the main body of the movement. In effect the introduction is like a gigantic two-part overture to the unfolding drama of the *Allegro* proper.

But once again, Brahms gives us no metronome markings, either at the beginning or along the way, no obvious clues to basic tempos or tempo relationships. In this case Brahms's notation is certainly not precise or explicit; *Adagio* can be legitimately interpreted in a wide range of tempos—I suppose anywhere from ♩ = 25 to ♩ = 60—while *Più andante* is considered by most conductors a fairly ambiguous tempo indication, especially in this case since it is clear that it is to be related somehow to the previous *Adagio*; and last, the rather vacillating marking, *Allegro non troppo, ma con brio* (not too lively, and yet sprightly). As can be easily imagined, this lack of specificity has led to all kinds of interpretations, none of which seems to have seriously, analytically probed the inherent tempo possibilities, but some of which (Klemperer's, Abravanel's, Walter's, Wand's, Dohnanyi's, Kondrashin's, for example) have sheerly intuitively arrived at admirable solutions. But on the premise—on which this entire book is based—that truly penetrating, comprehensive analysis can lead to the most illuminating, imaginative, and inspired musical performances, let us see with what reasonable options Brahms's score does, in fact, present us. And indeed, upon closer scrutiny we do find several important clues in Brahms's notation, that provide crucial guidelines towards evolving a cogent, coherent interpretation.

The most important and precise clue for solving the tempo puzzle, I believe, occurs in mm.29–30 in the timpani part. But while it is a solid piece of evidence as to how those two measures are intended to *relate*, by itself it doesn't tell us unequivocally what the *actual* tempos ought to be. Yet m.29 and m.30, if they are taken literally, narrow the discussion—and the options—in very compelling and fascinating ways. What Brahms is clearly saying in the timpani part (Ex. 25)—and it is astonishing that he was inspired to be so explicit—is that the

Ex. 25

tripletized sixteenths in m.30, notated as equal the tripletized thirty-seconds in the latter half of m.29. In modern 20th-century notation we might now confirm this by a 'metric modulation' notation of ←♪ = ♩→ or ←♪ = ♪→. That much seems clear, for one cannot imagine why Brahms would have set down such a relatively complex and intellectual, mathematical and very modern notation, if by so doing he didn't mean something quite specific.

What this understanding of Brahms's tempo conversion signifies is that his *Più andante* at m.30 will be twice as fast as the preceding *Adagio*. But this is where it gets complicated and elusive, for if we were to assume for the moment that this section (mm.30–61)might be at a tempo, say, of ♩ = 72 or ♩ = 60, then m.29 would have to be ♩ = 36 or ♩ = 30, respectively. Now, that is a reasonable possibility for the opening *Adagio* and is in fact the tempo range within which the vast majority of conductors set the *Adagio*. (That most of them in turn do the second movement, which is marked *Andante*, just as slow as this *Adagio* doesn't seem to disturb them particularly.)

If we allow for the moment the reasonableness of a ♩ = 30–36 tempo range for the *Adagio*, we are confronted with the apparent *un*reasonableness of maintaining that tempo throughout the *Adagio* phase of the introduction, that is, through mm.22–28 as well. But there isn't a single conductor that I know of, past or present, who kept or keeps the original slow tempo in those seven measures. Why this acceleration occurs is not really clear; it seems simply to be a long-standing tradition with origins buried in the distant past. Brahms certainly does not indicate any accelerando or stringendo in m.22 or m.23, as he does, for example, in mm.8–11 and mm.18–19. Yet tremendous accelerations of tempo have become *de rigeur* in this passage, I suppose, because nobody seems to be able to countenance the idea of the thirty-second-notes in mm.24–26 dragging along at a speed of, say, ♪ = 60. And in truth that would seem to be a musical *im*possibility. Most conductors reach a tempo of around ♪ = 112–16, the slowest I have heard (Boult) at ♪ = 92, the fastest (Stokowski) at ♪ = 138. Needless to say, if the tempo is speeded up in mm.22–28, then the original *Adagio* tempo is certainly abrogated and with that, of course, the possibility of retaining the m.29/30 ♪ = ♩ relationship I postulated earlier. We might also note in passing that if the tempo reached in mm.25–28 was, for example, ♩ = 60 (♪ = 120), Brahms's sixty-fourth-notes in the timpani in m.29 would become clearly unperformable.

We seem to find ourselves thus far in an impenetrable interpretational thicket in which no tempo choices, either intuitively or rationally arrived at, seem to be structurally compatible. If we maintain the original *Adagio* tempo and adhere to Brahms's metric modulation in mm.29–30, the climactic passage in mm.22–29 will be clearly incongruous; if on the other hand we follow our intuitions and accelerate to a much faster tempo in the climactic thirty-second-note passage, we cannot perform Brahms's metric modulation in mm.29–30, for then the *Più andante* tempo would turn out to be somewhere around ♩ = 120, depending on how much accelerando we had made previously. And it must be very clear that mm.30–61 with their alphorn melody and brass chorale, cannot be at a tempo of ♩ = 120.

Obviously some adjustment, some compromise, must be made somewhere. My solution lies in the recognition that the agitated syncopated figures in the woodwinds in mm.27–28 (Ex.26a) appear also in mm.279–84 (Ex.26b) of the main *Allegro,* and that similarly, the descending thirty-second note-runs in

Ex. 26a

Ex. 26b

mm.24–26 (Ex.27a) reappear several times in mm.106–10, 234–42 (Ex.27b), 259–67. This then may be a clue as to how fast what I have been calling "the climactic passage" (mm.24–28, 29) might be. That, of course, in turn would presuppose that we know what tempo the movement's main *Allegro* ought to

Ex. 27a

Ex. 27b

be; and here, indeed, there seems to be general agreement among interpreters. Keeping in mind that Brahms's *Allegro non troppo, ma con brio*—note that Brahms does not say *allegro ma non troppo,* but simply *Allegro non troppo,* the subtle difference being between 'not too lively' and 'lively but not too much'—has to be significantly faster than m.30's *Più andante,* but also slower than the coda's ¢ *Più allegro.* Since there is a limit to how fast the sextuplet eighth notes in the coda can be sensibly played—probably around ♩ = 136–144—it suggests, by retracing our tempo steps back through the stringendo of mm.383–90 to the main *Allegro,* that the *non troppo allegro* of m.62 will typically range somewhere between ♩ = 104 (Järvi's tempo) and ♩ = 120 (Furtwängler's, Abravanel's, Abbado's, Kondrashin's tempo).[39] Transferring this range of tempos further back to mm.24–29 of the introduction, we find that both cited figures, the syncopations in mm.27–28, the thirty-second-note runs in mm.24–26 work perfectly well within that range (counted in eighth-notes, of course, i.e. ♪ = 104 to ♪ = 120).

But if that solves the problem of the relative tempos at m.1, mm.24–28, m.30,

39. Only a few conductors do m.62 slower (Giulini at a sluggish ♩ = 90, Bernstein at ♩ = 92, Ozawa at ♩ = 96), and only a few do it faster (Toscanini, Jochum, and Stokowski, ♩ = 144, ♩ = 138, and ♩ = 124, respectively). Karajan in some of his recordings can't make up his mind, starting around ♩ = 96, then speeding up some measures later to ♩ = 120.

and presumably m.62, it still leaves one loophole, namely, m.29. For if m.29 is somewhere between $\quad = 104$ and $\quad = 120$ (as it should be if m.28 is in that tempo range), then carrying out the implied metric modulation would cause the horn call at m.30 to be also at a tempo between $\quad = 104$ and $\quad = 120$. This we also know is impossible, because we know that Brahms took this horn passage from an alphorn call that is played—by the way, to this day—on the Swiss alps, as we learn from a postcard Brahms sent to Clara Schumann in September 1868, telling her about hearing this call and quoting it in musical notation on the postcard. That alphorn call is certainly not played at a tempo of $\quad = 96–120$ by the cow- and goatherds of Switzerland. My solution to closing that 'loophole' is a relatively simple one: make a gentle ritard in the last two beats (four eighths) of m.29, easing the music from the faster tempo into a slower one at m.30, which nonetheless respects Brahms's *Più andante* in that it is 'more moving' ('more walking,' literally) than the initial *Adagio*. Thus the whole tempo schematic can be rendered as in Fig.7, showing not specific determined tempos but approximate, meaning possible, tempo ranges.

Fig. 7

mm.1–7	mm.8–10	mm.11 (*a tempo*)–17	mm. 18–19
$\quad = 30–50$	accel. to $\quad = $ca.132	$\quad = 30–50$	accel. to $\quad = $ca.132

mm.20–21	mm.22–23	mm.24–28	m.29	m.30
$\quad = 30–50$	accel. to $\quad = $ca.100–120	$\quad = $ca.100–120	rit. in 2. half to	$\quad = 60–72$

As I indicated earlier, certain conductors have arrived intuitively at similar solutions to the tempo question(s) left unclarified by Brahms. My proffered solution provides, I believe, a logical and rational approach to the problem within certain objective parameters.

Having established the parameters within which the tempos may range, the conductor must still, however, determine a *specific* tempo or at least narrow it down to a more limited range than $\quad = 30–50$. Most conductors, as I have indicated, tend towards an excruciatingly slow tempo for the opening *Adagio*, I suppose under the mistaken notion that it will ensure the most 'profound' and 'anguished' expression. They ought to remind themselves that Brahms's marking is merely *Adagio*, not even *Adagio molto*, let alone *Largo* or *Grave*. At too slow a tempo the music loses all line, all tension and inner strength. It simply becomes lugubrious and empty. In the case of some conductors' recordings, Bernstein's or Böhm's, the music seems to come to a standstill in the first few measures. My preference—and that, evidently, of Toscanini, Klemperer, Walter, and Wand—is $\quad = 46–50$. That is quite slow enough, and with that tempo the essence of the music, residing in both its sustained chromatic lines and its pedal points, can be fully expressed. Figure 8 shows the tempo choices represented in the fifty-odd recordings of the Brahms First that were analyzed.

Fig. 8

♩=24 Bernstein[+]
♩=26 van Beinum, Tennstedt[*], Sanderling[+]
♩=30 Munch[+*], Ozawa, Suitner, Skrowaczewski, Szell, Karajan[*], Abbado[+*], Solti[+*]
♩=32 Rowicki, Kertesz, Böhm[*]
♩=34 Stokowski, Boult, Dohnanyi[*], Giulini[+], Leinsdorf, Furtwängler[+] (Vienna), Walter[+*], Herbig[+*]
♩=36 Abravanel, Kondrashin[*], Janowski[*], Muti[+], Ormandy[+], Haitink, Levine, Jochum[+*]
♩=38 Järvi, Chailly
♩=40 Horenstein, Klemperer (1928), Toscanini, Lehel[+]
♩=44 Wand[*]
♩=46 Klemperer (Philharmonia)

[+] identifies those conductors who make a fermata (⌢) on the fourth beat of m.1.
[*] identifies those conductors who make (made) further excessive ritards in mm.4 and 5.

Just as the tempo should not be exaggerated, so too the dynamics should be contained within reason. Too often all the *f*'s are turned into *ff*'s or *fff*'s, the crescendo in m.1 is tremendously exaggerated, and the first timpani trill is transformed into a terrifying cannonade, all out of proportion to the elemental simplicity and doleful mood of this introductory phrase (see Ex. 28a). The melodic line, with its poignant drop down a major third in m.3 (in the violins and

Ex. 28a

1.horn), will reappear slightly altered as the main theme of the *Allegro non troppo* in m.62 (Ex. 28b). At first glance this opening line may seem to be a two-bar phrase, but it is in fact a four-bar phrase, in which the third and fourth bars are a near-repetition a varied repetition of the first two bars (not counting

Ex. 28b

the anacrusis measure). Too often this opening is treated as two separate two-bar phrases, thus destroying the grand line which, in my view, is more important to preserve rather than to indulge in some painfully slow tempo. The almost two-octave drop and intervening rest (in m.4) in the first violin line is potentially disruptive enough, so that one must make every effort to create a line of four bars. Only three conductors managed to preserve the *Viertaktigkeit* (the hyper-measure, as it is called by modern musicologists and theorists) of this opening phrase: Furtwängler, Toscanini, and Norrington. Towards that end the chromatic winding thirds in the woodwinds and the violas can be of enormous help. At the same time the intervallic clashes in m.4 and m.5 of A♭ and F♯ against the pedal G must be made audible. If the musicians involved are made to hear and feel these harmonic tension points, the tragic feeling of the music is expressed much more powerfully than with some impossible-to-sustain slow tempo.

As for the dynamics of the first phrase, apart from my suggestion that the *fp* in m.2 not be too explosive, the remainder of the phrase should be contained within essentially a *p* dynamic range, receding to a *pp* only at the end of m.5. The crescendo/diminuendo of mm.2–3 is not much more than a phrasing nuance, as distinguished from a major dynamic modification. Brahms often indicates such subtle dynamic nuancing in his works as a confirmation of what a sensitive, expressive musician might do quite naturally. Brahms was quite precise about such matters in his notation, from which we should infer that when he does *not* indicate such phrase nuances, they are not wanted and we, as performers, ought not to impose them.

In this connection, I feel compelled to point out something that seems to be little known among conductors and instrumentalists, that it is a quite common practice on the part of composers as well as music engravers, copyists, and editors that an incidental 'hairpin' nuance (< >) crescendos only to the next dynamic level above the one from which it starts. Thus a marking like *p* < > is meant to go only up to *mp*. If a composer wants to have a bigger crescendo than that, he must indicate as much (*p* —— *mf* ——) or (*mf* —— *ff* ——). Admittedly, not all composers adhere to this notational principle, but many do, and it is a good rule of thumb to follow. It is also reasonable to assume that the first beat of m.3 ought not to be played *mf* or *f*, given the fact that the sustained G in the lower instruments is marked *p*.

There is a danger—and it is represented on numerous recordings—that the violins in m.3 (and violins and violas in m.14) make too great a diminuendo because of not 'saving the bow.' As I've mentioned before, this is a world-wide bad habit among orchestral violinists, and unfortunately most conductors do very little to correct it, or don't know *how* to correct it. The effect of this excessive immediate diminuendo is exacerbated when the crescendo in m.2 (and in m.13) is also overdone, bringing the E♭ in m.2 to a *f*, for example, with the immediate result that the violinists use up half the bow before the first beat is over. The further result then is that beats 2 and 3 are close to *p* and the fourth beat *pp*. Then what usually happens is that in m.4, after the quarter rest there, the violins, with a 'fresh bow,' unknowingly come in *mp* or *p*, which, of course, has the effect of destroying the evenness of the four-bar long line (or three-bar in the case of mm.13–15). Instead of sounding as graphically displayed in Fig. 9a, which is what Brahms's notation calls for, it sounds as in Fig. 9b, which is in no way what Brahms's notation indicates.

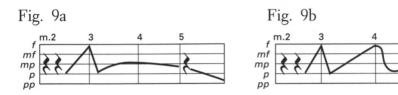

Fig. 9a   Fig. 9b

Since the end of m.5 is *pp*, m.6 must be a *subito p*, that is, one dynamic level *higher*. This nuance is hardly ever observed in performance and recordings. Indeed many conductors produce some kind of super *pianississimo* at m.6 that is totally misapplied, among them Bernstein, Dohnanyi, Giulini, Janowski, Ozawa, Abbado, the worst being Muti and Kertesz, on whose recordings I had to turn up the playback level *substantially* to be able to even hear the passage at all. Even more disappointing is that on the vast majority of recordings in the ensuing pizzicato passage the violas and cellos play much too softly, while the violins and basses play quite vigorously, with the result that the passage is distorted to sound as if Brahms had only written as in Ex. 29a, when in fact Ex. 29b is what

Ex. 29a

Ex. 29b

he actually wrote. The viola and cello parts are the most important; they are the binding that gives the whole passage its unity and line. How this sounds when well balanced can be heard to wonderful effect, for example, on Haitink's, Levine's, Skrowaczewski's, Norrington's and Furtwängler's recordings.

Another bad habit is an excessive accelerando in m.8 and m.9. First of all, Brahms writes *stringendo poco a poco* for mm.8–11, as opposed to *string. molto* in mm.18–19. This should alert conductors to the fact that Brahms definitely wants to make a difference between the two *stringendos*, with the first one being of a moderate and very gradual sort. Second, that first *stringendo* is stretched across four bars, but only across two bars for the later one. Instead, however, many conductors (like Tennstedt, Chailly, Munch, Janowski, Wand, Bernstein) have already made so much *stringendo* by m.8 and m.9 that they cannot accelerate any more in the remaining two measures, which simply end up being fast without any accelerando, thus completely subverting Brahms's idea. One conductor who gets it right is Toscanini, among the older conductors, and another, among more recent maestri, is Suitner.

At m.12 Brahms takes the opening five bars and, in effect, turns them upside down (although foreshortened by one measure), just as Beethoven had done in the last movement of the Fifth Symphony (see pp.214–15). The three anacrusis notes in m.12 are the upturned counterpart to the first three notes of the movement, Brahms having transferred them from the low register and the strings (primarily) to the high register and the woodwinds. Here again the vulgar fermata ($\curvearrowright$) on the fourth beat that, for example, Muti does, is to be avoided. The theme, first heard in the high violins, is now set in the lower middle register, very much as we shall see it in its 'main subject' form at the outset of the *Allegro* (m.62). One interesting modification Brahms adds to this permutation by inversion is the more pronounced sustaining of the pedal tones, now also moved from their erstwhile low register to the upper register. The trumpets' and flutes' C's must be well sustained, so that the momentary dissonantal clashes in mm.13–15—D♭ and B♮ against C—are fully realized. Only a few recordings exploit this particular Brahmsian (and Beethovenian) effect intelligently (van Beinum, Haitink, Furtwängler, Suitner, Abbado, Norrington, and Kondrashin).

What has been said earlier about mm.6–8 obviously applies as well as to mm.16–18.

I have already discussed the problems relative to the passage starting at m.22. I should like to add that in my view the accelerando here should not be overdone. We should recognize that Brahms has already quadrupled the speed of the audible rhythms in mm.22–26, compared with the opening. There we had movement in quarter-notes; at m.22 and onward we have movement in thirty-second-notes. The basic feelable pulse at m.22 is in eighth-notes (as it should be and, as it is incidentally, mostly conducted). Therefore, given this substantial accretion of tempo flow, already composed into the music, it is not at all necessary to make any enormous accelerando; the music speeds up significantly all by itself. My preference is to move from a basic *Adagio* tempo of ♩ = 46–50 to about ♩ = 60, that is ♪ = 120 by m.24. It is, however, also possible to stay at ♩ = 50 and render m.24 in a tempo of ♪ = 100, as, for example, Levine does. Boult's tempo and Skrowaczewski's tempo at m.24 are the slowest of all the recordings I have heard (92 and 98 respectively); Stokowski's is the fastest, clocked at ♪ = 138.

We should note in passing the intimate relationship between the melodic line in mm.22–23 and that of mm.31–32 in the second movement (Ex.30 a,b), material we shall see and hear again very prominently, somewhat transformed, in the main body of the last movement in m.156 (Ex.30c) and several other places.

Ex. 30a   Second movement

Ex. 30b   Fourth movement

Ex. 30c   Fourth movement

In most performances and recordings mm.24–27 go rather well,[40] but in m.28 far too often the dotted eighth-notes, especially in the brass, are played either too short or with sudden diminuendos (   ), or both, seriously undermining the effectiveness of this powerful climax. I have already suggested that it is best to conduct mm.22–28 in 'eight' (i.e. in eighth-notes)—indeed even the last two beats of m.21. I suggest further that m.29 also be conducted in 'eight,' so as to more readily control and pace both the long diminuendo and the ritard I have spoken of earlier at the end of m.29. The eighth-note beats are then converted directly to equivalent quarter-note beats in m.30.

Brahms was not generally given to tone painting, or extra-musical allusions, especially in his symphonies. But it is quite likely that, here and there, there are hidden or secret scenarios. I think this is the case in mm.30–51 which begins with Brahms conjuring up a beautiful cloudless sunrise morning on a Swiss alm, like the one he had visited in 1868, with an alphorn sending its simple song across the valleys and villages below. I say cloudless, because the driven, turbulent music of mm.24–28, but especially mm.27–28, surely evokes in our musical imagination a summer thunderstorm, particularly the *sf* crashes of m.28 (which must have their precedence in the 'storm' movement of Beethoven's *Pastorale* Symphony)—a storm that passes on as quickly and as suddenly as it came. The lingering, disappearing thunder in m.29 in fact led Stokowski in his in many ways remarkable Philadelphia Orchestra recording of 1927, to add a thunderous bass drum solo *(sic)* to the timpani part in mm.28–29.

The horn and flute intonings, followed by the reverential 'thanksgiving' chorale, comprise one of the most magical moments in all of Brahms. But its majestic, transcendent effect depends on a scrupulous observance of Brahms's score, especially in respect to the designated dynamics. The orchestral accompaniment (mainly trombones, timpani, and muted strings[11]) must be a mystical and misty *pp*—the morning mist rising from the valleys (if I may be permitted one more metaphor)—and the horn must declaim its call against this hushed background

40. A strange anomaly exists in one recording: while the thirty-second-notes in the violins are universally played 'on the string,' Walter had them played staccato 'off the string,' in context a very odd sounding, incongruous effect.

41. I recommend that *all* the strings be muted here, including cellos and basses, who can put on mutes one by one in the first few measures of the *Più andante*.

with nobility and rapture. Unfortunately the wondrous effect of a *f* horn against a *pp* impressionistic backdrop can be heard on only a very few recordings. On the great majority of recordings the whole dynamic level is spiraled up: a bombastic, blasting horn in *ff* or *fff*, playing with a bloated, oafish, obese tone, and an ordinary *p* or *mp* in the strings and trombones (which would be even worse if the violins weren't muted). I am also amazed to discover in systematically listening to all these many Brahms First recordings, on how many of them the horn players play woefully out of tune and also misrepresent Brahms's interesting rhythm in m.30, ♩♪. More often than not it sounds as if Brahms had written ♪. The only horn players who get both the intonation and the rhythm right are those in the Halle (Skrowaczewski), Chicago (Levine), Liverpool (Janowski), Concertgebouw (Haitink), Berlin Staatskapelle (Suitner), Berlin Philharmonic (Abbado), London Philharmonic (Jochum), and Philadelphia (Stokowski and Ormandy) orchestras.

The dynamic marking in the second horn in m.31 is apparently unclear to many interpreters, but it is very clear to me that Brahms intended the second horn player to enter gently in something considerably less than *f*, swell quickly and subtly to the level of the first horn, and then recede gradually to a lesser dynamic, say *mf*. This is to be repeated in m.33 and m.35.[42] The worst recording of this passage is Boult's with the London Philharmonic: the horns are out of tune, the second horn never holds his notes through, the trombones are much too loud, as is the timpani; and to make matters worse, there is a terrible editing splice at m.30.

An even brighter ray of sunshine now breaks through with the flute (m.38), emulating the horn call, and answered gently by a radiant solo trumpet. The counter-lines in bassoons, first and third horn in mm.41–44 must be treated discretely, with modest expressive dynamic swells. On Kertesz's and Abravanel's

---

42. It is disturbing to realize that Brahms's ingenious orchestration of the two horns, which, if it were played as written, is fail-safe, is still so often mismanaged. Could it be that many second horn players do not know how their notes fit into the over-all declamation of the alphorn melody? Brahms realized that if one horn alone were to play the entire eight-bar melody with a healthy *f*, there would inevitably be a big breathing gap, probably a whole beat's worth, at the end of the second, fourth, and sixth measures. To counteract that, Brahms brought in the second horn to in effect relieve the first horn, giving the first horn player a chance to take a big (and necessary) breath to continue the melody. But instead of having the second horn burst in with a full *f*, merely doubling the first horn, Brahms brings the second horn in under the first horn ♩ *< f >*,

then swelling to the level of the first horn, continuing its line, as it were, and then bowing out with a slight fourth beat diminuendo (more precisely notated as ♩ *mf < f > mf* or ♩ *mf < f > mf*. That some second horn players don't even hold through the full four beats, which is their only purpose in being involved in this passage in the first place, defies explanation; that their world-famous maestri allow them to do so is even more disturbing.

recordings the third horn is so overbearingly loud that it completely drowns out the solo flute. Also in several recordings—Cleveland (Dohnanyi), Los Angeles (Giulini), Philadelphia (Muti)—the third horn players are painfully sharp in mm.43–45 (Doesn't anybody hear this when recording?). On another recording, Furtwängler's 1952 recording with the Berlin Philharmonic, the third horn is missing entirely.

Stokowski, who always had to retouch everybody's orchestration, not only uses two bassoons to inflate the second bassoon countermelody in mm.41–43, but adds half the cello section as well.

Brahms's glorious organ-like chorale follows—a poignant prayer of thanksgiving—as simple as any in a Lutheran hymnbook, but transformed into something sublime by the addition to the trombones of a contrabassoon, like a 16–foot organ pedal. This chorale passage is usually played quite well, although I am often bothered by the excessive number of breaths trombonists take in these five bars—some trombone sections as many as three breaths, and in one case even four breaths (because of Bernstein's interminably slow tempo)—when in fact the whole phrase can easily be done in one breath—of course, only if it is played *p dolce* and in a reasonable tempo.

The horn returns, this time marked *poco f* (emphasis on *poco*), a notation and qualification which is ignored on almost all recordings, substituting instead a bellowing *ff*. As a result the succeeding flute and clarinet entrances (in m.54 and, m.55, respectively) are never heard. I am also surprised to find how often the interesting trombone interplay in mm.56–58—truly Schubertian trombone-writing, by the way—is not heard. It is an important new color specifically brought out here. But somehow this escapes most conductors and recording producers. Even more amazing—and depressing—is the fact that the simple G dominant seventh chord in horns and trombones at m.61, which closes the whole introduction, has hardly ever been played in tune. (What is so hard about tuning a simple G seventh chord?). On *only* one recording is this chord in tune and balanced in a beautiful *pp*: and that is Furtwängler's Vienna recording. In several others—Skrowaczewski's, Stokowski's, Ormandy's, and Leinsdorf's—the chord is almost in tune, in all the others painfully *out of tune*.

Before we leave this section, mention must be made of the unfortunate habit of many conductors to accelerate the tempo in mm.56–58 and, of course, to ritard in m.59—none of which is to be found in Brahms's score. The worst offenders are Bernstein and Ozawa, especially the former, who first of all turns m.52's *poco f* into a *più f*, goes suddenly faster at m.56, actually doubling (*sic*) the tempo at m.58, and then just as irrationally, relaxes the tempo into a very long fermata in m.60. Poor Brahms!!

The main theme of the fourth movement is one of the most well-known and most popular melodies in all of the symphonic literature, almost comparable to the popularity of the main theme of the Finale of Beethoven's Ninth Symphony, to which, incidentally, Brahms's theme was almost immediately compared by listeners to the early performances of the symphony in the 1870s. I suspect that Brahms's subtle allusion to the "Ode to Joy" theme was conscious and purposeful, a kind of tribute to the master he so revered and from whom he learned so much. This is the kind of thing that composers have occasionally done from time immemorial, sometimes in very subtle and hidden ways, sometimes in a more obvious or overt manner. That Brahms considered this reference to Beethoven's music to be fairly obvious was confirmed when, in response to one such early listener who pointed out to Brahms the similarity of the two themes, Brahms rather testily replied: "that's something any ass can hear."[43] And yet, for all of the theme's popularity and celebrity, it nevertheless eludes many conductors' understanding or, alternatively, provokes them to perform it in ways that do not correspond to what Brahms wrote in his score. There is, first of all, the question of tempo, which I have already alluded to, designated by Brahms as *allegro non troppo, ma con brio*.[44] To recapitulate, the range of tempo choices typically runs the gamut from the slowest (♩ = 100) to the fastest (♩ = 120), not considering a few extremists (Celibidache, for example, with a funereal ♩♩ = 80, and Jochum with a headlong ♩ = 138) on either side, a range of 20 points. Apparently those who are on the slow side emphasize the *non troppo* part of the tempo designation and pretty much ignore the *allegro* part, while those on the fast side emphasize the *con brio* as well as the *allegro*. That's the way it often is with unmetronomized tempo markings. My own preference is for a tempo of approximately ♩ = 100, give or take a few metronome points. I relate it to the opening *Adagio* of approximately ♩ = 50, half the *Allegro*'s tempo, but also to my *Più andante* tempo of approximately ♩ = 66, as it were, about one-third of the way between 50 and 100 and, in addition to the final *Più allegro* (m.391), which I do at a tempo of approximately ♩ = 100. In doing so, I am not just playing some number game or obsessed with the notion of mathematically relating all the tempos of the movement, but also that those tempos seem best to allow a realization of *all* aspects of the music.

In any case, the tempo of Brahms's *Allegro* is a legitimate conductor's choice which is left somewhat open by Brahms's reluctance to use metronome markings. What is less open to such a wide range of options are the dynamics that Brahms gives for the beginning of the *Allegro*: *poco f* in the violins. But apparently many conductors see only the *f* part of the designation and produce a heavy, thick, sensuous, overwrought string sound. They also fail to see the *p* in the horns, the *mp* in the cellos and basses, and the fact that the *two* violin sections are partnered at times (especially in mm.70–73) with only the *one* viola section. The latter point is important since Brahms's notation cannot leave any doubt that the violin and viola sections should match in a balanced *poco f*, which should be in my view something near or between a *mp* and *mf*—at most; and that balance is very difficult to achieve if the violins play too loudly. There are too many of them (violins)—in many orchestras it might be as many as 28

43. "*Das bemerkt ja schon jeder Esel*" (literally: that would be noticed by any ass.)
44. We know that Brahms altered the tempo marking from his original *Allegro con brio* first to *Allegro moderato, ma con brio* and finally to the perhaps even more explicit *Allegro non troppo, ma con brio*.

to 30 players—as against 8 to 10 violas. Obviously, if a dynamic match is to be achieved, the violins, playing on their richest, fullest sounding string (the G string), will have to moderate their dynamic level. Conductors should also keep in mind that at the recapitulation of this theme at m.186, Brahms varies the orchestration, as a kind of augmentation of the theme's sonoric amplitude, by adding among other things the cellos to the violins' theme.[45] The trick is to pay homage to all three of Brahms's tempo and character clues: *allegro, non troppo,* and *con brio,* that is, a basic 'lively' tempo, but 'not too much so,' and yet with a certain sprightliness. This can obviously be realized in several subtly different ways, depending on how much emphasis one might give to each of those three characterizations, and how one might interpret 'not too much.' For myself I see no contradiction in the 'not too lively' and 'but sprightly' notions, for a moderately lively tempo can achieve a certain *con brio* by the flow—its forward motion—and by a degree of lightness with which one inflects that tempo. I also see a major clue in the rather unusual staccato-dot marking of the pizzicato cellos and basses, a marking observed by hardly *anybody,* which signifies to me a certain briskness and lightness, as well as a clarity of articulation. It is doubly interesting that Brahms reiterates this marking when the violins and violas have the pizzicato sixteen bars later. There are further clues as to the somewhat restrained basic dynamic level in mm.62–77 (except for a midway crescendo and one *sf*), such as the bassoon's entry marking *p* (mm.70,72) and the return to *p* in the cellos and basses in m.74.

I should also point out another clue which seemingly no one has observed— primarily I suppose because it is not included in the score and parts usually used in performances, but which can readily be seen in Brahms's manuscript score (which has been available to be viewed for more than forty years and some years ago was published in facsimile)—namely, that articulation dots are situated above the two middle notes in first and second violins in m.64

. I know of no performance or record-

ing that has considered this phrasing articulation, which automatically lends a touch of lightness to the passage, as opposed to the gluey, syrupy, 'sexy' sound that most conductors favor here. As I have said before, no wonder many people have the impression that Brahms's music is turgid, thick, and heavy.

At m.78 Brahms indicates a uniform *p* in all the parts, a *p* which is only rarely observed. It is definitely meant to provide a significant contrast to the first

45. That is why I would disagree with Toscanini's decision to add a few stands of cellos to the theme already at m.62. It is an unnecessary enrichment of the string sound and anticipates—undercuts— what Brahms himself did 124 bars later.

One of the worst renditions of the *Allegro*'s main theme is to be heard on Abbado's recording, where there is an inept splice between m.62 and m.63, with the front end in a tempo of ♩ = 96, the continuation in a tempo of ♩ = 120. Again the question: How can a conductor or a record producer allow such an editing and performance bungle to be released to the public?

sixteen bars with a light transparent texture and crisp, nimble pizzicato in the strings and timpani. Instead this is usually regarded by the wind section as its chance to shine, to outdo the strings, and thus to 'play out.' Thereby not only is the mentioned contrasting effect gone, but Brahms's careful dynamic structuring, which will lead to a brilliant *ff* eventually in m.94, is subverted. Along the same lines, I believe the *sf*'s in m.73 and m.89 should be moderate in intensity, and the strings—not just the winds—should diminuendo in m.89 (probably to *p* in m.90) and thence begin a new crescendo.

Many conductors make an accelerando in mm.90–93 to arrive at the *animato* in m.94. I am opposed to this idea (a) because it is much more effective and in keeping with Brahms's structuring and periodization that the *animato* be achieved directly at m.94; (b) I am not absolutely convinced of the authenticity of the *animato* designation—it is not in Brahms's manuscript score (but it may have been added by him after the premiere but before the score's first printing); (c) I am not convinced that this *animato* (printed in small italic letters) signifies a tempo change, and if so, a radical tempo change, or whether it might mean just a subtle brightening, animating of the temper of the music—or a subtle combination of both ideas. I suggest this moderate cautious approach to m.94 because for many conductors this passage, especially the scalar runs in mm.106–13, has become an unbridled display of technical virtuosity, gaining a tempo of around ♩ = 144. This is, of course, superficially exciting but, again, destroys Brahms's fine sense of form, proportion, and balance of structure. The fact that Brahms doubles the speed of the notes in m.94—the actual perceived amount of rhythmic activity—from the more staid quarter-notes of mm.62–93 to clusters of eighth-notes in mm.94–105, and then doubles it again to sixteenth-notes in m.106, provides enough rhythmic intensity and excitement not to require any big tempo increase.[46] Also, the syncopations in mm.94–95 (in the bass instruments), which are unfortunately almost never heard, even on recordings, provide the essential 'animation,' i.e. increased motion and activity, so that, again, a big increase in tempo is quite unnecessary.

These syncopations are now followed up and extended in two different ways: the offbeat *sf*'s in mm.95–96 and mm.99–100, and in a polyphony of syncopations in mm.102–105. These last four bars are hardly ever realized correctly from a rhythmic point of view. Brahms compounds three unit levels of syncopations into a remarkable polyrhythmic composite. The high woodwinds and violas are written in very large syncopations (♩ ♩♩ ♩), the brass in syncopations of the next fastest unit level (♩ ♩　♪), the violins and bass instruments in the

---

46. It seems to be almost impossible to teach conductors as well as other musicians that (a) there is a difference between motion—sheer rhythmic activity—and tempo—the two concepts are separate and distinct; (b) great composers know how to create the illusion of increased momentum in the music by increasing the rhythmic activity *without* increasing the tempo; (c) therefore performers should not (without other overriding reasons) speed up the tempo when the composer has already composed into his music an acceleration of activity. In other words, sheer rhythmic activity creates a sense of acceleration, just as a decrease of rhythmic activity creates a sense of deceleration.

meantime being enmeshed in interlocking patterns

tion yet!—which are synchronized with the other two degrees of syncopations (see Ex. 31). Thus a rhythmic composite results in which every eighth-note in

Ex. 31

those four measures is articulated by some group of players. In order for that polyrhythmic complexity to become audible, it is necessary that all players perform their rhythms with very incisive articulations, not necessarily accents, but clearly articulated attacks.

We should note in passing how ingeniously Brahms transforms his main 'parent' theme (Ex. 32a) into three diverse variants (Exx. 32 b,c,d).[47] Let us also admire in mm.109–11 the ascending pattern of violin runs alternating between the firsts and seconds, constructed not in patterns of thirds, but of fourths (Ex. 33).

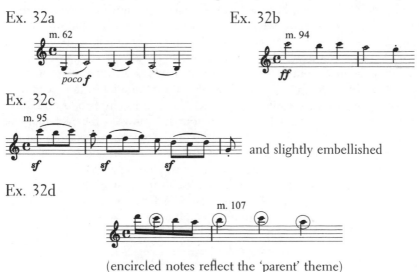

Ex. 32a

m. 62

poco f

Ex. 32b

m. 94

ff

Ex. 32c

m. 95

sf   sf   sf   and slightly embellished

Ex. 32d

m. 107

(encircled notes reflect the 'parent' theme)

47. Ex. 32c is, of course, reminiscent of several similar passages in the first movement of Beethoven's

Fifth Symphony   f   a thematic allusion which, if it

had been pointed out to Brahms by someone, would have undoubtedly again elicited his famous rejoinder "Any ass can hear that."

Ex. 33

Only one conductor—Stokowski in his 1927 Philadelphia recording—carries this passage out correctly, creating one single superbly balanced integrated line.

The next performance problem that arises occurs at m.118, where many conductors and orchestra do not achieve a real *p dolce*—instead we get the usual 'industrial strength' *mezzofortissimo*—often made worse (or perhaps even caused) by the fact that the violas, cellos, and basses, in an already inherently heavy three-octave unison, playing only four notes on one bow, use the full bow length, thus automatically raising the dynamic level considerably. These three sections should be encouraged to play very lightly: light in dynamic and light in texture—using very little bow. This little four-note motive is derived, of course, from the first four notes (in the bass instruments) of the Finale movement. Oddly enough, even the *f* in m.124 is in many recordings rather 'lukewarm,' odd because generally speaking musicians pounce on *f*'s with irrepressible relish.

I shall pass over the fact that most conductors rush through mm.130–31 with unrestrainable impetuosity to point to the next very interesting—and mostly mis- or underinterpreted—passage, mm.132–35. Most conductors manage some kind of *subito p dolce* at m.132, but then fail to make a significant crescendo and, even more, the all-important sudden *p/pp* at m.136. Only a handful of conductors have brought this off successfully on recordings: Tennstedt, Skrowaczewski, Janowski, Karajan, Haitink, Böhm, Kondrashin, and Walter.

The late Russian conductor Kyril Kondrashin makes an interesting point in regard to this oboe solo passage (mm.132–35) and its parallel in the recapitulation.[48] Noting that most conductors conduct this passage in 'two,' Kondrashin argues that it is more relevant to stay in four, pointing to the fact that the quarter-note triplets in mm.134–135—and I would add, the syncopations in the violas—retain much more of the typically Brahmsian rhythmic tension when played against a 4/4 beat and feeling. I think he is quite right. When the passage is conducted in 'two,' something may be gained in the flow of the passage, but the rhythms tend to get tensionless, very loose. A better idea of the rhythmic tensions, seen and heard vertically in three—and even four—levels (see Ex. 34), can be gained when feeling and conducting in 'four.'

Ex. 34

48. Kyril Kondrashin, *Die Kunst des Dirigierens* (Munich, 1989), p.37.

It is amazing to me in how many orchestras the violas—either tolerated or urged on by their conductors—pounce on the sixteenth-note tremolo passage at m.142 with a vigorous *mf* (or even *f*), when actually the prevailing dynamic is *pp*. Brahms writes only *marc*. The second violins in m.143 are asked to enter *p*, but, of course, won't do so if the violas have preceded them with a loud entrance. Also annoying is the bad habit of violins and violas accenting the third beats in every measure (the top notes in the pitch contour), when instead every attempt should be made to create a long six-bar tremolando line, from *pp* to the *ff* of m.148.

The next problem spot, rarely rendered correctly, is mm.156–57 (also mm.160–63) where the violas have to play eighth-note triplets on 'two' and 'four' without any follow-up notes on 'one' and 'three' (♪ ♫♫ ♪ ♫♫). This is indeed a difficult but not impossible rhythm to manage correctly. In most viola sections the rhythm turns into ♩♩♩ rather than ♪♪♪. Curiously, because they have been playing triplets all through mm.158–59, they usually play the triplets that follow in mm.160–65 more correctly. One of the few recordings where these triplets sound and feel absolutely right is the recent Berlin Philharmonic's with Abbado and the BBC Symphony's with Herbig.

In mm.164–67, many conductors are seduced by the temptation to urge the orchestra into too early (and too much) a crescendo. The excessive *ff* that then usually results in mm.168–75 contributes to a further obliteration of Brahms's superb contrapuntal workings here (see Ex. 35), including an expansion of roles between m.169 and m.173, combined with canonic imitations in mm.170–71. Most conductors and orchestras seem quite unaware of these important motivic relationships and their performance implications. Even the chattering triplet eighth-note figures in the woodwinds and horns (mm.168–69, 172–73), which look and sound like 'mere' passage work, turn out to be a variant of the violins'

Ex. 35

m.2 (or for that matter m.62) (see Exx. 36 a,b). It is so that wherever you look in the later pages of a Brahms symphonic movement, you will find that particu-

Ex. 36a   Ex. 36b

lar material to be a variant or extension of some previously stated elements—a theme, a motive, an innocuous accompanimental line, a bass line. And so too, for example, the descending lines in the oboes, clarinets, and horns in mm.170–71, which are a re-use of a similar descending quarter-note line in m.2 in the woodwinds.[49]

In m.176 (and mm.178,179) the abuse most commonly visited upon these measures is an enormous crescendo in the timpani into the fourth beat, some-

thing like [notation]. Needless to say, Brahms's score has no such dynamic indication yet conductors either tolerate or encourage this aberration because, presumably, it makes a grand 'effect.' But it is in fact a vulgar, primitive effect which simultaneously renders inaudible Brahms's excellent harmonic progression and the important woodwind triplet runs.

In mm.183–85 another form of abuse occurs with most conductors: an excessive alteration of the tempo. A subtle relaxation of the flow of the music is certainly permissible, indeed desirable. It is, after all, a major structural cadence and a modulation from the relative key of E minor back to the base key of C major. But more often than not what happens here is anything but subtle. There is usually a mighty wrenching back of the tempo, because the conductor has let the previous lengthy *allegro* section, especially the aforementioned *animato*, run out of control, picking up an enormous amount of exciting speed. But then, facing the return of the stately main theme (at m.186), he suddenly realizes that some tempo correction is in order. The various treatments given this passage—all wrong and unnecessary in my view—are shown in Ex. 37. Conductors who subject these measures to such dramatic tempo convulsions fail to realize that Brahms has already slowed down the motion of the music by reducing the speed of the notes from strongly articulated eighths and triplet eighths to quiet, smoothed-out quarter-notes. All one needs to do, assuming one has not inordinately rushed the *allegro (animato)*, is to gently relax the tempo *into* a *Tempo I* in m.186.[50]

49. The   only   reason   the   flutes   don't   have   the   complete   descending   line

—they have [notation] instead—is that

Brahms was reluctant to write high B's for the wooden flutes of that period. Substantiation of this can be found in mm.354–55, where the respective passage is located a third lower and the highest flute note, G, was technically within Brahms's reach.
50. Note the highly unusual omission in m.182 of B (the fifth) and D♯ (the third) in the two chords, respectively.

Ex. 37

A)	rit.	a tempo	rit.		a tpo
B)	rit.		piu rit.		a tpo
C)	no rit.	sub. meno mosso		molto rit.	a tpo
D)	no rit.	poco rit.		molto rit.	a tpo

The return of the main theme here in m.186 is a surprise, for it is not a true recapitulation (as in a proper sonata form) since there has not yet been any development section. Brahms here breaks with classical form and tradition by reversing the order of events: the development section develops out of an expanded return of the exposition. Brahms invents a fascinating new form—with a few wholly original elements added. The result was not only unprecedented but, to the best of my knowledge, has never been emulated by any composer. (I would be happy to stand corrected on this point.)

As Fig. 10 shows, the exposition contains not only five clearly discernible sub-sections but also its own mini-development section (C through E), so that when

Fig. 10

		A	62–77
		B	78–93
Exposition		C	94–117
		D	118–67
		E	168–85
		A¹	186–203
Re-exposition		B¹	204–19
		C¹	220–33
Development		F	234–84
		G*	285–300
Re-exposition cont.		D¹	301–51
		E¹	352–70
Transition		H	371–90
Coda		I	390–457

*The development section contains the climax of the movement at m.285 and a transition (G) from there back to D¹ for the continuation of the exposition.

it returns, like boxes within boxes, it is contained in a larger—the real—development section, which itself is encapsulated between the first and second parts of the re-exposition.

As we have seen so often, Brahms was virtually incapable of returning to some previously stated material without varying, refurbishing it in some way. Under his principle of 'perpetual variation,' the re-exposition starting at m.186 undergoes exactly such a conversion. The music is subtly enriched and elaborated in the most ingenious ways. We have previously noted the addition of cellos to the main theme here, giving the passage an extra sense of nobility and urgency. The previous bass line is now fleshed out with the addition of the timpani. Whereas the earlier version in cello and bass pizzicatos was

the re-exposition now offers (in basses and timpani, intermittently and abetted by bassoons):

(stems up = timpani; stems down = basses and/or bassoons)

It should be obvious that Brahms meant this to be heard (and played) as a single continuous balanced bass line, but judging by the many, many recordings in which the timpani is much too loud—in any case louder than the basses—it seems not to be obvious at all. It is useful and time-saving to rehearse these two (timpani and basses) alone, for the rest of the orchestra to hear. Meanwhile, the horns' previous purely harmonic accompanimental function has been reassessed. In alternation with the bassoons (see Ex. 38), the horns now participate in and enrich sonically the main melody. Although most currently available scores (and parts) say *poco f* in the bassoons, Brahms's manuscript clearly says *mf*. I think this is not only an important clue as to Brahms's desired dynamic level for this passage, i.e. the actual meaning of his *poco f*, but it also suggests to me in combination with Brahms's marking of *mp* for the horns, that bassoons

Ex. 38

M = main theme; A = accompaniment; bassoons upper staff; horns lower staff

and horns should alternate *mf* and *mp* dynamics here: *mf* for the thematic material (M), *mp* for the accompanimental material (A), so indicated in the example above. Finally, the reworking of the main theme in the re-exposition is completed by the addition of flutes, clarinets, and trumpets in a light byplay of quasi-pizzicato staccato notes (see Ex.39), all of which form a counterpart to the basses and timpani.

Ex. 39

I have gone into considerable detail regarding this passage not only to show the extent to which Brahms varies the recapitulation of his *Allegro* main theme, but to suggest how much will be lost if all these diverse elements and differentiated articulations are not represented fully and in appropriate balance. Incidentally, the *largamente* at m.186 is not to be found in his original manuscript. If it is Brahms's indication, it is meant, I think, to cancel the previous *animato*'s, and should be interpreted more as a subtle feeling of broadness than a substantial tempo change.

Brahms's constantly inventive ways of revisiting previously stated material show at almost every point along the way. Take, for example, the very sophisticated dynamic nuancing in mm.200–204—all of this roundly ignored in nearly all recordings and performances. The matter is complicated by some errors and omissions in the usually available scores and parts. A crescendo wedge is missing in the basses in m.200, while in the other strings the dynamics should

be [music], not [music]. As Ex. 40

shows, three different separate dynamic lines proceed simultaneously, all resolving in m.204 into an airy light *p*.

Ex. 40

The problems that arise in the pizzicato passage of mm.6–9 in the beginning of the Finale—problems of balance, we recall—are generally to be found again in mm.208–11 (and mm.216–19). The violas are usually too weak against the rest of the strings. One conductor who manages this entire passage, including mm.212–15, exceedingly well is Kondrashin. The pizzicato accompaniment is crisp and clear, well-balanced, and nicely flowing.

For a mere ten bars (mm.220–29) the recapitulation contains an exact repeat of the exposition, before Brahms veers off in a different direction. The earlier sixteenth-note passages—those that so often are played too fast for an obvious 'exciting' effect—are this time delayed and combined with a four-note motivic fragment, which first made its brief appearance in mm.97–98 and mm.101–102 (and, of course, in the just mentioned recapitulatory passage, mm.223–24 and mm.227–28) (see Exx.41a,b). This material is now developed in a most

Ex. 41a

Ex. 41b (its inversion in the basses)

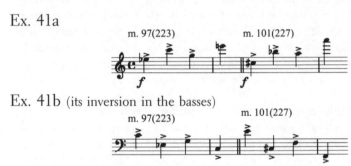

remarkable way, first in a striking upward leaping passage in the upper strings in mm.232–33 (Ex. 42a) and simultaneously, in contrary motion, in a downward bounding line in the low strings, consisting miraculously of the same pitches as in the upper strings (see Ex. 42b, diagonal connecting lines).

Ex. 42a

Ex. 42b

Incidentally, I am convinced that Brahms arrived at his for the time extraordinary ninth chords later in life, as in, for example, the Double Concerto for Violin and Cello, Op.102 (see Exx.43a,b) by way of these earlier melodic/harmonic/contrapuntal experiments.

Ex. 43a

(ninth and seventh chords are marked x)

Ex. 43b

Whereas in our present example from the First Symphony (the bass line in Ex. 42b), the pitches are still strung out horizontally, i.e. in succession, it did not take much to gather them together in a harmonic simultaneity, for exam-

ple  or or . It was still too early for Brahms to venture further into eleventh (or thirteenth) chord territory, but, of course, many composers within a few years of Brahms's death did just that; and thus were born not only the eleventh and thirteenth chords of the early 20th century, but also bitonality and polytonality. Brahms's descending bass line in mm.232–33, when gathered into a single chord, can be heard as a complex of several triadic or seventh-chord formations (Ex. 44)

Ex. 44

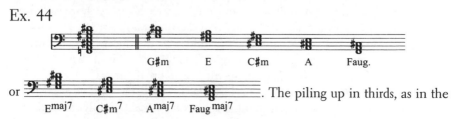

G#m   E   C#m   A   Faug.

or . The piling up in thirds, as in the

Emaj7   C#m7   Amaj7   Faugmaj7

first chord displayed directly above, became a favorite device of composers like Schönberg and Stravinsky as early as the first decade of the new century, and within very few years became a commonplace of 20th-century harmonic writing. I point this out primarily to underscore once again the astonishing modernity of Brahms's musical conceptions, of "Brahms the Progressive," as Schönberg put it in his famous essay.[51] Brahms was in so many ways always on the cutting edge

---

51. Schönberg, *Style and Idea* (New York, 1950).

of musical developments, despite his avowed conservatism and adherence to earlier classical models and values. Unfortunately, Brahms's modernity is more often than not either underappreciated, taken for granted or, from constant over-use (over-performance)—and even misuse—entirely ignored. It is also worth mentioning that Brahms had an ongoing fascination—a love affair—with me-lodic and harmonic construction in thirds, particularly falling thirds. The pas-sage just referred to is, of course, developed out of a series of thirds, clearly visible and audible in the bass line in mm.232–33, but disguised in rising sixths (the inversion of thirds) in the upper line. How consistently Brahms returned to such thirds constructions can be gauged by the fact that one of his most famous and beloved melodies, the opening theme of his Fourth Symphony, is built exactly on that principle and a similar series of thirds (see pp.381, 416).

The second way in which Brahms uses the aforementioned four-note motive (Ex.41a) is in a powerful contrapuntal/canonic passage, mm.234–43 (Ex.45), rhythmically foreshortened in stretto fashion in the later measures, at the same time pitting the motive against the sixteenth-note runs first heard in m.106 (which in turn were derived from the thirty-second-note runs in the introduction

Ex. 45

of the fourth movement). This passage and its even more sophisticatedly elabo-rated recapitulation in mm.257–67 are hardly ever played with any understand-ing of their content or structural conception. They are simply played in a per-functory "run-it-down" manner and in too fast a tempo,[52] fairly well emasculating the music. Conductors and orchestras who have, however, done well by these passages are Walter, Sanderling, Herbig, and Kertesz.

The cascading f sixteenth-note runs are suddenly subdued in m.244, turning into lacy, intermeshing p *leggiero* accompanimental figures in a delicate wood-wind interplay based on the head tones of the main theme (C B C A). The passage is cast in two phases (see Fig.11), first in a five-bar phrase, then in an eight-bar extension in which ingeniously the sixth through eighth measures are a re-orchestrated repetition of mm.3–5 of the first phase. It is a passage which offers few problems and normally goes quite well, as long as the p and mf dy-namics are respected. Surprisingly, however, the entrances of the clarinet and viola (in m.253) and the second horn (in m.255) are often covered. The prob-lem may be that Brahms gave no dynamics for the clarinet and viola—and most conductors don't offer one to the players, unless possibly when asked to do so—

52. Klemperer in his 1928 Berlin recording, for example, hits the amazing speed of ♩ = 156. Szell also pushes the tempo nervously forward and as a result the playing becomes rather ragged.

Fig. 11

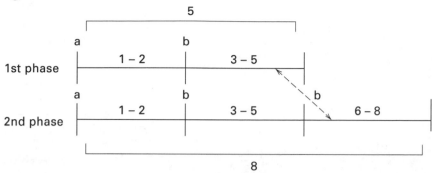

and the horns in m.254 are marked *pp* by Brahms, rather than *p* as in the other instruments at that point. It would be advisable to change that dynamic to *p* or

even *mp,* lest the second horn's lovely arpeggiated

line is lost in the over-all polyphony. In mm.257–67 Brahms recapitulates mm.232–43 in an elaborated form. He begins by inverting the original structure: the upper strings' leaping passage is now in the lower strings, and the descending jumps in thirds in cellos and basses are now in the violins (Ex. 46).

Ex. 46

Some of those pitches, C-A♭-F-D, are then isolated to begin a highly complex contrapuntal elaboration and enrichment of the earlier parallel passage. I use the word "complex" advisedly, for not only are the descending sixteenth-note scales now recast in three-part or two-part harmonizations (for example, the parallel triads in m.259), but Brahms produces a remarkable close canonic structuring, two beats apart, which includes shifting the leaping quarter-note figure one beat earlier than expected (in m.260, for example, in the low strings). The whole passage thus becomes a spectacularly complex and concentrated grid of contrapuntal lines (Ex. 47).[53] Unfortunately, in most performances (and the majority of recordings sampled) orchestra musicians are blissfully unaware of any of this intricate polyphony and simply plow through the music as best they

53. I am certain that Brahms was here influenced or inspired by the equally remarkable fugal/ canonic peroration in Mozart's *Jupiter* Symphony near the end of the last movement.

Ex. 47

can, with no sense of how all the parts of this multi-layered texture relate to each other.

Brahms now continues the strettoization canonically at even closer (i.e. eighth-note) range in m.264 in the strings—right out of late Beethoven!—(Ex. 48), while in the woodwinds a chain of sixteenth-note runs (Ex.49) leads to a

Ex. 48

major climactic section at m.268. This woodwind passage is, however, some-what difficult to bring off correctly. The problem is that the flute/clarinet line is in a favorable projecting range and has four instruments in octave unison, while the other line, in oboes and bassoons, is in less projecting registers as well as being divided in thirds (rather than unison). There is not a single recording in which the oboes and bassoons are heard as well as the flutes and clarinets,

Ex. 49

thereby destroying the single-line trajectory Brahms obviously had intended. It *can*, of course, be balanced properly by simply urging the oboes and bassoons to project well and, if necessary, by moderating the dynamics of the flutes and clarinets. Balancing this passage sectionally in rehearsal, as well as with (and against) the strings, is well worth the trouble.

The great twenty-bar climactic passage (mm.268–88), in two major phases, offers a number of performance problems. The strings' sixteenth-note runs are easily buried under the weight of the rest of the orchestra's sustained half-note chords and the timpani rolls. Overly loud timpani-playing, moreover, obscures the unusual bass line here featuring third inversion minor seventh chords

$$
\begin{array}{cc}
5 & 7 \\
(\text{G } 3 & - \text{ C } 5 \,). \\
7 & 3
\end{array}
$$
Also, conductors have for decades altered, i.e. modernized, the horn parts in mm.268–70, doubling the first horn in mm.268, raising both horns an octave higher in mm.269–70, and changing Brahms's original horn

parts 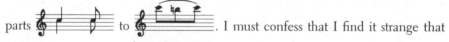 . I must confess that I find it strange that

Brahms did not write for the horns in mm.269–70, the B♮ being easily produceable on the natural horn for which Brahms always wrote (but which, by the way, no self-respecting horn player played any more by the 1870s). Brahms used much more difficult-to-achieve chromatic alterations throughout the First Symphony. Why he shied away from the above-mentioned easy solution is hard to fathom. In any case, my suggestion would be to keep the horns in the lower octave in mm.269–70 to preserve Brahms's initial intentions.

Many conductors also indulge in exaggerated tempo fluctuations here, most commonly a headlong accelerando in mm.274–77 and a big pull-back in m.278, all of which is quite unnecessary and a vulgarization of Brahms's already sufficiently 'exciting' music.

Measures 279–84 are one of Brahms's most original and daring conceits. Evolving quite organically out of the previous eleven-bar cumulative intensification of the movement's main theme, they also refer back to mm.27–28 of the introduction (in m.27, in m.279). It is also one of the more difficult passages in all of Brahms, since for six entire measures no one plays on the strong first and third beats—except, in effect, the conductor. It is one of the most totally syncopated passages in all music, and as such is unfortunately almost always played incorrectly, namely, as if written:

etc. when, of course, it is written

etc. It is another one of Brahms's remarkable rhythmic inventions, shifting the phrasing to the weakest beats of the measure. But I wish Brahms had notated the passage as follows: etc., for this would help players psychologically to anchor the syncopations rhythmically in the 4/4 metric frame. One of the few recordings on which this passage is right is Skrowaczewski's with the Hallé Orchestra: steady and well-paced, unfrantic, in which one can feel the underlying basic pulse and the fact that the rhythms are all syncopated off-beats.

These twenty-four propulsive syncopations culminate finally in an even more powerful outburst in m.285, a syncopated exclamation of the alphorn theme of m.30. It has long been a tradition to distort this passage with a tremendous slowing down and further ritarding in m.206–88. This is clearly wrong and unwarranted. Brahms could certainly have indicated a *ritenuto*, had he wanted one; second, Brahms has already slowed down the momentum of the music by augmenting all the rhythmic units from the previous ♪ ♩ ♩♩ to | 𝄽 ♩ ♩♩ 𝄾 |, against, by the way, even more slowed-down sustained dotted halves and whole notes, all of this requiring no additional *ritenuto*; third, changing the tempo in m.285 negates the whole point of Brahms's rhythmic augmentation; fourth, the second violins' and violas' sextuplet rhythms are a strong clue that Brahms wants to maintain the original *allegro* pulse. I have heard only three conductors who correctly maintain the tempo here: Erich Leinsdorf, van Beinum, and Lehel. Toscanini almost does. Curiously, Klemperer and Stokowski both slow up in m.285 but then accelerate in the next two measures, which makes little sense. Many conductors make a big fermata on the first, the empty, beat of m.285. Although most orchestras are by now prepared for this aberration, it *is* dangerous, and in the heat of the battle players sometimes anticipate the *ff* second beat (as, for instance, a bass player does on Dohnanyi's Cleveland recording). Incidentally, I suggest a slight diminuendo in all instruments except the first violins in m.283, to let the theme come through against this otherwise massive orchestral wall of sound.

Most conductors, having slowed down in mm.285–88, now have the problem of determining and re-establishing the tempo in m.289, an altered recapitulation of m.30. Most conductors blithely assume that Brahms made a mistake here and that the tempo must be the same as in m.30. How little they know their Brahms! Again he could have written *meno mosso* or some such designation, if he had wanted a slower tempo. Also, the timpani's eighth-note triplets in mm.289–97 are a good clue, related as they are to m.30's *sixteenth*-note sextuplets, that a faster tempo is wanted here, especially since there is a *calando* in m.297 in most printed scores.[54] The point is, if one is already at a slow tempo in m.289, followed by a *calando* in m.297, the music has to come to a virtual standstill by m.300, which, alas, is exactly what happens in most performances.

The relatively obscure marking *pf* in the horns and oboe in mm.289–90, ignored in all but the rarest of performances, means simply *poco forte*; and in effect should be no more than *mp*, rather than the obese overblown *ff* that is usually heard here. The appropriateness of the softer dynamic is further substantiated by the written *p* in the accompaniment and the *mf* two bars *later* in the first violins (m.291), also by the fact that a general *f* is not reached until m.293. Although violas and cellos can play *p* in mm.289–90 with the indicated bowing, they mostly don't, unless challenged to do so by the conductor. It is best

to add slurs here as Brahms did 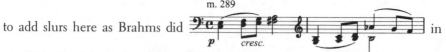 in

mm.291–92 and surely inadvertently did not in mm.289–90. One of the few conductors who handles this passage intelligently is Skrowaczewski.

I should point out that the diminuendo in the violins in m.299 is incorrect. The correct, and much more logical, placement of the hairpin dynamics in all the strings should be, as in Brahms's autograph

, and the *dim.* found in the printed

scores should be eliminated. The *animato* in m.301, corresponding to the one in m.118, is necessitated by the foregoing *calando*, but becomes unnecessary or self-evident if the *allegro* tempo has been maintained all along. What is much more important, it seems to me, and almost never realized, is the unusual, remarkably sombre voicing in the second cellos in mm.296–300, a sixth below the first cellos. Listen to Skrowaczewski's and Toscanini's recordings to savor how strangely beautiful this passage can sound when done correctly.

The conductors who make a huge ritard here, usually with a prolonged fermata on the bass B♭ in m.300, are, above all, Giulini, Abbado, Dohnanyi, Muti, Suitner, Bernstein, Böhm, and—with the longest fermata of all—Furtwängler (five seconds long!).

Measures 302–24 are for the most part a fairly literal repetition of mm.118–40, transposed of course (a fourth higher) and elaborated here and there, as in the addition of the four-note descending motive in the woodwinds in mm.303–310. The *animato* designation in m.301 leads many conductors to plunge through this section at enormous speed. When this occurs after an over-

54. This *calando* and *animato* in m.301, however, are not in Brahms's autograph, and are a later addition to the score, probably by Brahms himself—after the premiere.

extended fermata in m.300, it is an especially annoying rupture of the form. It is difficult to understand how an intelligent conductor, Rowicki, for example, can justify at m.310 the incredible tempo of ♩ = 160.

Brahms's dynamics in mm.316–20 are mostly ignored or misunderstood— only Skrowaczewski, Horenstein, Abravanel, Toscanini, Suitner, and Herbig seem to have paid close attention to them—even though similar to the parallel passage mm.132–36. It should be clear that the *poco f* in the violins in m.316 represents a *subito meno f*—something like a *mf*—followed in mm.318–19 by a crescendo, succeeded by another *subito meno f* in m.320, Brahms's notated *mf*. Be it noted that in Brahms's autograph the bassoons' dynamic in m.316 is *p*, not *mf*; also that Brahms has marked the first horn *p* in m.320 against the solo oboe's *mf*. This, again, is a dynamic nuance which is rarely observed by horn players and conductors.

Everything that has been said previously about the extended episode mm.142–83 should apply to its parallel, mm.326–67.

The cadence at mm.366–67, now in the base key of the symphony, C minor, is used by Brahms to initiate a transition to the coda, *più allegro* and C major. And what a magnificent transition it is! But also how fraught with all manner of performance problems, some of which are caused by engraving errors made by the Simrock editors over a century ago for the first edition. These errors, exclusively in dynamic markings, occur primarily in the first eight measures of the transition episode. In general Brahms's hairpin dynamics, especially in the strings (and bassoons in mm.371–74), should peak on the third beat of the second measure of each two-bar phrase, not the first beat as all printed scores have it. (The corrected version, perhaps presented here for the first time in the history of this symphony, can be seen in Ex.50.) What Brahms has in mind is that the rising arpeggiated melodic figures

 and their inversion in the

woodwinds, particularly in the bassoons,

 should crescendo—slightly, by the

way—to their final highest and lowest notes, respectively. The trombones, entering in m.371 after only very sporadic use since the end of the movement's introduction and thus a fascinating new color here, were marked *p* by Brahms, not *pp*. For me the most wondrous aspects of this eight-bar passage are (1) the remarkable harmonic progression in mm.367–74 and (2) the entrance of the contrabassoon on its low contra D♭ in m.373. I believe this harmonic progression is wholly original with Brahms, to my knowledge never composed or heard before in any music. I also believe that part of the beauty and mysterious effect of this sequence of harmonies lies in its remarkable symmetrical construction. Con-

Ex. 50

sider that the first two chords (C minor and E♭ minor) rise by a third, a minor third at that, while the two final chords (F major and D♭ minor) *fall* by a third, a *major* third. This, combined with the *rising* melodic figures over the first two chords and the *falling* figures over the final two chords—all so sophisticated and complex in its conception and yet so pure and simple in its outward effect— creates a magical mood of mystery that, it seems to me, is uniquely Brahmsian.[55] The passage fits superbly Schönberg's classic definition of superior creativity, paraphrased: "[In a great work] the profundity of the real meaning [does not] interfere with the elegance of the presentation and the polish of the surface."[56] How extraordinary this passage (mm.371–76) can sound can best be heard on Stokowski's and Haitink's recordings.

55. Further similar or analogous, primarily harmony-induced, magical moments occur in many of Brahms's other works, not only the other three symphonies but the *German Requiem* and the *Alto Rhapsody* as well.
56. Schönberg, *Style and Idea*, p.190.

The air of mystery, of ominous suppressed tension, continues in m.375, and now builds dramatically until its ultimate release in the brilliance of the C major stretto-coda (m.391). But this air of mystery is rarely well served in most performances. Again, the problem lies with too little attention given to Brahms's dynamic markings. Most seriously, the many *sf*'s are generally exaggerated and overblown—except curiously enough for the two important trumpet *sf*'s in mm.380–81, which in most performances and recordings are often woefully underplayed[57]—and, second, conductors either demand or tolerate a too-much and too-early crescendo. *There is no over-all crescendo until m.381.* In addition many conductors (Szell, Bernstein, Abbado, Celibidache, Ozawa, Ormandy, Järvi, Sanderling, Dohnanyi, Tennstedt, Furtwängler, Herbig, Weingartner, Muti, Norrington, and Skrowaczewski) start Brahms's *stringendo* of m.383 much too early. What results then is that both a *ff* and the *più allegro* tempo are reached four to five bars too early, making the real intended culmination of the crescendo and stringendo in m.391 an anti-climax. This is, of course, a much easier way of rendering—actually mis-rendering—this passage, for it makes the establishment of the *più allegro* much 'safer,' it having been reached several bars earlier. The harder, but also the better way, is to pace the stringendo at first less and then progressively more, hitting the new faster tempo only at m.391 (*più allegro*).

Moreover, if the *stringendo* is done too much too early, the second violin and viola parts in mm.385–87 become virtually unplayable and, in such performances, usually end up in rhythmic unison with the cellos and basses. Just as important, but seemingly unknown to most conductors, the aforementioned *sf*'s, are all *poco sf*'s, that is, in the context of a basic *p* dynamic level. This is especially crucial in the third trombone, cellos, and basses, but also in the violins. Be it noted as well that the bass instruments (including bass trombone) have no crescendo in mm.375, 377, and 379. (The diminuendos in the bass trombone in m.376 and m.378 are spurious, undoubtedly the work of an overzealous Simrock editor.) The same might apply to the woodwinds, who often, despite Brahms's differentiated dynamics, tend to make various phrase and over-all crescendos. The violins must always return to *p* in mm.377, 379, 381, which, of

---

57. I suspect that this may be due to the fact that Brahms's trumpet-writing is considered by most musicians, including conductors, not very 'exciting,' and as a consequence little attention is paid to the symphonies' trumpet parts. It is true, of course, that his trumpet-writing was extremely conservative and unadventurous, much more so than his horn- or trombone-writing. Brahms was the only late 19th-century composer who adhered to a use of the trumpet that goes all the way back to the very introduction of trumpets into the orchestra in the late 17th and early 18th century, that is, in permanent association with the timpani. This was a long-standing tradition harking back to the days when trumpets and kettledrums, on horseback, were the essential instruments in marches, parades, and military and royal festivities.

Brahms uses the trumpets exclusively as 'natural' trumpets—even though the valved instrument had been in existence since Haydn's day—and primarily partners them with the timpani. Still, he occasionally finds wonderfully imaginative ways of employing the trumpets, as, for example, the beautiful radiant entrance of the trumpet in m.39 of the last movement, or the powerful sustained *ff* in m.25 in the introduction to the first movement.

course, they are not likely to do if the whole orchestra is already started on a big over-all crescendo and a premature accelerando. Let the reader—and the prospective performer—be assured that the more the crescendo and stringendo can be held down or delayed, the more exciting and overwhelming will be the releasing climax at the *più allegro* (m.391). The conductors who most correctly and yet most excitingly realize this passage on recordings are Toscanini, and, more recently, Skrowaczewski.

A variant of the main theme

forms the basic material for the coda, which also incorporates a recapitulation of the chorale theme heard in the introduction (mm.47–51). Although it appears here (m.407) in a totally different context—a brilliant *allegro* rather than a moderate *andante*—most conductors unfortunately treat the coda chorale in a ponderous stentorian manner, as if Brahms had been incapable of writing *meno mosso* or *più maestoso* or *largamente* or some such terminology to indicate a slower tempo. Brahms surely meant the chorale to be rendered this time in a blaze of glory, in triumphant return—a kind of *Hosanna* or *Hallelujah,* not a sad melancholy prayer. I submit also that the combination of a slow- and heavy-paced chorale followed by a headlong charge for the finish line at m.417, which most conductors seem to favor, is one of the corniest and cheapest of effects. Keeping the tempo in the 'chorale' not only preserves Brahms's organic structuring but makes the remaining closing music of the movement even more exciting, more driving, because it is more logical, more organically generated.

I realize that conductors who have always done—or heard the chorale done —slowly, are not likely to be persuaded by the idea that it is much more effective in the bright tempo. But should anyone want or need evidence to that effect, a listen to Leinsdorf's and Skrowaczewski's recordings, to name just two of the very few that respect Brahms's score and wisdom, should be convincing. Toscanini stays almost in tempo, while Bernstein with his penchant for exaggerations drops a staggering 70 metronome points (*sic*) (from $\textbf{J}$ = 120 at m.39, to $\textbf{J}$ = 50 for the 'chorale'). Although Stokowski also makes a huge ritard into the 'chorale'—*and* has the nerve to double the orchestra with a full diapason organ—he is the *only* conductor who brings out the B♭ in the bass instruments in m.406, a most crucial note in leading the music into the brilliant A major of m.407. In the vast majority of recordings the B♭ is totally inaudible, obliterated by the timpani.

The sudden alternating waves of triplets, dramatically new and refreshing in this otherwise very much duple-oriented movement, find their release in the three powerful quasi-unison 'commands' in mm.431–43. But Brahms here is once again up to his favorite trick of seemingly shifting the beat. But it is only an illusion. Like the work of a great magician, the feat is anchored in logic and reality: the music's underlying pulse must not be destroyed in performance. It is, as always in such passages, the *tension* between the audible rhythm and the

underlying pulse that must be upheld. This cannot be done if the conductor
and orchestra simply convert the music to

(with 2 lower octaves)

. Matters are made worse when tim-

pani players are encouraged (or allowed) to make a murderous crescendo into
and a culminating accent on the last quarter-note of mm.434, 438,443. All that
needs to be done to perform this passage correctly is for the entire orchestra to
think and *feel* the downbeats in each measure, and above all not to diminuendo

on each note, as if written

—which

is the way it is most often played. In mm.439–43 there is, however, an orchestrational or balance problem, hard to resolve. It is a case of Brahms running out
of instruments to fully realize his idea. Looking closely at these five measures,
one will note that that the first two pitches, C and A, are reiterated canonically—and, significantly, on the strong beats in certain instruments (Ex. 51).

Ex. 51

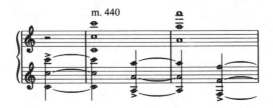

m. 440

However, when Brahms arrived at the F and D which complete this phrase, he
found that he had indeed run out of instruments to continue the canonic reiterations; for the F he found only one instrument, the first clarinet, and none for
the D. With these on-the-beat pitch alliterations, I believe Brahms wanted to
restore partially the metric equilibrium in the third and final of the three unison
'commands,' with an eye (and an ear) towards the full unequivocal re-
establishment of the pulse in m.444. It is a shame that Brahms did not have
enough instruments in his orchestra to realize fully this remarkable idea. I have
not heard a single recording among the fifty or so I have studied in which the
conductor seemed even aware of this problem, let alone attending to it in some
way. (In my own performances, so as not to tamper too much with the score, I
merely ask the first clarinet to play its F *fff* with a good accent in m.442—also
taking away the connecting slur—and ask the second clarinet to join in unison;
its A is already well represented by the first trombone. I leave the D unreiter-
ated, reluctantly.)

The final, final performance problem that remains in the closing fourteen
bars of the symphony occurs in mm.450–52, where the magnificent, imperiously

rising bass line ⟨musical notation⟩ is usually made completely inaudible by over-enthusiastic timpanists *and* conductors going for the ultimate cataclysmic applause-getting effect.

This account of Brahms First Symphony recordings has not been pleasant to write, although I have done my best to point out wherever possible the felicitous and occasionally grand interpretations of certain fine conductors. I would much rather have written a more positive report, in which respect for Brahms's magnificent score was much more evident. It is staggering—and I can well believe that many readers will read my findings in disbelief—to realize that a masterpiece as virtually perfect as this great symphony could be so consistently abused and misused, misunderstood and misinterpreted, its myriad notational details cavalierly ignored. It is indeed a sad commentary on the conducting profession. But all the more praise for those relatively few who have distilled from Brahms's score a moving, imaginative, compelling performance—without excessive distortions and deviations: Toscanini, Weingartner, Leinsdorf, Kondrashin, Suitner, Haitink, Skrowaczewski, and even Stokowski, who despite some of his capricious eccentricities, really loved and understood this work.

Brahms: Fourth Symphony

Although it took Brahms more than twenty years of painstaking, often struggling effort to complete his First Symphony, the Second Symphony appeared almost instantaneously and was premiered within a year of the First's premiere. The Third Symphony followed in another six years and the Fourth two years later; a mere 30 opus numbers separate it from the First. Thus the three last symphonies were created in less than half the time of the prolonged gestation period required for the First Symphony.

This is remarkable enough, especially if one considers that in the same nine-year period Brahms also wrote some ninety-five other compositions, including the staggering number of sixty-eight songs. But perhaps even more remarkable and interesting is the fact that his four symphonies are at once totally original, and in mood and conception completely different from one another. As uniquely distinctive as each of the four symphonies is, all are quintessentially Brahmsian and could, by no stretch of the imagination, have been written by anyone else, not even by Dvořák, Brahms's closest artistic colleague and musical soulmate.

This uniqueness is particularly appreciable in Brahms's Fourth Symphony, a work which, even more than the First, combines the most astonishingly 'modern' and intrinsically complex musical ideas with a fundamental adherence to classical and pre-classical models, as particularly manifested in the *Passacaglia* Finale movement. For all its complexity and frequent daring and radicality, the elegance and polish of its surface, its sheer naturalness and accessibility, are never disturbed. Brahms's extraordinary inventiveness and ingenuity in exhaustively exploiting every motivic/thematic feature—what one may aptly call the 'intellectual' side of Brahms's creativity—is not very much appreciated by the average listener, although he can dimly sense the logic and attractiveness of what he is hearing. Unfortunately, most orchestra musicians and most conductors have little more understanding and appreciation of Brahms's musical intelligence than the public. They relish the great tunes, revel in the obvious climaxes and rich harmonies, but rarely appreciate the extraordinary craft, skillful structuring, formal control of the musical materials that went into the creation of his mature works.

This should be evident from the preceding perusal of the First Symphony and its recorded performances. That minutely detailed account, if read with care, will save me—and the reader—from examining the Fourth Symphony in similarly exhaustive detail. All my previous inveighings against premature accelerandos and crescendos, against inattention to dynamics, against exaggerated modifications of tempo, against willful arbitrary deviations from the score—in short the whole petty paraphernalia of misguided musical conduct—applies as well to the discussion of the Fourth Symphony, saving us a lot of unnecessary redundancies. There are, however, many moments in the work that are uniformly abused, misinterpreted or in some crucial way ignored, that are, moreover, special and unique to the Fourth, and which, therefore, must come under discussion.

The first eighteen-bar statement of the first movement, one of the most beloved and popular theme expositions in all classical music, is nonetheless rarely played correctly. Leaving aside for the moment the always vexing problem of tempo—tempo considerations have already been ventilated too many times to warrant further reiteration now—there are subtler performance questions that are rarely even noted, let alone resolved. Take for example the violins' first eight bars, a theme which surely everybody thinks they know and know *how* it is to be played. The fact is that it is almost never played correctly; moreover, it is very difficult to play really correctly. What one usually hears is any one of four versions (Exx.1a,b,c,d).

What Brahms actually wrote can be seen in Ex. 1e. The problem with the four

Ex. 1e

misinterpretations (with the possible exception of Ex. 1d) is that they dissect what should be an eight-bar lyric thematic statement into eight tiny separate motivic fragments, strung together in succession, like so many link sausages. To maintain the long line Brahms obviously had in mind but made more problematic to achieve by the inclusion of the quarter-note rest, one must sustain each half-note the exact length—four eighth-notes' worth—without diminuendo or tapering off, and *feel* or *think* across the rest to the next upbeat quarter-note. To do that precisely and musically, that is, not mechanically but feelingly, is extraordinarily hard, but with care and concentration can certainly be done.

Most conductors, of course, conduct the violins here, a gesture that looks fine in the audience—especially since it offers the audience the conductor's handsome profile—but it does not necessarily do justice to the fullness of Brahms's musical idea. It leads rather quickly to an over-balancing of the violins, to the detriment of the woodwinds. But the woodwinds, who shadow the violins at a respectful distance of two quarter-note beats, are an integral part of the theme and must not be slighted. If the woodwinds are attended to at all by conductors, one usually hears only the flutes, while the clarinets and bassoons in less projecting registers remain virtually inaudible. (Disbelieving readers are invited to sample any available recording.) The bassoons especially add a slightly darkening color to the three-octave counterplay, a favorite orchestrational device of Brahms, as we have already noted several times.

Even the arpeggiated accompanimental figures in the cellos and violas do not necessarily 'play themselves,' as the saying goes. Each bar should comprise one single upward gesture—if we can imagine Brahms playing and composing this at the piano, we will come close to realizing how it should sound—each measure then connecting with its successor into a twelve-bar understructure on which the melody can float and sing its song. The violas should not approach their figure as if it were a new entry, but rather as a continuation of what the cellos have initiated. My suggestion to help towards this long-line effect is to bow the violas and cellos as in Ex.2. I can find only two recordings on

Ex. 2

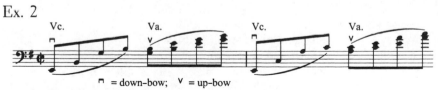

⊓ = down-bow;   ∨ = up-bow

which all aspects of this wondrous opening subject are treated intelligently, musically, and in balance. They are Toscanini's (NBC) and Walter's (Columbia Symphony).

The astute reader with a good memory will note a close relationship between the Fourth Symphony's main subject and a similarly constructed passage in the First Symphony in the last movement (see Exx.42 and 46, pp. 364 and 367). The fact is that Brahms loved to construct ideas out of falling thirds. As in the First Symphony example, the Fourth's main theme's first four measures are really derived from the following sequence of pitches in *descending* thirds

, while the next four bars are built

out of *ascending* thirds . Moreover,

as the English writer Ivor Keys has pointed out in his Brahms biography,[1] there is in this E minor symphony a remarkable concentration on and emphasis, both melodically and harmonically, of the minor-sixth C, which, however, almost always resolves back to the fifth, B. In the opening subject the C is brought out by inverting the descending major-third E-C into an ascending minor sixth, which by m.4 has resolved back to the B. The reiterated C's in mm.9–12, restated in varied form in mm.153–56, even more pronounced in the rhythmic augmentation of the opening subject in mm.246–58, and the final restatement of the theme in mm.394–406, as well as in severely concentrated form in mm.422–29—to point out only some of the more salient moments—continue the interplay with C throughout the movement. The play continues with the harmonic ambiguity of the unharmonized beginning of the symphony's second movement, which until E major is established in m.4, can certainly be heard in C major; or in the final measures of that movement the magnificent shift from E major to C (mm.114–16), which in turn prepares for the C major third movement, and the many C-emphasizing A minor diversions in the last movement. Perhaps it is this emphasis of the minor sixth—and frequently of F (the minor second of the tonic key)—that gives the symphony its melancholy yet consoling tonal flavor.

Great confusion exists regarding a Brahms marking, used often in this symphony, a short hairpin dynamic (**<>**) in mm.9–12, placed in such

a way notationally that it can lead to various interpretations. Does mean or or, as some conductors even suggest

. Actually, it doesn't mean any of these; it simply means a

certain bow or left-hand (vibrato) warming of the sound, in effect a subtle *espressive* accentuation. Violinists and conductors should avoid an automatic diminuendo in m.10, just because the phrase falls; m.10 is still part of the over-all crescendo which should not reach more than *mf* in m.12, as it also should not in the varied and expanded repetition (mm.19–38) of the main subject. Measure 31 should not rise above *mf* so as to be able to complete the crescendo to *f* in mm.33–37. More on that later.

A well-hidden problem exists in mm.17–18, where the first note of the oboe's entrance is hardly ever heard, even on recordings. I must confess that I am on a one-man campaign to restore this beautiful phrase to its rightful place in the melodic/thematic scheme of things. The point is that Brahms wrote

as a beautifully arching line. But

this is invariably destroyed by the first violins' over-enthusiastic crescendo in

1. Ivor Keys, *Johannes Brahms* (London, 1989), pp.188–89.

m.17 (usually on the G string with a thick heavy vibrato), completely covering the oboe's first B. My suggestion is to remove the violins' crescendo in m.17 (it will happen anyway with the octave rise), modify the woodwinds' dynamic to *poco f* or *mf* (especially the four horns), and urge the oboe to play its most projecting lower B. The only recording on which the oboe's initial B can be heard is Mehta's (probably accidental). Notice how the oboe's C resolves to the second violins' B—again that C to B resolution—initiating in turn a remarkable transformation of the main subject.

The violins' theme is now broken up into alternating eighth-note fragments, antiphonally divided between the two violin sections. And yet the fragmentation should not lead to more isolated 'sausage linking.' This is, however, what invariably happens when violinists indulge their aforementioned bad habit of chopping off—throwing away through inattention—their last notes in each little four-note grouplet. In order to make the main theme survive *as a theme*, the two violin sections must *not* drop their last notes, either rhythmically or dynamically, and, beyond that, should try to think/feel across the rests between them. All it takes is an awareness of the problem and a kind of responsive playing that two violinists would do automatically in a chamber music—say, a string quartet—situation. Meanwhile, in m.19 the former woodwind 'echoes' of mm.1–8 have been relocated in the cellos and basses, but a fifth lower. I see this not as a mere accompaniment but, as far as the dynamic level is concerned, as an integral part of the theme/main subject.

One of the most abused passages in the early pages of this symphony is the eighth-note figurations in the violas and woodwinds in mm.19–26. Almost no one seems to have the imagination to comprehend what Brahms means by the annotation *legg. Legg.* (*leggiero*), of course, means 'lightly' in Italian, and is Brahms's way of telling us that these descending lines are to be played lightly, as a kind of embroidering accompaniment to the main subject, filling in the rhythmic gaps left by the rests in the primary theme. But these lines are usually played by the respective players and sections as 'solo' passages, thereby obscuring and threatening the priority of the main theme (in the violins). We should note that even these *leggiero* lines are an embellished variant—embellished with passing tones—of the first three notes of the main subject (see Ex.3a), delineating an F minor triad. While the fourth theme note, C, appears prominently in the first violins (m.20), the remaining thematic notes appear in the viola part in

Ex. 3a

(encircled notes are thematic)

m.21 (Ex.3b). In the meantime the clarinet and bassoon in m.20 and the flute in m.22 echo in embroidered fashion the cellos' and basses' versions

Ex. 3b

of the theme (Exx. 3c and d, respectively), again at the respectful distanceof two or three beats. Thus ingeniously, these lines are doing double duty, as it

Ex. 3c   Ex. 3d

were: (1) serving as an embellished horizontalized variant of the woodwinds mm.1–4, and (2) simultaneously, as a variant of the cellos' and basses' recasting of that very same material. Talk about economical use of materials!

In m.23 Brahms converts the eighth-note garlands from single measure exchanges to half-bar alternations, all still marked *legg.* These are invariably played too loud, especially by the woodwinds, treated by the players as expressive 'solo' passages, rather than as the discreet accompaniments they really are. Nor should there be any crescendo in these four measures. (Brahms originally had a *poco cresc.* in m.23 but, according to the autograph, had second thoughts and moved it four bars later.) Moreover, every effort must be made to collect these eight four-note bundles into one *Klangfarben* line: oboes and violas,[2] alternating here with flutes, clarinets, and bassoons in Brahms's favorite three-octave distribution.[3]

The worst offenders in this passage are Furtwängler and Celibidache. The former makes an enormous crescendo in mm.23–26, surging to a full *f* in m.27, when in fact the score tells us that that bar should still be *p*. Celibidache's dynamic distortions are even more extravagant and peculiar. Like Furtwängler (whom Celibidache idolized and mimicked early in his career), he starts a big crescendo in m.24, continues it through mm.27–30, arriving at a full *f* in m.31, and then makes a big two-bar diminuendo (sic) to start m.33 in *p!*—all a reverse of what Brahms has written.

I have already alluded to a special performance problem in mm.33–36. Instead of an over-all four-bar crescendo in the violins, now in parallel thirds, one invariably hears the dynamic rendition of Ex.4a. Even more musical damage

Ex. 4a

2. Note how often in his symphonic works Brahms associates the viola, almost his favorite instrument (almost as much as the mezzo soprano or alto voice), with woodwinds.
3. I suggest, by the way, to shorten all group-ending quarter-notes to eighth-notes so that a better over-all line may be achieved and the harmonically destructive unpleasant dissonances resulting from a pedantically precise rendition of the quarter-notes avoided.

is done when the violins drop their final notes in each measure (dynamically and rhythmically). Between these two misinterpretations the result is four huffing and puffing short-breathed phrases instead of one grand over-arching crescendo line.This occurs through sheer inattention on the part of conductors, concertmasters and violinists, and not listening to the produced result—my 'third ear' (see p.17–19) would do well to come into play here—to see whether what is being played in fact relates to what is in the score. To help counteract any musical and physical/ technical tendency to diminuendo in m.34 and m.36, the bowing in Ex. 4b suggests itself. The only recording that manages to repro-

Ex. 4b

duce these four measures more or less correctly is Skrowaczewski's, while Mravinsky's and Furtwängler's fail utterly: Mravinsky with huge hairpins (◁▷) *per measure* but no over-all crescendo, Furtwängler in a full, virtually out of control *ff*, with no *possibility* of a crescendo.

The rhythmic problems in the famous transition theme in mm.53–56 (and its many later incarnations) are too well known to warrant much comment here. Brahms's vivid rhythmic imagination concocts a theme which contains in it five different rhythmic units ($\downarrow. \_ \downarrow \_ \downarrow \_ \downarrow \_ \downarrow^3$) in ingenious juxtapositions. To play it accurately is not altogether easy, but it is generally at least taken rather seriously by orchestra players, except, however, for the initiating F♯, a note which is almost always played too short, that is, as if Brahms had written:

Don't ask me why! Let us notice also that the unusual, odd character of this theme, apart from its rhythmic uniqueness, resides largely in the aforementioned C (minor sixth) to B (fifth) relationship, only here transposed up one tone to D-C♯.

At this theme's next appearance there is real performance trouble. It is absolutely astonishing that on not a single recording are mm.73–75 and mm.77–79 (and all similar parallel passages) played correctly, even Carlos Kleiber's, Skrowaczewski's, and Reiner's in so many ways splendid recordings. What is even more astonishing is the *manner* in which this passage is played incorrectly: the last notes of each rhythmic layer (see Ex. 5) in each measure are always

Ex. 5

played together! In this, one of Brahms's more brilliant rhythmic inventions—
once again pitting 'three' against 'two'—the duple rhythm players' fourth beat
must fall between the second and third notes of the conflicting triplet. But
somehow in all orchestras—I say in *all* orchestras—the two rhythms accommodate each other in some convenient way, ending up together and thereby, of
course, taking all the invention and tension out of Brahms's idea. And no conductor seems ever to notice this discrepancy and attempt to correct it. Astonishing and lamentable! (Once again, I invite any disbelieving reader to check
any recording of their choosing for verification of what I have just stated.)

It is a most disturbing example of the casualness—and sloppiness—with
which rhythms are generally treated in most orchestras, especially triplets.[4]
Granted, the time differential between the fourth beat and either one of the last
two notes in the triplet group is minuscule and not probably feelable or technically controllable in relation (or in reaction) to one another. At a tempo, say,
of ♩ = 72, a common tempo for this movement and this passage, the elapsed
time between the attack of the second triplet note and the fourth beat of the
measure is about one-sixth of a second (or equivalent to a beat at metronome
360), a time span not really controllable by the average instrumentalist. So one
can perhaps understand how easily the two rhythms can become congruent: all
that needs to happen is for the fourth-beat note to be played a hair late, and it
will coincide with the last of the triplet notes; or alternatively, if the triplet is a
tiny bit rushed, its third note will coincide with the fourth beat. (This is precisely
what happens on many recordings where the woodwinds and two C horns
play ♩. ♫ ♫ ♩ instead of ♩. ♫ ♩ ♩ ♩—as, for example, on
Weingartner's 1938 London Symphony and DeSabata's 1939 Berlin Philharmonic recordings; or, as on Celibidache's 1959 Milano concert recording, where
the same players play a convenient ♩. ♫ ♩ ♫.) Either or both of these
misreadings invariably occur, alas—mostly the former—as any number of re-

---

4. It is astonishing—and very depressing—that the vast majority of musicians do not know how to
play triplets precisely, especially so-called 'large' triplets, i.e. (a) | ♩ ♩ ♩ | or
(b) | 𝅝 ♩ | or (c) | ♩ ♩ ♩ |. Everyone does pretty well on the 'small'
triplets, i.e. | ♫♫ ♫♫ |, less well on | ♩ ♪♩ ♪ |. Almost universally, 'large'
triplets are played not as | ♩ ♩ ♩ |, but as | ♩. ♩. ♩ |. The quick solution
to this problem—and hardly anyone seems to teach this anymore—is to go to the next fastest levels
of triplets as needed, which means, in trying to play (a), above, correctly—precisely with the right
feeling—thinking | ♩♩ ♩♩ | or, if that doesn't achieve the right result,
| ♫♫ ♫♫ ♫♫ ♫♫ |. This is such an elementary rhythmic lesson that I am
continually surprised that it hasn't been learned (seemingly, it is not being taught).

cordings will attest. On a very few recordings one can hear a dim, accidental awareness on the part of the musicians—not the conductor, mind you—that those last notes in each rhythmic configuration should not be together. Curiously, the basses and cellos on those recordings enter the fourth beat almost on time in the first measure of the passage (m.73), but by m.74 have been seduced—or intimidated—into siding with the triplets, playing that measure's fourth beat late and in perfect congruence with the violins' last triplet note. And, shockingly, *nobody*—no conductor, no musician, no recording producer— ever notices this and challenges the conductor and/or themselves to work out the problem.

The only way this remarkable rhythmic idea can be properly realized is if the duple rhythm players concentrate on playing the fourth beat *exactly* in time—it may sound easy, but is actually rather difficult to do (as the countless inept recordings readily show)—and if the triplet rhythm players concentrate equally hard on playing their rhythm *precisely*. Remember that with even the slightest rhythmic deviation on either side—166 milliseconds are an infinitesimal instant of time—the two attack points will come together. If then in addition the players observe the *sf* on their last notes, Brahms's intended effect will be finally fully realized. And how extraordinary and unique it is! Most musicians and readers will not know what I mean, since it has hardly ever been heard properly played by anybody. I know that when I have exposed the problem to orchestras and rehearsed this passage very carefully, there is general amazement upon hearing it finally played as written—and consternation and chagrin that they have never before played correctly something so elementary.

What makes matters worse is that there are in the Fourth Symphony at least half a dozen similar rhythmic passages which, to the best of my knowledge, have never been played correctly, and certainly not on any of the thirty-odd recordings of this symphony that I have studied. (We shall return to these places at the appropriate time.)

Triplet and duplet rhythms battle it out in Brahms's symphonies all the time: so too in mm.91–94, where the triplets in the violas and cellos are almost always suppressed by conductors, as a kind of rhythmic annoyance against the 'ecstatic' violin and horn melody.

Skipping over various interpretive and performance questions, I turn briefly to mm.110–11, where the half-notes in the woodwinds, within the *pp* dynamic, must be firmly sustained and for the precise duration. This is easier said than done, as many recordings testify; for invariably what one hears is ♩·· or ♩ ⎯⎯ or ♩___|♪, but not what Brahms wrote: ♩ without ⏤. That Brahms was quite serious about these durational aspects is seen in the various succeeding versions of this 'fanfare' passage, where we see ⅄ ♫|♩· and ⅄ ♫♩ and ⅄ ♫♩__|♩·.

One of the most remarkable rhythmic passages ever written in the entire symphonic literature occurs in mm.128–32. It is also another passage which, as far as I know, is never played correctly. After another cross-accent passage (mm.127–29) similar to the one discussed above, combining triplet and duple-rhythms, Brahms tops it all with a multi-layered structure of such complexity that I dare say there is nothing like it even in the *Rite of Spring*; one has to turn to Ives's Fourth Symphony to find a parallel. Capitalizing and expanding upon the two most predominant musical ideas Brahms has been just working with— the little 'fanfare' motive and the triplet-against-duplet second subject—which until now have been kept separate and heard only in succession, they are sud-denly thrown together in a three-way collision of rhythms. Leaving aside the violas' fast sextuplets which almost sound like a tremolo, we have quarter triplets in the violins, ♩ ♩ ♩ ♩ ♩ ♩, in the bass instruments we have reg-ular quarter-notes ♩ ♩ ♩ ♩, and in the flutes, oboes, horns, and trum-pets the 'fanfare' motive ♫ | ♩. ♫ ♩.. There is even a fourth rhythmic layer in the timpani, in effect ♩ ♩ ♩ ♩ ♩ ♩, although written as | ♩ ♩ ♩ | ♩ ♩ ♩ ♩ |. Putting these rhythms all together—I have marked them a,b,c,—and relating them vertically we see that Brahms has cre-ated a composite rhythm 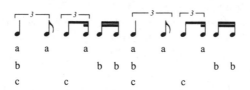 which, counting all the attack points, is astonishingly intricate and indeed difficult to render accurately in its multi-layered complexity. Notice that, as in a bell pat-tern, the three layers coincide rhythmically only now and then, in this case on the first and third beats (see Ex.6). The example does not include the viola

Ex. 6

sextuplets, but shows which notes belong to which layer (alphabetized). As if this were not enough, Brahms has shifted the phrasing of the quarter-note layer (c) one beat early, starting on the fourth beat of m.129 and stretching it to six beats, the result being two 6/4 entities stretched across three 4/4 measures.[5] This is borne out by the way Brahms continues this bass line in mm.133–34 (Ex.7), the rhythmic anticipation and shifting now reduced from a quarter-note to an

Ex. 7

_____
5. This is not unlike the famous 6/4 passage in the *Rite of Spring* at the end of the "Cortège of the Wise One," rehearsal number 70.

eighth-note. It should be obvious that all this rhythmic, polyphonic multi-layering takes great care in rehearsing and performing, which, alas, it almost never seems to receive. I have found only one recording in which these measures (mm.130–31) are played correctly, a quite obscure recording at that, by György Lehel and the Budapest Philharmonic. The Fourth's performance and interpretation as a whole is rather unspecial, and the orchestra's playing and ensemble are generally lax and rough (not to mention the recording's poor technical quality). Therefore, it is a little difficult to vouch absolutely for the accuracy of performance in these three measures, but as far as my ears can discern, it sounds correct. Without having been present at the sessions, there is no way to tell *why* these three measures were played correctly—uniquely so in the history of recordings of this work. Whether it was Lehel who saw the problem that no one else seems to have been aware of, and then rehearsed it carefully; or whether it was the first trumpet player who made his colleagues play the sixteenth rhythms correctly, I cannot tell for sure. But I tend to think the latter, judging by the forcefully secure manner in which that player seems to lead his colleagues. On another recording, Levine's with Chicago, one can hear the great trumpeter Adolph Herseth trying to fit the trumpets' two sixteenths correctly into the over-all rhythmic scheme, but to little effect since the rest of the orchestra's rhythms are played so imprecisely. One of the most anomalous and vexing renditions of this passage is DeSabata's with the Berlin Philharmonic. The orchestra in a technically virtually impeccable performance —the Berlin Philharmonic was in 1939 in its absolute heyday—plays the rhythms in mm.130–32 as accurately as I have been able to hear, except for the fact that someone (DeSabata?) changed the woodwind and brass parts to

(Yes, I didn't believe it either, and had to listen to it in disbelief several times to verify the fact.)

I know that when I have tackled this passage in my own performances of the symphony, it has required at least fifteen to twenty minutes of rehearsing time—sorting out and clarifying the three discrete rhythmic layers section by section—to finally arrive at the correct (desired) result, much to the consternation and amazement (and self-satisfied delight) of the musicians.

Of the many performance and balance problems that occur around the *Scheinreprise* (false recapitulation), mm.145–68, I single out: (1) the need to make sure that the E of the basses in mm.139–40 balances well and expressively with the B-major dominant chord in the winds—most conductors and recording producers seem to shy away from this 'dissonance'; (2) the need for the third horn player to know that he is to blend with the two clarinets in m.153, in effect a third clarinet, not the little 'solo' that most horn players think they have; (3) the need for the violin sections in mm.153–56 to shuttle their little motive, which first appeared in m.9 (and which is a close relative of the First Symphony's last movement's main theme), back and forth so as to create one interlocking line. What usually happens instead is that, through unawareness, the players—with conductors' blessings—drop the last note either rhythmically or dynamically or both, making any linking up of the alternating phrases impossible for the listener. The same will apply twelve bars later to the solo woodwinds. A similar dropping of final notes occurs almost always in the flutes and bassoons in mm.156–65, a transformed and transposed (to G minor) variant of the main

theme. From  etc., Brahms derives,

with passing notes added,

 etc.

As in the case of the beginning of the symphony, a way must be found to make one grand line out of these eight three-note bundles, and surely the way

to achieve that is *not* to play . It also

does not help when in m.161 the clarinets burst in too loud, as if they have some important 'solo passage,' when in fact they are merely the continuation of the *leggiero* accompanimental lines started in m.157 (which, of course, hark back to those analogous figures in mm.19–26). Brahms evidently forgot to mark m.157, m.158 (violins) and m.161 and (clarinets) *leggiero*; he did so two bars later—that is, two bars too late—for the violins.

Notice also the tiny melodic/harmonic clashes Brahms builds into this G minor episode: C♯ and A against C♮ and B♭ respectively (m.157), F♯ and D against F♮ and E♭ (m.158), and even more piquantly, C♯/A against E♭/C♮ in m.160, and so on. These dissonantal delicacies must be brought out, must be heard in a performance. Generally, however, the conductor and musicians involved—flutes, bassoons, violins—seem to be unaware of anything interesting going on at all; the notes glide by unnoticed, unheard, unsavored—and bland.

In m.168 we come to one of the more complex and motivically convoluted passages in all music, as two three-note (or six-note) fragments, generated in the previous *Scheinreprise*, battle it out contrapuntally, canonically, in a fierce *f -ff* fray, which is finally resolved in m.184. To do full justice to this extraordinary passage in performance is very difficult, because it presumes that every player in the orchestra would know exactly at every point which cog of this gigantic wheel he is representing, the problem being exacerbated by the fact that the roles initially assigned to specific instruments and players change and are given over to others at various times during the sixteen-bar passage.

Let me see if I can sort out these multiple motivic strands in an easily graspable format. To begin, let us isolate the two main motivic fragments: one, the flutes'/bassoons' variant of the main theme (mm.157–60) is further trans-

formed into (*a*) , appearing initially

in the violins and in canonic imitation two beats later in the bassoons, only to appear four bars later in the cellos/basses and canonically in two horns. Its inversion (a^1) [music] appears initially in the cellos and basses, but is then transferred to the violins (mm.173–78).

The other motivic fragment (b) [music] is derived from the violins' mm.153–54, also, of course, from

[music] m. 9 [music] or, as condensed,

[music] m. 155 [music]. It appears initially in two horns, but is later given over to flutes and oboes. Its inversion (b^1) [music]

appears initially in flutes and oboes, but is then handed over to clarinets and bassoons. Thus the following polyphonic structure (Exx. 8a and b) evolves, at least in its first two phases (mm.169–76). To help the reader through this labyrinthian complex in musical notation (simplified and reduced), the various fragments and strands are identified by a (the first one cited above), a^1 its inversion, b the second motive, b^1 its inversion. Notice again the dissonantal clashes (E♭ and E♮, m.169, 173; A♭ and A♮, mm.170,174).

Ex. 8a

Ex. 8b

Then, at the end of m.172 Brahms inverts the entire previous four-bar structure, somewhat as Beethoven had done at one point in the last movement of the
Fifth Symphony. We should also note that much of the structure is metrically off
kilter. The *b* motive has been shifted over to the weak fourth and second beats,
and even the *a* motive, with its canonic interlocking and uniform accentuation,
gives the impression of being in a constant state of ambiguous rhythmic flux. In
effect, the entire passage in all its component parts sounds strangely unanchored
rhythmically/metrically; and in fact, it needs to be played in a kind of agogically
'neutral' way.

At the end of m.176 the fearsome contrapuntal structure undergoes still another sea change. Brahms, heading for an eventual cadence and resolution of
the polyphonic struggle, simplifies the structure somewhat. But again, he does
so ingeniously and utterly logically. In mm.172–73 we find in the first violins

the following:  . The G♭ and F♮ are

the last two notes of the *a* (main theme) material; the A♮ is an extra note, not
part of any thematic reference, added simply to provide some harmony (the
third in an F major chord); the next B♭ is, as we have already seen, a part of an
inversion of the main theme. Brahms now takes these three pitches, F-A♮-B♭,
and fashions a new three-note motive from them, puts it in the woodwinds, and
uses it as a link between successive violin phrases (see Ex.9). This idea is then

Ex. 9

further expanded at the end of m.178 (along with a sudden *ff*), the woodwinds

reiterated (and transposed) in cellos and basses 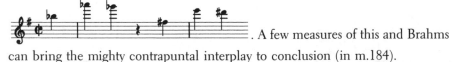 and im-

mediately canonically imitated in the upper woodwinds, while the violins,
bassoons, and horns take up the second three-note part of the theme

 . A few measures of this and Brahms

can bring the mighty contrapuntal interplay to conclusion (in m.184).

I have gone to some length to detail the construction of this interlocking,
multi-layered, timbrally delineated structure (mm.169–84) to indicate its complexity and difficulty of performance, a complexity of design which can be rendered meaningful in performance only if, as I suggested above, *all* the performers are aware of the inner workings and construction of the entire passage. This
is, of course, never the case. Orchestras usually just plow through this passage
as best as they can, not really knowing where even one note in their parts be-
longs, or why it is there. Musicians, no matter how good or how intelligent,

cannot just from their individual part glean anything of the intricacy of structuring of the various relationships that their parts have to all the other parts. The only way to make this aurally and intellectually clear to the musicians is for the conductor to separate out all the component parts, and then rehearse them all separately in a series of tiny 'section rehearsals.' That takes time, but is well worth the effort. I have often rehearsed the passage this way, and not only has the resultant performance been more 'intelligent' and 'understanding' on the part of the musicians—an intelligence which transmits itself to the audience in performance—but also the musicians seem actually quite relieved (as well as pleased with themselves) to have been able to bring some sense to a passage which they previously considered merely dense and obscure.

In mm.184–85 (and subsequent similar passages, like mm.202–203) the prob-

lem is that almost everyone plays

instead of Brahms's .

Skipping to m.219, I find that in virtually all performances (and recordings) too much attention is given by conductors to the clarinets and bassoons, rather than the recapitulation of the main theme in the pizzicato strings and flutes.

Measures 227–46 is one of those heavenly passages which only Brahms could devise. It is made up of a very beautiful and original chord progression and two little motives, first heard in m.9 and m.10. This sounds simple enough and, on the face of it, unproblematic; and yet I have rarely heard this passage played well. The problem is that the *Klangfarben* connections, both in the melodies and the underlying harmonies, are difficult to achieve, given that the players, just staring at their parts, have little or no idea whom they follow, whom they succeed, how the entire twenty measures function musically. It also takes a bit of painstaking rehearsing to make sure that *all* the hairpin dynamics are matched up between *all* the different players and sections.

The chord progression (Ex.10), distributed among alternating winds and strings, is in itself a thing of beauty. Highlighting this wondrous progression are

Ex. 10

the harmonies in m.229 and m.230 (and its sequence four bars later), the one a third inversion E♭ dominant seventh, darkened by its seventh in the bass, followed instantly by the other, a bright widely spread C major seventh chord in root position. The melodic line, meanwhile, colored with at least four different timbres, should sound as follows (Ex.11), that is to say, one continuous *Klangfarben* melody.

Ex. 11

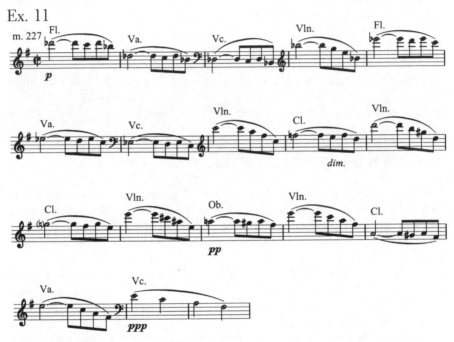

It takes enormous concentration on the part of the harmony players to sustain the whole-note chords exactly the right length, and at the same time maintain the right *p* dynamic so as to be able to hand the chord over, as it were, to the next succeeding group of players; also for *all* participants to produce the same degree and rhythmic timing of the hairpin crescendo-diminuendos in each measure. The same applies to the melody players, whether individual soloists or entire sections. On almost all recordings, they drop their last eighth-notes dynamically, making a musical connection to the next bar impossible. One recording on which these nineteen bars are beautifully played is Mehta's with the New York Philharmonic. Near the end of this episode, say from m.240 on, many conductors make a ritardando and, worse, even more so at m.243 where Brahms has already slowed the rhythm from eighth-notes to half-notes. As Strauss already complained (see p.90), "subjects which the composer has already drawn out, should not be drawn out further." The worst offenders here are, once again, Furtwängler, Celibidache, and Fischer-Dieskau (the latter two probably imitating Furtwängler), who slow to an incredible ♩ = 38 at m.243, after having shuttled back and forth between the mid-60s and lower 80s. But Chailly,

Mengelberg, and Sanderling are not far behind, slowing to half speed when Brahms has already done it in his composition.

A similar exaggerated tempo distortion occurs in dozens of recordings in the beautiful passage mm.246–58. Here Brahms takes the first eight notes of his opening main theme and elongates them to the following (Ex. 12).

Ex. 12

The long-held C and B are in effect written-out fermatas (six times the length of the original notes). Still, conductor after conductor does this already drastically 'slowed-down' music at a snail's pace, all but destroying Brahms's formal proportions and basic *allegro ma non troppo* pacing. Wagner's "imperceptible," "hardly noticeable" tempo modifications would be much more appropriate here.

Our discussion can skip the recapitulation (which *seems* to start in m.259 but which is actually already in progress as of m.246), except to note that the crescendo in m.277 this time reaches a full *f* by m.281, quite different from the earlier parallel passage in mm.9–12.

Another battle of 'three-against-two' occurs in mm.387–89, the twos in the winds, the threes in strings and timpani. Simpler than the other rhythmically complex passages I have referred to earlier, nevertheless even this one is rarely played correctly. On innumerable recordings, notably Walter's and Celibidache's, the winds and timpani play their notated duplet quarter-notes squarely with the strings' triplets. The three bars mm.387–89 are another place where a horrendously loud timpani has ruined many a recording.

The total effectiveness of the towering climactic passage at m.394 depends very much on two factors: (1) that the half-notes be well sustained, that is, not tapered dynamically; and (2) that they not be held too long, that is, as , which is the way they are unfortunately almost always played. (This is in essence the same problem as at the very beginning of the Symphony.)

There are two additional very special moments in the first movement of the Brahms Fourth that warrant specific mention, and particularly fascinate me as a composer and orchestrator. One is the remarkable and wholly original harmonic progression in mm.418–20, centered and voiced in open tenths in the lower instruments (Ex.13). It is itself an inversion of a similar but descending line in thirds (not tenths) in mm.414–15. This is in turn an elaborated recapitu-

Ex. 13

lation of m.45 and mm. 49–50 in the exposition, the latter in tenths in contrary motion (Ex.14). What makes this ascending progression so unusual is its wide

Ex. 14

'open tenth' spacing and the augmented seconds (bracketed in Ex. 13), which in earlier harmony manuals used to be called 'false relations.' Brahms is here reaching out towards a harmonic language and voicing, a parallelism that antici- pates Debussy and Ravel and other early turn-of-the century modernists.

The passage becomes even more remarkable—and would even sound *start- ling*, if its 'remarkableness' were ever appreciated by conductors and musicians who, through over-use, blithely glide through these measures as if they were very ordinary—when one realizes that the *rising* tenths are complemented by a line of *falling* tenths in the high register (Ex.15). The resultant harmonic

Ex. 15

clashes (marked x in the example) again are rarely appreciated and heard/felt (or brought out) by performers. This amazing passage must be played with a rich, vibrant, intense sound—especially in the cellos and basses—to attain its full power of expression. Of all the recordings sampled, the only ones that cap- tured this power and intensity are those of Maazel, Mravinsky, and Toscanini.[6]

Even more startling is a slightly later passage (mm.428–30) of shifting parallel 6/4 chords (Ex.16), a remarkable progression, which would have been consid- ered 'wrong,' 'inept,' and 'impermissible' only a few decades earlier—and was probably still considered so by conservatives at the time of the appearance of this symphony. But unfortunately even this uniquely Brahmsian passage has, through decades of repeated performances, lost all its novelty and modernity for most players and conductors. I have often been successful in rekindling some wonderment in orchestral players over this progression by isolating the two

6. Let us note in passing the frequent use of high B's and C's in the flutes in this symphony, an advance over Brahms's more restricted use (only up to A) in the First Symphony.

Ex. 16

m. 429

chords in m.429 in rehearsal, having the musicians hear it out of context in all its startling nakedness. The reader might kindle (or re-kindle, as the case may be) the same sense of wonderment for the passage by going to the piano and playing it to himself.

Finally, I feel obliged to protest the ubiquitous vulgarization of the first movement's final measures on the part of countless conductors, by either slowing the tempo inordinately, often as early as m.430, or by an overly bombastic timpani, especially in the last two measures—or a combination of both. (The way some conductors over-indulge in these mannerisms, one would think they were conducting the end of the Shostakovitch Fifth or the Mahler Third.) One of the worst offenders is Celibidache, who slows suddenly in m.430 to a pedantic ♩ = 76, after surging along at a bright ♩ = 92 for most of the coda, and expands further into a ponderous timpani bombardment in the penultimate measure. DeSabata, Mengelberg, and Maazel are not far behind in this type of tasteless (but audience-grabbing) trivialization of Brahms's noble and stately ending.

The Fourth Symphony's second movement is also full of wondrous, uniquely Brahmsian moments, a few of which I would like to dwell upon here. The first is the very introduction of the movement and its transition to the main theme at m.5. Apart from the unusual harmonic ambiguity of the opening—is it C major or E minor phrygian?—let us note and appreciate the organ-like instrumentation of mm.1–4. Brahms was not unfamiliar with the organ and its literature—in his younger years he had composed and performed (conducted) a fair amount of choral music, often for just female choir, with organ accompaniment—and it seems to me that he was here in m.2 playing the organist, that is, bringing in the four-foot stop with the oboes and the two-foot with the flutes, gradually eliminating them again in m.4.

I think what happens in m.4 is most extraordinary. Having teased us for the first three measures with a harmonically undefined theme statement, Brahms suddenly resolves the ambiguity by the unequivocal establishment of E major in the second half of m.4 (in the clarinets). Unequivocally, yes, but also in typically Brahmsian fashion, quietly and subtly. I take Brahms's *very* precisely marked dynamic notation in m.4 to mean that the unison E's, especially those reiterated as late as the fifth eighth of m.4 by the horns, should envelope and subtly mask the E-major entrance of the clarinets. By that reading the entry of the clarinets (and first bassoon) must not merely be truly *pp* but unobtrusive and barely audible. The effect then is of the E major rising quietly out of the previous harmonic vagueness, just as an island might rise out of the disappearing morning mists in a lake or shoreline scene (Hear how wonderful this can sound on Celibidache's recording.) It takes impeccable control on the part of the players, especially the three entering voices of clarinets and bassoon, who must enter as a perfectly balanced and blended trio. In most performances, alas, one hears a too loud first clarinet—thinking it has a 'solo'—and a too soft and self-effacing second clarinet, and an either too loud or unblending bassoon. Admittedly, this is all very difficult to manage, as, by the way, any second bassoonist will tell you, having to enter pp in m.6, taking over from the second clarinet.

The entire ensuing, quiet, march-like procession must not be disturbed by either the flutes entering too obtrusively in m.8—flutists tend generally to overplay, that is, over-compensate in their low-register passages—or, for that matter, the horns in m.13. These should merely continue the softly flowing clarinet melodic line. (Many horn players see this as an important 'solo' entrance, and one seldom hears a true 'clarinet' pp here.) A beautifully executed transition from a clarinet to a horn sonority can be heard only on a few recordings, notably Levine's, Reiner's, Barbirolli's, and Kempe's. On Mengelberg's recording, on the other hand, the horns mindlessly invade the quiet with an implacable, obese mf.

As soft as the horns and bassoons must play here, they must nonetheless sustain every one of their notes, especially in the important harmonic suspensions, to achieve the full effect of Brahms's beautifully melancholy harmonization. Consider the typically Brahmsian daring of the following key points, extracted and thus isolated from the complete passage in Ex.17.

Ex. 17

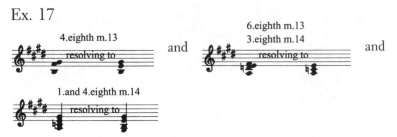

Another detail that calls for special mention is the exchange of overlapping phraselets between first bassoon and second clarinet in mm.20–21, both together providing the 'bass line' for the entire phrase. A caution also for conductors to not allow the first clarinet to diminuendo too early or indeed to play the passage too softly: it is far too often played in a mere (though mellifluous) mf, instead of full f. Nor should the typically Brahmsian sophistication of grouping the sixteenths in mm.20–21, in both pitch organization and articulation/phrasing, in six groups of three be allowed to undermine the basic underlying 6/8 march-like pulse.[7]

In mm.22–26 Brahms conceives one of the most remarkable—and original—

---

7. The same admonition must be voiced in regard to the unusual clarinet phrasing in mm.11–12.

musical ideas, for which I have not been able to find a parallel in any music, either by Brahms or any other composer (at least in any absolute, text-less music).[8] I am referring to the gradual fragmentation and distension to which Brahms subjects his main theme in mm.22–26. The phrase starts out as expected in m.22, but already in m.23 Brahms excises a beat and a half's worth of theme—compare m.6 and m.23—only to continue later in the same measure as if the phrase had been fully carried through. In the process, however, Brahms subtly drops the horn, which had been temporarily in the lead, allowing the melody to revert back to the clarinet, the pizzicato strings in turn having, almost unnoticeably, filled in the gap left by the missing melodic fragment.

But then suddenly, in mm.24–25, the most extraordinary thing happens: Brahms interrupts the clarinet's phrase again, and whereas in m.23 the incision in the melody is momentary and does not impair the original structuring of the phrase, this time the theme comes to a complete halt—in a kind of *phrasis interruptus*. At first thought Brahms seems to want to liquidate the theme; but surprisingly he picks it up exactly one bar later and continues with it as if there never had been any interruption at all[9] (Ex.18).

Ex. 18

m. 24   m. 25

This uncanny—and subtly humorous—bit of musical legerdemain is accomplished with such sly skill and ingenious artlessness, that one is apt to not even take note of it. Just as Brahms rid himself of the horn in m.23, he now goes about eliminating (in m.25) the pizzicato accompaniment, continuing the phrase in that measure solely with a quintet of winds, without benefit of strings. It is interesting to note in the autograph score a *più p* in the strings on the sixth beat of m.24, as well as a diminuendo wedge in m.23, both markings confirming his intention to liquidate the strings gradually. Brahms, however, crossed out both markings for his final draft, probably reasoning that the *pp* at m.22 was soft enough not to require further dynamic moderations. But I think his original impulse was right—and I usually ask the strings to play even softer here, or at least not let them play louder, that is, soloistically.

I have no documented proof to corroborate my just-related scenario, but from

8. Such exists—as, for example, in the text- and scenario-oriented interruptions in Beckmesser's serenade to Eva in the second act of Wagner's *Die Meistersinger*; or the so-called echo aria in the Cantata, "Flösst, mein Heiland, flösst dein Namen," of Bach's *Christmas Oratorio*. But I have not found any in purely instrumental music.

9. It reminds me of a hilarious routine by the great comedian-musician Victor Borge in which sentences are constantly interrupted by various distractions and interpolations, only to be blithely continued seconds later as if there had never had been any interruption at all.

all internal evidence of the music, it seems to me at least a highly plausible one. Furthermore, in the autograph facsimile p.56, which begins with m.25 there is clearly to be seen—unmistakably in Brahms's hand—a caesura sign ‖, used by musicians to indicate a break in a phrase. In any case, I do not recall in the many times that I played this symphony as a horn player *any conductor* ever offering such an explication of one of Brahms's most ingenious devisings, or for that matter any explanation at all or ever stopping even to rehearse this passage. Nor can I find any recording in which the performers seem to have understood that something highly unusual is going on here (possibly Carlos Kleiber's, Chailly's, and Suitner's).

To make my interpretation of this passage clear in performance, I ask the quintet of winds in m.25 to breathe after the quarter-note (or at least to subtly interrupt the phrase), even though they, of course, have no need for a breath. But if Brahms had not interrupted and delayed the phrase, that is, left it intact, then a quick breath would certainly have been needed, as is the case in the original main theme at this point (m.8). I also ask the clarinets and bassoon as best as possible to think across the rests in m.24–25—to 'hold the thought,' as it were—and to continue the theme (in terms of attack and dynamics) *exactly* as they had left it in m.24, in other words as if the phrase had never been interrupted.

Before we leave this passage, one other aspect calls for particular mention, namely, the crescendo all the way to *f* in m.27 and immediately back down to *p* in m.28, so different from the initial statement of the same theme.

In the glorious, warmly singing passage of mm.30–36, care must be taken to bring out (or at least not underplay) the highly unusual A♯'s in m.32 and m.33,

and to not ignore the remarkable dissonance   in m.33 (fourth beat).[10]

Similarly, the tension-producing harmonic suspensions in the woodwinds in m.34 must be well sustained—sung out, as it were—to assure the full effect intended by Brahms. And one other performance problem in this passage must be noted, namely the crescendoing 'too-much-too-early' syndrome. To crescendo prematurely takes away from the glowing, quietly ecstatic beauty of the beginning of the second subject (mm.30–31). Many conductors stoke the crescendo fires immediately (Masur, Janowski, Szell, Mehta, Toscanini, for example, and, worst of all, Furtwängler, who reaches a full *ff* by m.32). It is even worse to reach a climactic peak at the beginning of m.34, then diminuendo as the violins'

10. I have for many years theorized that Stravinsky, with his remarkable harmonic ear, in his neo-classic works learned to take chords, such as the one just displayed here, which he heard and discovered like so many 'found objects' strewn all through the great classical and Romantic litera-ture, and used them without resolving them harmonically—as Brahms certainly still had to do.

line dips down registrally, only to come up again at the end of m.35—an inadvertent 'misinterpretation' which unfortunately happens far too often. Even a modest knowledge of and sensitivity to acoustic matters and relative registral projection of instruments ought to alert interpreters to the reality that, for example, on a violin, the notes on the E string project more readily than notes on the thicker G string. Thus, in the passage in question (mm.34–35) the first few notes in m.34 project well and brilliantly, but as the line descends in the second part of the measure, its projection also lessens, while at the end of m.35, as the line skips back up into the high register, the projection increases again dramatically.

The result in notational terms is | f ⏵ | mf ⏵ | f , certainly not what Brahms intended. This dynamic roller-coaster effect is easily avoided if the initial crescendo, starting in m.31, is held to a modest increase, arriving at no more than *mf* in m.34 and then making sure that a further crescendo is maintained. Brahms's phrasing and bowing in mm.34–35 lend themselves excellently to such a gradual augmentation of the dynamic. It is sheer laziness and inattention that allows this passage to be 'misinterpreted' in the manner described. On recordings, those conductors who solved this problem well are Maazel, Suitner, Haitink, Reiner, and, above all, Skrowaczewski, whose beautiful pacing, dynamic control, and warm singing sonorities in this passage contribute to making his recording of the Fourth Symphony one of the finest ever.

In the soulful second subject 'cello episode' of mm.41–49 care must be taken that (1) the violas, while ceding priority to the cellos, must nonetheless be clearly audible as the 'bass line' throughout; and (2) that the second violins in m.42 sustain their final notes well—there is no diminuendo!—so as to render in its full daring the extraordinary clash of pitches with the violas and bassoons on the last eighth—actually a cluster 🎼. We should note that as soulful and poignant as this cello melody is, it is derived from its opposite, a somewhat rhetorical, stubbornly rhythmic passage that enters the fray in m.36 (compare the two in Exx.19a and b).[11]

Ex. 19a   Ex. 19b

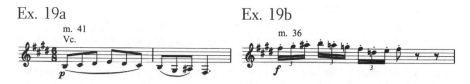

One of my special interpretive *bêtes noires* is the manner in which mm.57–59 are usually performed, namely as if Brahms had written:

11. At   the   end   of   this   episode   (m.51),   I   wish   that   Brahms   had   written
as he had done in the three directly previous presentations of this motive (clarinet and first violins, m.50; bassoon, m.51). It would have made a more logical and smoother over-all line.

alternating Vln. + Fl.   alternating Va., Vln.

that is, shortening each second note in the descending chain of note pairs. Apart from the fact that Brahms was quite capable of putting staccato dots over notes that he wanted played short—his scores are full of such examples, but here in mm.57–59 there are no shortening dots—it is silly and mindless to perform the beginning of the passage as I have described and then have the violas, who have the full continuous pattern starting in the middle of m.59, play in a sustained manner, as invariably happens, absent any conductorial suggestion to the contrary. It may help those who adhere to—or inadvertently permit—this type of disjunct phrasing to think of Brahms composing and playing the passage at the piano, as was his wont. It is highly unlikely that Brahms would have played it in such a disjointed fashion. In any case, if a conductor feels some compelling need to interpret mm.57–59 in that erroneous manner, then that conductor should also extract a similar interpretation from the violas and all the other strings in mm.59–63, and beyond that in the ensuing wind accompaniments (mm.64–71). Or is that too logical?

I pass quickly on to the recapitulation of the second subject in mm.88–97. This time the melody is in the first violins. Again there is the 'cluster' clash in the second bar (m.89), B-A-G#, with the G#, however, now in a lower non-adjacent octave. But in Brahms's re-casting of this already harmonically and contrapuntally rich second theme, he dispenses with the earlier descant garlands in the violins, raises the dynamic level from *p dolce* to *poco f espr.*, and enriches the texture by expanding the number of voices in the accompaniment from three to seven, at times even eight. This especially copious and dense musical weave is made all the more remarkable by Brahms's wholly original use of low divided cellos, with the second cellos adding a particularly 'dark' coloring to the entire passage. Unfortunately this typically Brahmsian sound is suppressed or ignored in recording after recording; only a handful of conductors (Haitink, Skrowaczewski, Maazel, Szell, Reiner, Barbirolli) sees the special dark beauty of this unusual instrumentation.

I will assume that conductors and performers know by now that the special magic of the woodsy, impressionistic passage near the end of the movement (mm.106–10) depends crucially on everyone involved playing a real *ppp*—not just 'mezzo softly.' The rustling string tremolos and softly throbbing timpani thirty-seconds combined with veiled horns and bassoons provide a gossamer sonic carpet for the delicate *pp* clarinet and oboe solos. If the diminuendo and *smorzando* in m.109 are observed, presumably causing the phrase to end in *pppp* in m.110, then the full *p* of the recurring clarinet theme appears to bloom forth with the intended dynamic and textural contrast. The point is that the clarinet theme should not have to be played more than *p*, lest it lose its nostalgic 'sad farewell' mood. It also should not lose its feeling of momentary reanimation after the near-stillness of the previous pastoral passage; but that cannot happen if the latter is not played at that heart-stoppingly soft level. The few conductors who realized this passage (mm.106–12) beautifully on recordings are Chailly, DeSabata, Skrowaczewski, and, above all, Carlos Kleiber.

After the pensive resignation of the clarinet phrase, Brahms's next move comes as a total surprise, if not actual shock. The very opening of the movement is recapitulated, but now unequivocally in E major and harmonically tremendously enriched. Indeed, the particular harmonization is very special in Brahms's canon. I know of nothing like it in the rest of Brahms's *oeuvre*; in fact, I cannot quite explain the unmistakable Spanish Andalusian flavor of this coda (m.113). The reader may well ask: "Spanish? What Spanish?" Compare, then,

Brahms's

and to a similar quintessentially Spanish Andalusian

theme, from the "Córdoba" movement of DeFalla's *Nights in the Gardens of*

Spain  , to cite just one strikingly

similar example. (There are hundreds if not thousands of other examples in the Spanish orchestra or guitar literature with the same harmonic progression.) Perhaps it is a mere coincidence, resulting from combining a phrygian melody with an E major harmonization, the standard key for flamenco guitar music. But if it is coincidence, it is certainly a highly unusual one, and one to be especially savored in performances.

What makes this coda passage even more striking is the galvanic, puissant

clash of the cellos' E against the horns' F on the third beat

of m.113 and m.114. And if one may wonder whether the movement's opening is in E or in C, Brahms tells us in the movement's conclusion it could have been in either. His sudden move to C major in the middle of m.114, coordinated with the entrance of the bright sounding high-register 'two-foot stop' flutes, makes a dramatic harmonic shift. Brahms, of course, resolves back to the tonic, although not without letting a bassoon and viola F♮ rub against a sustaining E in m.116: Brahms's final comment, as it were, on the movement's ongoing fascination with the juxtaposition of E and F, both melodically and harmonically.

I will limit my observations regarding the third movement to a few of the most grievous misrepresentations of Brahms's score. In this movement we encounter once again what many consider to be Brahms's 'shifting of the beat' or his 'metric ambiguity.' Indeed, much of this movement sounds rhythmically off-kilter, but that is mostly because of the way it is usually performed—in my view erroneously. The problems begin almost immediately: if one did not know that the movement was in a lively 2/4 throughout and heard the average tradition-bound performance of the first nineteen bars, from m.5 on one would assume the music to have been composed in a metric pattern of one 3/4, two 4/4 bars, seven 2/4 bars, and one final 3/4 bar (see Ex.20). Clearly this is not what Brahms wrote or intended (see Plate IX), and it puzzles me mightily why almost all conductors consider the rendition (as in Ex.20) of Brahms's score in this

Ex. 20

manner appropriate or stylistically authentic. I have several times stated my reasons why such an 'interpretation' cannot be in the spirit and intentions of Brahms. And, indeed, if conductor and players merely keep Brahms's 2/4 meter continually in mind, feeling the weight of the downbeats (not necessarily accenting them), Brahms's cross-accents (as in mm.10–16) turn out to be powerful syncopations. The effect, once heard, is infinitely more exciting and more original. What I find particularly annoying is the universal, senseless habit of chopping off the held-over notes in mm.11,13,15, and 17, as if Brahms had written ♩ ♩ ♩ ♪ ♪ ♪ ♪ ♪ ♪ ♪ ♪ ♪ . Again, had he wanted such an effect he certainly could and would have written as much, as indeed he does in the timpani part in m.11 and m.13, for example. When these tied-over quarter-notes are sustained into the succeeding eighth-note triplets, at the same time feeling the weight of the downbeat pulse, the effect is electrifying. I know of no performance or recording that attempts to preserve the integrity of the basic 2/4 pulse, that is, attempts to do justice both to the underlying meter and the shifted cross-accents. This then is another passage in which Brahms's intentions have been continually and forever subverted and misrepresented.

How wonderfully Brahms immediately re-uses his expositional thematic material is seen in mm.19–31 in several instances. The four-note lead line in mm.6–8 of piccolo, oboe, first violins is re-interpreted lyrically in mm.19–23, but with the harmonization of mm.8–10. The furtive eighth-note figures in the strings in m.23 (Ex.21) are obviously derived from the first three notes of the movement,

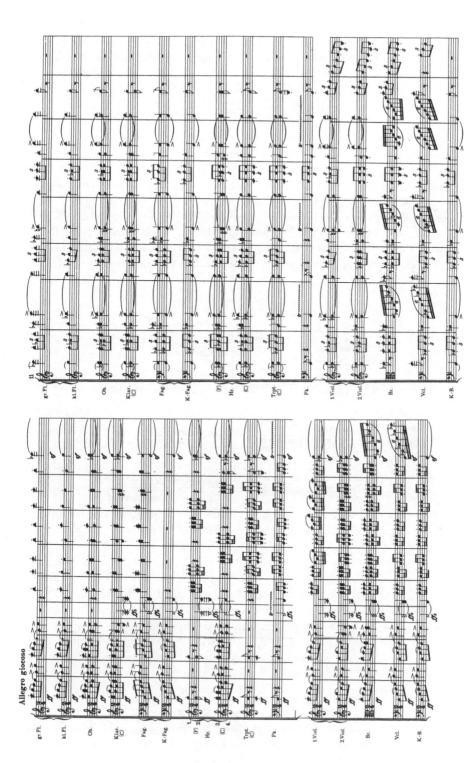

Plate IX   Brahms Fourth Symphony, opening of the third movement

while the sinuous violin figures in mm.21–23 and mm.27–31 have already appeared in mm.8–10. And finally the pseudo-recapitulation in mm.35–38 is sim-

Ex. 21

ply mm.1–4 turned upside down: what was in the violins is now in the bass instruments, what was in the basses is now in the violins.

The brilliant, hurtling E major passage in mm.133–38 works best when the two competing and orchestrationally/registrally separated motivic fragments and are reconnected, spliced together again, as it were, by the players. Fascinatingly, there are two ways of doing this; both work, but one—the first cited—is probably preferable over the other. Looking at Ex. 22, a reduced transcription of the passage, we can see that two lines

Ex. 22

are constructed of the same material, the 'bass' line representing the original formation (as in m.23), the 'treble' line representing its converse. Because the phrase jumps, so to speak, across sections of the orchestra, the individual motivic components need to be reunited. But how this is to be accomplished is impossible for the players to glean from merely their printed parts. A conductor has to sort it out for the orchestra in rehearsal, and once heard by the musicians, the passage plays in a more coherent, cohesive way.

Because of the particular interrelationship of the two motivic fragments, there are, as mentioned, two ways of making the desired connections. The one is to make the first violins and woodwinds aurally aware of their interconnection (the treble line in Ex.22), and, as well, the other strings of theirs (lower line in Ex.22). The other way, which I have tried on occasion—and which also works

12. These fragments are derived, of course, from (m.22) and further elaborated (even fugally) in the first part of the development section (mm.117–32).

very well—is, on the one hand, to interconnect aurally all the sixteenth-note

fragments,  and on

the other hand, the dotted- eighth- sixteenth material,

. We should note that, in any case,

the two fragments are further cross-related by virtue of the fact that they operate in octave unisons: first and last notes of the four-sixteenth fragments coinciding octavally with the other fragment.

As I say, the first cited approach is probably preferable since it preserves more the thematic integrity of the original phrase. But either approach produces a cohesive, incisive, clearly audible result, which is preferable to the rather mindless helter-skelter way in which this passage is usually rattled off.

Speaking now to a small detail, I am puzzled by the notation *dim.* in m.175 (see Plate X), for if one follows Brahms's dynamic markings strictly and if the dynamic in the winds is still assumed to be the previous *pp*, one would have to either arrive at a *ppp* in m.177 (with then a further diminuendo in m.179), or interpret the *pp* in m.177 as a suddenly slightly louder dynamic. But both of these propositions seem illogical and overly fussy. I suspect that Brahms simply forgot to indicate a fresh dynamic for the winds in m.174: *p*, or possibly even the strings' *mf*, from which, of course, a "*dim.*" to *pp* could easily be made. I have taken that approach (i.e. *p* in m.174) in my own performances of the symphony, which is what most orchestras seem to do anyway, quite naturally.

What is much more bothersome, however, is that nine conductors out of ten completely ignore or otherwise re-compose Brahms's music here, including the tempo indication(s). There is only one tempo modification indicated: the unusual but very explicit marking *poco meno presto* (a little less fast). What traditionally happens is that this *poco meno presto* is turned into an *adagio Romanza*. Brahms's *molto p sempre* (see Ex.23), a very special marking, is also completely

Ex. 23

ignored, third and first horns usually interpreting their lines as expansive 'solos,' usually around *mf.* But this takes all the mystery, all the pathos and sentiment, out of the passage. Here we are suddenly in the key of Db—mind you, in a C major piece—a very special moment indeed in the over-all scheme, and it is turned into a very ordinary, maudlin *chanson.* Furthermore, when the horns and bassoons play too loud here, the incoming oboe (m.187), marked *mf* and

Plate X   Brahms Fourth Symphony, mm.168–188 of the third movement

which is therefore meant to top the horns dynamically, cannot possibly do so, and Brahms's carefully calibrated progression of gradual intensification is completely thwarted. Be it noted in passing that this lovely D♭ horn theme is ingeniously a lyric recasting of the explosive cross-accent music that forms the second half of the movement's expository statement, specifically mm. 10–17.

But even before the horns' entrance in m.181 there is usually already trouble. If conductors wonder about how to get from the basic *Allegro giocoso* tempo to the *poco meno presto*, Brahms having not indicated any ritardando or calando anywhere, and decide therefore to make a ritard between m.177 and m.181, that is one thing. (How *much* ritard is, of course, another question—the answer to which should be: 'very little.') But when conductors—and most do this (again one of those allegedly 'venerable' traditions)—start a huge ritard as early as m.175, that is quite another matter, and one for which I can find no justification. I have only been able to find two recordings that more or less navigate the interpretive reefs of this interlude section well: Barbirolli's and Maazel's, although the latter ritards too much in m.175 (but at least observes Brahms's *molto p* after that).

But the best solution in my view is to follow Brahms's score precisely (see Plate X), namely, not to make any ritard, either at m.175 or at m.177. Once again, Brahms has already slowed down the momentum of the music, coming from eighth-notes to tied half-notes (in effect *whole notes*) in mm.177–80. Clearly, no further ritardation of the tempo is needed. And if one postulates that the basic *allegro giocoso* tempo (taken by most conductors) is around ♩ = 120–32, then dropping to, say, ♩ = 90–104 at the *poco meno presto* is a very reasonable assumption. It retains a nice flow in the music, in keeping with the idea that this movement is, after all, a kind of lively Scherzo, and is thus much preferable to the portentous '*adagio Romanza*' that is most of the time made out of this lovely, innocent transitional interlude—transitional to the movement's recapitulation.

Before we leave this section, it is well worth pointing out that this Scherzo, constructed rigorously in a combination Rondo-and-Sonata form (see Fig. 1) is one of Brahms's tersest symphonic movements, and that therefore any exaggerated distending, bloating of the form—anywhere—is detrimental. The formal plan exists on two interlocking levels simultaneously: one, a Rondo form, the other a Sonata form. The movement thus partakes of the most salient features of both forms.

In the recapitulation of the second subject's extension (mm.258–81)—in effect a brief development section within one of the Rondo episodes—there arise some serious performance problems. The gently lilting grazioso second subject has here been transformed into a powerfully rhythmic affirmation. But there are problems, mostly of balance and of sorting out the intricately interrelated thematic/motivic materials. The passage can only be played really correctly, that is, as originally conceived by Brahms, if every musician—and, of course, the conductor—understands how his or her particular part fits into the over-all

Fig. 1

Sonata	Rondo	
	A	mm.1–18
	B	mm.19–34
Exposition	A	mm.35–51
	C	mm.52–88 (also Sonata second subject)
	A	mm.89–117
	B	mm.118–81
Development section	D	mm.182–98 (Interlude)
	A	mm.199–207
	B	mm.208–23
Recapitulation	A	mm.224–46
	C	mm.247–82 (Sonata second subject recapitulated)
Coda		mm.283–357

scheme of things. Again, as so often in Brahms, it is deceptively easy and simple looking, when in actuality it is astonishingly complex.

Let us begin with the primary voice: the third horn, in a staccato variant of the erstwhile *grazioso* second subject. It is embellished in near-unison with interpolated eighth-note triplets in the violins (see Ex. 24). But this violin figuration is also derived in part from the fanfare-like cross-accented figures in the exposition (mm.10–17). In addition, the upper strings' triplets are anticipated in each measure by similar triplets in the cellos and basses—and timpani. And

Ex. 24

therein lies the rub. The timpani, unable to double all of the basses' pitches—

it is relegated to just three pitches 𝄢 —is left to merely hammer out repeated notes, and for six measures of this 18-bar segment (as the music modulates to keys like D and E major) is forced to lay out altogether. The first problem then is to balance timpani with cellos and basses in such a way that (a) the latter's important notes are not obliterated by the timpani; and (b) that when the timpani stops playing (mm.266–71), it is not so blatantly noticeable. Unfortunately, as most recordings will attest, such sensibility is in rare supply; this passage is regularly ruined by over-played or over-recorded timpani. (The timpanist in Fischer-Dieskau's Czech Philharmonic completely mars this passage—and most of the recording—with his unmusical, hard, loud, constantly over-accented playing.) This in turn makes whatever pitch relationships Brahms has composed into his triplets (see Ex. 25) inaudible and pointless. Finally,

Ex. 25

the regular reiterated quarter-notes have as their running partner, as it were, the eighth-note triplets. They, too, must be heard as a continuous chain

 as shown in Ex.25 by the diagonally connecting lines.

The coda (m.282) must begin with a hushed, controlled, suppressed excitement. Dozens of brief figures, all heard previously in the movement, are scattered over a G pedal point, darting out of the subdued texture like so many tiny snake tongues. It is as if the music is slowly gathering force again after its previous exertions—there is no crescendo for at least fourteen measures!—and Brahms is gathering all the little scraps of motivic material together for a final consummation in the elaboration of the coda. The sudden interruptive *f* in the horns in m.294 is usually well managed, but the answer in low clarinets and bassoons (see Ex. 26), both an inversion and a retrograde of the horns, is rarely heard, particularly if timpani and strings are too loud here.

Ex. 26

In m.317 a most astonishing thing happens, which I have, however, never heard consciously brought to the fore by any other conductor: the lead notes in mm.317–25, A-B-C-D-E♭, are identical to the first five lead notes of the Passacaglia theme at the beginning of the last movement, transposed. One might be forgiven for not hearing or seeing the relationship for, in the Scherzo, the melodic line skips back and forth across two octaves (Ex.27). But the notes (pitch classes) are identical by transposition. The question then arises: which came first, the Scherzo or the Finale theme? The question is complicated by the fact

Ex. 27

that we know that Brahms drew his Passacaglia theme in part from a Bach chorale (from Cantata No.150). So, was that theme already in his mind while writing the Scherzo, or was it chance that these notes came to him during the writing of the Scherzo coda? Or was it the coincidence of Brahms writing these notes and then, by chance, coming upon the Bach chorale, much the same way that Alban Berg late in the writing of his Violin Concerto came upon a Bach chorale whose first four notes happened to be identical to the last four notes of the Concerto's tone row? Or did Brahms early on know that he would be using the Passacaglia theme and simply found a way to sneak it into the Scherzo movement? We will probably never know. But in any case, it is well worth alerting the orchestra to this wonderful Brahmsian touch and to bring the theme out in the relevant instruments.

What I have suggested for the earlier cross-accented tied-over quarter-notes should obviously apply as well to mm.329–31 and, above all, mm.337–46.

One final suggestion for the Scherzo movement: the quick run in m.352

, rather under-orchestrated (in only the first violins against the whole orchestra) and therefore usually inaudible, can be helped by having half or even a majority of the section play the notes separate-bow. The extra articulation makes the passage project much better.

The last movement of Brahms's Fourth Symphony, although the clearest of the four movements in its over-all form and continuity—thirty variations on an eight-bar Passacaglia theme plus a coda—nonetheless contains innumerable performance problems, many of which are very serious and basic, and have rarely if ever been resolved appropriately.

I believe that basic to an authentic and informed performance of the movement are the two following premises: one is the audible presence of the Passacaglia theme at all times, not, of course, necessarily at the expense of other primary materials, but still a continuously felt presence; the other is the preservation of the integrity of the 3/4 meter and its pulse. Unfortunately these are precisely the two elements of Brahms's in some ways perhaps finest symphonic achievement that are the most neglected, the most ignored or misunderstood. It baffles me why these so very elementary aspects of the Fourth's Finale are so thoroughly disregarded. One can hear performance after performance, recording after recording, in which nary a shred of the Passacaglia theme is heard, except perhaps in the most obvious places; or in which the innumerable rhythmic and metric wonders of the movement remain unheard and unfelt.

I believe that a fully comprehending, intellectually and emotionally representative performance, reflecting the fullness of Brahms's genius, is not possible without every player in the orchestra knowing the Passacaglia theme and keeping it in mind throughout the performance, not only to be immediately aware of it when it appears in their own parts, but, perhaps even more importantly, to know when it is in *someone else's part* and to respect that. I'm afraid that such an approach is an ideal which will rarely be attained, especially in the major orchestras who mostly think they know the Brahms symphonies thoroughly and feel they have really nothing new to learn about them. One is more likely to have success with the allegedly 'lesser' or 'less famous' orchestras, who often are still ready and eager to learn something about the music they are performing. In any case, playing the Fourth Symphony's Finale without knowing the Passacaglia theme and without being constantly aware of it in all its myriad incarnations, is like trying to drive a car without knowing where the steering wheel is and how to use it.

Since the Passacaglia theme (Ex. 28) is at times partially hidden or disguised or set in unexpected places—as well as being, of course, at other times clearly prominent—it will be best to plot its course as it wends its way through the

Ex. 28

movement. In the first eight bars the theme is obviously in the lead voices (flutes, first oboe, first trombone), but in the first variation (mm.9–16) it is tucked away in the first violins' pizzicato, which is easily covered and rendered inaudible by overly boisterous horns and timpani. In the next variation it is even more 'hidden,' given an octave lower to pizzicato violas and first cellos (and temporarily two horns). The problem here is that most conductors concentrate entirely on the woodwind lines, leaving the Passacaglia theme to fend for itself—with the result that it is simply not present. The reader can confirm this on dozens of recordings.

In variation 3 (mm.25–32) the theme is distributed among a number of instruments. In the first four measures it is to be found not only in the first violins' pizzicato chords, but also in first horn and trumpet as well as the solo woodwinds, while in the remaining four measures only in the violins, first oboe, second flute, and, for part of the phrase, first clarinet and trumpets. Note, too, that the first four bars of variation 3 are played, except for a few sustained notes, very staccato, whereas the woodwinds and horns in the second group of four bars are marked *marcato* and by implication more sustained.

In variation 4, most maestri conduct the first violins—as if they needed the conductor's help and would otherwise not be heard—in the meantime neglecting the Passacaglia theme, which in its powerful octave leaps should be energetically intoned by the cellos, basses, and bassoons. The basses retain the theme in the next variation in heavy *ff* pizzicatos. I have found that bass sections are so thrilled to find out that they have the principal voice and that a conductor will occasionally pay some attention to them, neglected 'underdogs' that they are, that they respond enthusiastically with mighty pizzicato strokes, propelling the entire accompanying structure forward.

The basses' leading role continues in variation 6 as they now sing out the theme in a deep *espressivo* (mm.49–57). In the next variation they, along with second bassoon and contrabassoon, retain only the slightest hold on the theme (Ex. 29, circled notes),[13] but in variation 8 and 9 they come back full

Ex. 29

m. 57

force (Ex. 30) with, remarkably, some for that time very unusual and daring double stops, marked *sf*. The basses finally relinquish the Passacaglia theme in

Ex. 30

m. 65

variation 10, handing it over to a series of diverse instruments, virtually hidden in the chorale-like chordal progression (Ex.31), but nevertheless to be brought out or at least hinted at in performance.

Ex. 31

m. 81   1. Vln.   1. Cl.

Vc.   2. Cl., Bn. Vc.   2. Cl., Bn.   Vc.

Now cellos and violas divide the theme up among themselves (Ex. 32), at least for five measures. From there on via a cycle of fifths, the cellos hand the phrase

Ex. 32

Va.

Vc.   Vc.

over to the solo flute and a change from 3/4 to 3/2. Set over a quietly pulsating tonic pedal point, the flute variation, expanded to twice the length of the earlier variations by way of the 3/2 meter, is one of the most masterful (and memora-

13. We should, however, take note in passing of the striking dissonances, of virtually Stravinskian modernity, that Brahms has squirreled away in several places, including the following gems in variation 7. The second-beat sixteenths in mm.58, 59, and 60 comprise the following three chords, respectively: . Try these on the piano to fully appreciate their shock value.

Since these startling discordances, in sixteenth-note durations flit by in about an eighth of a second, nobody takes notice of them—not conductors, not musicians, not audiences. They remain one of Brahms's little hidden treasures, tiny precious details that he lavished on so many of his works.

ble) thematic elaborations in the entire symphonic literature. It is, along with the flute solo in Ravel's *Daphnis et Chloé*, the most notable celebration of the flute as a lyric instrument in the orchestral repertory. It is also a moment of repose, of calm, not the 'dramatic climactic' declamation many flutists and conductors like to make of it. In that connection, I believe that the peak of the solo is in m.101 (the theme's fifth measure) and that Brahms forgot to indicate *dim.* in m.102 (as he did, however, in the accompanying instruments) and that, therefore, the solo must gradually recede dynamically towards its final *p* cadence, the dynamics in the final two measures being merely *within p* phrase nuances.

Other soft woodwinds now take over in variation 13, the music still floating over an E pedal point (with slyly insinuating 'lower-neighbor' D♯'s), and the Passacaglia theme's notes subtly disguised in the melodic and harmonic fabric. In the meantime, from the plaintive flute solo, still in E minor, the music has suddenly turned to the major and a lighter, airier mood.

The sarabande-like next two variations return the Passacaglia theme to its rightful bass position—it is heard in the second bassoon. Trombones, *tacet* since the movement's first four theme statements, now return to intone, like a congregation in prayer, a solemn chorale. The quiet elegance of these two variations, clearly a kind of sublime, stately dance music, is unfortunately almost always marred by the lack of tempo control of most conductors. The reader will find it hard to believe, but on thirty out of the forty-odd recordings the two silent beats in mm.114,116,122, and 124 (see Ex.33) are completely disregarded. It is shocking to find famous maestri like Dohnanyi, Chailly, DeSabata,

Ex. 33

Carlos Kleiber, Haitink, Suitner, Max Fiedler, and Janowski bring the instruments in m.115 (or 123) a whole or a half a beat (or worse yet, four fifths of a beat) early, in other words already in the previous measure. As I have mentioned before, if an orchestral musician decided to come in one beat early, there would be hell to pay. But when conductors do it willy-nilly, unconsciously, incomprehendingly, nothing is said—at least not in public. (Unfortunately such misdemeanors are never even noticed by critics, managers, board members, or other taste- and decision-makers, who collectively determine who our orchestras' next conductors will be.)

In variation 16, a close relative of the original theme statement, the Passacaglia theme is clearly in the lead voices, but in the next variational phase (mm.137–44) it is heard, half-hidden, in soft sixteenth sextuplet tremolos in the cellos. For variations 18 through 21 the theme is transformed—and therefore for most listeners disguised—into melodically altered variants, for example, in variation 18    or in varia-

(the original primary pitches indicated with x)

ation 21, chromatically altered as:

(simplified, condensed)

In variation 22 (mm.177–84) the theme is sacrificed, retaining only its harmonic implications. We find instead in the first four bars, over a chatty spiccato E pedal point, a note pattern in falling thirds (Ex. 34) which in essence is a reminiscence of the first movement's main subject.

Ex. 34

Variation 23 puts the theme squarely in the first horn and first violins, while in the next two variations it appears in dissected *marcato* triplets in the high register lead instruments (violins, flutes) or first horn (in variation 25).

In variation 26 the music shifts to C major, just as it had done in variation 3. Along with the new key, each variation now goes more and more melodically afield, leaving it mainly to the harmonies to carry the Passacaglia message. In variations 29 and 30 the falling-thirds patterns return, in a hushed, furtive *p* in the former, in the latter in stentorian canonic formations. Even here the Passacaglia theme is ingeniously embedded in the series of thirds, shown encircled in Ex. 35. But these falling thirds, already anticipated eight bars earlier in variation

Ex. 35

29, are again nothing more than a transposed, rhythmic re-working of the pitches of the symphony's opening main subject, transposed (see pp.380–81).

At the end of variation 30, in the *poco ritard*[14] we hear the old ambivalence

14. This *poco ritard* is frequently exaggerated by conductors (Skrowaczewski, Carlos Kleiber, Fischer-Dieskau, for example) into a huge *molto ritard*, in my view to rather trivial effect. I believe one should think of this ritard not so much as a holding back of the tempo as an *expansion*, an *enlargement*, of the beat and pulse, just before the oncoming driving *Più allegro*.

between E and C that we have encountered so often in this symphony. The ritard stems the relentless tide of the music only temporarily, bounding forth in the coda (m.253) with renewed energy (*più allegro*)—and a brand-new slant on how to further exploit the Passacaglia theme. After a powerful, clear, although re-harmonized and melodically truncated statement of it (mm.253–60)—in effect the thirty-first variation—Brahms takes the last four notes of the theme, three of which he had not used in this last variation, and constructs a remarkable contrapuntal design, by which means he also modulates the music to the remote key of F major. As Ex. 36 shows, the complete contrapuntal grid is of considerable complexity, an intricacy of design, which is made all the harder to bring out in performance because a number of instruments enunciate different segments of the motive (rather than complete statements of it), while others share in only two of the four notes. Thus it is imperative, once again, for the performing musicians to know precisely what part of the contrapuntal web they are representing and—just as important—when they are *not* a part of it. This can be achieved in extensive, detailed rehearsing (if there is time) or by marking the parts in some elucidating way. Unfortunately, neither happens normally, and orchestras generally just plow through the passage, energetically, but also mindlessly, with Brahms's ingenious contrapuntal design left to chance.

Ex. 36

Brahms begins the passage with a B♮, the expected pitch which had been withheld in the previous eight-bar variation. This gives him a B⁄C⁀C⁄F motivic shape and, by transposition upward, the succeeding iterations of the motive. But the choice of transposition is hardly arbitrary—nothing in Brahms ever is—but rather is linked to the canonic response of the first motivic fragment, a response in the bass instruments beginning in m.263 at a distance of two bars. The pitch chosen by Brahms for the entry of this second voice is C, in other words in unison with its canonic partner at that point. Once this linkage is established, the whole passage modulates its way upward—crab-wise, if you will—until in twelve bars it has fulfilled one of its major functions, namely, to modulate to some remote key, in this case F.

But Brahms is apparently not satisfied with the resultant dearth of harmony in his canonic construct. Rather than harmonizing it in some intuitive way, he invents another voice (seen in the third stave in Ex.36 above), a B-A-C-H-like motive which also, crab-wise, wends its way upward. Considering the multi-layered structuring involved here—two of the three lines multiplied into forceful three-octave unisons—Brahms's orchestration, with a limited resource of instruments, can only be called masterful. And yet bringing it to correct acoustic life doesn't necessarily happen by itself. One would like to think that if every musician would play his or her part in a well-balanced integrated way, the passage would play itself, so to speak. But this rarely happens. All that has to occur is that one player or one section plays a little too soft—or a little too loud—and the whole contrapuntal house of cards breaks down. And, as proven by the available recordings, it is indeed a rare performance that renders Brahms's wondrous polyphonic achievement accurately.

If we ask ourselves where before have we seen the notes in the declamatory trombone theme pronouncement in mm.273–76, the answer is in the *third* movement in that remarkable premonition of the Passacaglia theme I referred to earlier (see Ex.27, p.411).

Finally, two more remarkable theme variants call for particular mention. Having exhausted—presumably—virtually all potential variants of the Passacaglia theme, Brahms nonetheless, on the verge of the ultimate dynamic climax of the movement, as if in triumph, finds one more transmutation of the theme. Bifurcated, the one part the transpositional inversion of the other (see Ex.37) and further harmonized in thirds, Brahms is apparently so happy with it that he

Ex. 37

repeats it immediately (of course, slightly altered), thus also completing the expected eight-bar phrasing. Notice that Brahms manages to sneak in the first three notes of the theme (as here reconstituted) in the viola and trombone in mm.291–93.

The final triumphant statement of the much-used theme is also the first one that, instead of dipping down the octave in its seventh bar, strides confidently upward towards the tonic (Ex. 38). A simple idea, in a way, but the timing and

Ex. 38

placement of it at the very end of the movement—meaning among other things that Brahms was able to resist for a remarkably long time the temptation to round out the theme in this ultimate and fulfilled form—are a mark of his perfect craftsmanship, his disciplined creativity—in short, his genius.[15]

The lengthy exploration above of the arduous path traversed by the Passacaglia theme was, unfortunately, necessary given that in most performances and recordings this seems to be unexplored territory for conductors and orchestra musicians alike. While that theme's course is a factual and objectifiable matter, and therefore hardly arguable, the other major fourth movement performance/interpretation issue I raised earlier is perhaps more subjective—more in the realm of pure interpretation, exegesis, and connotation. Nonetheless, for the reason I have given several times previously, I firmly (though modestly) believe, if not in the 'absolute correctness'[16] of my theory of metric integrity in Brahms, in its possible rightful applicability. At the very least, I believe it offers another viable interpretational performance option. Certainly the results of such a realization are startling—as we shall see—especially to anyone who has never heard such a 'rightful' interpretation. The premise is really a very simple one: to do full justice to Brahms's extraordinary rhythmic inventiveness, specifically fully to honor both the basic metric design and pulse, as well as the actual rhythmic configurations and their particular feeling. And let me emphasize that this analytical exercise is not about mathematics, mechanics, abstractions, or a mania for precision. On the contrary, it is about feeling, expression, about the sensation the music creates in the listener; it is, therefore, about the actual expressive, communicative content of the music, not merely its artifacts of construction. But that fullness of expression cannot be achieved without a complete understanding—and complete rendering—of the creative/intellectual construction, the techniques of composition, without which composers could not create their great and enduring works.

The interpretational problems I refer to begin—in their subtlest form, to be

15. It is interesting to note in the autograph score that at some point in the creation of the work Brahms had considered adding the marking *sost.* to the rising scalar theme in m.297 and then an *accel.* four bars later (m.301). But evidently he soon changed his mind and crossed out both markings—wisely so, I think.

16. There is no such thing as 'absolute correctness' in music, just as there is no such thing as a— or *the*—'definitive performance,' as I have stated earlier.

sure—immediately in mm.9–16. Here the single quarter-note interjections in trombones and strings should be played with that particular feeling that is unique to a second beat in a 3/4 meter.[17] The difference between that suggestion and the way it is universally played, namely, simply as a loud *f* stressed note, is subtle but crucial. To simply have eight heavy pizzicato 'plunks' in this phrase (the same problem occurs in the next eight-bar variation, mm.17–24, as well) is to deprive the passage of its true feeling and meaning. The problems intensify in mm.16–24. Here Brahms has superimposed on the Passacaglia theme melodic woodwind lines which in terms of phrasing and articulation start always on the second beat of a measure. This 'construction' has led musicians for over a century, I am sure, to play the passage as if Brahms had written

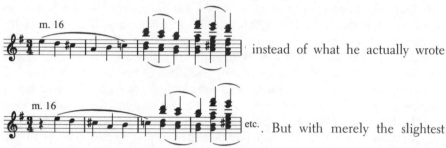 instead of what he actually wrote

etc.. But with merely the slightest

stress on the downbeats, the slightest sense of the expressive primacy of those first beats, the passage not only takes on a whole new feeling, but in fact does full justice to Brahms's beloved sense of rhythmic compositeness and ambivalence. There is another way to argue the point, namely, that if the passage is played in the usual off-kilter manner, i.e. simply shifting the bar lines one beat later, then there occurs of necessity at the end of the eight-bar phrase a disturbing metric shift (see Ex. 39 for a hypothetical notation), which Brahms could certainly not have wanted.

Ex. 39

We encounter another category of Brahmsian rhythmic ambivalence—and ingenious inventiveness—in mm.41–48 in the woodwinds, a passage which, once again, is hardly ever played correctly, that is, as written. Only in Reiner's and Mravinsky's recordings can one hear a semblance of what Brahms actually wrote. For,

some woodwinds have the following rhythmic phrase

while at the same time others have |—once again Brahms's

---

17. See my discussion of the hierarchical distribution of beats within certain basic time signatures, 3/4 or 4/4, and their disposition in terms of strong (stressed) or weak (unstressed) beats (p.110).

fascination with 'three against two.' Obviously most of those notes should not sound together, and yet that is exactly what happens in recording after recording, performance after performance. As we saw in certain rhythmically unusual passages in the Fourth's first movement, the musicians simply accommodate themselves in some casual unthinking way to each other, making the passage come out in a bland rhythmic unison that is not only wrong but infinitely less interesting than what Brahms actually wrote.

These single-bar phraselets would have been easy to play (and would probably be played mostly correctly) if Brahms had included in each line a downbeat first note instead of the eighth-note rest. Given the imprecise, casual way in which most musicians generally play rhythms, and given the closeness of timing between the respective first notes in each rhythm pattern, a matter of about 100 milliseconds (or one-tenth of a second),[18] and given the almost unavoidable natural time lag between a musician's impulse to play a note and its actual acoustical appearance—musicians, after all, are not machines—it is easy to see why this passage is rarely if ever played correctly. But, of course, it *can* be done. It takes two things: an awareness on the musicians' part of the problem at hand, and a bit of extra rehearsing, separating the triplet figures out from the duple figures. With a little attention to the passage of that sort on the part of the conductor, it quickly yields the right results, as I have been able to experience many times in my own performances of the work.

Before we leave this fifth variation (m.41), I must point out by way of reminder that the hairpin nuances $<$ $>$ in the strings extend over two measures at a time, but in the woodwinds are contained in single measures. When performed correctly this makes for a wondrously rich dynamic diversity, a kind of polyphony of dynamics.[19]

18. At a basic conventional natural tempo of $\quad = 108\pm$.

19. Various constraints prevent me from dealing with the many dynamic-related questions and problems in this Passacaglia movement. Some comments on a few special places are nonetheless in order and will have to stand for all others. Brahms's score is in this regard at times not entirely clear. For example, after a crescendo in mm.38–39 (fourth variation) from a previous *f*, the beginning of variation 5 is lacking in dynamic explicitness. Should the strings continue at their just newly arrived at *ff*? Surely not. The woodwinds' marking of *poco f* leads one to think that the melodic strings should start their two-bar phrases *mf*, with the basses a full *f* or *ff* pizzicato. The ensuing crescendo *sempre più* (m.45) will bring the strings to full *f* or *ff* at m.49. This implies in turn that the *più f* of m.57 represents the ultimate climactic dynamic thus far in the movement. But this means that strings and woodwinds must recede to a mere *f* and *mf* respectively, not the bow-breaking *ff* with which most conductors charge into m.65, variation 8. The majority of conductors unfortunately arrive at a *ff* by m.33, and stay at that peak level for the next three minutes of music, resulting in an unrelieved acoustic orgy which renders the music both boring and brutalized.

The other spot that is almost never conducted or played right is the two-bar scalar phrase mm.175–76. Rarely does an orchestra come down to a true *pp* after the three previous trombone 'explosions,' but worse yet almost everyone makes too much crescendo in m.176 and no *subito p* in m.177. Admittedly, Brahms's notation is somewhat ambiguous here, for the crescendo wedges in m.176 fail to indicate how much crescendo is to be made. Is it to *f* by the end of the measure, followed by a sudden *p* on the next bar, or is it in the brass and strings a crescendo from *pp* only to *p*? No one knows; and we will probably never know Brahms's true intentions. But is it too much to expect conductors ocassionally to try some of the alternative realizations just mentioned, rather than the very ordinary, '*mostly ff*' solution.

No rhythmic/metric problems occur until variations 24 and 25. But these have remained unresolved and ignored forever, judging by the evidence of recordings. Measures 193–200 represent a gigantic augmentation and intensification of the first Passacaglia variation (mm.9–16), especially in terms of orchestration. But rhythms are also intensified. The second beats, originally simple staccato quarter-notes, are now transformed into—one can almost guess what— a complex mix of duple eighth-notes and eighth-note triplets—another skirmish between 'three' and 'two': eighths in the woodwinds, eighth triplets in the strings (Ex. 40). At even a moderately lively tempo, let alone the inordinately fast

Ex. 40

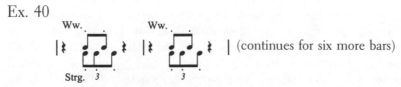

(continues for six more bars)

ones some conductors take starting with variation 16 (m.129), keeping those two rhythms apart, playing them exactly right, is not easy at all. It requires a degree of rhythmic precision on the part of *all* the musicians involved that is almost never demanded. And the conductor can do very little except to lend his ears and monitor the results of rehearsing. All that is not to say that what Brahms asks for cannot be achieved, but it will require painstaking rehearsing, certainly choir by choir. I have in my own conducting of the work always been able to achieve the desired result, but only by dint of very tough, extensive rehearsing, sometimes taking as much as ten to twenty minutes of precious rehearsal time.

The precise problem is, as in some of the similarly complex rhythmic figurations in the first movement, that if the duple players are even the tiniest bit late with their *second* eighth-note and/or the strings are, similarly, early with their *third* triplet note, the two notes come together. In most orchestras the strings play their triplets too fast, something like ♫♪ or ♪♪♪. It is relatively easy to err in this way, because the strings have no third beat to play which, if they did, would help to firmly anchor the rhythm. It is *infinitely* harder to play | ♪♪♪ | correctly than | ♪♪♪ |.[20] What I have often done in rehearsing this passage is to ask the strings to play a note—any note, it doesn't matter—on the third beat, which after a few tries helps *them* to hear what the correct speed of their triplets should be. After a few rehearsings in this fashion, the crutch of the third beat can be taken away, and then—with perhaps a few relapses now and then—the triplets begin to be played correctly. At the very least the players are now intensely aware of a very difficult problem, of which they were previously totally *unaware*.

This unawareness is, of course, the real problem. And in truth, why should string players, unless they are score readers and/or somehow know exactly in every detail what Brahms committed to paper, know that the woodwinds have

20. This is the identical problem as in the last movement of the Brahms First Symphony in mm.156–63, where the violas have isolated triplet figures, unanchored in any surrounding beats.

plain eighths in mm.193–200. Similarly, why should wind players know that the strings have competing, conflicting triplets. They certainly cannot glean this information from their instrumental parts; and having never heard the passage played correctly, they have no inkling that there is something wrong in what is usually played, that there is a problem here. But, my goodness, isn't the conductor, who has the score, supposed to notice this? How can one explain that not a single conductor—not Toscanini, not Furtwängler, not Dohnanyi, not Ozawa, not Muti, not Carlos Kleiber, not even the rhythmically meticulously finicky George Szell—at least judging by the recordings, has ever addressed this difficult but fascinating performance problem.[21] It *can* be solved, of course; and I can vouch for the fact that once the orchestra musicians have heard how the passage sounds when played correctly—inevitably their first time—they are amazed, and delighted with themselves for achieving what at first they didn't feel could ever be done, what at first they didn't even know was wrong. As I say, there is not a single recording where this rhythmic detail, so quintessentially Brahmsian and in the scheme of things so important, is dealt with correctly.

The problem is exacerbated in the next variation, a close relative (by way of augmentation) of variation 2. Here (mm.200–208) the violins, violas, oboes, and bassoons play an intensified version of what the woodwinds had earlier. And as there, here too these instruments must firmly re-establish the 3/4 meter and downbeat pulses to counteract the strong second-beat attacks of the entire rest of the orchestra. Moreover, to complicate matters—but really to heighten the intensity of expression over the preceding variation—the contesting eighth triplets and duplets are redistributed: flutes, oboes, horns, trumpets, and timpani in triplets; trombones, bassoons, cellos, and basses in duples. Again, to sort all this out to truly reflect what is in Brahms's score, is very difficult and will take time to rehearse. But it is well worth the effort; the results are startling and exciting!

We end this critical traversal of the Fourth Symphony and its recorded performances on a happier note by considering a very interesting possible notational error in both the printed score and Brahms's autograph. In m.233 he gives the flute the line in Ex. 41 to play (note, by the way, in a hemiola configuration). Obviously the second pair of measures is an imitation and variation of the

Ex. 41
m. 233

first two measures, and as such its sequence of notes is perfectly logical and 'correct.' But if we notice that the pitches in these two bars, except for the third beat G in m.235 and second beat C in m.236, are identical to the first six notes of the symphony's main theme (see Ex. 1), we may be forgiven for speculating

21. There is a precisely identical problem in Stravinsky's *Octet* where in the second movement there are triplet eighths in the two woodwinds against plain duple eighth-notes in the trumpets, confined to one beat. There too, on all recordings and performances I have heard—and I admit I have, of course, not heard them all—this passage is played incorrectly; in some mysterious way the musicians accommodate each other and play the rhythms together rather than untogether.

that this G and C should have been E and A respectively. Thus the phrase
would have been as in Ex. 42; it would have been just like Brahms to play this

Ex. 42

kind of note game. When one then finds, as if to confirm our conjecture, that
eight bars later (mm.341–44), secreted away in the first viola part, the identical
pitch sequence as the flute has in mm.233–36 (except for the two anomalous
notes), it gives one pause for thought. In other words, in the viola part in
mm.243–44 the pitches of the symphony's opening melody are exactly replicated—which in turn *does* make one wonder if Brahms meant to create the
same effect in the flute, and if not, why not. It is a curious matter and, I suspect,
will always remain an 'unanswered question.'

Finally, I wish to point out to the serious student of Brahms scores that in a
half-dozen places in the last movement Brahms contemplated, sometimes between the initial finalization of the score (in ink) and its engraving for publication, certain tempo modifications (in heavy pencil). A *sost.* and *largamente* were
added in m.193, a *tranquillo* in m.217; another *sost.* in m.249; yet another *sost.*
in m.297, followed by an *accel.* in m.301. Before a conductor of the free-
wheeling, capriciously variable tempo school jumps with the joy at this discovery, assuming that these annotations by Brahms give justification for taking those
and other tempo liberties, I should point out quickly that Brahms, before printing and publication crossed all these changes out, substituting *poco rit.* for one
of the crossed out *sost.* markings (in m.249).

These facts ought to serve as another reminder to us that Brahms did not
relish tempo modification in his works beyond those he specifically indicates in
his scores. That ought to be sufficient proof that he was most intent on preserving the classical symmetry, integrity, and logic of his forms.

Strauss: Till Eulenspiegel

Strauss's *Till Eulenspiegel* is unquestionably one of the composer's most popular works, and has been so since its premiere under Franz Wüllner's direction in 1895. Indeed, both *Till* and *Don Juan*, that wondrously exuberant, slightly earlier 1889 masterpiece, were not only outrageously popular right from the start with audiences and critics alike, but had a profound impact on the course of music in that both works, having more or less abandoned the classical sonata and variation forms, successfully explored the new freer narrative form of the 'tone poem.' While Liszt was the real inventor of the tone poem and Wagner contributed—for all his respect and love for Beethoven's symphonies—enormously through his operas to the break-up of the classical symphonic forms, it was Strauss, in *Don Juan, Till Eulenspiegel*, and some of his other tone poems, who brought this literary-oriented genre to world center-stage. The tone poem clearly stood in direct opposition to Brahms and his classical forms, and led the revolution, along with other brilliant break-through works, like Debussy's *L'Après-Midi d'un faune* (1892–94), and the virtual decimation of the symphonic form(s) by Mahler in his symphonies and *Das Lied von der Erde*, that eventually brought down the entire house of classical forms.

It is of more than passing interest to us, in view of the foregoing extensive discussion of two of Brahms's symphonies, that *Don Juan* was composed only five years after the older master's Fourth Symphony, and that *Till Eulenspiegel* was completed two years before Brahms's death, when Brahms was writing his two elegiac clarinet sonatas.

Strauss was only thirty-one when he composed *Till*, which may very well be, along with the final scene of *Salomé*, his most masterful work, certainly its most perfectly constructed work, with an economy of means that, interestingly, rivals Brahms in this respect. As in Brahms's best works, there is in *Till* not an iota of extraneous material. If even the 'least significant-looking' material is exploited for all it can yield, what can we say about the work's two principal themes:

Ex. 1a

Es war ein - mal (ein Schelm)

and the famous leaping horn theme:

Ex. 1b

Strauss's constant re-invention of the main thematic materials results in one of the most tightly (and, perfectly) constructed works in the entire 19th-century repertory. Not since Mozart's *Jupiter* Finale and Beethoven's Fifth Symphony had there been anything quite as succinctly integrated as *Till*. Nor—looking forward towards the 20th century—was there to appear anything as rigorously developed as *Till* until some of Webern's scores of the 1920s.

It will be countered, undoubtedly, that *Till Eulenspiegel* does pay some allegiance to one older form, the Rondo. But this occurs more in name than in actual practice. Strauss is himself playing a roguish game, a play on words, when in his full title of the work, *Till Eulenspiegels Lustige Streiche, nach alter Schelmenweise in Rondeauform für groses Orchester gesetzt,*[1] he refers to the 'rondeau form' used by troubadours and poets in French medieval and Ars Nova times, a form which bears only a minimal resemblance to the much later Italian rondo with its rather strict alternation of a principal theme and various secondary episodes. In the work itself Strauss is—if he ever intended a true classical rondo—so free with that form as to leave little evidence of its possible initial inspiration. I suspect that whatever Strauss's original intentions may have been, as he worked on the piece, using his own personally devised Eulenspiegel scenario, in turn culled from many different extant versions of Till's (Tyl's in the original old German) tale, Strauss found himself following those narrative impulses rather than adhering to any strict classical form. Even *Till's* short-lived 'recapitulation,' slightly varied and split into two episodes (mm.430–33 and mm. 466–85), is more representative of the narrative 'tone-poem' conception than of the sonata-rondo form.

Our admiration for *Till Eulenspiegel* must to a large extent also derive from Strauss's spectacular and, for the time, very daring use of the orchestra. And finally, it has endeared itself to musicians, audiences—and even critics—for its restoration of humor to music. As Busoni once pointed out about Strauss's *Till*, not since "Papa Haydn" had any composer "handled lightness and humor so masterfully." Yet for all its popularity and consistent adoration by the musical world, performances (and recordings) of *Till Eulenspiegel* have on only the rar-

---

1. *Till Eulenspiegel's Merry Pranks, after an Old Rogue's Tale in Rondo Form for Large Orchestra.*

est of occasions done full justice to the work. To this day many of Strauss's most brilliant ideas and explicit notational instructions remain largely ignored or unrealized, as we shall see in the ensuing discussion.

Since so much of the work centers on and derives from *Till's* horn theme, our discussion must of necessity begin with it and an analysis of its correct rendition. The point is that on none of the forty-odd recordings I have sampled is this famous passage played really correctly, nor are its many variants and permutations throughout the piece rendered as intended by the composer. The problems—apart from the horn solo's sheer technical difficulty[2]—are two-fold in that Strauss devised a theme whose first two phrase segments comprise seven eighths, overlaid, however, on an underlying 6/8 meter; and, second, that in addition, the two parts of the solo are prescribed in very explicit, contrasting tempo terms. This startlingly novel idea—even for today, let alone 1895—has resisted proper interpretation for the same reason that metric ambivalence in Brahms is generally inaccurately rendered. It must be obvious to any thinking reader—and musician—that if a 7/8 pattern is overlaid on a 6/8 meter, the pattern will on each successive repeat appear one eighth later in the metric structure (see Ex.1b: the 7/8 phrase is bracketed). Thus, once again resorting to the commonly used 'stressed' and 'unstressed' markings, ╛ and ╰, the first half of the horn theme should be phrased as in (Ex.2a), thereby preserving the 6/8

Ex. 2a

feeling and dance-like lilt of the music. What one invariably hears instead is one or the other of the following versions, i.e. rhythmic distortions (Exx.2b, c). Far too often—as can be confirmed in literally dozens of recordings—the G# is

Ex. 2b

2. The story is told, although it may be apocryphal, that when the younger Strauss showed his *Till Eulenspiegel* horn theme to his father, Franz Strauss, the most famous and reputedly best horn player in Germany at the time, playing often at Bayreuth in Wagner's operas, Strauss senior complained bitterly about the difficulty, even 'unplayability,' of the passage. Strauss is reputed to have countered with, "But, my dear father, I have heard you practise similar passages almost every day of your life; it is after all based on the horn's basic harmonic series." Strauss senior, we are told, was neither amused nor placated!

I remember in my younger days as a horn player that the *Till Eulenspiegel* theme was still considered technically very difficult, and we young players practised it many times every day. The horn player who in those days could play the *Till* theme securely and with relative ease was considered almost phenomenal. Ironically, though nowadays many horn players can perform this solo with consummate technical control, and even high school students 'polish it off' several times a day, almost nobody plays it correctly conceptually, i.e. rhythmically/metrically.

Ex. 2c

etc.

held only two eighths or two and a half eighths, but not three. Under the circumstances, the conductor and the violins (with their accompanying sixteenth-notes) are relegated to vaguely following along, hooking up, it is hoped, with the horn player at m.10 (the high D).

Indeed, in the early days, conductors were quite content to merely follow along no matter what the horn player did, because in most cases the conductors were relieved to have a player who could get all the notes. Any kind of rhythmic distortion was acceptable, and horn players liberally recomposed Strauss's theme into whatever was most comfortable or seemed safest. Early recordings show — for example, Furtwängler's 1930 recording with the Berlin Philharmonic — that in those days horn players took a breath in the middle of m.8, adding a whole eighth and still more rhythmic/metric distortion to the passage. (Nowadays horn players usually play the entire passage in one breath.) Strauss himself, as is evident from his 1944 performance with the Vienna Philharmonic, had to accept the considerable restructuring (including an altogether wrong note) of the horn solo.[3] The point is that Strauss's horn theme in *Till* is one of his great strokes of genius[4] Moreover, there is a real significance to Strauss's 7/8 phrasing, for what better way could a composer find to describe in music Till's ribald character. In one fell swoop, Strauss depicts all of Till's basic characteristics: his anti-authoritarian, free-wheeling, irreverent nature, always at odds with society and especially with the pomposity of bureaucratic authority — the 7/8 pattern at odds with the basic 6/8 metric design; his nimble quick-footedness, heard in the bouncy staccato of the horn's theme; his "quick-witted and elusive manner" depicted, as Norman del Mar has pointed out,[5] in the syncopated shifting 7/8 pattern; and finally, Till as a folk hero represented by a horn, so often cast in the late 19th century as the ideal instrument associated with heroes (especially, of course, in Wagner's works). Thus the 7/8 patterning is not some arbitrary, meaningless gesture, even less a creative lapse on Strauss's part,[6] but rather a

3. However, this may not have bothered Strauss particularly, since he tended to be rather casual in his conducting of his own works, as is testified to by many observers, from Arnold Rosé, long-time concertmaster of the Vienna Philharmonic, to Fritz Busch, a close associate of the composer during the many years of successive Strauss opera premieres (starting with *Salomé*) at the Dresden State Opera, as well as the evidence of his own recordings of his works.
4. I would add to this the final scenes of *Rosenkavalier* and *Salomé*, and the opening thirty bars of the *Alpine* Symphony, a remarkable musical depiction of the clouds (or mists) descending upon and blanketing an Alpine valley.
5. Norman del Mar, *Richard Strauss, a Critical Commentary on His Life and Works*, Vol. 1 (London, 1962), pp.125–26.
6. I recall giving a talk for the New York Philharmonic Friday afternoon Pre-Concert Lecture Series many years ago on the occasion of a performance of my *Symphony for Brass and Percussion*, which features in the last movement a repeated 9/8 pattern in the horns, overlaid on a 4/4 (i.e. 8/8) meter. I allowed that I had been inspired in this idea by Strauss's *Till Eulenspiegel* horn theme, and

master stroke, portraying in music in the most succinct way the many complex aspects of Till Eulenspiegel's character and his pranksterish exploits.

But there is more that the score tells us, all points, alas, that are regularly ignored or subverted in performance. First, Strauss indicates quite clearly that the beginning tempo of the horn solo shall be in the exact relationship of a 6/8 dotted quarter equaling the eighth of the previous 4/8 (Ex.3). In nine out

Ex. 3

of ten recordings (and performances) this instruction is completely disregarded, the horn solo usually starting in much too slow a tempo. Strauss himself is one of the few conductors who does the tempo conversion right (in his Vienna Philharmonic performance). Second, Strauss indicates *allmählich lebhafter* (gradually livelier, meaning a gradual accelerando) in m.8, arriving at the full tempo by m.13 (*Volles Zeitmass, sehr lebhaft* [full tempo, very lively]). Here Strauss pictures Till a little hesitant at first, as if slyly making sure that the coast is clear before venturing forth. But at m.14 he is fully confident, with the horn accordingly now in full tempo. Strauss's dynamics underline this scenario: the first 'call' is *p* for four bars, then crescendos a little (to *mf* or *mp*), while the repeat at m.14 starts *mf*, crescendos two bars earlier than in the first phrase all the way to *ff* at m.19. Neither the dynamics nor the tempo indications are generally observed, the first 'call' usually (as mentioned) starting too slow but also crescendoing immediately, and reaching the full tempo several bars too early, the second 'call' usually *not* reaching the brash *ff* Strauss calls for.

For all the reasons given, it is imperative that this theme be played absolutely correctly in all respects. Instead, between the rhythmic/metric distortions and the failure to honor the tempo and dynamic indications, in the vast majority of performances and recordings the entire meaning and essence of the brief underlying scenario of mm.6–19 are lost.

On recordings, only a few players came close to capturing both the spirit and the letter of this passage: Chambers (Bernstein, New York Philharmonic), Rolf Götz (Masur, Leipzig Gewandhaus Orchestra), Satoru Umeda (Paternostro, NHK Symphony, Tokyo), Greg Hustis (Mata, Dallas Symphony). Even so, on *no* recording is the fascinating relation between the horn theme's 7/8 pattern and the underlying 6/8 meter brought out.

As I already indicated, *Till's* horn theme functions as the central thematic/motivic material throughout the entire work—I count some 55 appearances of

---

began to wax lyrical, almost ecstatic, about Strauss's ingenuity in creating this remarkable theme, so extraordinary in its invention, particularly for 1895. I was suddenly brought up short when a little old lady in the front row interrupted me and with a steely voice irritatedly demanded to know: "Well, what's so great about that?"

it—in just about as many guises as Till in real life is alleged to have assumed. Therefore its correct rendition is important not only in its four solo settings, but wherever it appears throughout the work.

Turning now to the five-bar prologue, which parallels, according to Strauss himself, the opening line of any folk or fairy tale—as in "Once upon a time . . ." ("Es war einmal") (in German)—we find the other *Till* theme (see Ex. 1a) set in the simplest folk-song manner. The theme's gentle character and lovely turn-of-phrase need to be handled with the utmost care and taste. Unfortunately this theme often sounds strained and edgy, as the violinists use too much bow and too much bow pressure with a resultant heavy *mf*, instead of Strauss's *p*. Most annoying is the strident accent on the B♮ one hears in many performances. This is usually the result of aural carelessness, combined with the following bowing

 , particularly

when too much bow is used on the first up-bow, thus landing too low on the bow for the B, in turn causing a heavy mezzofort-ish sound. Just as annoying are the stretching and distorting of the rhythm and tempo in the first measure (particularly an over-long C♯), which goes against Strauss's *gemächlich* tempo marking and the intended simplicity of the phrase. The ideal bowing is as fol-

(mid-bow)

lows:  etc.

The reader might note here the specific slur notation

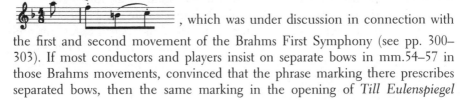 , which was under discussion in connection with

the first and second movement of the Brahms First Symphony (see pp. 300–303). If most conductors and players insist on separate bows in mm.54–57 in those Brahms movements, convinced that the phrase marking there prescribes separated bows, then the same marking in the opening of *Till Eulenspiegel*

ought also to result in separate bows, i.e. [musical notation] , which,

of course, no one in his right mind would consider, nor do I know of anyone's ever thinking of such a phrasing. This just confirms the inconsistency and confusion in interpretation which this particular phrasing notation seems to generate.

I count as the most beautiful renditions of *Till*'s opening phrase on recordings those of Fricsay, Blomstedt, Furtwängler (1930), Karajan (Vienna Philharmonic), Busch, and Mata.

In m.2 clarinets and bassoons have a *sfzp* marking, which, however, is almost always ignored, flattened out into a mild non-committal pseudo-accent. In m.4 all violins and violas play ♪ instead of ♪, thus ignoring the intended over- lap with the entrance of the clarinets. For their part, clarinets rarely make the (admittedly difficult but not impossible) diminuendo Strauss calls for. And why do almost all clarinetists make an accelerando in m.4? Is it because the phrase rises in pitch? Hardly a valid reason. The *sfzp* in the flutes in m.5 has a slightly better success rate for some reason than the same marking in m.2, but could still stand improvement on most recordings.

Thus we find in just five simple uncomplicated measures half a dozen points of negligence in most performances/recordings, detracting seriously from the beauty and inviting enchantment of this most memorable and loveliest of intro- ductions.

What has been said about the horn theme applies, of course, to the first variants of that theme in the oboes (m.21) and clarinets (m.26). These passages invariably sound as if they had been rhythmically notated as:

rather than Strauss's:

The integrity, the feeling, of the underlying 6/8 pulse must be preserved at all costs; otherwise Strauss's particular notation is rendered meaningless and his conception nullified.

The subtle—I would add sly, Tillian—metric ambivalence Strauss creates here often upsets orchestra and conductor in the entrance of bassoons and lower strings in m.30. Indeed, this passage, in which Strauss readjusts the horn theme to fit into the 6/8 meter, once gave orchestras tremendous problems, as many shaky performances in early recordings of the work attest (although not in Albert Coates's excellent late 1930s London Symphony Orchestra recording). Though eminently manageable nowadays, orchestras still have to be on their best alert to execute mm.30–33 without a rhythmic hitch.

Having negotiated these particular musical rapids, many conductors now drive the orchestra inordinately to achieve, presumably, some kind of 'audience- exciting' climax at mm.39–45. Once again, had Strauss wanted such an acceler- ation, he was quite capable of indicating as much and writing another *allmäh- lich lebhafter*. But it is in fact much more 'exciting' to let this climax accumulate through the *orchestration*, through the constantly added voices as the phrase rises into the upper register. The passage rises chromatically to a *f* C^7 chord in m.39—be it noted *only* to *f*.[7] Yet virtually all conductors push the orchestra (or

7. In Karajan's 1973 recording mm.39/40 are simply left out—cut *(sic!)*. So much for Karajan's and producer Hans Weber's musical acumen and artistic integrity.

allow the orchestra to race) prematurely to a full *ff*. The real *ff* should not come until two bars later, signalling the true climax of the 'introduction.'

In m.44, a totally different problem arises: intonation. The woodwind's four-octave unison is indeed hard to tune correctly, although it can be done, of course. It takes a little more attention than most players or conductors are apparently willing to give it, judging by the recordings. On twenty-six of the more than forty recordings analyzed, this measure is painfully out of tune. At the same time, hardly any orchestras/conductors play m.44's rhythm correctly. Strauss wrote

⟨notation⟩ , but (almost) everybody plays ⟨notation⟩ .

In m.46 we hear for the first time *Till's* other theme, a variant of the very opening introductory phrase of the work, here turned into an irreverent, mocking, nose-thumbing gesture, played on the D clarinet (nowadays on the E♭ clarinet). The problem here is that nine out of ten clarinetists fail to play this, and all similar places in the piece, as Strauss wrote it, in 6/8 (Ex.4a), but as if notated as in Ex.4b.

Ex. 4a

Ex. 4b

Measures 51–54 present no enormous problems, other than that many orchestras play this jaunty passage—Till at his most carefree, sauntering in the countryside—much too loud, in lieu of Strauss's *p*. In mm.55–62, Strauss gives us *Till's* 'mocking' theme in the horns, but grossly fragmented and rhythmically dissected, torn to shreds, as it were—a remarkable idea which the conductor must ensure is prominently heard (Ex.5). On many recordings it is virtually inaudible through sheer neglect.

Ex. 5

In mm.63–66, Strauss calls for the distinction in the strings between primary motivic material (*f*) and harmonic accompanimental material (*mf*), which again, however, is largely ignored by players and conductors. But these dynamic shadings are very crucial in making Strauss's relatively complex and fleet-moving polyphony texturally clear (Ex.6). In m.62 and m.68 we encounter the first of many places in *Till* where woodwind runs, especially in bassoons, clarinets, and bass clarinet, with which Strauss loved to flesh out his *allegro* pas-

Ex. 6

sages,[8] are seen in the score but never heard in performance. We shall encounter many of these in *Till Eulenspiegel*. In m.62, the E♭ clarinet's chromatic run *can* be made audible if the clarinetist really plays out and the other instruments (other woodwinds and upper strings) hold back just a little, saving their *ff* for the downbeat of m.63. Similarly, the bassoons' and E♭ clarinet's runs in m.68 (Ex.7) can be

Ex. 7

clearly heard if they are encouraged to play out and if Strauss's dynamics—*mf*'s, mostly—are not over-played, and if the horns in particular restrain themselves to a tempered but healthily exuberant *f*, not the all-out *ff* one usually hears. Attention to such details—and there are thousands in *Till* and, for that matter, in all of Strauss's mature works—is of paramount importance. Realization of them fully gives this work an entirely different look and feel. As Mies van der

8. Strauss has often been accused of overloading his scores with 'gratuitous' passage work, all manner of decorative filigree and ornamental clutter. There is some truth to the charges, although we should recognize that this 'instrumental clutter' is the result of an extraordinarily fertile musical imagination, one that often was reluctant to leave an empty spot on a score page. And indeed, much of this ingenious and dazzling passage work usually remains unheard in performance, buried in the always intricate orchestration and complex textures. On the other hand, it is also true that performers (conductors, orchestras) could do much better in making these secondary ornamental passages audible. It takes care and attention: the players involved in such passages—usually fast running figures—must be made aware of their relative importance and encouraged to play out (not ignore, as they usually do); and, second, the other players in and around these passages must restrain themselves dynamically. For if the latter play one iota too loud, the secondary materials will, of course, be covered and rendered inaudible. But these are again things the conductor must hear and control. Unfortunately most don't.

Rohe said, "God is in the details," an aesthetic message most conductors would do well to take to heart.

In m.73, we encounter the first of many ensemble problems related to the timpani parts in *Till Eulenspiegel*. Strauss doubles the last four notes of *Till*'s 'irreverent, nose-thumbing' theme (Ex.3a), set here in low strings and bassoons, in the timpani, marking the latter with the same *ff* as the other instruments (Ex. 8). The problem is that if the timpanist really plays his full *ff*, especially

Ex. 8

with hard sticks, he will completely overbalance not only the other thematic instruments, but in fact the whole orchestra at that point. I find it curious that Strauss, the master orchestrator and meticulous notator, would throughout his life mark timpani parts in a doubling situation with the same dynamic as the other instruments, a practice which many late 19th-century composers had begun to abandon by the 1890s (Mahler, Tchaikovsky, Rimsky-Korsakov). Perhaps for Strauss it was still a relevant notational convention (as in Beethoven), or perhaps timpanists around the turn of the century, still playing on calf-skin heads, simply did not produce such thunderous, all-obliterating sounds[9] that modern timpanists nowadays not only *can* produce but seem to enjoy unleashing. Most conductors also don't seem to mind that in a passage such as just

cited (Ex.8), after the first three theme notes , one usually hears only timpani and nothing of the primary thematic instruments. In any case, it is wise—if not imperative—to mark all such timpani parts at least two degrees lower, i.e. softer, with the result that the timpani notes will blend naturally into the thematic context. The same will apply to m.75 and m.79 and dozens of similar places in *Till*. Indeed in m.75, a loud timpani roll will simply blot out the all-important primary thematic material in the strings (Ex.9).[10] Here not only the timpani's dynamic must be moderated, but all woodwinds

Ex.9

and horns must diminuendo in that measure—perhaps to *mf* or at least to *f*— for they too will otherwise cover the strings, as any number of recordings will

---

9. This may also be the explanation—and solution—for Beethoven's timpani parts, for example, in many places in the Seventh Symphony (see pp. 246, 248, 254, 272).

10. Let the dynamic adjustments in these two timpani-related passages stand as solutions for the balance problems in all similar situations (especially mm.149, 243, 279, 477–84, and, above all, mm.560–72).

attest. The only recordings on which the strings can be heard at all are Bern-stein's and Toscanini's.

Very much the same problem occurs four bars later in mm.79–80. Here Strauss's fertile musical mind devises an ingenious polyphonic construction (again in the strings) (Ex.10), all forged from the same 'mocking' *Till* theme. Here the horns had better join in the other brass's *diminuendo*.

Ex. 10

The next problem spot is mm.89–90, where the fleet figures in third flute and three clarinets are consistently covered. I defy any reader to find one re-cording (or identify one performance) in which the four woodwinds (Ex.11)

Ex. 11

are audible. If, as in most performances and recordings,[11] the remaining instru-ments do not observe Strauss's *p* and bowings, all is lost. I have, for example, never conducted an orchestra in *Till Eulenspiegel* in which (a) the first violin, viola, and cello parts weren't divided into two bows (instead of one, starting mid-bow), and (b) the second violins didn't fail to observe Strauss's *pp* in m.88 and didn't avoid a crescendo in the ascending run of that measure

m. 88

. Strauss's crescendo wedge (———) in the

flutes, first violins, and violas, I must admit, does not help matters, and is best suppressed. It has usually taken me five to ten minutes of rehearsing—often much resented by some of the string players—to restore balances and bowings to the point where every precious detail of those two wonderful measures can be heard. It is certainly worth the time and effort. Similarly, care must be taken that the wonderful third flute part in mm.93–95 is clearly heard.

11. Including, alas, Strauss's own recording; also Toscanini's, and even that of Fritz Busch, who was one of the finer Strauss conductors.

In mm.97–102 almost all orchestras rush—especially the flutists—and in addition lose all sense of the 6/8 meter. Rarely does one hear anything but a series of 4/8 measures, as if Strauss had merely writtten

, leav-

ing no sense or feeling of the fact that the passage is set in a 6/8 meter. One of the few conductors who kept this passage in check was Fritz Reiner.

A quite different (and deplorable) rhythmic subversion of Strauss's notation and musical intentions can be heard in many recordings in mm.113–22. Strauss's

is usually rendered as  (Coates, Blom-

stedt, Karajan (Vienna), Krauss, Maazel, Mackerras, Marriner, Masur, Slatkin, Stokowski, Fricsay, Stock, Sawallisch, and Strauss himself), or even more oddly

(Bernstein, Ormandy, Solti, Stein-

berg, Toscanini), or simply (Reiner,

Karajan (Berlin), Haitink). The passage is also rarely played in a true *p*, and anything but *grazioso*: 'spooky and lumbering' might be a more appropriate description. On only five recordings can this fascinating passage be heard correctly played in all respects: Furtwängler's 1930 recording, as well as those of Kempe, Henry Lewis, Mata, and Busch. (Munch, Dorati and Szell get the right rhythms, but no *grazioso p*.)

What I have written earlier about Strauss's timpani parts can be applied to virtually all the percussion parts in *Till*. A case in point is the cymbal crash in m.135, unfortunately marked *fff*. The problem is that if the cymbalist really plays *fff*, with his largest crash cymbals at that, one might as well send the E♭ clarinet player and the rest of the woodwinds and violas home. The point is that all three successive clarinet entrances in mm.133–35 (Ex.12) and their ascending runs must be well-heard. This depicts Till's charging into the town's

Ex. 12

marketplace on horseback, overturning all the carts and tables, with the women screaming (flutes, oboes: mm.135–36) and scurrying around in startled fright. A forceful, short cymbal crash (with medium-sized plates) not only allows the E♭ clarinet to be heard, but produces a much more dramatic, incisive and surprising effect than the all-obliterating, cataclysmic explosion that has become standard practice here. On Haitink's CD the cymbal crash actually causes a total distortion.

With care and attention of the sort mentioned earlier (in reference to m.62 and m.68), all the runs in the bassoons (m.138) and clarinets (mm.141–42, 145–46) can be made audible. One usually gets to hear only the horns in m.141 and m.147 and the ratchet and flutter-tonguing trumpets.

Note the dynamics in m.153: *p* in the clarinets and upper strings, but a whacking big *ff* pizzicato G in the basses. It is Strauss's humorous way of signaling the abrupt unceremonious end of that episode, as Till beats a hasty retreat from the demolished marketplace.

The basses' G is slyly continued in the contrabassoon in m.155, marked *p*. Note that all other instruments here are marked either *pp* or *ppp*, dynamic differentiations that appear not to be observed on even a single recording.

Every time I have conducted *Till Eulenspiegel* I have had to caution the cellos (in mm.159–60), the second violins (in mm.163–64), and first violins (in mm.167–68) to play their pizzicatos more forcefully. Strauss's *pp* marking is misleading, especially since these notes are set in three-part divisi,[12] meaning very few players on each of the three pitches.

Even more problematic is the eight-bar transitional modulatory passage, mm.171–78. I am applying the term "transitional" because Strauss's *Till* is very much like a short orchestral opera with scenes and quick scene changes, each describing different episodes in Till's life. Scene I starts at m.51, for example, Scene II at m.133, Scene III at m.155, Scene IV at m.179, and so on. In between, there are transitional 'scene changes,' such as mm.111–32 and, as just mentioned, mm.171–78. The performance problem in the last is one of balances and dynamics, and to make Strauss's main melodic line clear (see Ex.13)—which it is unfortunately in only three or four recordings: Busch's and (barely) Masur's, Haitink's, Kempe's and DeWaart's.

I first learned *Till*, after having heard it a number of times in performance and recording (particularly Fritz Busch's early 78 rpm recording with the BBC Orchestra), in my early teen years from the composer-approved piano reduction made by Otto Singer. There one can clearly see and hear the six-bar *Klangfarben* chain of running sixteenth-note figures (Ex.13). The performance problems here are two-fold: (1) to produce, despite the half-bar segmentation, one long connected instrumental line, at the same time making in effect an over-all diminuendo from *f* (m.171) to *p* (m.176); (2) to prevent the accompanying instruments from covering this line once it is in place. Unfortunately, Strauss's dy-

---

12. A famous German-speaking concertmaster of the Berlin Philharmonic (who shall remain nameless) once argued with me that Strauss's *dreifach* marking indicates a triple-stop and that, as such, the pizzicato is automatically louder, the implication being 'loud enough.' This would in fact be the case, except that *dreifach* does *not* indicate a triple-stop, but, rather, quite the opposite: divided

into three parts. This is proven also by the fact that in the cellos the three notes 𝄢 cannot in any way be played as a triple-stop!

Ex. 13

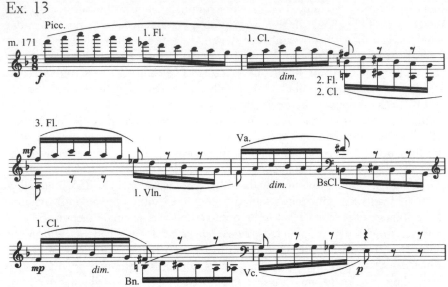

namic notations here are not much help, and need to be slightly adjusted. Since
the players cannot easily divine all this from their individual printed parts, even
after they have been adjusted, this passage requires considerable rehearsal time
to bring off correctly, that is, to achieve a clear balance between the (primary)
melodic line and the (secondary) harmonic accompaniment. The outline of the
latter can be seen in Ex.14.

Ex. 14

Rather than listing the dynamic adjustments I'm speaking of individually and
verbally, it is more practical to show them in score, as in Ex.15. Note that all
sf's have been changed to *poco sf* and, in any case, need to be played in the
context of the prevailing dynamic. If (1) the dynamic adjustments here cited are
adhered to, and (2) if every player involved in this passage understands his specific function(s) in it, I can almost guarantee that this passage will become
structurally and expressively absolutely clear, as opposed to the ambiguous, arbitrary renderings one hears on virtually all recordings.[13]

13. The reader, aware by now that I am vigorously opposed to re-orchestrating the masterworks of
the literature, may wonder why suddenly I am making an exception here, 'retouching' Strauss's
score. For me to say that I am *in principle, fundamentally* opposed to retouching is not a cop out,
because I do hold strenuously to that principle. But every rule, every principle, does have somewhere
along the line an exception. Furthermore, what I am suggesting here is not the wholesale re-
touching/re-orchestrating that has become common in Beethoven and Schumann, even Brahms—
while generally ignoring tempo and dynamic indications—but simply a mild (cosmetic) adjusting

of some dynamics (upward, by the way) to make Strauss's five-and-a half *Klangfarben* sixteenth-note line become clear, automatically audible.

Of course, some readers and conductors will say that that's exactly what *all* conductors contend: they are simply trying "to help the composer's intentions become clearer." That certainly was what Mengelberg, Mahler, Klemperer, and others claimed. But ultimately, in my view it should be the exception, not the rule; also it is a matter of degree, that is (a) a matter of *what* one is retouching, (b) *how much*—many of the retouchings of the past constituted complete re-composings of the passage, changing notes, re-instrumentating the work—i.e. is it micro-surgery or major surgery?— and (c) automatically accepting the retouchings without ever questioning them.

In the exceptional example I have cited here, a few dynamics have been *slightly* adjusted. This hardly comes under the category of re-orchestrating.

It defies understanding as to why almost no orchestras and conductors respect Strauss's dynamics in m.179 (*p!*) and m.187 (*mf*). (Perhaps it is because Strauss himself did not do so in his own recordings.) Only Fricsay, Blomstedt, and Kempe (all basically fine Strauss conductors) do, although in Kempe's Dresden recording, the dynamics prematurely sneak up in mm.183–86 to the upcoming *mf* (m.187). It *is* important to observe the *p* dynamic in m.179 not only because Strauss wrote it and there is no over-riding reason to ignore or adjust it, but because the dynamic here best expresses the mood and character of this scene (Scene IV in the over-all scenario), which depicts Till, dressed as a monk, joining a procession of clerics, spouting religious platitudes. Till's sermonizing is at first unctuously devout, thus *p*, then eight bars later more brazenly burlesquing, thus *mf*. The low-register line  in contrabassoon, bass clarinet, tuba, basses, and two low horns, depicts the real Till slyly peeking out from his monk's cloak and laughing "roguishly" ("schelmisch") in m.191 (E♭ clarinet).

Till's snicker is now imitated in the solo violin (mm.194–96)—a quirky caricature of Till's 'mocking' theme (Ex.16)—as Till, ever bolder, taunts the real monks, sneering at what he considers their holier-than-thou fatuousness and

Ex. 16

clerical hypocrisy. It seems to me that Till's sarcasm is superbly characterized by Strauss in the solo violin passage. And yet, unfortunately, almost nobody plays, i.e. interprets, the passage in the suggested satiric manner. It is usually played as a romantic, sentimental, vibrato-y solo, completely at odds with the intended sardonic characterization. Nor is it played with the right rhythmic inflections. Furthermore, there seem to be very few persons who recall that in some of the earliest editions of *Till Eulenspiegel* the E's in mm.194–95 were marked with a small circle, meaning to play the note on the open E-string. When this is done, it immediately adds a grating, sneering expression to the passage which is perfectly in keeping with Strauss's musical-descriptive intentions. The only concertmasters on recordings who played this solo rhythmically correctly and with the open E string are Coates's and Stock's leaders, and Rafael Druian on Szell's Cleveland recording. Oddly enough, Druian's successor, Daniel Majeski, plays the open string on only the first two E's—not on the third one—on Maazel's recording with the same orchestra.

As for the rhythmic aspects of the passage, it is sad to report that they are rarely rendered correctly. What one usually hears is as if Strauss had written

, in effect completely subverting and turning around the beat, and turning Strauss's purposely twisted phrase into something quite ordinary and boring. (See Ex.16 for what Strauss actually wrote.) A tiny accent on the high D and a careful counting of the sixty-fourths and dotted sixteenths in each eighth-note will easily rectify the problem.

I truly don't like the accelerando so many conductors make in mm.197–98. It is sufficient to observe Strauss's *doppelt so schnell* (twice as fast), without escalating that with another tempo acceleration, especially since these six measures (mm.196–202) represent in Strauss's scenario an ominous premonition of the solemn judgment and death sentence handed down eventually by the authorities. Hurrying the phrase just makes it sound a bit silly and trivial. Strauss's instrumentation here is rather unusual: muted brass and four muted solo violins. What is curious is that the brass are marked *mf*, while the strings are marked *pp* and muted. If these dynamics are taken literally, given the inherent projection capacities of the two types of instruments (brass and violins), the result will be a tremendous dynamic imbalance. Could this be what Strauss had in mind? Or could the violins' *pp* be an error? Later, in m.605, the violins' notes—both full sections now playing and unmuted—are marked *mf*. One of the more interesting and logical solutions to the seeming dilemma thus presented in m.196 is one that Fritz Busch turns to, namely, to have the trumpets and horns play *p*. This makes sense because, as I just pointed out, the initial appearance of this passage is in the nature of a foreboding, a tentative hint of things yet to happen.[14]

The score is, by the way, in error in m.202. The annotation *wieder noch einmal so langsam*—a curious way of putting it, literally: again once as slow—should have been placed in the middle of m.202, not at the double bar of m.203. For, if taken literally as printed, the three violin notes in m.203

would have to be played in the 'twice-as-fast' tempo of m.197—a clear impossibility technically and a bizarre idea musically.

The transitional episode (mm.209–22)—after the famous chromatic descending violin solo—is usually played well enough, although perhaps a bit loudly, especially for the delicate instrumentation in mm.215–18 of one flute, three oboes, four solo strings, all marked *pp*. However, in the next 'scene'—in which Till falls in love (mm.229–52), but is peremptorily rejected, throwing him into a violent rage (m.253, *wütend* (furious)—many small and large performance/interpretational problems abound. The entire scene is musically quite challenging, in terms of its myriad instrumentational details, dynamic balances, and structural complexity. As usual, most of the problems arise from conductors' and musicians' inattention to what the score actually prescribes, while a few problems—a very few—are the result of ambiguous notation.

Dealing with the latter first, I would cite the crescendo in trombones, tuba,

14. For the record, Strauss in his own recording with the Vienna Philharmonic has the four solo violins play *f* to more or less match the muted brass.

and timpani in m.224. It would have been useful for Strauss either to have added *poco* to the crescendo wedge (—◁) or—even better—to have indicated a terminal dynamic, like *p* or *mp*. But it would seem to me that any intelligent conductor ought to be concerned about the degree of crescendo in m.224. As it is, on most recordings the crescendo, especially in the timpani, usually goes all the way to *f*, completely covering the other instruments: flutes, bassoons, string trills. In the same measure the oboes and English horn should have been marked *mf*—not *p*—to match the flutes, E♭ clarinet, and second violins. The

solo-viola trill  should also be marked *mf*. It is audi-

ble on only two recordings. It would perhaps also have helped if Strauss had given some dynamic indications for the 'love theme' in the first violins, E♭ clarinet and third flute. Given the accompanimental context in mm.229–44, I suggest that the dynamic in m.229 should be *p*, m.233 *mp*, followed by Strauss's specified *mf* in m.237 and the further crescendo to *ff* in m.243—the increasing dynamic levels representing Till's mounting ardor and passion.

The real problem in this entire episode is that hardly anybody wants to accept Strauss's *p*'s and *pp*'s, scattered throughout the entire scene.[15] I am convinced that Strauss thought of the string pizzicatos (see Exs.17a and b) as a delicate

Ex. 17a

15. Let us note the special subtlety of Strauss's dynamic nuancing here: *p* when the 'love theme' appears in G minor, *pp* when it appears in G major!

Ex. 17b

guitar: Till serenading his *amour*. The only recordings on which these dynamics can be heard correctly are Blomstedt's and Kempe's, both with the Dresden Staatskapelle. (I suspect that this excellent orchestra, so well trained in the Strauss tradition, had these soft dynamics ingrained in their playing of these passages). As I have mentioned before, for me there is nothing quite as exciting as a whole 85-piece orchestra playing a subtle, refined *pp*. And a true collective *pp* in mm.245–52 is especially ravishing, capturing the intimacy of Till's ardor and the voluptuous, dance-like lilt of the pizzicato accompaniment, making Till's sudden furious *ff* outburst at m.253 all the more exciting and dramatic by contrast.

As Till's fury mounts, swearing revenge on the world, Strauss builds to an extraordinary climax in which *Till's* 'mocking theme' is elaborately expanded in rhythmic augmentation, supported by massive, wildly modulating chords (Ex. 18). Here a special effort is needed in m.276 and m.280 in the woodwinds to make the ends of both descending runs audible—they are *in*audible on all but one or two recordings. The flutes can't help much here, descending as they do, to their weak middle register; but the clarinets certainly can by crescendoing as much as possible to their final notes.

In a brief transitional interlude (mm.287–93), Till quickly forgets his anger,

Ex. 18

tripping gaily off in search of further mischievous adventures. A new theme, first heard in a quintet of three bassoons, contrabassoon, and bass clarinet (Ex.19),

Ex. 19

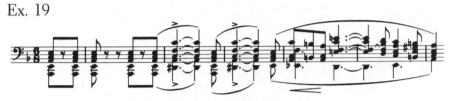

is juxtaposed with a new variant of *Till*'s horn theme. In this scene (Scene VI, mm.293–370), Till is in an encounter with the most learned pedagogues of the land, plying them with incomprehensible fact-distorting questions (the horn theme variant), to which the academic philistines can find no logical answers. Their confusion and dumbfoundedness are depicted brilliantly by Strauss in various rhythmically 'confusing' passages, the professors all arguing among themselves in ingenious double canons (mm.319–29) and helpless stutterings (mm.335–46). At m.308 Strauss writes into his score above *Till*'s theme in the violins *lustig* (merry, comical), depicting Till laughing heartily as the learned men—in the 'bassoon quintet' now doubled by cellos and basses—in all earnestness consider Till's ludicrous questions. This passage, with its conflicting rhythms, takes a very steady, poised conductorial hand, which it evidently doesn't often get, judging by the many recordings in which this passage is very shaky—as confused as the bewildered professors it describes. The entire episode shows Strauss at once at his most daring creatively and his most brilliant and imaginative in the handling of the orchestra. This makes this section also extremely difficult to perform, especially if one is intent on realizing every one of the myriad details of the score. The two canonic episodes mentioned above are particularly difficult. As the musical examples (20a and 20b) show, the canons are one beat apart in the first instance, but only an eighth apart in the second instance. This daring rhythmic/metric dislocation of the pedagogues' theme—

from 𝄽 to 𝄽—goes beyond what Brahms was do-

Ex. 20a

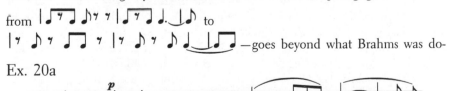

Ex. 20b

ing in this respect in his symphonies, and presages in 1895 what Charles Ives was to do a decade or so later, taking this idea of metric dislocation much, much further.

One wonders how musicians—and conductors—coped with these difficulties in 1895, for they are still challenging to this day[16]. While most orchestras can now deal effectively with the rhythmic aspects of this section, for the most part, dynamics and balances remain elusive and unrealized, especially the three *subito p*'s/*pp*'s (mm.318, 335, and 358). On only a very few recordings—notably Blomstedt's, Kempe's—are these dynamic differentiations realized and the various intricate contrapuntal lines clarified.

At m.344, Strauss again over-marked the timpani and bass drum. If they really play their indicated *ff*, the trombones and tuba—let alone the low strings— might as well go home for all they're going to be heard. Testimony for this can be found on innumerable recordings. (Hear especially the ridiculously thunderous percussion on Mata's and Karajan's recordings, making even the trombones in effect inaudible.)

The next problematic passage occurs near the end of the scene (m.362), in which an amazing series of triadic harmonies (Ex.21) leads to the climax of the scene. Because each chord is set in a different instrumentation, the specifics of which cannot be gleaned from a player's individual part (only from a score), it is not easy to achieve a balance among the discrete chords and produce an overall musical line. That is where the conductor has to step in and, by rehearsing and balancing, create a continuity in which all the instruments involved are properly heard, to arbitrate, as it were, between trombones and bassoons, or trumpets and clarinets, for example. In most recordings the trombones and per-

---

16. Some orchestras and conductors still had troubles with these rhythms in our own time, as witness Steinberg's 1970 Boston Symphony recording of *Till Eulenspiegel*.

Ex. 21

cussion are by far the loudest, the low trumpets the weakest, and the woodwinds fend variously for themselves in the *melée*. The problem of balancing and actually hearing the disparate harmonies is made more difficult by the speed with which all of this is happening: a lively tempo of somewhere around ♩. = 120. I have found only a few recordings in which this remarkably daring and inventive passage reflects what Strauss actually wrote: Szell's, Solti's, Blomstedt's (almost), and Ormandy's (however, at a sluggishly slow tempo).

The scene builds to a *ff-fff* climax with Till derisively taunting the perplexed academics (four times) , culminating in a strident trilled chord , which Strauss called "die grosse Grimasse" (the big grimace): Till thumbing his nose at the outraged professors.

Till now flits away (mm.371–74), dashing down the nearest alley and whistling a common popular tune—a *Gassenhauer*, as Strauss puts it. It is amazing to me how many conductors ruin this delightfully witty scene by ignoring its most salient features: the accelerando in m.372, the *p* at m.375, and—even more important— the *pp* at m.378. The accelerando depicts Till's quick escape, the *pp* his disappearance off into the distance—all perfectly realized on Fricsay's recording.

The music evaporates, as it were, before our very eyes and ears into a shadowy, purposely tenuous passage (Ex.22) as Till contemplates his next move. The music is fragmented, made up of distorted bits of *Till*'s 'mocking theme.' Care must be taken to distinguish clearly in m.389 and m.393 between the extremely short clarinet sixteenths and the longer fully sustained quarter-note F's—a distinction made on almost no recordings. Nor is Strauss's tempo/character admonition—*schnell und schattenhaft* (fast and shadowy)—generally respected, most conductors slowing down to a bland *adagio*. Talk about missing the 'spirit' of the music—as well as the 'letter'! What these conductors also completely miss is the idea that this music is already distended rhythmically—from the previous eighth-notes to three-times-as-slow half notes—and therefore does not need any further slowing down of the tempo, and that the entire 'shadowy' episode is structurally a bridge passage to what will eventually become the recapitulation of *Till*'s exposition. The point is that the *poco rit.* in mm.408–409 should lead to, i.e. elide into, the *etwas gemächlicher* [17] of m.410, not ritard beyond it, and

17. *Gemächlich*, Strauss's 'tempo' marking for the Prologue and used in m.410 as an important tempo reference point, is a virtually untranslatable word meaning something between leisurely, comfortably, and easy-going.

Ex. 22

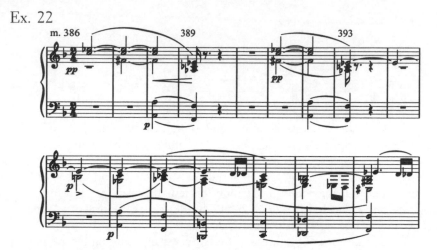

thereby causing a kind of *a tempo* at m.410. Nor should the basic tempo prior to m.410 be so slow as to require an *a tempo* there. Moreover, if the bridge passage is taken too slow, the ingeniously twisted, distorted oboe phrases (mm.403–409)

lose their wonderful grotesquerie, and become instead some sweetly romantic solos that are quite out of place in this scene. Last, these oboe solos should not be *gemächlich* (leisurely), for then they anticipate and annul the effect of the real, wonderfully amiable *gemächlich* of mm.410–28. Strauss's *etwas gemäch-licher* in fact refers by inference to both the immediately preceding section *and* to the tempo of the work's opening Prologue. In the first instance, *etwas gemäch-licher* (a little more leisurely) clearly implies that the previous section must be at a faster tempo—we shall see in a moment how much faster—and in the second instance has the clear implication that the tempo at m.410 should in-deed be that of the opening Prologue. Stated in rhythmic notation, the ♩. of the m.410 6/8 equals the ♩ of the *gemächlich* Prologue.

By any accurate, respectful reading of the score the tempo in mm.375–85 (the *Gassenhauer* tune) should be quite fast—for several reasons. If we assume that the basic 6/8 tempo for *Till Eulenspiegel* is anywhere between ♩. = 100 and ♩. = 120, the tempo range adopted by most conductors, then—obviously—the three-bar accelerando mm.372–74 will push the tempo up to a somewhat faster speed level. At m.375 (2/4) Strauss says "*leichtfertig* ♩ = ♩. *des 6/8*" (frivo-lous, 2/4 ♩ = 6/8 ♩.). This clearly means that the 2/4 tempo shall equal (via a ♩. to ♩ conversion) the tempo reached at the end of m.374. This could be anywhere between ♩ = 112 and 132. If we now respect Strauss's *schnell und shattenhaft* (m.393)—i.e. avoid slowing down—and observe the *poco rit.* in m.408 (not *molto rit.*), then we have a tempo unit relationship between m.409

and m.410 of ← 𝅗𝅥 = 𝅘𝅥. →. If the tempo at m.375 were, for example, 𝅘𝅥 = 132—
more or less the practical upper tempo limit considering the fast triplet figures in
bassoon and oboe—or, stated in half notes, 𝅗𝅥 = 66, then the little *poco rit.* in
mm.408–409 might retard the tempo effectively to 𝅘𝅥. = 60 at m.410. Such a sequence of tempos not only fulfills all the expectations of Strauss's score and notation, but has the further virtue of positioning us in exactly the right tempo to allow
for the return of the main horn theme (m.429) to be a replication of its original
appearance in the exposition—the recapitulation I mentioned earlier.

There is an excellent way from a baton-technical point of view to achieve all
of this, namely, to conduct m.375 through m.385 in a fast 'two' (having, to be
sure, made the previous accelerando), go into 'one' at m.386 (maintaining the
same speed, of course), that 'one' converting (via the little *poco rit).* into the
'two' of the new, slightly slower tempo at m.410. In mathematical-metronomic
terms the sequence would be stated as in (Fig.1).

Fig. 1

mm.386–407 𝅗𝅥 = 66	m.408 (rit.) 𝅗𝅥 = 66, rit. to 𝅗𝅥 = 60	m.410 ← 𝅗𝅥 = 𝅘𝅥. → 𝅘𝅥. = 60

In mm.410–28 there are two aspects that require some special attention. One
is to bring out (or at least not allow to be under-played) the lovely cello count-

ermelody 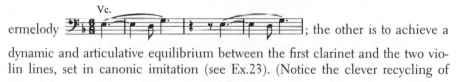; the other is to achieve a

dynamic and articulative equilibrium between the first clarinet and the two violin lines, set in canonic imitation (see Ex.23). (Notice the clever recycling of

Ex. 23

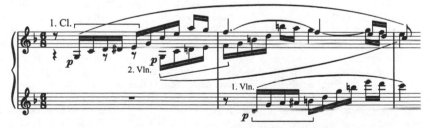

Till's horn theme (bracketed in the musical example), set against *Till's* other

main theme in a *gemächlich* augmentation . The clarinet

has by nature, even when played *legato*, a clearer note-by-note articulation,
while the strings played legato tend inherently to be much smoother. I believe
if Strauss had put the second and first violin in second clarinet and flute respectively, the canonic interplay between the three lines would be automatically

clear. As it is, one has to admonish the violins to play with a certain kind of left-hand articulation to make the intimate relationship between the three lines explicitly clear. (These correspondences between clarinet and violins become critical again a little later in mm.451–64.)

In the short-lived pseudo-recapitulation (m.429), all the points—dynamics, tempo—made earlier regarding *Till's* horn theme are applicable again, including, of course, the third horn solo: *Till's* theme, now in the key of D.[18]

Horns continue to dominate the entire next section (mm.449–63) (see Exx. 24a,b,c,d); and while horn players have long ago mastered these 'solos' techni-

Ex. 24a

Ex. 24b

Ex. 24c

Ex. 24d

cally[19], it is depressing to note that none attempt to play the four passages in the correct rhythmic-metric configuration; nor evidently do any conductors insist on any correct rendering. It is as if Strauss had written

 in m.449.

18. As an example of how traditions—not necessarily valid traditions—are formed, I might cite my own recollections from my years as a young horn player in New York. The New York Philharmonic's horn section all through the Mengelberg-Toscanini-Barbirolli years was dominated by a marvelous horn player, Bruno Jaenicke: not the flashiest technician on the horn, but the consummate expressive artist on the instrument. Two brothers, related by marriage to Jaenicke—Adolf and Robert Schulze (the latter my teacher)—and an Italian, Luigi Ricci, completed the horn section. Stylistically Ricci, originally hired by Toscanini, did not fit into this otherwise all-German-trained horn section, although it must be said that Ricci tried his best to blend tonally and dynamically with his three horn partners. Ricci, like many Italian horn players of that era, had a fantastic virtuoso technique, an incredibly fast tongue, and a flashy, peppery style of playing that was really light years removed from Jaenicke's conception. Throughout my young years I heard *Till Eulenspiegel*, a big favorite in the New York Philharmonic repertory, many, many times; and for all those years Ricci played the D horn solo (mm.436–42) faster and flashier than it had ever been played before. Ricci became legendary for his technical prowess—he never missed a note in this solo (or in any other, for that matter)—and single-handedly set a tradition, at least in New York, of playing the third horn solo in *Till* as a dazzling virtuoso display, a 'tradition' which was followed in New York orchestras for decades, and which no conductor dared to oppose. The moral of this anecdote is that a kind of exceptional technical virtuosity determined the interpretation of the work, while the work itself—its character, its musical essence, its form and continuity, its stylistic integrity—became interpretationally irrelevant, alas.

19. Nonetheless, it is surprising how many third horns miss the C ♪ in m.450, as any number of recordings will attest.

In the second part of the recapitulation (mm.465–84)—a recapitulation some-what modified in orchestration and elongated by two measures—all previous comments and suggestions should obviously apply again.

One can quibble about missing or ignored details in the accompaniment to the exuberant, triumphant transformation (mm.485–500)

 of Till's horn theme—details like

the trumpet fanfares (not audible on the vast majority of recordings) and some of the frolicking woodwind runs and trills—but, in general, this passage is played reasonably well. I suggest, however, that every effort be made to bring out the trill in basses and low woodwinds in m.499—a uniquely Straussian touch.

In the next section (mm.500–573), leading to Till's capture and arraignment by the authorities, many wonderful details are unfortunately ignored or misinter-preted. To begin with, the deliciously witty trumpet part in mm.500–508 (Ex. 25) is rarely heard (but hear it splendidly on Fricsay's and Haitink's recordings).

Ex. 25

Strauss lavishes all of his dazzling orchestrational skills on the final climactic section of the work. It is masterful in conception and refinement of detail: noth-ing, no instrument, no motivic idea, escapes Strauss's attention and imagination. Indeed, it is so rich in detail that one wonders whether any performance can ever do full justice to it all. I shall limit myself to some of the more important and consistently ignored or misrendered aspects.[20] In m.516 through m.527,

20. I raise a question as to the accuracy of the high F in the bassoon in m.514. I have always

m. 512

wondered if this shouldn't have been a high C,

as in the otherwise absolutely identical solo viola part. Arguing for the high C is the fact that it makes a more logical musical line, and that dropping suddenly down to the F makes an awkward jump and very poor 'voice leading'; it also better parallels the viola part. Arguing against the high C

is the fact that in m.514 the lowest 'bass' note is that very same F . It is possible that Strauss wanted to sustain that F, otherwise represented in only two tiny eighth-notes in the tutti violas. Moreover, the bassoon's F was perhaps considered by Strauss a dominant resolving to the

Bn.   1. Hn.   BsCl.

upcoming (temporary) tonic of Bb major (m.516)  .

in virtually all recorded performances, all one can really hear are the flutes and the horns, and nary a sound from the all-important 'horn' main theme

 in the oboes and violins. It is typical

of the general problem just mentioned, namely, that Strauss's fertile imagination creates an abundance of polythematic layers, *all of which* often have equal status and must therefore be correctly balanced and integrated. This takes a very sharp, caring, and constantly monitoring conductorial ear. In this particular passage all three thematic layers are marked either *p grazioso* (flutes), or *pp* (horns, oboes, violins) and the remaining accompanying instruments (trumpet, violas, etc.). In nine out of ten performances, however, the flutes play—and are allowed to play—*mf*, the horn somewhere around *mp*, and the poor oboes, oddly enough playing a real *pp*, recede into the distance.[21] Flutists, by the way, do not help matters when they rush through their thematic figures or, in their anxiety

 to get all the

notes in, enter one eighth early, i.e. on the downbeat (as can be heard on innumerable recordings: Toscanini's, Maazel's, Mata's, Steinberg's, Paternostro's, among others). Horn players sometimes also do not play *Till's* theme correctly, adding to the rhythmic confusion. For example, on Mackerras's recording with the London Philharmonic, the first horn ingeniously converts Strauss's

 in m.520 into his own personal ver-

sion: which, if any composer actu-

ally wrote such a thing and asked a player to render it accurately, would immediately elicit a storm of wrath and grumbling disdain.

   Problems of this sort abound in this climactic section of *Till*. There are far too many to enumerate here, but some of the more salient derelictions are worth mentioning, to serve as examples. The *fp*'s in mm.532–38, followed by crescendos to *ff*, are ignored on all but one recording: Kempe's. On all others one gets at best a kind of grudging *f - mf* or an accented *f*. Nor can one rely on the upper strings to make the diminuendos in their parts in mm.532–39, diminu-

---

For an interesting parallel case, where Strauss allows the ascending line to rise to its expected high point, see mm.169–71 in the violas and English horn:

21. There are also many recordings on which the whole orchestra here simply plays loudly—*mf* or *f*—notably Bernstein's, Karajan's, and Solti's.

endos of great importance since they go simultaneously against the rising *cre-scendos* in the lower strings, woodwinds, and horns. As a result of such care-lessness in the dynamics, the fantastic high woodwind figurations are all but lost in the shuffle. Very few conductors and orchestras make the crucial dynamic differentiations in mm.544–48 between *fff*, *ff*, and *f*. Again, only Kempe, Dorati, Reiner, and Haitink render these correctly, while Blomstedt, Marriner, Masur, Karajan, Maazel, for example, do not.

One of the most miraculously inventive passages in all of late 19th-century music is the climactic section mm.558–65, a chain of forty-two contrapuntally organized, wildly divergent chords, played by the entire orchestra. This most extraordinary harmonic tour de force is, however, rarely—if ever—done full jus-tice, most often because the timpani, entering in m.560, marked *ff* and playing thirty-five reiterated D's—notes which for the most part do not fit into any of the chords—simply obliterates the rest of the orchestra. Nothing of the won-drous harmonic progression is then heard. Two things need to be done in order for this extraordinary passage to be performed as originally conceived by Strauss. First, the timpani needs to be marked down to at least *mf*, particularly with the hard wooden mallets that Strauss calls for (for rhythmic clarity). Second, all the accented thematic notes—variants of *Till*'s 'mocking theme' (see Ex.26a)—

Ex. 26a

should be brought out *ff*, while all other non-thematic notes need to be played *substantially* softer, in order for the main line of canonic-motivic interplay (Ex.26b) to become aurally clear. I would also suggest that the conductor has the orchestra play through the entire passage in slow motion so that everyone can hear and savor—even tune and balance—this most remarkable chord progres-sion.[22] It is then all the more exciting when this tonally completely unanchored, almost cacaphonous, music is released into a pure, brilliant D major (m.567). Among the better realizations of this entire passage is Toscanini's, who takes the timpani way down, and Blomstedt's, Reiner's, Fricsay's, Sawallisch's and Henry Lewis's. By comparison, Barenboim's, Solti's, and Maazel's 'readings' of this

22. I have done this with great success with a number of orchestras, when even the most blasé of musicians have been amazed at what they are hearing for the first time. In truth, rattled through at full speed, especially with a cannonading timpani, even the best and most interested musician can-not possibly hear what Strauss has wrought here. It is a pleasure to report that, once played through very slowly chord by chord, the passage makes more sense to an orchestra and, when then played at full tempo, attains a harmonic clarity that is startling and otherwise unachievable.

Ex. 26b

passage are among the worst: not only raucously loud and insensitive to the harmonies, but (in the case of the latter two) almost falling apart rhythmically (in mm.563–65)—a real mess!

There follows the famous scene of Till's being apprehended and sentenced to be hanged on the gallows, replete with the tribunal's judgments in heavy F minor chords and the executioner's muffled drums. I am opposed to the long-standing tradition of adopting a slower tempo at m.577—for a number of reasons. Again, the composer has already considerably slowed down the momentum of the music—from the fast eighth-notes of the previous section to heavy dotted half-notes. Second, the fermatas in mm.581,589 and the fermata-like elongated chords in mm.596–97,600–601 are portentous and threatening enough so that further tempo deviations are quite unnecessary. Third, Strauss calls for an *etwas breiter* (a little broader) at m.602, a clear indication that the preceding section should *not* be played at a broader tempo. I am also vehemently opposed to the 'cheap trick' of inserting a *subito p* at m.602, as one can hear, for example, on Blomstedt's recording.[23]

One of the worst instrumental abuses, either tolerated or aided and abetted

23. Blomstedt's *Till* recording is actually, all in all, one of the best, one of the most intelligent and notation-respectful, well played and beautifully recorded. But it is also a bit humorless and rarely captures the *capriccio* character of the piece.

by countless conductors, is the trombones' and tuba's bad, lazy habit in the entire 'judgment' episode (mm.577–603) of dropping their sustained dotted half-

notes:  . The word *drohend* means

threatening, ominous, and it ought to be self-evident that the more firmly sustained these 'judges' pronouncements' are intoned, the more effective, the more dramatic their musical impact. There is simply no excuse or reason for playing

A descending seventh in low brass and woodwinds signals Till's death sentence. He is hoisted up on the gallows (m.615), dangling in mid-air as his last shuddering breath (in the flute) escapes him. Strauss's graphic depiction knows no bounds. The grisly 'gallows' chord, first heard in squealing high woodwinds (m.618) (incidentally the same chord that is part of *Till's* E♭ clarinet theme—see Exx.4a,b—except inverted), then, after Till's body is brought down, three octaves lower in low woodwinds and muted horns , is resolved to F major as Till expires. These two horn chords (mm.623–24) are not easy to tune and balance, especially as they are muted. The F chord, moreover, is not set in the usual manner but in an odd voicing , emphasizing the third of the chord, the lower A, in an unusually low position, and is practically inaudible on most recordings.

The Epilogue (m.632), all of it in soft orchestral colors, is an expanded variant of the Prologue, one of Strauss's most gently lyrical creations. It speaks clearly of the affection we have for Till despite his roguishness—or perhaps *because* of it: a kind of German Robin Hood. In performance it should have the feeling of reminiscence, of a fond farewell and a poignant memory. The Epilogue's tempo marking *Doppelt so langsam* (twice as slow) is very important, serving as an unequivocal reminder of what the relationship between the basic *lebhaft (allegro)* — the main 6/8–2/4 body of the piece — and the Prologue/Epilogue is to be.[24] Since Strauss does not use metronome markings in *Till Eulenspiegel*, nowhere does he tell us anything precise and specific about *what* tempo — we have to divine that from the Prologue's *gemächlich*, applied to the 4/8 meter. But the tempo conversion at m.6 and the "twice as slow" indication at m.632 tell us at least — and unequivocally so — that whatever tempo we do commit ourselves to at the beginning, it should determine the *lebhaft* tempo as well, i.e. twice as fast as the Prologue. The point is that Strauss's *Volles Zeitmass (sehr lebhaft)* in m.13[25] is not a precise tempo indication. We only find out near the end of *Till*, at the Epilogue, that it means twice as fast as the Prologue tempo.

Several small, typically Straussian details in the Epilogue are often neglected or misinterpreted. For example, the single triangle note in m.644 — such a splendid touch — is inaudible on numerous recordings. In mm.645–46 Strauss plays around with *Till*'s 'horn theme, now in a 2/4 context in subtly different variants. As Ex.27 shows, the final two notes of the motive are rhythmically one

Ex. 27

sixteenth apart. Strauss's cautionary ⟩ in the horns is, unfortunately, rarely observed — it *is* hard (but not impossible) to make a diminuendo into a high note on the horn — and thus, on most recordings, while the clarinets maintain their *p* or even diminuendo into their last note, the subtle rhythmic play between the two pairs of instruments is generally lost: one hears almost only horns. How this should and can sound can be beautifully heard on Kempe's recording.

The sublime diminuendo into m.640 is best achieved and enhanced in the first violins by staying on the A string.

I am opposed to the big ritardando that most conductors make in mm.647–48, as if Strauss's diminuendo were not enough. The delicate suspended surprise A♭ chord — just before the rowdy, boisterous final eight-bar coda — is much

24. Toscanini evidently did not understand this relationship. Although his program note writer refers to the identity of tempi between Epilogue and Prologue and claims that Toscanini observes this, the Maestro in fact does not, taking an inordinately slow tempo in the Epilogue.
25. Full tempo (very lively)

more of a surprise when it is not prepared for or hinted at by an obvious ritard; it should come out of nowhere.

The other—and final—vulgarism visited upon *Till Eulenspiegel*, perpetrated by many conductors, is a *ffp* ————— in mm.653–54—a real corny 'Hollywood show business' effect. Hear it, for instance, on Marriner's recording. Strauss has made *Till's* sneering chord in its final appearance more interesting by way of instrumentation: the trombones and tuba drop out in the middle of the measure, the second horn quartet at the end of the measure (Ex.28). But this means that the remaining instruments must maintain their *ff*

Ex. 28

at full strength. It also means that the instruments that drop out must do so accurately and audibly, i.e. no diminuendo fading. In any case, the corny *ffp* effect is completely out of place.

Why any conductor would distort and re-compose Strauss's 6/8 rhythms in the last four measures of the piece—or allow them to be distorted—is beyond my comprehension. On quite a few recordings, most shamefully Karajan's and Stock's, the music, written by Strauss (Ex.29a), is re-composed (Ex.29b). Unbelievable!

Ex. 29a   Ex. 29b

Even in this relatively exhaustive perusal of *Till Eulenspiegel* and its performance/recording history, I have been able to touch upon only the most serious interpretive problems and abuses. Given the extraordinary fertility of Strauss's musical imagination and inventiveness, there is a wealth of other detail that might be discussed. But even these brief deliberations may give a mild idea of not only how many hundreds of musical details a performance, a recording, should have to address, but also how far most performances stray from the intended path, from the notated score.

Ravel: Daphnis et Chloé
Second Suite

It is hard to name a composer who was more meticulous and detailed in his musical notation than Maurice Ravel. Debussy, Stravinsky, Prokofiev, Scriabin, Schönberg, Berg, and (especially) Webern are possible contenders, but I doubt that they surpassed Ravel in the exactitude with which he normally finalized and refined his scores. The increasing complexity of music in the first decade of the 20th century—in the French school the subtle use of varied timbral mixtures, of multiple divisi in the strings, in the Viennese school the dramatically increased use of polyphony, a more fragmented continuity combined with a more instrumentally individualized chamber music approach, in the Russian school the greater use of massive harmonic and rhythmic structures, carefully differentiated orchestral/registral layerings—all this virtually compelled composers to scripturally 'fix' their musical conception more completely and precisely than had ever been necessary earlier. Composers such as Bach, Mozart, Beethoven—even Liszt and Wagner—could rely on a high level of established performing traditions, in which their notation stood for a great deal more than actually appeared on the printed page. But with the ever more rapidly developing advances in musical conceptions and the emergence of highly differentiated nationalist styles, stylistic awareness on the part of orchestral musicians (and conductors) could no longer be taken for granted. Composers, virtually in self-defense, resorted to ever greater precision of notation, hoping to pin down as many minutiae and details of composition as notation permitted. Ravel's *Daphnis et Chloé* is a remarkable example of such notational *raffinement* and precision, several major scriptural discrepancies not withstanding (more on these later). And yet, considering that the Second Suite from Ravel's ballet is one of the most popular and most performed early 20th-century scores, it is staggering to discover that it is rarely played correctly, rarely played as it was written, too often performed in considerable disregard of the hundreds of details of instrumentation, dynamics, tempo relationships Ravel lavished on this, one of his most brilliant, perfectly realized creations.

459

As just mentioned, the Second Suite score, as printed by Durand in 1913, is not without errors and misprints. Some are quite serious, such as the wrong metronome indication at m.106 (it should have read ♪ = 66); or the intent of Ravel's *retenez* in m.77. Less serious, because Ravel's intentions can be readily deduced from the prevailing context, are the missing dynamics in the strings at m.60 and in the lower strings in m.76, in the pizzicato strings in m.124; the missing *pp* dynamics in horns, alto flute, and harp at m.20, and numerous others of the same sort.

To reconcile and clarify such notational omissions and ambiguities, the musicologist and conscientious conductor normally turn to the composer's autograph or, alternatively, when such does not exist, to a piano reduction, usually published a short time after the initial full score edition. By an ironic twist of fate neither of these sources is of much help in the case of *Daphnis et Chloé*. Ravel's autograph[1] turns out to not only contain almost all the same errors as the Durand score, but is full of omissions and discrepancies which *were* later rectified in the printed Durand score. Viewing the autograph, as I have, two things become clear: the score shows signs of great haste, a fact confirmed by what we know about the circumstances during the final months before Diaghilev's Ballets Russes premiere in 1912. This presumably might explain the surprising number of errors and omissions. For example, much of the percussion parts is entirely missing in the autograph, evidently intended to be filled in at a later stage in the process of producing the parts, editing and printing the score. (The percussion parts are remarkably detailed and near-perfectly represented in the Durand score.)

Second, it is clear from the autograph manuscript that it was used by the editor and engraver as the basis for the engraving of the published score, since every page contains indications superimposed in another hand in red pencil of how many staves would be required for each printed score page. This means in turn, at least conjecturally, that either another more complete score by Ravel existed and was used supplementarily to arrive at the final engraving, or Ravel himself completed and refined the engraved score in the process of proofreading. But this is truly conjectural on my part since, despite years of research in this matter, I have been unable to find any evidence that such a supplementary score ever existed, or any absolute evidence that Ravel actually proofread the engraved score. That is, however, a reasonable and generally accepted assumption.

As for the piano reduction of *Daphnis et Chloé*, made by Ravel himself and published in an unusual turnabout in 1910, nearly two years before the premiere and three years *before* the publication of the full score, it contains many of the same errors and indeed has a few of its own, not in the orchestral score (see, for example, the *andante* tempo marking at the first entrance of the *Danse générale* music, p.92 of the piano score).

1. Ravel's *Daphnis et Chloé* autograph is located in the Harry Ransom Library and Humanities Research Center of the University of Texas in Austin, having been sold some years ago to that insitution by Durand, Ravel's publisher.

As a result we are left with a conductor's score which for the most part is remarkably detailed and precise, but which at the same time contains a number of irresolvable discrepancies and ambiguities, regarding which no one can say with absolute certainty what Ravel's intentions may have been.[2] (We shall deal with these questions at the appropriate moment in our analytic perusal of the score.) This also leaves us with the mystery of how, by what process, by whose hand, the printed score became a much more complete and detailed realization of Ravel's work than his own autograph score.

Measure 1 (rehearsal number 155) shows a tempo marking of *lent* ♩ = 50. This metronomization turns out to be, upon closer scrutiny of the score, of crucial relevance since many other tempos in the Second Suite are, either explicitly or implicitly, mathematically related to this opening tempo, a fact most conductors are seemingly unaware of. In the hands of many conductors and orchestras, Ravel's *Daphnis et Chloé* Second Suite has become primarily a virtuoso showpiece, in which super fast tempos are seen as a sure-fire means of 'knocking the audience out' and gaining their thunderous ovations and screamed bravos.

Although the famous *Danse générale* is most conductors' favorite part for pushing the tempo to the edge of the impossible, the tranquil opening of the Suite is no less immune to tempo distortions. Most conductors ignore Ravel's ♩ = 50 and, as Fig.1 shows, the gamut of tempos ranges all the way from an immobilizing, desultory ♩ = 36 (Dervaux) and a too deliberate, too slow ♩ = 42 (Munch and Mengelberg) to a nervously driven ♩ = 62 (Thomas and

Fig. 1

♩ = 36	Dervaux
♩ = 42	Munch, Mengelberg, Muti, Mehta, Barbirolli, Levi
♩ = 44	Haitink, Casadesus, Barenboim, Martinon (Chicago), Slatkin
♩ = 46	Toscanini, Ansermet, Commissiona, Previn, Boulez, Schwarz, Gaubert, Nagano, Nowak, Rattle
♩ = 48	Abbado (LSO), Skrowaczewski, Dohnanyi, Paray, Ozawa, Sinopoli, Mata, Inbal, Jansons, Levi
♩ = 50	Solti, Dutoit, Rosenthal, de Burgos, Slatkin, Ormandy
♩ = 52	Mackerras, Tortelier
♩ = 54	Kondrashin, Levine, Abbado (BSO)
♩ = 56	Bernstein, Monteux, Stokowski (1970), Koussevitsky (1944), Karajan
♩ = 60	Maazel
♩ = 62	Koussevitzky (1928), Thomas

2. It is unfortunate that the several errata lists published in the *Conductors Guild Newsletter*—Vol. 6, No.4 (1985), Vol. 7, No.1 (1986), Vol. 11, Nos. 1 & 2 (1990)—although dealing with an immense amount of valuable notational detail and minutiae (including wrong pitches), fail even to allude to, let alone resolve, some of the *major* questions and ambiguities referred to above. The same, alas, needs to be said about the newly published score by Clint Niewig.

Koussevitzky (1928)). The livelier tempo, apart from being disrespectful of Ravel's conception, tends to make the flute and clarinet passage work in the opening measures sound like virtuoso exercises, rather than the evocation of the quiet beauty of early dawn and the gentle murmuring rustling sounds of nature. Ravel's score reads "Aucun bruit que le murmure des ruisselets amassés par la rosée qui coule des roches" (No sound but the murmur of rivulets of dew flowing down over the rocks.)

In m.2 we encounter the first relatively serious performance problem. There seems always to be considerable confusion as to what notes the violas' and cellos' harmonics are to produce, even though the score is quite clear on this point.[3] Second, on almost all recordings these viola and cello harmonics are to all intents and purposes inaudible; and third, almost no conductor seems to realize that the horns (muted) are to blend and balance with those string harmonics. On only a very few recordings (Mackerras, Karajan, Skrowaczewski, Abbado) can some of the viola and cello harmonics be heard at all, and then not really blended with the muted horns.

There are four notes in the violas, sounding [musical notation], all in harmonics. [musical notation] produces [musical notation], [musical notation] produces [musical notation], [musical notation] produces [musical notation], and [musical notation] produces [musical notation]. (All these harmonics are produced on the C string). The lower staff is often interpreted (and is so rendered in the printed viola part) as [musical notation], which would produce [musical notation]. This is definitely wrong,[4] but is what one hears clearly on the recording by Dutoit, a French-speaking conductor who should really know better. Why? Because Ravel clarifies the potential ambiguity by writing *2nds* above the C and *1rs* below the G, verifying that there are two notes here (not one) and that the c^1, though written above the G, is to be played by the "seconds" on this staff, actually the fourth part in a total four-part divisi. Having no stems available in whole notes, this was the only way Ravel could indicate which viola plays what note. This disposition of the four parts is further confirmed in m.6, where—because the meter is 3/4, thus allowing for note stems—the matter is unequivocally clear. But more important, in further answer to the question why, it is because the four pitches in the viola, along with the two harmonics in the cellos and three pitches in muted horns, combine to make the following beautifully voiced nine-part chord [musical notation]. The cello notes are [musical notation], [musical notation] sounding [musical notation]

3. The confusion is compounded by the fact that the printed parts do not correspond to the score, leaving the concerned, inquisitive violist in even more of a quandary.
4. That all this was misunderstood by the original editor/engraver is clear from the "*en 3*" in the viola and cello parts in m.2, *not* in Ravel's autograph score.

𝄢 , 𝄢 sounding 𝄢 , both played on the C string. (Note the respective *1rs* and *2ds* indications again.) The horns have three notes , pitches which could not be produced as easily in string harmonics. The fact that the two *top* notes of the string chord are is confirmed by the presence of these same pitches in the clarinets' top notes, i.e. the first pitches in each 12-note *gruppetto*.

This understanding of m.2 should make conductors (and players) realize that the violas, cellos, and horns should blend into a single balanced nine-part chord. This is rarely the case, and usually left to chance by most conductors, most being unaware there is even a problem. Usually the horns, but particularly the first horn, overbalance the violas and cellos. It would seem obvious that string harmonics, having a basically soft 'muted' veiled sound, have to be dynamically adjusted. Ravel, like most composers, here indicates the resultant and desired dynamic level: *pp*. It is likely—disregarding for the moment such other variables as acoustics, the size and strength of the viola section, the skill of the players in playing harmonics, the quality of their instruments in producing harmonics clearly, and, finally, the different dynamic levels at which the four different harmonics in the violas respond—that each violist will have to play *p* or *mp* rather than *pp*. But this is also subject to another variable, namely, the type and quality of mutes used and the horn players' ability to produce an evenly balanced three-part *pp* chord. It should also be obvious that the harmonics in mm.2–6 should be played without mutes, which, alas, is made ambiguous or confusing by the indication to remove mutes in m.7 and m.8, an indication which, however, is valid only in the context of the entire ballet. I strongly suggest that the violas and first and second cellos play without mutes from the beginning of the Suite. In any case, the final result should be that the clarinets and viola/cello/horn ensemble match each other, all in *pp*! Example 1 clarifies how the harmonics in m.2 proceed to the next measure.

On only a few recordings (Mackerras, Skrowaczewski, Ansermet, Stokowski, Kondrashin, Tortelier, Abbado, Levine) can some of these harmonics be heard at all (usually just the viola G), and then not really blended with the muted horns. On only *one* recording are all four viola notes audible: Karajan's.

One final point: these delicate harmonics are hard enough to bring out, even when not muted, and even when the rest of the orchestra plays really *pp* (as prescribed by Ravel). But the harmonics have *no* chance of being heard when the flutes, clarinets, and harps play too loud, as they do in so many recordings, for example, Bernstein's, Mata's, Maazel's, Slatkin's, Kondrachin's, even Toscanini's. As mentioned, in Dutoit's recording the violas don't even play the right notes, a clearly audible high G ( ) was not intended by Ravel.

From the foregoing it seems logical that the cellos in the first seven bars should also be divided into four parts: third and fourth for

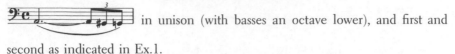 in unison (with basses an octave lower), and first and second as indicated in Ex.1.

Ex. 1

It is distressing to hear on the majority of recordings the lower cellos and basses dragging their eighth-note triplets (in mm.1–7). As a result, at the downbeats of each measure, the woodwinds, in full flow and unable to wait for the low strings—nor should they have to—are ahead of the latter, causing considerable ensemble and harmonic disarray, as one can hear, for example, on Karajan's, Abbado's, Solti's, Schwarz's, Mackerras's, Rattle's, Nagano's, and Janson's recordings. Even worse are the performances by Dervaux—who actually slows up at the end of each measure—and Muti, with whom the cellos and basses play the triplet eighths in the speed of regular eighths, forcing the flutes and clarinets to loiter uncomfortably on *their* last three or four thirty-second-notes.

The opening of the Second Suite is, of course, not only about correct tempo and the proper realization of the viola's and cello's harmonics. It is in fact one of the most ravishingly beautiful and evocative moments in all music, and I dare say nothing like it had ever appeared in music before. It is all shimmering colors and atmosphere. It seems to me almost fail-safe in its extraordinarily imaginative orchestration. Even so, not every conductor manages to capture the magic of this opening. Toscanini allows the flute and clarinet 'rivulets' sound like a technical exercise, rather 'strangling' the music, while Bernstein's musicians play with an exhibitionistic virtuoso *mf* (rather than Ravel's delicate *pp*), thereby making any magic or atmosphere quite impossible. In many recordings the flutes' and clarinets' figures are poorly coordinated. In Ozawa's, for example, the flutes are both untogether rhythmically and slower than the clarinets in m.2. In Rosenthal's and Kondrashin's recordings, the clarinets are substantially louder than the flutes, playing also with an edgy, brittle sound, quite inappropriate for this passage. Conductors who manage this opening beautifully are Boulez, Haitink, Skrowaczewski, and Previn.

The next problem arises at m.8. Almost all conductors allow this measure to sound *p* or *mp*, rather than *pp*. Granted, the horns are now open, and the 1. and 2. bassoon parts are hard to play really *pp*. Some conductors even encourage the first bassoon and first basses to 'play out,' as if it were an important 'solo.'[5] However, this all goes against Ravel's intentions of having the thematic material enter unobtrusively out of the misty sonorities of the first seven measures of the introduction, eventually rising to the first *f* climax (at m.15). Thus, m.8 represents a moment of elision, not an obvious interruption by the main theme.

In m.9, care must be taken that the three solo violins balance into three-part chords. One often hears only the top violin. Moreover, this wonderful bit of Ravelian onomatopoeia, evoking the first tentative bird calls at dawn—the score says "On perçoit des chants d'oiseaux"—set in three solo violins, piccolo, and flute, does not always come off so well in performances and recordings, mainly through lack of attention on the part of conductors. As any number of recordings attest, some parts of this passage are usually inaudible, be it the violin trills in m.11 and m.14, the piccolo in mm.12, 15, and 16, or the violin harmonics in m.12. The problem is that most conductors are unable or unwilling to keep Ravel's very gradual crescendo in mm.8–14 from peaking too early, thereby drowning out the 'bird calls,' especially in the three solo violins. It is a shame, because Ravel's dynamic markings are unequivocal and right, pacing the crescendo ever so precisely: mm.8–9 *pp*, mm.10–11 ——— to *p*, mm.12–13 ——— *mf*, m.14 ——— to *f* in the 'peak' measure 15. It is surprising and disheartening that even on recordings, where balances can so easily be adjusted, this wondrously imaginative passage is so rarely rendered correctly. Only a few recordings do full justice to it: Skrowaczewski's, Jansons's, Ormandy's, Dohnanyi's, Ansermet's, Previn's, Rosenthal's, Gaubert's, and, above all—nearly perfect—Martinon's Chicago recording with Walfred Kujala, one of the very few to play the piccolo part really correctly.

In m.15 over-all balance can easily be a problem. Horns and trombones must not be allowed to play more than their written *f*. The indiscriminately 'loud' sound so often heard here completely drowns out the first violins, which are divided *a 4* (Ex.2), meaning that in most orchestras there are only three, at best four players, per individual part, which even *one* overly loud trombone can easily render inaudible.

Another frequently encountered problem is that here (and in many similar passages, for example, mm.38–49) the harpists pay no attention to Ravel's pre-

---

5. I have even seen and heard conductors double the first bassoon in m.8 with the second bassoon.

Ex. 2

cise glissando markings. The glissandos should encompass only two octaves (in the upper register). Most harpists play large sweeping four-octave glissandos, with the result that through the greater speed of the sweeping hands *and* the fact that now all the most projecting thicker mid-range strings are activated, the glissandos become inordinately loud and soloistic, thereby obscuring the other important primary thematic/melodic material.[6]

Too many conductors disregard Ravel's *pp* in the clarinet and violas in m.20, causing a big 'dramatic' entrance here. The result is vulgar and obvious. How beautiful this can sound when played *pp* may be heard on Schwarz's altogether excellent recording. Three bars later a slight crescendo to *p* (or at most *mp*) will suffice, the higher dynamic and slightly greater expressivity being dissolved in the collective diminuendo of m.25. Accordingly, the phrase-ending dynamics in m.20 in the harp, horns, bass clarinet, and alto flute should be *pp*.

Regarding the piccolo part (m.26–27), Ravel intended this to be played in staged ballet performances on-stage or off-stage, (i.e. not in the pit or the orchestra), with the direction "Au loin, un berger passe avec son troupeau" (In the distance, a shepherd passes with his flock). It is not unequivocally clear whether Ravel's marking of *mf* is the resultant intended dynamic for the audience, or whether the piccolo player, playing off-stage, for example, should play *mf* with the actual effect in the audience being more like *p*. (The same applies to the E♭ clarinet passage at m.31).[7] Arguably the latter result represents Ravel's true intentions, that is, an audible *p*, slightly above the *pp* of the orchestra. But if the piccolo part is performed on stage, it should be played *p*, not *mf*.

There are three notational errors in the piccolo part in m.26 in all three sources: autograph, printed score, and piano reduction, namely, the double dot in the third beat should obviously be a single dot, the first ascending run should

6. Although the harpists in Rosenthal's recording play their parts quite correctly, the second harp is seriously *under*recorded, while the first harp is badly *over*recorded, almost turning the work into a 'Harp Concerto.'

7. I have in my own concert performances on several occasions had these piccolo and E♭ clarinet passages played off-stage with wonderful effect.

have a modifying "9," while the last six sixty-fourths in beat 4 should have a

modifying "6":   . This is a moot point, however, since, to

my knowledge, only a few rare players have ever played the third and fourth

beats rhythmically correctly. Instead of   every-

body plays something like   , except the piccolo

players in Barbirolli's Halle Orchestra, Gaubert's Straram Orchestra, and Sinopoli's World Philharmonic Orchestra.

There is scarcely a recorded performance that avoids the self-indulgent crescendo in mm.27–28 in the violas and clarinets. Ravel was perfectly capable of
writing ————— or *cresc.* if that is what he had wanted. Absent such an indication, we have the clear obligation to perform these bars without crescendo—
until Ravel himself calls for one at the end of m.29. Indeed many conductors
and violists change Ravel's phrasing and bowing from

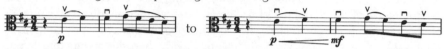

with, of course, a resultant heavy thick *mf.* The problem then is that Ravel's
thirty-second-note 'rivulets,' which run all the way through the first scene *(Lever
du jour)*—some 75 measures!—become quite inaudible. Again, Schwarz in his
Seattle recording finds the perfect balance for this magical passage, and so does
Gaubert in his 1928 recording with the Concerts Straram Orchestra.

In both Ravel's autograph and the printed score a *cresc.* is inadvertently missing in the first violins in m.30. By the same token, the first violins' thirty-second-
note passage at m.31 should be marked *mf,* or else it remains simply inaudible.

A fairly serious error by omission occurs here (m.31) in both autograph and
the Durand score. Fortunately here the piano reduction comes to our aid. Since
there is a crescendo in m.35 to *p* or *mp* in m.36 (see the oboe, horn, and violin
parts), it is obvious that something is missing between those measures and the
last previous dynamic, namely, *mf* at m.31. What is missing is a one-bar diminuendo in m.31 to a basic *p* for mm.32–34 (this is clearly indicated in Ravel's
piano reduction). This notational oversight has caused conductors to come up
with all kinds of makeshift 'solutions,' the strangest of which is Rosenthal's,[8] who
maintains a vigorous *mf* in m.31,32 and the first two beats of m.33, then makes
a sudden *one-beat* diminuendo (≫—) to *p* in m.34.

8. Manuel Rosenthal, ninety-one in 1996 and still active as a conductor, studied with Ravel and
became the composer's leading disciple, protegé, and musical confidant. As a composer he is best
known for his symphonic suites *Joan of Arc* and *Musique de table,* but also as a brilliant orchestrator
of some of Ravel's songs and works for piano.

It is important that the ascending chromatic line in the cellos and alto flute in mm.32–35 is well heard, both as a counter-line to the violas and clarinets and as a modifier of the underlying harmonic progression. (In m.34 the flute's G should obviously be a G♯.) Measures 36–37 are a clear instance of the scriptural lacunae in Ravel's autograph score, mentioned at the beginning of this section. The dynamic < > we all know so well and so indicated in the Durand score, exists only in the horn parts in Ravel's score. This is clear proof that some other document (another working score) or process (proofreading) intervened to achieve the final notation of the printed score.

Up to this point the basic tempo of ♩ = 50 should have been maintained. A *slight* harmony-resolving ritardando (*kadenzierend* in German) is possible at the end of m.37, followed by an *a tempo* at m.38, of course. But here, alas, too many conductors allow themselves extraordinary tempo liberties. The worst of these offenders are Bernstein and Mengelberg. Both in m.14 and later in m.37, Bernstein elongates the final quarter beats to such an extant that their eighth-notes are as slow as the basic quarter beats, so that these 3/4 measures become in effect 4/4 measures. Both conductors make matters worse in the second passage mentioned, by indulging in a huge—in the case of Mengelberg abrupt—accelerando already in m.35, followed, of course, by the aforementioned huge compensatory ritard in m.37. These vulgar excesses not only totally distort the music, destroying Ravel's superb sense of pacing and the wonderful aubade-like serenity of the introduction, but make much of the filigree passage work completely unplayable. (Pity the poor violinists in mm.35–37 when the conductor's tempo climbs to around ♩ = 96 in those measures!) In addition, many conductors disregard the seductive *p* at m.38, allowing (or urging) the violins to play much louder, following it with a big crescendo in mm.39–40. Again, as in mm.26–27, this passage is much more expressive and poignant when played without a crescendo. On the other hand, on some recordings, like Levine's (Vienna) and Abbado's (Boston), the chorus and most of the orchestra sing and play so loud— way beyond Ravel's *p*—that one can barely hear the beautiful main theme in the violins.

The problem we encountered in mm.27–28 recurs in its reorchestrated variant in mm.45–46. There are very few recordings that avoid the immediate crescendo in the strings, a crescendo Ravel withholds until m.48. Again Bernstein and Mengelberg completely ignore Ravel's *p*, substituting an unsubtle heavy molasses-like *mf.* Among the few recordings in which this passage is beautifully realized are Rosenthal's, Gaubert's, Previn's, Schwarz's, and Paul Paray's altogether superior rendition with the Detroit Symphony in 1961. Since the strings

will start 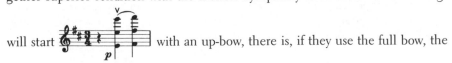 with an up-bow, there is, if they use the full bow, the

automatic, virtually unavoidable result of a crescendo on those two notes. This is then usually followed by a full down-bow on the F♯ in m.46, and another crescendo on the succeeding four eighth-notes. However, all of this unwanted increased sound is easily avoidable, if the string players realize that they must *not* use heavy full-length bows here, with which the aforementioned crescendo bulges are virtually inevitable.

In mm.43–44 Ravel devises one of the most sublime and yet most simple of harmonic progressions in all of music: the move from $D^{maj.7}$ and D^6 to C^6 (Ex.3). Part of the magic of this passage derives from Ravel's imaginative and

Ex. 3

highly original instrumentation, for the *softer* dynamic (p) at m.44 is achieved with a *larger* orchestra: trombones and chorus enter here as new additional voices. Unfortunately, on numerous recordings this supernal effect is ruined by poor ensemble. The beauty of the passage is all the more striking when conductors do not indulge in a ritardando in m.43; there is an added sense of surprise and revelation when the C major harmony arrives *in tempo*. A ritard in m.43 complicates matters considerably, because now the four woodwinds have to slow their thirty-second-note figures to conform to the ritard as they do clearly in the Jansons recording (forced also to adjust again in the next measure). Most woodwind sections, however, do not adjust to the conductor's ritard, with the result that they arrive at their C major well before the rest of the lingering orchestra. (The worst example of this can be heard on Ozawa's and Levi's recordings.)

Ravel composed a suddenly surging crescendo from p to f in one bar (m.48), a much more interesting, inventive use of dynamic coloration and phrase-building than the gradual crescendo one usually hears in this passage. Ex.4 shows the desired bow positions for the entire passage (mm.45–48).

Ex. 4

In m.49 the brass are easily too loud, covering both the strings and woodwinds. On many recordings (including Boulez's and Martinon's) the latter two sections are virtually inaudible. The second beat particularly, in the strings, is usually lost completely. The standard bowing

 does not work well here, especially

since 90 percent of all string players use almost the whole full bow on the first quarter-note alone, making the second beat automatically much too weak. Saving bow on the first beat will help, of course, but a still better solution is the following bowing (first violins, Ex.5a, 2nd violins, violas, cellos, Ex.5b). This bowing/phrasing is certainly justified, since Ravel's long slur over three bars

Ex. 5a

Ex. 5b

is clearly a *phrasing* indication of a long legato, *not* a bowing.

Grievous sins are committed in mm.52–54 by almost all conductors. Very few recordings sampled reveal any respectful treatment of Ravel's dynamic indications here. Violas and clarinets are marked *p* with slight hairpin $<\ >$ nuances per beat, not the commonly heard nervous *mf* or *f*. These three measures (mm.52–54) should evolve subtly (though, of course, clearly) out of the previous three-bar diminuendo passage—more in the sense of an elision than an abrupt new start.

The other distortion visited upon this passage is that it is often played too fast, that is, with a sudden *più mosso*. Apart from the fact that Ravel was perfectly capable of indicating *plus vite* or *pressez* if that's what he wanted—there are dozens of secondary tempo changes in the Second Suite alone, and hundreds in the entire *Daphnis et Chloé* ballet—the thirty-second-note groups in the first violins are a clear indication that the tempo should remain at the basic ♩ = 50, in keeping with all the other previous thirty-second-note runs. Furthermore, Ravel has ingeniously given the three bars (mm.52–54) a faster *feeling*—but in the same tempo—by the subdivision of the music into six eighth-note beats per bar, in effect doubling the tempo. At an eighth-note of ♪ = 100, this passage is fast and anguished enough; no additional tempo increase is needed. Finally, if a *plus vite* is taken here, m.54 *has* to ritard in order to get back to the original tempo at m.55. All of this is unnecessary and unwanted. True, Ravel writes *angoissé* in the score; but that is a stage direction for the dance pantomime, hardly a specific indication of a tempo change. Nor is it a license to play the whole three-bar passage *f* or *mf*, as so many orchestras are encouraged (or allowed) to do. Of the recordings sampled, only Haitink's (with the Boston Symphony), Skrowaczewski's, Paray's, Rosenthal's, Ansermet's, Jansons's, and Abbado's bring this somewhat difficult passage off quite well, while Bernstein's,

Mengelberg's, Barenboims's, and Mackerras's are rough, loud, and messily played. But even those performances that observe Ravel's *pp* at m.52 and maintain the correct tempo are almost all flawed by that already often-mentioned careless habit of dropping last notes before a rest. Ravel writes

in the violas and

in the violins. But on all recordings the last notes in each phrase segment (circled in the examples) are *totally* inaudible — in Kondrashin's recording with the Concertgebouw as if Ravel had written

$\qquad$.[9]

Having traversed the first fifty-plus measures of the Second Suite in some detail, I believe it is necessary to show in a summary sequential manner how some conductors are ignorant of or choose to ignore what the composer's score indicates. If Muti's and Karajan's recordings of *Daphnis et Chloé* were exceptional, one might be able to disregard them as singular aberrations. The sad truth is that their performances are typical of the customary misrepresentations of Ravel's score. In both recordings cellos and basses drag the triplets in the opening measures, in Muti's, as already mentioned, to the point where the flutes and clarinets have to slow up drastically their final three or four thirty-second-notes, obviously seriously interrupting the intended flow of the music. In m.6, as in almost all recordings, the low F♯ in cellos and basses is sharp, as the musicians, not understanding the bitonal harmony of C^7 over F♯, inadvertently reach high to fit their F♯ into the G of the C major chord. Muti now takes a much slower tempo in m.8, and both conductors crescendo much too early, thereby drowning out the all-important bird-song material. Karajan reaches *mf* as early as m.10 — see Ravel's indicated dynamic sequence (p.465) for comparison — while Muti reaches a full *f* by m.13.

In m.20 Muti and Karajan ignore Ravel's *pp*, entering with a beefy thickish *mf* in the violas, that completely misses the quiet early-dawn awakening-of-nature atmosphere of the music. Muti makes matters worse by stretching his already painfully slow tempo to a lugubrious ♩ = 32. All upcoming *p*'s and *pp*'s (mm.27,32,38,46) are ignored, favoring instead a heavy *mf*, with the result that in Muti's recording it is impossible to hear any of the rippling thirty-second-note figures in the violins. At least Karajan avoids Muti's huge vulgar ritards in mm.37,43,48 — the last 3/4 bar re-composed by Muti into a 4/4 bar.

Measures 52–54 must count as possibly the most maltreated three measures in the entire work. Muti's and Karajan's distortions are typical. Both conductors,

9. When this theme returns in augmentation, rhythmically slightly altered, in the *Danse générale* in m.269, the same performance problem occurs (see pp.492–93).

ignoring Ravel's delicate *p* and *pp*, lunge into m.52 with an aggressive, offensive *f* and at a suddenly faster tempo. Muti's Philadelphians struggle through these (complex and difficult) three measures more or less successfully, but Karajan's Berliners, completely thrown by his 'interpretation' and his vague beat, produce an incredible mess.

To continue now where we previously left off (around mm.52–54), two useful changes (improvements) can be seen in Ravel's autograph. According to the manuscript, only one oboe was intended to play in m.53 (an excellent idea to achieve the desired *p* dynamic), and Ravel had marked the horns (in m.53) handstopped (+).

Great care must be taken to make accurately and naturally/organically the transition from a duple division of the beat in m.54 to, in effect, a triple division of the beat in Ravel's 9/8 in m.55. (It is very shaky or unclear in many re-cordings.) In addition, to give full meaning to Ravel's E minor$^{7(9)}$ chord at m.55, the D in the third horn and second trombone—the only instruments sustaining the important 7th of the chord—must be clearly projected. This becomes even more important when the C#'s in the bassoons, strings, and third trumpet in m.56 clash momentarily with the sustaining D. Just as the transition from duple to triple beat-division is critical at mm.54–55, so the reverse at m.57 is just as crucial.

For some reason many conductors (and first violin sections) like to make a huge crescendo in mm.60–61, ending *f* at m.62. This is quite wrong, for the following three reasons: (1) The dynamic at m.62 in all other instruments is *pp*, clearly indicating that the alto flute and first violins' dynamic must relate to this underlying soft sound. Thus the solution is certainly not a beefy *f*, but at most a *mp*, more likely a *p*. (2) The first violins are doubled in the alto flute. There is no alto flutist in the world who can produce a *f* F# (concert) comparable to a whole violin sections' *f*, and certainly not without taking a huge breath before the previous B♭. Therefore, logic compels one to conclude that, if Ravel's alto flute doubling is to have any meaning (and any chance of being heard), then the violins' crescendo must be kept to a moderate increase. (3) The usual violin *f* at m.62 completely covers the *pp* entrance of the bassoon, second cellos, and first basses with what is unquestionably an important primary theme.

The first entrance of any percussion occurs here: timpani in m.62, suspended cymbal in m.66. A relatively high-pitched shimmering cymbal sound is wanted here, so that the two percussion instruments—the cymbal high, the timpani low—embrace the entire orchestra in a gradually encroaching wave of sound which breaks and peaks at m.70.

Ravel mounts his biggest climax at m.70 with a *ff* in all instruments. This should make clear that the previous dynamic peaks (at mm.15, 49, and 55) must never be allowed to over-reach, lest the effect of the ultimate climax at m.70 is reduced to an anti-climax. Here it is often necessary to exhort the celesta and glockenspiel (Ravel preferred the keyboard *jeu de timbre*, a rare instrument in the United States) to play really *ff*, to preserve, along with the triangle and two harps, the full brilliant, scintillating, iridescent effect of this climax. (For the record, on not one recording are the celesta, *jeu de timbre*, and triangle audible.) There is also at m.70 a fantastic sense of release and affirmation, as Ravel confirms the music's basic tonic tonality of D major.

The tempo distortions that usually occur here (especially m.69) are extreme. Most conductors turn the 3/4 into a 4/4 by stretching the last two eighths into quarter-notes, among them Bernstein, Dutoit, Rattle, Jansons, Barenboim, Dervaux, Koussevitzky and Boulez. Maazel and Slatkin distort the measure even more grotesquely. Instead of [♪ ♩ ♩ ♪♩], Maazel contorts the measure into a [rit. ♩ ♩ ♪♩], while Slatkin deforms it even further: [rit. ♩ ♩ ♪♩]. But Mehta's disfigurement of this measure is the most blatant of all: starting m.69 at ♩ = 34 (instead of ♩ = 50), he makes two huge fermatas on the two eighth-notes of the third beat, the second of which is *four seconds long*, i.e. longer than *one entire measure* at Ravel's designated tempo!

Care must be taken at mm.70–71 that the whole orchestra does not diminuendo, that it sustains the full intensity of the climax, for at least two and a half measures. At the same time, some attention must be paid to not covering the main melody in the first violins, violas, alto flute, oboes, and English horn. As the main theme [musical notation] subsides, falling to ever lower octave positions and reaches the third violas in m.76—the *divisi en 3* of m.74 continues into m.78—it is well to mark the violas *p* and *dim.* (missing in score and parts).

Ravel's *retenez* in m.77 is one of the most consistently misinterpreted markings in the entire score, for it has been erroneously assumed by several generations of conductors that this *retenez* refers only to m.77, and that somehow Ravel forgot to indicate an *au Mouv* at m.78. The resultant *a tempo* (or even *più mosso*) in m.78 which one can hear on many recordings (Munch's, Dervaux's, Ozawa's, Dutoit's, Jansons's, Rosenthal's, Martinon's, Solti's, de Burgos's, Rattle's Kondrashin's, and Barbirolli's, for example) has always sounded awkward and abrupt to my ears, and seems to me to be a complete misconstruction of Ravel's intentions. A much more plausible interpretation is that Ravel's *retenez* indicates the beginning of a subtle, eleven-bar graduated ritardando which leads eventually to the 6/8 ♪ = 104 at m.88. There are several reasons that argue for such a realization; most important (1) the four-bar repetitive thematic phrase in the clarinets starting m.76; (2) the lack of a tempo modification in m.87 to lead to the explicit ♪ = 104 in m.88; and finally (3) the meaning of the French term *retenez*.

To take the first point, it makes little sense to have the clarinets play one measure slow (m.76) and the next measure fast (m.77). Indeed, matters are made worse in that somehow many conductors feel compelled to jump to a much faster tempo at m.78, instead of at least retaining the basic ♩ = 50 (♩. = 50 in 9/8). They do so probably because they see m.78 as a brand new episode in the over-all continuity, and feel impelled to make that audible with a brand new tempo. But it is not a new phase; it is instead the final outgoing phase of the Suite's *Lever du Jour* (Daybreak) introduction. In that view, Ravel uses one of his favorite devices of elision in m.78, where one primary theme (in the clarinet) overlaps and elides with a new incoming motive (in the oboe). His intention was not to make an obvious structural break here, but rather a smooth, subtle transition from one musical idea to another. It does not help matters either when the oboists are allowed (or encouraged) to play mm.78–85 as a grand *solo*, often at a *mf* dynamic level, while at the same time the at least equally important main theme in the clarinet is ignored or suppressed. On several recordings (Haitink, Munch, Solti, Dutoit) this clarinet melody is literally inaudible; it is as if it had never been written by Ravel. The lilting lulling oboe motive [10] in this (my) interpretation is gradually slowed down until it elides with a rhythmically augmented version in the clarinet. At this point the music's movement comes almost to a standstill—almost in the sense of a cadential fermata—to be quietly reanimated in the three oboes at m.88.

A second strong clue as to the intended eleven-bar ritard is given by Ravel in the last full beat of the first and second oboe parts in m.87, where the final sixteenth-note (in 9/8) is obviously intended to match the sixteenths in m.88.[11] If one accepts the usual interpretation of a 9/8 ♩. = 50 at m.78, a sixteenth in that tempo would equate with 300 on the metronome. It makes little sense to have this sixteenth at 300, but the ones at m.88 ♪ = 208. Most conductors sense this, of course, and get around the problem by making a ritard around m.86, a ritard, needless to say, not indicated by Ravel. This substitute ritard, furthermore, seems quite gratuitous and inappropriate, since Ravel has already composed a relaxation of the music's pace by means of rhythmic augmentation in the clarinet (m.86)—the clarinet's note values are more than twice as slow as the preceding oboe's notes—making an additional sudden belated ritard quite unnecessary and unjustified.

The third reason for my suggested gradual relaxation of tempo between m.77 and m.88, is the term *retenez* itself. Note that Ravel uses the imperative form, not the past participle *retenu* or *ralenti*, clearly indicating an instruction to 'hold back,' implying a continuing, ongoing holding back. In this conception one ritards imperceptibly per measure to arrive in m.87 at the ♩ (♩.) = *ca.*34 tempo, thus allowing a subdivision of the last beat of m.87 into three eighths in 9/8, each eighth equalling ♪ = 104, the tempo given by Ravel for the next formal episode.

A handful of conductors interpret the *retenez* more or less in the manner suggested here, including Dohnanyi, Schwarz, Boulez, Barenboim, Paray, Tosca-

---

10. Gaubert's 1928 recording (made under Ravel's supervision) is the best proof that this tempo destortion is wrong and unwanted. This oboe motive is derived from the first measure of the Suite, being a distillation and simplification of the first flute part's thirty-second-note opening runs, essentially preserving the flute's original range and contours in the oboe.

11. The English horn part should have been rewritten as [musical notation] or as [musical notation].

nini, Stokowski, Mackerras, and, suprisingly, Mengelberg, who can so often be idiosyncratically expressive and willful in his tempo modifications, but who here (mm.78–87) maintains his basic slow tempo ♩ (♩.) = 40. Unfortunately, he undermines his own concept of the passage by irrationally jumping suddenly to ♩ = 50 at the entrance of the clarinet in m.85. Boulez does exactly the same thing, as do many other conductors. Koussevitzky in his 1928 Boston Symphony recording, one of the first ever of *Daphnis et Chloé*, takes a different approach. He adopts a tempo of ♩ (♩.) = 52 at m.78 (after a subtle ritard in m.77), holds that tempo throughout the succeeding ten bars, and then simply doubles the tempo at the 6/8 ♪ = 104 in m.88. His interpretation is, however, marred by the curious anomaly that the oboes and English horn play an eighth-note pick-up in m.87, rather than Ravel's sixteenth.

Finally, I offer yet another rationale for my interpretation of the score's *retenez* as an ongoing eleven-bar ritardando. In addition to the elisional, transitional character of this passage, there is an important harmonic progression here, which carries the music via three ingeniously chosen harmonies from D major to the new implied key of B minor at m.88. This tonality will basically dominate the entire next section of the work, the so-called *Pantomime* episode, including the magnificent flute solo of Chloé's dance (mm.106–30) on the secondary dominant F♯ minor, resolving eventually down a fifth to the tonic B major at m.207.[12] The three harmonies Ravel uses to modulate from D major (m.78) to B minor (m.88) (on a second inversion pedal point F♯) are C$^{7♯4}$ (m.80)—A minor (m.82)—F♯^{9} (m.83), all harmonic stations prominently encountered in the first (*lent*) section of the Suite. Further tying this passage to the opening Daybreak music is the reappearance of the four-note motive (at mm.79–

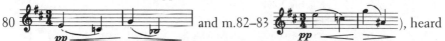

80 and m.82–83), heard

earlier in mm.60–62, now slightly varied and re-orchestrated. Note, too, that the harmonic progression here (mm.78–83)—D - C^7 - F♯^{7}—is a condensed microcosmic variant of the two basic alternating harmonies of the Suite: D major^9 (m.1) and the bitonal mixture of C and F♯ major[13] (m.2). It is by means of that C/F♯ mixture that Ravel arrives ultimately at the 'dominant' F♯ pedal point, mentioned earlier, in m.84. However, this F♯ sounds virtually inevitable to our ears, since it is constantly present and reiterated in the oboe parts mm.78–85, the third of the key of D thus being transmuted into the fifth of the new key of B minor. All of this combines to confirm unequivocally that the passage

12. The Suite's final tumultuous *Danse générale* is basically in A major, thus yielding an over-all four-part harmonic/formal scheme for the entire Suite of: D major—F♯ minor (flute solo)—B minor/B major—A major.
13. This bitonal chord is often referred to as the "Petrushka chord" because of its prominent use in Stravinsky's famous 1911 ballet—although it should be noted that Franz Liszt used the same bitonal harmony, distinctive and unresolved (in F and B major), as early as his 1831 *Malediction Concerto* when he was just twenty years old(!).

mm.76–88 is (1) transitional, comprising a crucial harmonic modulatory function; (2) was intended to relax—to 'cadence,' as it were—gradually into the new section at m.88.

The flute's F♯ in mm.85–86 is often played too loud with a luxuriant vibrato, as if it were some important solo passage. It is instead a subtle, quiet pedal point extension of the clarinet's F♯, and should be played more in the manner of a string harmonic.

It is amazing how conductors have ignored and continue to ignore Ravel's tempo marking ♪ = 104 at m.88, most of them by a very wide margin, as Fig.2 shows. Clearly, they can't all be right!

It has long been a tradition variously to distort and dress up the four measures, mm.91–94, with exaggerated rubatos and gratuitous glissandos. Especially annoying and unmusical is the silly, uncalled for *Luftpause* hesitation between the last two sixteenths of m.92. This phrase, obviously a harmonic, though *not* a rhythmic, variant of the oboe/English horn passage four bars earlier, should not

be phrased any differently. Ravel's separation of ♫ into

two slurs gives no license to distort the given rhythm. It is a cute, 'sexy' seductive idea that someone (was it Koussevitzky?) started and, like weeds in an untended garden, has been impossible to stamp out. Again, Ravel was capable of writing some hesitating *Luftpause* rhythm, if he wanted to, as he demonstrates handily

Fig. 2

♪ = 64	Nagano, Levi
♪ = 70	Stokowski (1970)
♪ = 72	Koussevitsky (1928), Mackerras, Sinopoli, Skrowaczewski
♪ = 74	Bernstein, Barenboim
♪ = 76	Boulez, Jansons
♪ = 78	Abbado (Boston), Ormandy
♪ = 80	Muti, Dohnanyi
♪ = 82	Ozawa, Previn
♪ = 84	Levine, Maazel, Haitink, Martinon, Inbal, Rattle
♪ = 86	Karajan, Mehta, Abbado (London Symphony), Nowak
♪ = 88	Koussevitsky (1944), de Burgos, Slatkin, Jansons
♪ = 90	Mata
♪ = 92	Tortelier, Schwarz
♪ = 94	Munch, Toscanini, Paray
♪ = 96	Slatkin, Mengelberg, Kondrashin
♪ = 100	Solti, Dutoit, Monteux, Casadesus
♪ = 104	Commissiona, Gaubert
♪ = 106	Rosenthal
♪ = 112	Dervaux
♪ = 128	Ansermet
♪ = 138	Barbirolli

enough in m.97  —surely a first in musical his-

tory in rhythmic *raffinement* and meticulousness of notation. He could easily have done something similar in m.92, but didn't. Nor is there any justification for the long-standing aberration and misinterpretation of Ravel's *très ralenti*

in m.93. The passage is usually played

This common distortion of Ravel's notation is hard to explain in any rational terms, for what Ravel wrote is quite clear and simple (Ex.6). Notice, by the way,

Ex. 6

the past tense (*très ralenti*), meaning "very held back" (not "holding back"), in effect a *meno mosso*—indicating a slower tempo for the entire three bars of mm.93–95. Ravel did not forget an *au Mouv^t* in m.94, as so many conductors seem to have arbitrarily assumed. He *did* put it two bars later in m.96. What possible reason is there to doubt and change this conception? In addition, Ravel's $\Longrightarrow$ in the latter half of m.93 brings the dynamic back to *pp* at m.94, where it should remain until the last eighth of m.95. All of this makes totally inexplicable the *subito pp* usually heard in m.94.

That all these tempo, dynamic, and phrasing distortions were established in the first decade of the work's existence—by whom I do not know—is proven by the fact that they can already be heard on one of the two earliest recordings of the *Daphnis et Chloé* Second Suite, Koussevitzky's 1928 Boston Symphony recording. They can *not* be heard on Gaubert's recording of the same year, recorded under Ravel's supervision. It is a remarkably faithful, intelligent, clear and (for the time) stunningly well-played performance.[14] How unmannered and warmly expressive this beautiful *Pantomime* passage sounds, when rendered without all the aforementioned distortions, can also be heard on Toscanini's, Stokowski's (1970), and Mengelberg's recordings (the last's, however, marred by the Concertgebouw's piccolo and flute players moving to their D♮ in m.95, three whole beats (♩) early!).

14. Phillippe Gaubert (1879–1941), probably best remembered today by flutists for his many beautiful flute compositions (he himself was a leading flutist in France around the turn of the century), was also one of France's finest and most active conductors in the decades between the two world wars, leading both the Paris Conservatoire and Paris Opera orchestras. His many fine recordings all made in the late 1920s and early 1930s comprised mostly French repertory (Debussy's *Nocturnes*, Dukas's *L'Apprenti Sorcier*, Franck's D Minor Symphony), in many cases first recordings of such works.

The *au Mouv*[t] in m.96 refers to the tempo at m.88, i.e. ♪ = 104, a
fact pretty much ignored by most conductors. (One of the odder interpretations of this *a tempo* is Skrowaczewski's, who, after having selected a very
slow tempo for the oboe/English horn trio of m.88–91 (♪ = 72), at the *au
Mouv*[t] of m.96 returns not to that slower tempo, but to a much faster one,
♪ = 92.)

Many conductors (for example, Previn, Dervaux, Kondrashin, Solti) do quite
the opposite, taking a fairly lively tempo at m.88, then slowing dramatically for
m.92, some (like Dutoit) adding a huge silent fermata (⌢) in the middle of
m.91, causing the music to come to a dead stop. In Mehta's Los Angeles recording, in m.96 the oboist manages to recompose the phrase from

apparently with the blessings of both the conductor and the recording producer.
The ritard in m.97 is also often exaggerated, particularly by Bernstein, Previn,
Dohnanyi, Rattle—even Paray, who as a Frenchman should certainly have understood Ravel's very special marking *cédez très peu*, meaning in essence "very
little ritard" (literally: yielding very little [in tempo]). Indeed, the two score
pages encompassing mm.96–105 are among the most performance-abused of
the entire work, this in spite (or in defiance) of Ravel's most detailed, precise,
and unambiguous notation.

Apart from the frequently exaggerated ritard in m.97, the tiny pause at the
end of that measure is often held much too long. (Can anything be clearer than
Ravel's *très court?*—over a thirty-second-note rest!) How arbitrary and illogical
conductors can be in their 'interpretations' is shown by comparing the passage
here under discussion in two recordings by two conductors presumably expert
and knowledgeable in the French repertory, Boulez and Martinon. The former
seemingly can't read French, because he makes a very long fermata 𝄐 in m.97
when the score clearly says *très court*. Martinon, on the other hand, does a
very short fermata, but then ignores Ravel's *cédez très peu* in the same measure,
making a *huge* ritardando. In m.98 rarely is the *au Mouv*[t] really *a tempo*;
equally rarely is the *p* dynamic in m.99, contrasting with m.96's *mf*, respected.

The *pressez* of m.100 is generally misinterpreted to refer only to the oboe run,
which is additionally taken too fast, and then followed by some kind of slower *a
tempo* at m.101. Ravel's *pressez*, signifying a gradual 'pushing the tempo,' is
eventually cancelled out in m.105 by the complementary *retenez*. This means
in simple language that, starting at the second half of m.100 there is a *poco a
poco accel.* until the ritard in m.105. A further more precise refinement of that
idea suggests that, in effect, the *pressez* of m.100 cancels out the immediately
previous *cédez*, thereby restoring the tempo to *a tempo* (♪ = 104) in m.101.
From there the tempo continues to press forward. A good thing too, for in the
muted strings it is all but impossible to maintain a strong *f* in m.102 on the

long $\overset{\frown}{\Gamma}$ $\overset{\cdot}{\varrho}$ note, even at an accelerated tempo.[15] The oboes' C♯ major arpeggio is not only generally played too fast but with an enormous crescendo, instead of Ravel's final *pp*. To match the oboes' near-*forte* high notes, conductors then arbitrarily change the first violins' *pizz.* to *f*—as if two wrongs would make a right. Although this oboe passage *is* difficult, it *can* be played correctly with Ravel's dynamics, especially if not played too fast.[16]

15. Typical of the disregard of composers' detailed notation by conductors is the recording of *Daphnis et Chloé* by the late Kyril Kondrashin, a conductor often praised for his "precise and faithful reproduction of the score." He was also notorious with orchestras, so one is told, for demanding pianissimos and diminuendos from his players. His book, *Die Kunst des Dirigierens*, is, as I have mentioned earlier, a valuable and insightful manual on "the art of conducting." But evidently Kondrashin was, like many conductors, not one to practice what he preached. The dozens of important *pianissimos* in the *Danse générale* or the crucial *p*'s in the *Lever du Jour*, say, in mm.20, 27, 45 are anything but *p*, more like *mf*, even *f*!

In the *Pantomime* section Kondrashin is anything but 'precise' and 'faithful' to the score. As one instance, having taken a tempo of ♪ = 96 for the 'oboe trio' mm.88–91, he now drops the tempo precipitously to ♪ = 88 at m.92, causing some confusion and poor ensemble in the orchestra (the Concertgebouw, not generally a rattleable orchestra). Then, in the middle of m.92 Kondrashin makes a huge ritard, suddenly snaps back into tempo in the beginning of m.93, followed by *another* ritard in the second half of that measure. In m.94 he continues the ritard (Ravel's *très ralenti*), but then irrationally indulges in a huge *accelerando* in the piccolo and flute in mm.94–95, jumping to ♪ = *ca*.100. A quick glance at Ravel's score will show that none of these yoyo-like tempo distortions can be found therein.

Incidentally, why do those conductors who indulge in rhythmic distortions in m.92 refrain from doing the same in m.102. (Actually, Bernstein and Sinopoli do, making matters even worse.)

A word about Bernstein, one of the most famous and revered conductors of all time, is perhaps in order—a word that many of the millions of Bernstein's fans will undoubtedly reject summarily. The truth is that on the one side Bernstein was at once one of the most remarkable musical talents ever to be born in this country (or anywhere else for that matter) and yet, on the other side, an often confused and even tragic figure, who was in many ways torn apart by his many talents. Although a multi-gifted pianist, composer, conductor, musical television evangelist, he seriously misused and abused his talents through his ego-driven distortions, exaggerations, podium histrionics, ignorings of the composers' texts—not always, but most of the time.

Recent biographies of Bernstein have dealt with his drinking, his often outrageous behavior, his pursuit of the trendy and chic, his bisexuality, as well as the genius, the brilliance, the charm. And what emerges is a picture of a life of frustrations, non-fulfillments, of a man with almost too much talent, too profusely squandered in too many directions.

Bernstein had very little discipline and no shame. He seemingly needed to be loved passionately by millions of music lovers. And insofar as he continuously catered to these extra-musical demands—a moth drawn to the flame—he was much the lesser artist for it. As someone who knew Bernstein extremely well—since 1943 when we first met—I can attest to the veracity of these comments. At his best he was a brilliant all-around virtuoso; at his worst his influence was dangerous and damaging.

Bernstein constantly vacillated in his praise and recollections of his two main teachers, Koussevitzky and Reiner, depending on expediency, place, and occasion. He sometimes claimed Reiner as his most important teacher, saying among other things, "Reiner is responsible for my own very high standards" (no modesty there!). Yet Bernstein's podium antics and athletic exhibitionism, wonderful for audiences and television, give the lie to that myth. Indeed, after one of Bernstein's telecasts in which he had given one of his more physical conducting displays, Reiner was heard to remark, "He didn't learn *that* from me" (quoted in Philip Hart's *Fritz Reiner: A Biography* (Evanston, 1994), p. 66).—Levi unmutes the upper strings at m.101. Why?

16. The G♯ (pizzicato) in the first violins in m.101 is spurious, nonexistent in the autograph score; it should be eliminated.

Ravel was, of course, one of the supreme orchestrators of all time, and *Daphnis et Chloé* is replete with one remarkable orchestrational innovation after another. It is beyond the scope and intent of this book to deal exhaustively with this aspect of Ravel's *oeuvre*, except where especially relevant to the structuring and continuity of the music and, therefore, particularly relevant to the conductor. A case in point is the passage presently under discussion, in which Ravel ingeniously prepares for the arrival of the key of F♯ minor and the famous extended flute solo. Over a C♯ pedal point (the dominant V of the upcoming key) Ravel weaves a series of seventh chords (Ex.7). To reinforce the C♯ pedal

Ex. 7

point, Ravel adds three further C♯'s in harps, woodwinds, and cellos, dropping from the highest piccolo C♯ over four octaves to the cello's C♯ (Ex.8). It is an

Ex. 8

idea Ravel was very fond of, using it in many different formats in many of his orchestral works, and first used in the Second Suite in mm.88–91 (F♯ pedal point), reiterated in mm.92–93 (violas and harps), and now varied and extended in mm.102–105. While Ravel's orchestrational devisings usually work by themselves, requiring little help from the conductor or the players, the C♯ passage under discussion requires a little extra attention. The problem is the cello's C♯ harmonic (marked with a small circle in Ex.8) in mm.104–105. The effect of descending octave leaps in different overlapping instrumentations should work perfectly well, except for the fact that (1) the cello's C♯ harmonic projects rather feebly; and (2) most cellists in any given orchestra (as well as their conductors) are usually unaware of the crucial linking function of that note and that they must take that note over from the flute in the middle of m.104. In most performances and recordings one can usually hear the piccolo, oboe, and harps well enough, but in the next bar the flute and English horn are often underplayed, these players not realizing that they are the middle link in a three-way chain of descending C♯'s. But, as mentioned, even less audible usually is the cello harmonic in m.104, which must emerge from the flute's final C♯ in *Klangfarbenmelodie* fashion. In order for this passage to come off as intended, the cellos may have to play *mf* rather than the score's *mp*. Only on Stokowski's London Symphony, Paray's Detroit Symphony, and Nagano's recordings is this cello C♯ effectively projected.

Chloé's dance—Ravel's inspired flute solo, one of his crowning achievements—is also, alas, one of his most misinterpreted, both by flutists and conductors. The most conspicuous misrepresentations occur in the realm of tempo and rhythm. There are the usual interpreters—flutists and conductors—who righteously believe that since it is a major solo, akin to an entire slow movement in a mini-concerto, *any* interpretive liberties and rhythmic freedoms are permissible. Happily there are also those who rightly interpret the whole episode as a stately dance (in tempo), perhaps in the manner of ancient Greek drama and dance, in keeping with the basic source of the ballet's origin, the legend of *Daphnis and Chloé* as told by the third-century A.D. Greek poet Longos.

A degree of interpretive uncertainty and divergence of views can undoubtedly be laid at Ravel's feet, on two accounts. First, all three sources, the autograph, the printed Durand score and the piano reduction, contain a major error in the tempo marking at m.106, namely, the metronome indication ♩ = 66, which even at a first glance seems discrepant with Ravel's tempo marking *très lent*. Second, in m.108 the score reads *suivez le solo* ("follow [or accompany] the soloist"), an admonition that has, of course, given many a flutist and conductor leave to perpetrate any license that might come to mind.

As for the metronomization at m.106, it is easily proven to be wrong by deduction and a little sleuthing in the score, although oddly only retroactively from the vantage point of a later passage in the work. In m.139, as mentioned earlier, there is a tempo indication *vif* (lively), and ♩ = ♪ *précédente*, meaning that the quarter beats in m.139 shall be equivalent to the sixteenths of the preceding tempo. This, however, is clearly technically impossible if that previous tempo is ♩ = 66; and, since all the prior tempo fluctuations, mostly slowing down the pulse, are cancelled in each instance by *au Mouv*ᵗ (*a tempo*), we must assume that m.138 is in the original m.106 tempo. We also already know that ♩ = 66 is not a possible tempo for the immediately preceding eight measures, for the various sixty-fourth note runs in the flutes in mm.132, 134, and 138 (see Ex.9a, 9b, 9c) are also unplayable at that tempo (it is impossible to

Ex. 9a

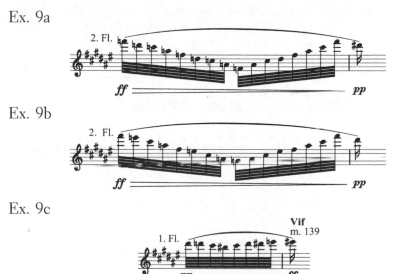

Ex. 9b

Ex. 9c

play eight notes in the time of an eighth-note at 138, even more so in mm.132, 134, where the score says *pressez*, i.e. accelerando). Something, therefore, is drastically wrong. But if we assume that the score's ♩ = 66 in mm.106–38 should have been ♪ = 66, everything falls into place and everything Ravel wrote becomes possible and playable. It, of course, means as well that the metric modulation in mm.138/139 ←♪=♩→ is in fact quite correct.

Incidentally, one can easily see how Ravel was misled into thinking that the flute's eight sixteenths in m.139 could neatly match its eight sixty-fourths in m.138, as long as at that moment he had in mind the slower ♪ = 66 tempo. Turning the metric equation the other way, if the tempo in m.138 were indeed ♩ = 66 and one held to the ♪ = ♩ metric modulation, then the entire ensuing section of flute runs becomes absolutely impossible to play, since then the tempo at m.139 would have to be ♩ = 264; and try playing septuplet sixteenth runs at that tempo, not to mention dealing with the several *plus animés* and *en animants* that are scattered throughout the next several score pages, culminating in the spectacular piccolo-to-alto flute descending four-octave run in mm.180–82.

Thus by retrograde calculation we can safely establish that m.106 and the ensuing flute solo was meant to be set in a tempo of ♪ = 66. This is confirmed by Ravel's indication *très lent* in m.106. For if in his nomenclature he calls ♩ = 50 (the Suite's opening tempo) *lent*, then *très lent* could easily be ♩ = 33 (or ♪ = 66). That tempo also suits well the entire flute solo conceived, as I suggested, by Ravel as a very stately dance of Greek antiquity, and which also contains several sixty-fourth-note groups. I am convinced, too that Ravel meant it to be played *as* a dance, that is, with a steady pulse. The evidence for this is in the score, both explicitly and by inference: (a) in mm.124, 126, and 127 the composer clearly indicates tempo variations, in turn signifying that at other points he does not expect any; and (b) the flute solo already has all the rhythmic variation it can possibly contain. Thus any further tempo deviations are not only *not needed* but would actually cancel out and subvert Ravel's extraordinary rhythmic invention. A passage which has in it sixteen different divisions of the

beat hardly requires, or can tolerate, any further rhythmic additions or liberties.

It is astounding that this major notational error was never corrected by Durand in score and parts,[17] considering the enormous success, innumerable per-

17. It is my understanding that a new revised score (and parts) is now being produced by Durand, to be available in the near future.

formances, and therefore tremendous royalty income, *Daphnis et Chloé* has enjoyed. As already mentioned, the recent Nieweg score and the various Conductos Guild Newsletters unfortunately offer no new or valuable information on this major notational error.

In 1922, Ernest Ansermet evidently sent Ravel a telegram asking for clarification of the tempo question at m.106. Ravel responded indirectly to Ansermet in a letter to Jacques Durand, dated March 23, 1922,[18] allowing that "complete lunacy" [une loufoquerie intégrale] reigned over the tempo indications. "At the *très lent* it should read ♪ = 66," then adding two additional revisions: an *a tempo* at m.136 and *"le double plus vite"* (doubly faster) at m.136. "Therefore at m.139 (*vif*) the ♩ equals the preceding ♪." Somehow—and unfortunately—Ravel's clarification to Ansermet never was widely published and never had much currency. Thus the confusions throughout this section continue to this day.

As for the indication *"suivez le solo"* in m.108 (incidentally, not in Ravel's autograph), it is, admittedly somewhat confusing, and has given many a flutist license to take Ravel's written rhythmic variants rather lightly. It is conceivable that Ravel—in 1912—realized that this solo might cause its early protagonists some performance problems and thus felt inclined to allow for some *colla parte* freedom. This goes along with his further annotation *"expressif et souple."* But I can't believe that either or both of these indications were meant to completely annul or re-arrange Ravel's richly imagnative and ingenious rhythmic invention.

As for the dynamics, Ravel marks the solo initially *p*, meaning, of course, a 'solo *p*.' I don't think it is too much to ask that flutists observe this dynamic, instead of the usual heavy-breathing *f*. For, playing the solo at a higher dynamic level than indicated creates considerable breathing problems, that is to say, the louder a wind player plays, the more he will tend to run out of breath. This fact has through the years forced many a conductor to speed up in tempo to accommodate the flutist (especially at mm.115–116, 117, 118, and 124–25). At m.115 a slight crescendo—or at least full maintenance of the sound—is suggested, not only leading to Ravel's clearly indicated *mf* at m.116 but paralleling the crescendo in the strings.

The accompaniment in strings, harp, and two horns is almost never realized in its full harmonic beauty and ingenuity. As we have seen, two seventh chords (B^7 and $C\#^7$) in mm.104–105 lead chromatically to a D^7 chord in m.106 in the upper strings (arco *and* pizzicato), suspended over a double pedal point of C#s and F#s. After six bars of D the harmonies progress in subtle chromatic alteration through one of Ravel's most inspired harmonic creations (see Ex 10). The C# in the second cellos, second horn and second harp in mm.106–111 must be clearly heard, clashing with the D and C directly above and below. But beyond that most of the above chords clash with the pulsating bass F# and C# pizzicatos. It is very useful to rehearse these string harmonies (mm.106–26) slowly, tempo-

---

18. I am indebted to Arbie Orenstein for this information, contained in his *Maurice Ravel: Lettres Écrits, Entretiens* (Paris, 1989), p.217.

Ex. 10

rarily playing all pizzicato notes arco to hear better the poignantly dissonant clashes.[19]

There has been over the years considerable controversy or confusion over the opening scale of the flute solo (m.108). There are those who insist and claim to know that the scale should contain a D#, some others argue for a D# *and* an E#. The fact is that nobody really knows what Ravel intended, for all three sources offer a different version. The piano reduction, the earliest published source, has both the D# and E#. Ravel's autograph score has only the D#. The Durand score has neither D# or E#, D♮ and E♮ instead. The D# of the autograph, making it a melodic minor scale (in F# minor) is a possibility, except that it clashes with the many very clear D♮'s in the accompaniment. It is for that reason that some conductors have argued for the "harmonic minor" scale (D♮-E#). This latter choice would seem also to be confirmed by Gaubert's 1928 recording, recorded as mentioned with Ravel in attendance. It seems that through the various stages of composition and publishing, Ravel revised and refined the passage, finalizing it in the last proofs for the printed edition as we see it in that score (D♮- E♮).

There have been a number of fine recorded performances of the flute solo, notably Moyse's with Gaubert, Kincaid's with Stokowski and the Philadelphians, Colin Fleming with Tortelier and the Ulster Symphony, Peter Lloyd with Previn, James Pellerite with Paray, and an unidentified flutist with Sinopoli's World Philharmonic. But there have also been renditions full of willful tempo distortions and idiosyncratic liberties. The most common deformation is the accelerando in mm.115–16, where many flutists, not preparing for this long C#, run out of air. The conductor then obliges them by making a big accelerando: Haitink and Doriot Dwyer (in an otherwise well-played though overly intense solo),[20] Sidney Zeitlin

19. G♮ against F# in mm.116–17, D against C# in mm.117–20, E# against F# in mm.121–22, G and C against F# and C#, respectively in mm.124–25. In m.124 a *pp* is missing in both score and parts in the first violins, second violins, first violas, and second cellos.

20. Changing the opening run from  to

is also hardly a justifiable or necessary 'interpretation'.

with Skrowaczewski, Dwyer (again) with Munch, and Joshua Smith with Dohnanyi (the last makes matters worse by making a huge ritardando in m.114).
Though Bernstein rushes mm.115–16 tremendously, his flutist (Julius Baker),
still cannot sustain the C♯ for its full written duration.[21] But the strangest and
most inept rendition of all recordings sampled occurs on Mengelberg's 1938
performance. Mengelberg apparently wanted to believe the erroneous ♩ = 66
metronome marking in m.106. His flutist bravely embarks upon the solo, however—in self-defense—starting one eighth early. But by m.110 he has pulled
back the tempo, realizing that m.111 is quite unplayable at ♩ = 66. Thence the
solo lurches and wobbles forward and backward—the flutist basically playing
each first beat of a given measure in tempo, the second beat held back. The
result is a completely distorted, nay re-composed, flute solo, which makes no
musical sense. It must have had the poor bass players, trying to stay with the
erratic beat, on tenterhooks. The pizzicato violins and violas also had no idea
where to place their notes. Actually, conductors starting the dance at a relatively
lively tempo—although not as lively as Mengelberg's—is a fairly common occurrence. The flutists inevitably are forced to pull the conductor back in m.110
or m.111, as can be heard on, among others, Muti's, Mehta's, Rosenthal's, Dutoit's, Sinopoli's recordings. Karajan's attempt to liven up the flute solo is almost
comical. He starts at a brisk ♪ = 76, but his flutist, Karlheinz Zöller, quickly
pulls him down to ♪ = 58. So does Christina Smith in Levi's recording.

Another one of the more amazing distortions can be heard in m.130 in Bernstein's recording (with Baker), in which after a making a big gratuitous ritard in
m.122, but at the same time ignoring in m.124 Ravel's *retenu légèrement* , m.130

is deformed into  .

In m.129 horns are often rather lazy and slow about quieting down to
p immediately after their entrance, most horn sections doing so much later
than indicated. In that same measure the Durand score has misplaced the
f in the flute part; it should be at the beginning of the measure

The *au Mouv*[t] of m.131 refers—it should be obvious—to the flute solo's basic
tempo (♪ = 66). I mention this because almost all conductors take this passage
too fast, more or less in the wrong tempo of ♪ = 132, some as fast as ♪ =
150, one (Rosenthal) even at ♪ = 180. It is, of course, quite impossible to play
the second flute part in m.132 (Ex.11) at such fast tempos. There is no flutist
on earth who can play sixteen sixty-fourth-notes in one ♩ = 70 or ♩ = 90 beat.

---

21. Laurent and Koussevitzky 'solve' the problem in their 1928 recording by having the second flute
hold the C♯ in m.115–16 for the first flute (also the A♯ in m.124), which is not an unreasonable
idea, except when it is done as unsubtly and noticeably as in this recording.

Ex. 11

But the fact that, in consequence, flutists are forced to fake their way through the passage, usually leaving out at least half the written notes—hear Cleveland's flutist completely swallow the last six notes—doesn't seem to bother Maazel, Nagano, Chailly, Levi, Slatkin, or Mehta at all. In only Muti's and Mata's recordings, taken at too fast but at least a playable tempo can one hear all the notes Ravel wrote. More trouble at m.138, where again Ravel's eight sixty-fourths cannot be played at the fast tempos conductors take here. The result is a mess of one kind or another: the sixty-fourths changed to thirty-seconds (i.e. twice as slow) or even to sixteenths (Ansermet), garbling the notes badly (Commissiona, Slatkin, Karajan, Levine), or leaving some out (Jansons, Rattle, Mata: four of the eighth-notes totally missing). Boulez 'solves' the problem by making a big ritard in mm.137–38 and is also wrong. How wonderful this entire passage sounds when done at the right tempo can be heard on Toscanini's and Paray's recordings.[22] The slower (correct) tempo of $\eighthnote$ = 66 also works beautifully for mm.136–37, giving time for the two echoes, f - mf - p, to be controlled and well heard. In those two measures the flute should obviously also have a two-bar diminuendo.

The next episode, marked *vif.* is usually rendered fairly correctly, at least until mm.148–51. Here flutists pay no attention to Ravel's very special dynamics, playing the passage the same way as in mm.142–45 (assuming, I suppose, that Ravel 'must have made a mistake'). Whereas the dynamics of the earlier four measures naturally suit the flute's lack of projection in the lower register, therefore *p* in m.143 and m.145 (*f* in m.144 and m.146), Ravel asks the absolute opposite in mm.148–51: the high notes (mm.148,150) soft, the low notes (mm.149,151) loud. I have not heard a single recording where this remarkable idea was correctly realized, although apparently Haitink and his Boston flutists at least tried to achieve the desired effect, without, however, fully succeeding. (Ozawa in his innumerable performances of the *Daphnis et Chloé* music has consistently ignored this typically Ravelian *raffinement.*)

Much of the next (mostly flute-led) episode offers no serious interpretational/ re-creative problems. We should only quickly note that a *p* dynamic is missing in score and parts in m.154; and indeed hardly any winds and strings play a *p* here, forcing the solo flute in self-defense to also play *f*. In mm.184 and 188, we often have the reverse problem, namely, that the muted horn cannot be heard at all. Following the accelerating *animez* and *en animant* of mm.170 and 174, respectively, Ravel follows with a *moins animé* at the climax of the passage

---

22. In m.132 there is also a danger of covering up the runs of the second flute. I suggest that the harp's dynamics be changed to p $\diagup$ mf.

(m.180). It is a good idea to think of a specific metric modulation in preparing for this sudden and difficult tempo change. If, for example, one has arrived at a tempo of ♩ = 175 at m.179, and if one thought of the septuplet sixteenths in mm.180–82 as being the same speed as the previous plain sixteenths, then the *moins animé* tempo would be ♩ = 100.[23] This flute run is impeccably played by the four flutists on Schwarz's recording.

The tempo in m.183 (*très lent*) in Ravel's autograph and piano reduction is ♩ = 40 (missing in the Durand score). Also missing is a diminuendo wedge (>) for the second beat of m.182 in the alto flute, and a *f* at the beginning of m.183, both items in Ravel's manuscript score. In m.184 and m.186, it is a

rare occasion if the flutes play what is written:

What one usually hears is  . Also on many recordings in m.184 the D♯ of the alto flute (written G♯) is not fully held out to the end of the measure, that is, *beyond* the release of the piccolo and flutes. In Bernstein's recording the alto flute is inexplicably not held through (m.186) and over into m.187.

Apart from the fact that almost no one observes the *pp* in m.187 and m.193, the next big performance problem occurs in mm.191–92 and its parallel place, mm.197–98. Here for some reason no one—absolutely no one[24]—pays any attention to Ravel's unusual dynamics, despite the fact that the score's indications are unequivocally clear—and, I might add, remarkably inventive. As Ex.12 shows, the orchestra is divided into two separate and distinct renderings of these two-bar phrases, both rhythmically and dynamically. As intended by Ravel— although ignored by *everyone*—woodwinds and two horns have two quick unusual dynamic swellings, while the strings and second and fourth horn have

a *long* six-beat diminuendo from *f* to *pp*, not the usual | ♩ ♩ | ♩ ♩ | .

Also, most conductors make matters worse by imposing a ritardando on these two measures, in no way intended by Ravel; also by making two one-bar phrases, rather than one two-bar phrase.

Just as universally ignored is Ravel's very clear *retenez—peu—à—peu* in mm.201–206. With but few exceptions (Koussevitzky (1928), Toscanini, Paray, Stokowski (1970), Mackerras, Casadesus) the majority of conductors and concertmasters make a ritard in mm.201–202 (Ravel's *retenez*), often very substan-

---

23. At ♩ = 175, ♪ = 700; at tempo ♩ = 100, ♪⁷ = 700; therefore m.179 ♪ = m.180's ♪⁷.
24. Gaubert does not quite succeed either, although there is evidence on his recording that he and the orchestra tried to realize Ravel's unusual dynamics. Gaubert's rendition of these measures is, in any case, the least distorted of all extant recordings.

Ex. 12

tial, but then make a considerable accelerando (*sic*) in the rising violin line, just where the score clearly says *peu à peu*, meaning unequivocally to continue the ritard 'little by little.'

Ravel's *au Mouv^t* in m.207 is marked ♩ = 80 in the autograph score and the piano reduction, but is inadvertently left out of the Durand score. All kinds of tempos have been attempted here, ranging all the way from Dervaux's ♩ = 50, Koussevitzky's ♩ = 63, and Munch's ♩ = 68, to Solti's ♩ = 108, and the most ridiculous (because virtually unplayable in flutes/clarinets and completely out of context), Bernstein's ♩ = 120.

The *lent* metronome marking at m.212 is an important one, for it not only refers back to the opening tempo of the Suite and forward to the next *lent*, m.221 (also marked ♩ = 50), but, I firmly believe, it also offers a strong clue, if not absolute proof per se, that the *animé* of the *Danse générale* was intended to be taken at ♩ = 150 that is, three times the speed of the various *lents*. (In purely metronomic terms the equation is: *lent* ♪³ = *animé* ♩, or *lent* ♩ = *animé* ♩.). There is additional substantiation of this concept in mm.290–303, where the climactic passage of mm.55–59 is recapitulated, although transposed, slightly varied rhythmically, and reconstituted in a different meter (See Exx.13a and b). To be an exact replication of the original phrase in the new meter and tempo,

Ex. 13a

Ex. 13b

it would have had to have been written as in Ex.13c, but the resemblance is obviously close enough to justify the assumption that the two passages should

Ex. 13c

be more or less identical in tempo, which they will be if the *Danse générale*5/4 is taken at $\quarternote$ = 150 ($\dottedhalf$ = $\halfnote$ 50, $\halfnote$ = 75). Another clue, suggesting such a tempo relationship, are the last eighth-note triplets in m.216, which equal the quarter-notes in m.217 (4/4 $\eighthnote^3$ = 5/4 $\eighthnote$).

The *lent* of mm.212–16 is generally played very well,[25] as is indeed the entire rest of the Second Suite. Perhaps the reason is that the *Danse générale* is technically very demanding, a virtuoso tour de force, leaving very little, if any, room in its inexorable rhythmic drive for any interpretive deviations. Generally, the only major departure from the score is in fact in the realm of tempo. For most conductors take a much faster tempo than the one I have suggested is inherent in Ravel's conception, undoubtedly thinking that the fastest possible tempo is the most 'exciting,' the most likely to 'thrill' the audience into a rousing standing ovation. I think that conception actually sells Ravel's magnificent score short. I would argue that while the faster tempo does create a kind of superficial excitement, it also leads to a fair amount of 'faking' in the orchestra, whereas at a tempo closer to $\halfnote$ = 150 any fine orchestra can play Ravel's dazzling passage work with crystal-clear clarity, incisive exciting articulations, and a greater contrast of texture and dynamics.

One possible reason for the adoption of this faster *animé* tempo may be the tempo marking *animé* $\halfnote$ = 168 at m.217 in the piano reduction for four hands (although I am not at all convinced that many conductors have looked at that publication). Paradoxically, the same measure in the *two-hand* piano reduction has the completely irrational marking *andante* $\halfnote$ = 68, obviously a complete misreading by the engraver and not detected in proofreading (by Ravel or the editor).

The slowest *Danse générale* tempos on the fifty-five recordings I sampled were Barbirolli's ($\halfnote$ =144), Ansermet's and Paray's ($\halfnote$ =152), and Previn's ($\halfnote$ =156), while the fastest were Bernstein's and Schwarz's, a ridiculous $\halfnote$ = 204 and $\halfnote$ = 200, respectively, with most conductors more in the middle to upper

---

25. Mengelberg is the grand exception. For some implausible reason, he felt the need to strip the trumpet parts here of all their slurs, offering instead a harsh, quite out-of-context, heavily tongued *marcato* on every note.

170s. At least those are the tempos with which these conductors *start* the *animé*. Most (Koussevitzky, Munch, Toscanini, Jansons, Rattle, Monteux) settle down by m.241 to a less frantic ♩ = 156–62, a tempo much closer to Ravel's intentions. Bernstein's super fast ♩ = 204, mostly unplayable even by the remarkable virtuosos of the New York Philharmonic, gradually bogs down to a still very swift ♩ = 184. Boulez, often touted as the paragon of conductors in the repertory of Debussy and Ravel, is oddly wayward in tempo in the *Danse générale*. Starting out with a good and reasonable ♩ = 156 at m.221, he then slows to ♩ = 144 in the second half of the *Danse* (m.304), pushes up to ♩ = 150 by m.218, but at the end (around m.362) goes all out for a 'flash finish' at ♩ = 176. So much for inexorableness of tempo!

These fast tempos cause another immediate problem in that, if the conductor wants to adhere to the ♩ = 50 *lent* in m.221, he will have to ritard into that measure at the end of m.218, a ritard which, of course, is not in Ravel's score and surely not envisioned by him.

Most of the other performance derelictions or misinterpretations are in the realm of dynamics, as can be heard on virtually all *Daphnis* recordings. Ravel's important dynamic nuances are so universally ignored that they can be summarized and listed as follows:

1) the crescendo, especially in the flutes and oboes from *mf* to *f* in m.223

2) the *subito pp* for all players in m.224

3) the clearly marked *pp subito* in m.229 (Solti, Previn, and Rattle are the only ones to observe this wonderful *pp*).

4) the *pp* in m.250 (second beat)

5) the *subito p* in m.258 (second beat), sometimes done sort of half-heartedly

6) the *p* in m.269 (on virtually all recordings this is played with a vigorous, insensitive *mf* or, at best, *mp*).

7) the *pp* at m.306; the *p* at m.311 (Casadesus and Jansons are among the few conductors who observe these dynamics)[26]

8) the *p* at m.316, as well as the ensuing *mf* in m.317

9) the *subito pp* in m.318 (again, Casadesus does this well)

10) only *p* at m.321

11) the *p* in m.326 (second beat), sometimes done but half-heartedly; similarly m.330

---

26. I can remember as a young teenage horn player noting that in all recordings and performances of *Daphnis* I heard in those years the muted horns in m.308—a passage I had already diligently practiced—were always frustratingly inaudible. I ingenuously reasoned that Ravel had marked the horn parts too soft and that we horn players simply had to play the passage louder in order to be heard. What I didn't realize until I studied the score more carefully years later was that Ravel's horn dynamics were quite perfect, and that it was conductors and the other instruments (bass clarinet, bassoons, and lower strings) that were always too loud, ignoring Ravel's *pp*—their actual *mf* or *f* automatically obscuring the muted horns.

The same problem occurred—and still occurs to this day—in the next measure where the alto flute cannot be heard if the other instruments do not observe Ravel's *p* dynamics. Instead of paying attention to that, many conductors simply change Ravel's orchestration and put the alto flute part in m.309 in one or two clarinets—a quite unnecessary revision.

12) *subito p* and *pp* in m.362, usually only approximated (except on Skrowac-zewski's recording)

Additionally, on many recordings the E♭ clarinet and first trumpet (in mm.245–47, 248–49, respectively) fail to make the per-bar diminuendos (Ex.14a). Solti's disciplined Chicagoans are very good here, whereas with Hai-

Ex. 14a   Ex. 14b

tink's Bostonians it remains a split decision: the E♭ clarinet makes the ⟩, the first trumpet doesn't. In Bernstein's recording a uniform *f* prevails, while on Karajan's recording, his E♭ clarinet player rearranges the part as in Ex.14b.

Another common failing, present on all but a few recordings (Sinopoli's, for example), is the total suppression of the eighth-notes in the basses in the entire passage, mm.304–33. All one can hear in the vast majority of performances and recordings are the quarter-notes (a reverse bowing is very helpful here).

In my discussion of the first movement of Brahms's First Symphony I referred to the confusion that surrounds the marking ⌢♩ ♪ (see p.300). Here in Ravel's *Daphnis* the problem surfaces again—in m.312 (trumpets), mm.313–14 (woodwinds)—for many conductors erroneously make their players play the last note in ♩♩♩ tongued, while others (correctly) have them play the last note staccato but slurred into.

A curious anomaly is the omission of a timpani part in m.286 in score and parts. It is probably impossible to know definitively at this late date whether the omission was intentional, or to know how this 'error' (if it is one) came to be made, given the fact, as mentioned earlier, that Ravel's autograph score is virtually devoid of timpani and percussion parts. One possible reason for the empty measure is that Ravel may have thought it was necessary in order to tune from the E♭ and G♯ of m.285 to the D and G of m.287. However, this is rather remote since (a) Ravel was writing for chromatic timpani and (b) the two seconds that m.286 lasts do not offer all that much extra time anyway. To repeat m.285 in m.286 is entirely possible on modern timpani, and I strongly urge that this 'correction' be made. On only a few recordings (Mehta's and Karajan's) does the timpani play in m.286.

Ravel's magnificent fanning-out harmonic progressions (Exx.15a and b) are often poorly balanced, especially in the brass, because each of the twelve brass players cannot easily tell from his part how he fits into the over-all crescendo and harmonic scheme. (Bernstein's recording typifies how conductors and orchestra can sail uncomprehendingly through these passages.) I have found that playing through these harmonic sequences slowly in rehearsal helps the orches-

Ex. 15a

Ex. 15b

tra tremendously in understanding what is going on, to really hear and feel these remarkable chord progressions.

There is often among conductors and brass players some confusion regarding certain muting indications in the *Danse générale*. The first trumpet should take the mute out in m.250 (the printed score is in error here). Some conductors, like Muti and Levine, have the trombones (not the tuba) mistakenly play muted in m.277, presumably because the trumpets are marked muted, as if that were sufficient reason. The marking *ôtez la sourdine* in m.314 refers only to the third trumpet, leaving the first and second trumpet muted through mm.318–22.

One of the worst intonation problems often occurs in the trombones in mm.256–57. Along with the other brass and woodwinds, the trombones play an A♯ minor (or B♭ minor) chord here against an A major bass—a striking example of Ravelian bitonality. Unless the trombonists know and hear this in its true harmonic function, there is hardly any way that they can play their notes in correct intonation (except by accident).

Finally, there is on almost all recordings a blatant dropping of final notes in the strings in mm.269–70 (violins, violas), mm.273–74 (cellos), a sloppiness of phrasing that dissects what should be two-bar phrases into fragmented one-bar

units. Ravel writes  but what we usu-

ally get to hear is , with the last G's

virtually inaudible. Only on Gaubert's 1928 recording can these two passages be heard as Ravel actually wrote them.

Let us close this discussion of the *Daphnis et Chloé* Second Suite by pointing to the most bizarre performance aberration so far imposed upon this piece. Stokowski, like Bernstein, always felt compelled to rearrange, recompose, reor-chestrate, and 'improve upon' the works he conducted. In his 1970 recording of the Second Suite, Stokowski adds *after Ravel's last bar* a trashy Hollywood ending: a long-sustained *ff* A major chord in the chorus *a capella (sic)*. As the saying goes: "Folks, I don't make these things up; I just report what I hear." Hear it for yourself if you don't believe me.

Schumann: Second Symphony

The performance and interpretational problems in Schumann's symphonies—
or better said, *alleged* problems—have been a matter of discussion and controversy almost ever since the works were written in the 1840s, continuing well
into our own day. Such discussions have generally centered on two aspects of
Schumann's symphonic *oeuvre:* his orchestration—mostly negative—and the influence of Beethoven's symphonies—mostly positively.

Through the years many conductors (as well as critics and composers) have
contended that the great flaw in Schumann's orchestral music was his "poor,"
"weak," even "inept" orchestration, premised on primarily the notion that his
music was intrinsically pianistic and in his symphonies was not translated into
truly orchestral music. As a result of such thinking, Schumann's symphonies
have been instrumentally re-touched, re-written, re-orchestrated not only by major conductor-composers such as Mahler and Weingartner, but by virtually every
significant late 19th- and early 20th-century conductor. Whether such alterations
really constitute 'improvements' remains questionable, or at least arguable.

As for the question of Beethoven's influence on Schumann's symphonic
works, although in general positively viewed, I feel it is on the one hand over-
stated—Bach's and Schubert's works had much greater impact on most of Schumann's music—and, on the other hand, curiously disregarded in precisely such
works where Beethoven's influence was quite specific and profound.

In view of the unavoidable redundancy in the analysis and evaluation of the
six previously discussed major works and their recordings—unavoidable mostly
because of the endless recurrence of the same interpretational transgressions
and bad habits—it should not be necessary to inflict upon the reader a blow-by-
blow account of how the Schumann Second Symphony has suffered (or benefited) at the hands of generations of greater or lesser conductors. I will limit

myself therefore to a discussion of the two topics mentioned above: the one in need of a more thorough airing than the cliché of Schumann's "poor orchestration" generally receives, the other (Beethoven's influence) a matter of very specific concern in regard to the Schumann Second's Finale movement—and only that movement.

The reader may have gathered from the above that I do not entirely agree with the commonplace 'wisdom' that Schumann's orchestration is 'inept' and in constant need of 'improvement.' It is too sweeping a generalization, and one particularly inapplicable to the Second Symphony. Consider the ravishingly beautiful and imaginative use of woodwinds in the Symphony's slow movement, its idiomatic use of horns and trumpets, and above all the luminous, ecstatic quality of the string writing (especially in the slow movement's two great surging climaxes); the fine orchestrational symmetries and contrasts in the two Trios of the second movement;[1] the inspired dialogue between clarinet and piano in Schumann's Piano Concerto; the daring adventurousness of the horn writing in the Konzertstück for Four Horns and Orchestra; the solemn beauty of the trombones in the slow movement of the "Rhenish" Symphony—to name just a few outstanding examples.

I do not believe in all instances, especially in his orchestral works from 1845 on, that the homogeneity and uniformity of orchestral sound we hear in many Schumann works are the result of 'ineptitude,' but rather I find them to be an integral part of his particular sound world, sound vision. That this timbral uniformity was in part a reflection of the coloristically limited palette of his favorite instrument, the piano, can hardly be denied. But I think this was not merely— and certainly not always—due to an inadequate knowledge of the orchestral instruments, but rather a purposeful choice to set off by contrast those special orchestral colors and textures that are, if one be honest about it, scattered throughout his works, even his earliest ones. I am convinced that Schumann often strove for a certain rich and consistent instrumental palette in place of the more differentiated contrasting colors of the earlier classical, say Mozartean, orchestra. I believe this because he tried to reflect the singular yet overtone-abundant sonority of the piano, and also because that sound to him was the rich mixed-color sound, which embodied the Romantic fantasy aesthetic that he helped to advance and establish in the early 19th century. Schumann's concept of an orchestrally blended sound is the sonoric analogue to his notion of the new Romanticism, unifying and combining the poetic, literary, fantastic, and humanistic. Indeed, he referred to this concept as "tiefcombinatorish" (profoundly combinatorial), and it is the instrumental conbinatoriality in much of his orchestral writing that is in fact—right or wrong—Schumann's distinctive sound world. This also answers directly to Schumann's oft-criticized—I believe over-criticized—tendency of multiple doubling of individual melodic or contrapuntal lines.

1. In Trio I the primary voices of winds are answered by strings, in Trio II the exact reverse, both trios in turn contrasting with the full-orchestra whirlwind of the Scherzo proper.

In any case, that *is* part and parcel of Schumann's sound aesthetic, very special and characteristic in its own way; and it ought not to be automatically dismissed or tampered with. For, although it may not need "improvement" in re-orchestration, it may indeed need improvement in performance and realization. (More of that below.)

I am not so foolish as to argue that Schumann's use and knowledge of the orchestra were always the most accomplished and natural, let alone inspired. In his earliest orchestral works [2] we can discern the signs of struggle as he is trying to wean himself away from the 'mother's milk' of the piano. Clearly, some of these first forays into the orchestral realm constitute a fairly routine note-for-note orchestration of very pianistically idiomatic piano music. It should be noted, however, that, paradoxically, Schumann's piano music, even his earliest, is replete with instrumental colorations and associations. Obviously it took him some years and some experience to translate these sonorities successfully into an orchestral garb. As with almost any composer striking out in new directions, some experiences in Schumann's case proved misleading. He was not the only composer in the history of music who changed the instrumentation of this or that passage in a given piece, misled by an inept first rehearsal or a poor performance. We know this happened to Schumann, even in his mature years, as his friend Brahms testifies. The latter, writing about the first version of Schumann's D minor Symphony, of which he owned the score, recalls how upset Schumann was "by a first reading, which went so badly" that some time later he felt compelled to re-orchestrate the work. Brahms continues: "the original scoring has always delighted me. It is a real pleasure to see anything so bright and spontaneous expressed with corresponding ease and grace. Everything [in it] is so absolutely natural that you cannot imagine it different."[3]

By distinguishing between the first and earlier orchestral works and the later, more mature ones, I also want to remind us that around 1845, the year of the creation of the Second Symphony, Schumann, according to his own diary, began "to create and work out everything in my head, developing thus a whole new way of composing."[4] The greater latitude and innovative spirit in the instrumentation of Schumann's later works are undoubtedly ascribable to this new development.

As I suggested earlier, the 'alleged problems' with Schumann's symphonic

2. Symphony No.1, Overture, Scherzo, and Finale and Symphony No.4 (both written in 1841 but later substantially revised), as well as the incompleted earlier 1832 G Minor Symphony and several unfinished piano concertos.

3. Contained in a letter by Brahms to Heinrich von Herzogenberg, October, 1886. In another letter (dated two months later), this time written to his close friend, the conductor Franz Wüllner, Brahms expresses himself even more emphatically: "I find it enchanting how this lovely work sounded [originally] in its loveliest, most fitting garment. That Schumann was later induced to bedeck it ["behängt"] so heavily has undoubtedly to do with the bad performance by the Düsseldorf orchestra. Unfortunately all the symphony's beautiful, unfettered and graceful freedom of movement has become impossible in the new unwieldy garment.

4. Berthold Litzmann, *Clara Schumann, Künstlerleben; Nach Tagebüchern und Briefen*, Vol.1 (Leipzig, 1902) pp.372–73.

works may, in fact, not lie with their orchestration but with *how* they are performed. Given the special nature of Schumann's instrumentation, as detailed above, it takes a special sensitivity in matters of timbral blending and balancing to realize fully Schumann's intentions. I have never felt the need actually to re-orchestrate or re-voice any of Schumann's symphonies. I also find the re-touchings, even those of Weingartner, Mahler, Klemperer, Szell, Masur, Schuricht, Mehta, Celibidache, and others not so much 'improvements' as changes, which drastically alter and often subvert the very nature and character of the music. While such re-orchestrations may in a purely technical sense (and from a point of view wholly different from Schumann's) constitute "improvements," "enhancements," "corrections" of the material, they are really a negation and refutation of the very essence of Schumann's music. What I have done—and what it seems to me is sometimes necessary to do—is to adjust in minute and subtle ways balances, dynamics, inner voices, and harmonic details, in order to bring out instrumental colors, important harmonic shifts, polyphonic lines, as the case may be. This is not to rewrite or re-orchestrate the music (by doubling or adding instruments), but merely subtly to clarify what is already there.

But beyond that it is absolutely essential that in Schumann's music performers, especially in the woodwinds and brass, not only blend and balance in the usual sense but, when needed, merge their sounds in a symbiotic "combinatorial" rapport which, as a result of the merging, produces entirely new and distinctive sonorities.[5] This is not as easy as it sounds, and takes sensitive and caring musicians, not the so often uninterested bored-with-Schumann players one finds in most famous orchestras. Incidentally, none of these things can or will happen if conductors conduct only the top or most obvious melodic lines in Schumann's works (usually the violins). The conductor must be aware of and elicit every subtle adjustment and calibration, constantly monitoring the results—again, my "third ear."

To give a more precise idea of what I mean, let me briefly elaborate on two passages in the first movement of the Second Symphony. (I will deal with the last movement more extensively further on.) The first excerpt comprises mm.73–104 (see Plates XI, XII).

In m.73 the degree of *sf* must be well-considered and balanced in the flute and upper strings, the former's *sf* in its most projecting highest register and the latter's *sf* in a more flexible range must be adjusted to each other. In m.74 the other woodwinds line must be balanced dynamically and sonorically with the cellos and basses (or vice versa), while the horns must subtly under-play their implied *f* so as not to interfere with the main melodic material. All of the above will apply, of course, to all the succeeding pairs of measures, with the addition that the tremolos in violins and violas starting in m.78 must be downgraded to *mf*, or, in any case, must be played in a lyrical, melodic, non-

---

5. This is not only necessary and true for Schumann, but many other early Romantic composers, such as Mendelssohn, Cherubini, Bellini, Donizetti, Hummel, Clementi, and Chopin (in his orchestral works).

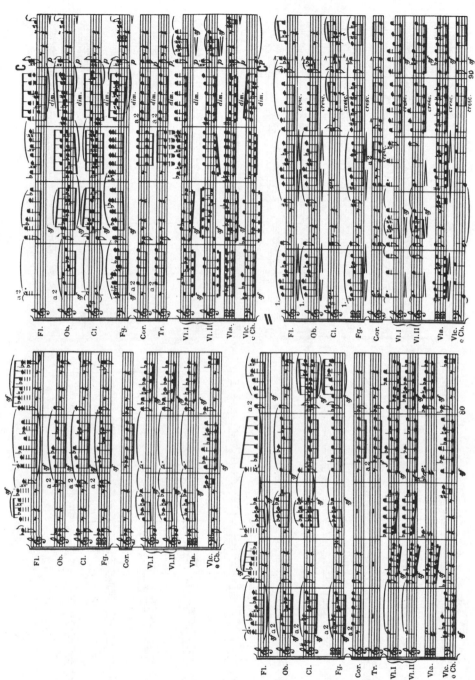

Plate XI Schumann, Symphony N° 2, first movement, mm.77–90

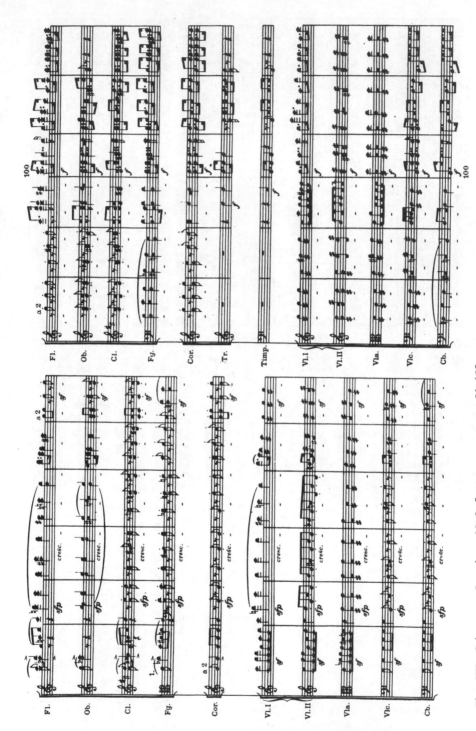

Plate XII   Schumann, Symphony N° 2, first movement, mm.91–102

boisterous way, lest they overpower the woodwinds. The trumpets entering in m.79 must moderate their *f* to fit into the over-all dynamic context, paired as they are (in m.79) with two clarinets.

To skip to m.85, again the *sf* in the violins has to be thought about. In my view it is a moderate *sf*, in the basic dynamic of, say, *mp*. That will allow the three solo woodwinds to be well heard, even in *p*. Care must be taken that the clarinets and horns balance/blend into a single clearly heard harmony

. The string *sf*'s in mm.90–91 must be graduated within the over-

all crescendo, i.e. in m.90 in *mp*, in m.91 in *mf*.

In m.92 things get even more complicated. The first violins must not be allowed to dominate or overpower the *legato* woodwind quartet of two flutes and two oboes, while at the same time the conductor must keep an ear on the secondary staccato counter line in the second violins. In m.96 the bassoons must be allowed to come to the fore, bringing out a new color, which will only happen if the entire rest of the orchestra—winds and tremoloing strings—carefully moderate their *f*'s (and *sf*'s). In mm.100–104, it seems to me that it is well to emphasize and bring out the entire wind section, while making sure that the timpani blend with cellos and basses, and not overpower them.

In the second passage (mm.126–42, see Plates XIII, XIV)—and its transposed partial recapitulation (mm.150–*ca.* 166)—our concerns must be slightly different. Here the conductor must make sure that the eight woodwinds are not only all dynamically balanced but are alike in their phrasing, so that Schumann's fine alternating colorations are faithfully represented, and that out of the mosaic-like structuring, long eight-bar *Klangfarben* lines are created. Meanwhile, the seemingly fragmented line in the violins must also be pulled together into a single musical idea. Again, precise identical dynamic balancing and phrasing will produce the desired result. In mm.134–42, both sections of violins must merely (softly) color the woodwinds, and their *fp*'s (as well as those of the woodwinds) must be gently expressive *in p*, not the hard, harsh *fp*'s one hears here so often.

The reader can now well imagine the extent of the performing challenges Schumann's Second Symphony presents. I have here dealt with only some fifty-odd measures; but there are 391 in the first movement alone. These challenges have nothing to do with 'poor orchestration'; they merely represent the subtler interpretive refinements one must bring to this piece—in actuality, of course, to *any* kind of fine, special, and original music. That process begins with understanding and *really* caring![6]

The influence of Beethoven in Schumann's work can be seen most explicitly in the Finale of the Second Symphony, in particular the direct influence of the first movement of Beethoven's Fifth. Having herein previously examined and analyzed that work, it will be interesting to see how it served as a model for

6. Anyone wanting to hear how magnificent Schumann's orchestral music is and can sound, *without* re-orchestrating it, should listen to Levine's superb recording of Schumann's Symphony No.1 (the "Spring" Symphony) with the Philadelphia Orchestra.

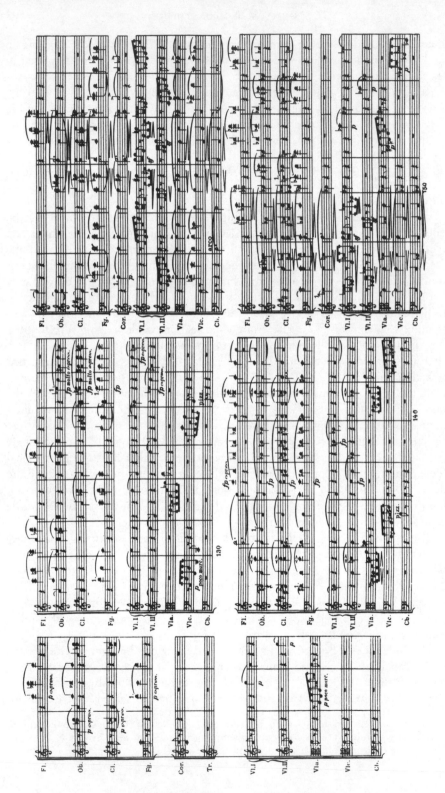

Plate XIII   Schumann, Symphony Nº 2, first movement, mm. 126-53

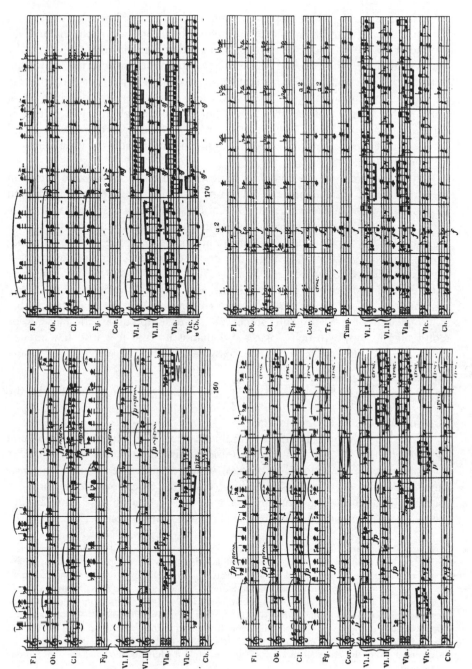

Plate XIV   Schumann, Symphony N° 2, first movement, mm. 154-78

what is arguably, in certain respects, Schumann's most Beethovenian work. I am
referring, on the one hand, to the movement's basic phrase structuring and, on
the other, to the character of some of its melodic/thematic material, especially
the lyrical second subject. And just as the specific phrase structuring in Beethoven's Fifth is generally disregarded and subverted, so it is in most performances
of the Schumann Second. At best, conductors and players are dimly aware of
some sort of four-bar phrase structuring, but are appallingly unaware of exactly
where the four-bar phrases fall. The result, as any number of recordings will
attest, is that the weights and stresses within phrases are all—or mostly all—in
the wrong place.

I think this problem arises here for basically the same reason it also occurs in
performances of the Beethoven Fifth: a very fast tempo notated in a meter and
note values which make whole stretches of measures look absolutely alike, providing no instantly obvious way of distinguishing phrase structuring, phrase beginnings, and phrase endings. Beethoven's first movement, the reader will recall,
was written in 2/4 ♩ = 108, *Allegro con brio*. Schumann's Finale movement is
written in 2/2 (¢) 𝅗𝅥 = 85 (actually ♩ = 170), *Allegro molto vivace*. The only
difference in performance is that the Beethoven is conducted in 'one,' the Schumann in 'two'—although certain sections of it can also be (and should be)
conducted in 'one.' Schumann's somewhat slower tempo allows the music to be
conducted in 'two'—in many places it demands it—and it is just slow enough
to admit the use of sixteenth-note passages, as in the very first measure

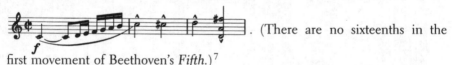

. (There are no sixteenths in the

first movement of Beethoven's *Fifth*.)[7]

There are no serious interpretive problems regarding phrase structuring and
periodization in the first 137 measures, these being constructed, with but three
exceptions, entirely of four-bar phrases, which at times accumulate variously
into eight-bar or even sixteen-bar entities. The three exceptions are two nine-bar
phrases (mm.13–21 and mm.109–17) and one five-bar phrase (mm.22–26),
which, combined with the first one in nine bars, simply restores the periodiza-

7. The two tempo markings in question here are illustrative of the variability with which different
composers specify relatively similar tempo situations, depending on the particular rhythmic and
melodic content (see discussion of this matter in Part I; also Rudolf Kolisch's aforementioned article
on Beethoven's metronome markings). On the face of it, Schumann's *Allegro molto vivace* ♩ = 170
looks quite a bit faster than Beethoven's *Allegro con brio* ♩ = 108. But in reality the Schumann
sounds at times much faster than it looks, for the metronomization (170) refers to the two-beats-per-
bar basic pulse of the music. On the other hand, the feeling of the music when conducted in one
beat per bar is considerably more relaxed, i.e. slower, than in the Beethoven, showing again that
what counts is not the nomenclature per se, but the actual rhythmic content and inherent pulse of
the music.

For the record, most conductors take the movement close to Schumann's metronome tempo
(usually around ♩ = 160), but some are off (slower) by a mile: Klemperer ♩ = 120, Kubelik, Ansermet ♩ = 132, Mehta, Marriner ♩ = 136, Patane, Zinman, Commissiona ♩ = 144, Sawallisch, Semkow ♩ = 148.

tion to binary structuring, putting the next 86 bars firmly back on the four-bar phrase track.[8]

The second nine-bar phrase constitutes the closing of the first part of the exposition and leads directly into what appears at first hearing to be a recapitulation of the very opening of the movement, only now on the dominant rather than the tonic. But Schumann quickly steers this 'deceptive' recapitulation in a quite different direction: rapid canonic imitation in the strings (Ex.1), while above them woodwinds 'sing' in beautiful sustained chorale-like harmonies. The

Ex. 1

contrast between the stately winds and the agitated string scales—whole notes against sixteenths—is quite wonderful when fully exploited in performance. On many recordings (Janowski's, Mehta's, Pfitzner's, Celibidache's, and Klemperer's, for example) the winds are barely present, and all one hears is strings.[9]

I digress to point out that in this symphony, as in so many early 19th-century, early-Romantic works, horns and bassoons must be absolutely matched in performance, dynamically and, as much as possible, sonorically. In the per-measure alternation of quarter triplets starting in m.134, for example, horns and bassoons must form a continuous rhythmic chain. This takes on especial importance be-

8. Again there is one exception, a six-bar phrase (mm.75–80), which is merely—very much as in Beethoven's Fifth—a two-bar extension of a four-bar phrase.
9. Masur, Schuricht, Mehta, and Patane don't help matters here at all by adding trombones. That helps to bring out the winds' 'chorale' all right, but adds a dark low-register coloring to the sound which fills in (and muddies up) the orchestral mid-range, and thereby completely vitiates Schumann's conception of registral and timbral contrast: luminous winds against darksome strings. How beautiful this passage can sound may be heard on Zinman's altogether excellently played Baltimore recording.

ginning in m.142, when horns and bassoons form an unbroken rippling line of
reiterated D's (Ex.2). Unfortunately, in the vast majority of performances the

Ex. 2

horns seriously overbalance the bassoons, so that no continuous line is heard.
How wonderful and exciting this can sound can be heard to excellent effect on
Haitink's, Mehta's, Sawallisch's, and Kubelik's recordings.

In m.139 Schumann breaks into the by now well-established binary (four-bar)
structuring with a three-bar phrase. This is brought about by one of Schumann's
favorite devices: elision and overlapping of phrases. As Ex.3 shows, the violas
and cellos enter in m.140 with a marcato half-note motive (derived from mm.2–
4 in the exposition), forcing, in effect, a truncation—an incompletion—of the

Ex. 3

violins' and basses' intended four-bar phrase. That is to say, the latter's phrase
would have finished in m.142. But with the violas' and cellos' 'early' entrance
in m.140 overlapping and eliding with the outgoing phrase, the whole structure
is shrunk by one measure. It is imperative, in order to do full justice to the
music, that conductor and players understand the (in effect) intricate polymetric

structuring here. This is especially urgent for the violas and cellos as they cut across the phrase construction of the rest of the orchestra.

Furthermore, if this structuring is not appreciated, in all likelihood Schumann's next remarkable idea in the succeeding measures will also be completely missed and misunderstood. I refer to the successive *sf*'s, starting in m.150 and continuing for over twenty measures. One is startled to discover that all these *sf*'s are on 'twos' and 'fours' in the four-bar structuring, acting therefore as powerful syncopations, not the heavy 'downbeat' accents as they are almost always universally played.

The interlocking canonic, stretto-like construction here (see Ex.4) produces a

Ex. 4

ten-bar edifice (not counting the anticipatory viola/cello entrance of m.140), which Schumann elaborates into a prolonged climactic passage, relentlessly spiked with *sf* syncopations and cross-accents (Ex. 5a). That these are truly

Ex. 5a

syncopations can perhaps be better seen when re-notated with smaller note values (as in Ex. 5b).

Ex. 5b

This type of syncopated cross-accents was, of course, not original with Schumann. Beethoven was the first to consistently use them as a powerful, expressive—not merely rhythmic—device. Examples abound, but, suffice it to mention only a few outstanding instances: Beethoven's Seventh, Eighth, and Ninth symphonies.

Let me emphasize that in the Finale of Schumann's Second Symphony we are not as likely to see these *sf*'s immediately as off-beat accents, because of the particular *alla breve* notation of half-notes and whole notes stretched across four measures, rather than in eighth- and quarter-notes contained in a single measure (as in Ex.5b). It is only when one engages in the kind of structural analysis I have undertaken here that one can discover the true character and meaning of these *sf*'s. And what a difference it makes to hear these *sf*'s played as 'two' and 'four' syncopations. It gives an entirely new and fresh feeling to the music, much more interesting than the ordinary 'one' and 'three' stresses one usually hears here. How wonderful this passage can sound when played correctly can be heard, alas, on only a few recordings, those of Ansermet, Masur, Mehta, Semkow, Haitink, Paray, and on Pfitzner's rather excellent old 1928 Berlin recording.

Schumann continues to play with such off-beat syncopations in the most remarkable ways. For example, in m.165 we see—and this should be made clearly audible—an idea first explored by Beethoven in the exposition of the Seventh Symphony's Finale, namely, what I called in the discussion of that work 'double syncopations.' We can see the similarity clearly by comparing the three following music examples (Ex.6a, Beethoven, and Exx.6b,c, Schumann), reduced here to their essential rhythmic/metric configurations),

Ex. 6a

Ex. 6b

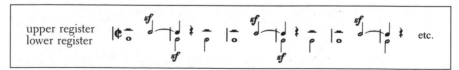

or Ex.6b reduced to Beethoven's rhythmic units:

Ex. 6c

Note, by the way, how Schumann pointedly asserts the idea of these double syncopations conceptually and instrumentally by reinforcing flutes and oboes with trombones, who—to further emphasize the point—make their first entrance in the movement here, Schumann having saved them until now.

Second, the *sf*'s in m.172 and m.176 turn out to be 'twos,' off-beat syncopations in the phrase structuring—not 'ones,' as they are usually rendered. These accentuations are then developed in diminution, when accents—in effect the former *sf*'s—now appear every bar on the 'weak,' i.e. second beat, a further rhythmic/expressive intensification of the music as it reaches the end of the exposition. Whereas in mm.149–72 the cross-accent syncopations occurred every two bars (see again Ex.5a), now in m.179 they occur in *every* measure (Ex.7),

Ex. 7

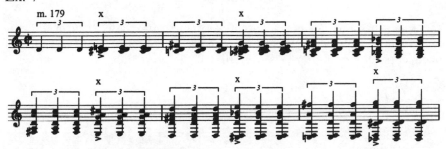

as the harmonies fan out in in contrary motion. Note, too, the extraordinary Beethovenian dissonances along the way (marked x in Ex.7), further energizing the cross-accents.[10]

In m. 191 Schumann introduces the first of two 'second subjects.'[11] It is presented here as the primary thematic material, dominating the entire next phase of the movement (mm.191–279). But we have heard this elegantly elliptical melody before, not only earlier in this movement (partly hidden in the contrapuntal fabric), but long before that as the main theme on which the

10. Celibidache, who is fond of 'improving' Schumann scores, allows himself the terrible and tawdry idea of adding a crescendoing twelve-bar timpani roll (on D—a note to which Schumann did not have access in this symphony).

11. Remarkably, the Finale of Schumann's Second Symphony has no development section, but instead two lyric second subjects. It also honors the cyclical conception of thematic linkings between movements by recapitulating, in (what amounts to) the coda, the 'signal' theme, first heard in the brass in the very opening of the symphony.

beautiful slow movement is built. There the melody appears in its primary form in C minor at the head of the movement (Ex. 8a), then immediately transposed to E♭ major in the oboe (Ex. 8b).

Ex. 8a

Ex. 8b

In the Finale the head notes of this theme appear quite suddenly and unexpectedly in m.63 in the violas, cellos, bassoons, and low clarinet (Ex.9), of

Ex. 9

course, in rhythmic augmentation and yet significantly faster than in the slow movement. We also hear the theme's second-through-fourth notes in a kind of hocquet patterning in mm.80–89 (Ex.10, reduced in outline).

Ex. 10

A little bit later (m.191) Schumann presents this same theme in the solo clarinet in inversion (Ex.11). While the clarinet's presentation of the theme is

Ex. 11

obviously positioned so as to conform to the underlying four-bar nexus, when Schumann gives it to the violins, flutes, and oboes twenty bars later (m.211)—answered in its prime (uninverted) form by cellos and basses in

m.213 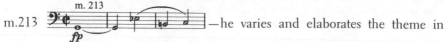 —he varies and elaborates the theme in such a way as to provide a recapitulation of the off-beat syncopations heard earlier. Again, their off-beat-ness is not readily discernible from a casual perusal of the score. One has to analyze the over-all substructure to realize that the *sf*'s starting in m.220 in violins and cellos are once again 'twos' and 'fours' (respectively): again syncopations against the beat, not the 'downbeat' stresses one usually hears in performance.

Here again that universal bad habit of dropping or chopping off last phrase notes comes to the fore (mm.211–25), especially in the strings. It makes the music sound nervous, fragmented and disjointed, when in fact—even with the bars rest—we need to create long, arching melodic lines (spanning sixteen to twenty-four bars), in which, indeed, the silent bars are as much a part of the melody (the music) as the notes in the played bars. How fulfilling and singing this passage sounds when played correctly can be heard on only a few recordings, notably Barenboim's, Pfitzner's, Celibidache's, and also, partially, on Masur's. But on the last-named only the upper-range instruments sustain the last phrase notes, while the answering cellos and basses do not.

It is truly beautiful how Schumann now manages to work the final measures of the *Adagio*'s main theme and its oboe variant into the Finale. Measures

227–30  come directly from the violin theme

(compare with mm.5–6 of Ex.8a), while the next four measures (mm.231–34)

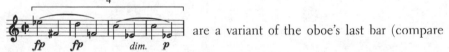 are a variant of the oboe's last bar (compare

with Ex.8b). In all this time, the *Viertaktigkeit*, as in Beethoven's Fifth, has never been disturbed. Comparing the last two music examples directly above, each comprising a four-bar phrase, we can readily see how Schumann shifted his *fp*'s around within a phrase in order to achieve a syncopated effect (mm.227–30), while at other times (mm.231–34) he shifts the stressed notes to their normal hierarchical position.

If one follows the four-bar structuring systematically through to the end of this section of the movement, one will see that the C minor resolution in m.274 falls squarely on a 'one,' with only one six-bar phrase extension (mm.267–73) needed—as in Beethoven—to arrive at the tonic (Ex.12). If this is not understood, in all likelihood the entire rest of the movement will be metrically,

Ex. 12

structurally out of kilter. On not a single recording have I heard any performance that seemed to understand where these four C minor chords fall phrase-wise. They are usually played vaguely, amorphously, indiscriminately, made worse on numerous recordings by an enormous ritard (Sawallisch, Karajan, Muti, Kubelik, Semkow, Szell, Mehta, Janowski, Masur, Barenboim, Celibidache)[12]—not asked for by Schumann—as if the movement had come to a final cadence. It is anything but that; it is instead a mere temporary halt in the music (Exx.13a,b), to be immediately followed by the *second* 'second subject' theme (m.280). The best way to realize the intent and feeling of this interim cadential resolution is to use the bowings indicated in Ex.13a. An easier way to

Ex. 13a

catch that 'intent and feeling' is to see it notated in 4/4 as in Ex.13b.

Ex. 13b

Now we come to the most fascinating, unusual, and original part of the Finale—and, I must add, the most consistently misinterpreted. For, the phrasing from here to the end of the movement—some 300-plus measures—is shifted one bar off from where, at first glance, the periodization appears to be. The oboe's phrase positioning (mm.380–83) is not

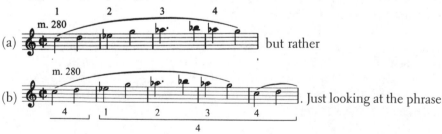

12. Many conductors are quite confused by the empty "G.P." (general pause) measures and some (like Mehta and Karajan) even make huge fermatas on the empty measures or on the C minor chords—or both—not realizing that the empty measures are an integral part of the music and its continuous flow. Hear how wonderful this C minor cadence sounds when played in tempo on Paray's excellent recording.

in the score, a phrase preceded by pause measures and seemingly ambiguously placed tonic chords, also encompassed by a single phrase slur, one is apt quickly to assume that the structuring is as in example (a) above. But closer scrutiny and analysis of the metric substructure reveal that it is in fact as in example (b) above, making m.380 an anacrusis (upbeat) measure.[13]

That this is so is further confirmed in my mind by the close resemblance of this lovely theme to the 'second subject' theme in the first movement of Beethoven's Fifth. I am firmly convinced that Schumann was directly influenced by Beethoven's theme, just a little short of reverent plagiarization. (For another reference to Beethoven, see below p.516). If we compare the two themes (Exx. 14a and b)—the Schumann example (14b) altered notationally to be in the

Ex. 14a

Ex. 14b

same metric/rhythmic format—we see that both have the same general shape and contour, with the melodic high point coming in the third bar (the second in the phrase structure). And if we hypothetically change the E♭ in the first measure of Beethoven's theme to a C (Ex.14c), the similarity becomes even

Ex. 14c

more striking—a similarity emphasized still more when Schumann's 'second subject' theme is conducted in 'one,' which it almost always is.

In any case, seen from this point of view, not only is m.280 an anacrusis measure, but so are mm.284, 288 (in the strings), m.292 (flute and clarinet), and m.296. In m.301 and m.309, Schumann recapitulates the double syncopation passage of m.165, re-orchestrating it in the process (Ex.15). It is crucial to follow (and observe in performance) the four-bar structuring, and probably to annotate the orchestra musicians' parts accordingly. For it is otherwise impossi-

13. Only a few conductors (and oboists) seem to understand the phrase structuring here. In fact, of all the recordings I sampled, only on Haitink's, Patane's, and Paray's is this passage played correctly. Pfitzner phrases it right, but ruins it all by dropping to a lugubriously slow (♩=108) tempo, the only major aberration in what is for the most part a very fine interpretation and which contains much fine playing by the orchestra (Neue Sinfonie Orchester Berlin), especially for 1928. It was probably the first recording of the Schumann Second Symphony.

Ex. 15

ble for the violins to know just from their parts that m.316 (*sub.p*) is a 'four,' or for the brass to know that m.324 and m.328 are 'upbeat' measures. It is curious that most conductors and orchestras feel this passage (mm.301–32), with its syncopated cross-accents, correctly, even those who interpret mm.289–99 *incorrectly*. They manage this—probably unknowingly—by adjusting to a five-bar phrase (mm.296–300). This in turn puts them on the right phrase footing in m.316, the aforementioned *subito p*—a dynamic which, unfortunately, is not observed by the majority of conductors, who opt instead for a beefy, inelegant *mf.* (Score and parts, by the way, are obviously wrong in the horns in 316–17. The *p* should be in m.316, clearly a printing error. And yet on recording after recording the error is perpetuated for posterity.)

The temptation to crescendo in m.331 is enormous, especially with the rising melodic line. But that should be strenuously resisted—as well in m.339—for the real crescendo does not come until m.343. It is a temptation very few orchestras and conductors are able to resist, judging by the recordings. Only Barenboim and Paray manage to keep the *p* more or less under control.

In m.359 and the entire succeeding forty-odd measures the performer (conductor) is faced with very serious scriptural problems, unfortunately, I believe, of Schumann's making. The *cresc.* markings in mm.367–68 and m.376 make little sense, if the earlier crescendo starting in m.343 is to bring the music to a *f* at m.359. How can there be two more extended crescendos after that, starting from a *f* base, and arrive only at a *ff* in m.391? The answer obviously is that there cannot be and, accordingly, the score must be wrong at m.359. I think the error lies in the fact that the *fp* we see in the horn, trumpet, and timpani parts was mistakenly left out in all the other parts. This is borne out for me in that the flute, oboe, and violin parts have no *f*, which presumably would be the terminal dynamic after the previous fifteen-bar crescendo. If one interprets the *sf*'s in the scale passages as *sfp*'s (Ex.16) or, better yet, starting in m.360 as *sf*'s

Ex. 16

in p, then the two later (aforementioned) crescendos can make sense, and in fact can be used to bring the music to its first major mid-movement climax (Ex.17). It would also allow the many successive *sf*'s to be interpreted in the context of the ever-escalating crescendo. This would mean, for example, that the *sf*'s in m.375 would be in *mp*, reach a *mf* level by, say, m.383, and the full *f sf* only in m.387. The beauty of this approach is that the climax at m.392,

Ex. 17

when it finally explodes, so to speak, is all the more meaningful and exciting after a prolonged twenty-five-bar build-up.[14] Let us note, too, that the *sf*'s in mm.374–80 are all on 'twos' and 'fours,' thus again phrase syncopations in effect, following very much the same pattern Schumann presented earlier, in mm.149–86.

I think the music that now follows is some of the most beautiful Schumann ever achieved. It is simply masterful the way the music builds gradually but inexorably to its ultimate climax and affirmation. The major performance problems here are (1) not to allow the music to rise to its highest dynamic levels too early; (2) to preserve the integrity and feeling of the four-bar phrase structuring. As for the first point, conductor (and orchestra) should not be carried away by Schumann's—perhaps *too many*—crescendos and *più f*'s, which, I suspect, were added (understandably) in the sheer excitement and white heat of creation. As seen in the score, they can be misleading. It is best to note that the first sign of a crescendo does not come until m.426, and that the first real *f* does not occur until m.474—and a *ff* not until thirty-five bars *after that*, in m.508! The dynamic pacing must therefore be very well controlled, especially in the brass, whose dynamic markings, it should be noted, are always slightly under—i.e. softer than—the rest of the orchestra.[15] (It is only in the coda, which I place at m.537, that the full brass and timpani begin to dominate, with the timpani having the 'final word' in the last five bars of the movement.)

After the great dominant-seventh climax—on the dominant (G)—the soft, gently swaying melody in m.394 (Ex. 18a) has a remarkably calming and serene

Ex. 18a

---

14. I present this suggestion as at least as interesting and viable an alternative as simply playing the whole passage unrelievedly loud for thirty-three measures. Evidently Zinman and Ansermet are of the same mind—the only ones—judging by the recordings sampled.

15. To clarify this very gradual rise in dynamic level for the brass players, I suggest marking the entire sequence as follows: m.445—*p*; m.461—*mp*; m.469—*mf*; m.481—*poco f*; and m.489—*f* (as written by Schumann).

effect.[16] (It should therefore be conducted in 'one.') It is even more beautiful when the performers understand that m.394 is an anacrusis (upbeat) measure. More often than not it is played as if it were a 'downbeat,' a 'one' (Ex.18b), even changing Schumann's phrasing and bowing. In string players' partial de-

Ex. 18b

fense it must be said that Schumann's phrasings, i.e. slurs, which readily can be taken as bowings, can be misleading if only viewed from the part, rather than from the score and the music's phrase infrastructure. A phrase such as

will obviously be seen at first glance as starting on the 'ones' of four-bar phrases. But the fact is that in this work phrasings/bowings and the metric understructure do not always coincide; indeed, they are often totally at odds. Similarly, flutists

might be forgiven if, seeing  —espe-

cially after twelve bars' rest—in their parts, they assume this to be a typical four-bar phrase starting on 'one.' And yet it isn't; it starts on a 'four' with a lightly syncopated upbeat feeling.

In any case, very few conductors attain the right phrase feeling here (mm.394–417). To my ears, only Mehta, Pfitzner, Szell, Ansermet, Schuricht, and Janowski do, but the latter five partially spoil things by disregarding Schumann's p dynamic. The accented E (m.402) and the woodwinds' fp's (mm.406, 410) are 'fours,' slight accents—not aggressive fp's—in the prevailing p dynamic.

Two six-bar phrases intervene in the otherwise systematic four-bar structuring. The extra two bars are needed for the brasses' 'signal' theme, borrowed from the very opening of the symphony, to modulate the music in the first instance to the supertonic (D minor), and in the second instance back to C major.

Without disturbing the four-bar structuring, Schumann now overlays a wood-wind theme in triple meter (Ex. 19), again borrowed from the first movement

16. Many musicians have observed the similarity of this phrase to a line in the last song of Beethoven's *An die ferne Geliebte* :  . Whether this was consciously a near-quotation by Schumann or whether the melody was, as I suggest, inspired instead by the Fifth Symphony's first movement 'second subject'—or both—is hard to prove or, for that matter, to disprove, there being no documentary evidence to support either thesis. Both ideas will have to remain conjectural.

Ex. 19

(m.15), on top of the basic continuing 2/2 (all obviously conducted in 'one'). When the 'triplet'[17] overlays are abandoned and the *alla breve* second subject returns in the woodwinds, in m.474, it is answered canonically in the violins, the latter starting on a 'two,' the former on a 'four.'

After several more six-bar interspersions,[18] the mounting waves of sound have reached *f*, marked *con fuoco* (with fire). This exhortation does not mean to 'play as loud as possible'—as, unfortunately, it is usually interpreted—but merely to play in *f* with great intensity and passion, not necessarily a matter of sheer volume. (Conductor and orchestra must save something for the *ff* at m.508.) This triumphant passage occupies more or less the same place in the over-all form of the movement as the very similar passage near the end of the Beethoven Fifth's first movement (m.439). Melodically/thematically very similar, it also hints at a similarly fragmented instrumentation (Ex.20a), although not as contrastingly varied as in Beethoven's example (Ex.20b).

Ex. 20a   Schumann

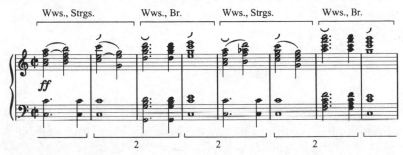

Ex.20b   Beethoven

17. Although written in triple meter (3/2), the effect is that of triplets over a duple base:

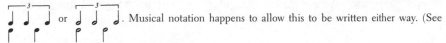

or . Musical notation happens to allow this to be written either way. (See Schumann's use of the triplets in mm.560–70.)

18. Measures 455–60, 469–75, 487–92.

A vulgar tempo distortion is perpetrated here by many conductors (mm.514–15): a huge ritard and, even worse, an immense fermata on m.515 (Semkow, Kubelik, for example). Apart from not being called for in Schumann's score—which ought to be reason enough not to do it—such a holding back of the tempo destroys the whole momentum and forward thrust of the music, which has been building for many minutes, in fact—if the equally unwarranted ritard in mm.273–79 is not taken—since the very beginning of the Finale.

This tempo pull-back also serves the next passage (mm.516–36) very poorly, for it takes all the surprise out of the sudden *subito p* in m.516 and the gloriously inventive chain of abrupt modulations that ensues (Ex.21a, b). It also kills

Ex. 21a

Ex. 21b

The same passage seen in harmonic abstract.

the surprising effect of the entire passage being off-kilter structurally/rhythmically. Both the phrase beginnings and harmonic shifts are on 'fours'—not on 'ones,' as unfortunately virtually all recordings interpret it. Indeed, this glorious passage seems to be a signal for conductors to indulge in various interpretive excesses. Even if they have observed the *sub. p* of m.516—so startling and exciting after the preceding eighty-bar crescendo build-up to *ff* (most conductors simply ignore it)—they tend to crescendo too-much-too-early, so that the next *ff*, to be reached only in m.537 (heralding the arrival of the coda), climaxes about sixteen bars too early, rendering the real *ff* a meaningless anti-climax.

Another terrible distortion of this passage is perpetrated by Szell, who makes eight successive two-bar crescendos (*p* ⟨ *mf*), instead of one long nineteen-bar graduated crescendo. The other favorite indulgence is to acclerate the tempo tremendously, I suppose, to achieve some audience-electrifying race to the finish line.[19] All that it really achieves is to degenerate this striking and

19. Many conductors—such as Barenboim, Sawallisch, Celibidache—rush the tempo even much earlier, as early as m.359. Marriner outdoes everybody else, however, by taking an unbelievable tempo of ♩ = 220 as early as m.394.

inherently noble passage to a cheap 'effect' (what Beethoven had already called *Effekthascherei*). Here again the *sf*'s starting in m.526 must be graduated, logically as follows: m.526—*sf* in *mp*, m.530—*sf* in *mf*, m.534—*sf* in *f*.

The four-bar structuring continues all through the coda. But rather than letting that be a constraint on his rhythmic inventiveness and imagination, Schumann keeps things lively and full of surprises (as he has all along in this movement) by means of cross-phrasings and off-beat accentuations and syncopations. These appear to be confusing or incomprehensible to most conductors and orchestras, who generally react arbitrarily, intuitively, with whatever seems to be the easiest or most comfortable interpretation. To be sure, there is not much an orchestra musician can glean from the instrumental part alone; as the bars rush by, at breakneck tempo, they all look structurally alike. But the conductor should have no such excuse, as any basic analysis of the work will readily reveal its intrinsic structuring. And yet in recording after recording one hears

the desired result in the first violins, the following bowing is advisable (Ex. 22).

Ex. 22

At point ⊕ (m.550) care must be taken to remain more or less at mid-bow; in any case, the thrust and direction of the phrase must go to the whole notes.

These tied whole notes in mm.552–53 would certainly, at a quick glance, lead one to think that they are 'one-two' in a phrase (or at least a 'three-four'), but they are neither. They are a 'four-one.'

Schumann's final inspiration is to place what looks like a phrase beginning (m.560) not on a 'one,' but on a 'two'—another phrase syncopation, analogous to an accent on the second beat of a 4/4 bar. (Measures 553–58 are a six-bar phrase.)

We should note that the timpani's final bar means to indicate,

in all likelihood, that the trill (or roll) should stop on the second beat as is the case as well in m.393. (The fermata is there *pro forma*, as it were, because the entire orchestra has a fermata for the whole bar.)

I have shown at considerable length and in some detail the true structuring and periodization of Schumann's Finale movement. If performed with an awareness *throughout* of this basic infrastructure,[20] the entire movement will sound completely different from the usual ordinary renderings, infinitely more interesting— virtually like a brand-new piece—and in fact like the remarkably original masterpiece it is.

---

20. Of all the 28 renditions sampled, only two conductors—Paray in his Detroit recording and Bernardi in his Calgary recording—get almost the entire movement right.

Tchaikovsky: Sixth Symphony

There are certain aspects of Tchaikovsky's Sixth Symphony that are very special, if not unique, to this work and to this composer's conception, that in turn require very special care and attention in performance, in 'interpretation.' These unusual interpretive problems come in addition to all the more conventional, general interpretational/notational issues at stake in any major orchestral symphonic work. That being the case, I will not reiterate all those fundamental and customary performance issues already extensively examined herein, but, rather, address only those which are peculiar to Tchaikovsky's Sixth Symphony.

These performance and notational aspects are, once again, primarily in the realm of tempo and dynamics, the notation of which Tchaikovsky had by the end of his life refined and expanded to a remarkable degree, to my knowledge more than any other prior composer or composer of his generation. This may surprise many readers, because Tchaikovsky is generally regarded as a highly emotional, unreservedly expressive, even at times unsubtle composer, and not given to tidy, fastidious subtleties of thought, let alone to intellectual precision. It is common knowledge that many 'intellectuals' in music—historians, critics, theorists, and such—consider much of Tchaikovsky's music inclined to vulgar emotionalism, thus irresistible to a mass audience, and therefore, almost by definition, a music devoid of any discipline, refinement, and discriminating intelligence. Indeed, most performances—interpretations—of Tchaikovsky tend in that direction, wallowing indulgently in every expressive effect, wearing their emotionalism ostentatiously on their sleeve, and disregarding those aspects of his music that reflect the workings of a brilliant, disciplined mind. In fact, many conductors have made entire careers of exploiting the more obvious superficial emotional effects of Tchaikovsky's music, some (Bernstein, for example) through incredible exaggerations and distortions, finding new depths of vulgarity that are nothing more than an offense to the music and to Tchaikovsky's genius.

Bernstein's Tchaikovsky renderings represent an extreme, however. Still, most conductors, past and present, approach Tchaikovsky performances—especially of his last three symphonies—as a kind of emotional 'field day,' where all whims, liberties, and indulgences can be safely indulged. Such performances are in general stubbornly blind and deaf to the intellectual and notational rigor with which Tchaikovsky not only *sought* to balance and temper the emotional content of his music but, in my view, brilliantly succeeded in doing so—if only we would perform faithfully what he wrote. Indeed, I would argue that precisely because Tchaikovsky's music is so inherently emotionally highly charged, that it is all the more necessary, not so much to restrain ourselves expressively, as to unequivocally respect those very markings and indications with which Tchaikovsky himself refined and disciplined his art.

Let me lay out the main points of contention here. In the realm of tempo and tempo modifications Tchaikovsky, in his Sixth Symphony (and many other late works, like his *Sleeping Beauty* and *Nutcracker* ballet scores) evolved a highly subtle and explicit code of notation, in turn supported and complemented by metronome markings of great precision and variety. Analogously, in the realm of dynamics, Tchaikovsky refined and elaborated his notation to comprise eleven and twelve dynamic gradations, not the usual six or eight used by the vast majority of 19th-century composers. The sad truth is that in these two fundamental respects, Tchaikovsky's notations are either roundly ignored or, by many conductors considered a typically Tchaikovskian notational excess, which in turn is re-interpreted to allow the performer any license he may wish to take.

In his ever more subtle tempo distinctions Tchaikovsky was merely elaborating on what Beethoven in his late works had begun to explore: a way of notating in an increasingly precise manner—but still with conventional Italian-language indications—the subtle variations of tempos that early 19th-century composers were hearing and incorporating in their music. It was nothing more than an attempt at defining for the performer through notation Beethoven's "Tempo des Gefühls." And just like Beethoven, Tchaikovsky used metronome markings to confirm and give substance to his Italian-language verbal annotations. The wonderful variety and subtlety of his tempo indications can be appreciated, virtually at a glance, in the following table (Fig.1a, b).

Notice how logical and consistent is the relation between the verbal tempo indications and their respective metronome markings, as well as the use of modifying terms like *poco, molto, mosso, quasi* (as in *quasi adagio*), *non troppo, non tanto*. Taken all together, Tchaikovsky provides us with a network of tempos and tempo relationships which is not only intrinsically clear but, when its components are seen in relation to one another, is consistent and logical, and thus should leave no room for misunderstanding. We should also note that Tchaikovsky is consistent in nomenclature and metronome marking from movement to movement. *Moderato assai* ($\boldsymbol{\downarrow}$ = 88) is used both in the first and last movements, as is *andante* ($\boldsymbol{\downarrow}$ = 69). While the fact that both *allegro vivo* and *allegro con grazia* are defined by a metronomization of $\boldsymbol{\downarrow}$ = 144 may at first seem

Fig. 1a

vivace
allegro molto vivace (152)
allegro vivo (144), allegro con grazia
un poco più animato (132)
allegro non troppo (116)
moderato mosso (100)
moderato assai (88)
andante mosso (80)
andante (76), andante giusto (76)
andante (69)
andante non tanto (60)
quasi adagio
adagio mosso (60), adagio poco meno che prima (60)
adagio (54), adagio lamentoso (54)

Fig. 1b

stringendo molto	ritenuto
stringendo	rallentando poco
affrettando	rallentando
un poco animando	ritardando molto
animando	
incalzando	
più mosso	

strange, it is really quite logical, for the former is a lively *allegro* 4/4 with a pulse of fast quarter-notes, while the *allegro con grazia* (Ex.1) of the 5/4 second

Ex. 1

movement is actually felt in a much more leisurely pulse of half-notes and dot-ted half-notes (♩ = 72 and ♩. = 48, respectively).[1]

1. This is the same interesting issue, namely, the same tempo indication allowing for different tempo *feelings* and, conversely, different tempo indications permitting *similar* tempo feelings, as discussed at some length in Part I (pp. 40–43).

Let it be noted, finally, that Tchaikovsky's tempos for the Sixth Symphony are in one sense all quite conventional. The basic tempos range from *adagio* to *allegro* and *allegro vivo*, the metronome markings from ♩ = 54 to ♩ = 152. In other words, we do not find in this score any indication at all of extreme tempos—fast or slow—or of any of the excessive super slow and super fast tempos taken in this work by the majority of conductors. They seem to feel that there is some hard evidence—evidence certainly not in the score—that Tchaikovsky was intent on expanding the range of tempos, something that actually can be said of Beethoven but clearly not of Tchaikovsky. The question then remains: why do most conductors feel justified in taking tempos thirty or more metronome points slower than Tchaikovsky's markings (in slow tempos) and similarly proportionately faster in fast tempos? Typical examples are: for the *andante* (♩ = 69), first movement, m.89 (see Ex.2) ♩ = 38 (Furtwängler, Bychkov, Celibidache, Martinon, Ormandy, Giulini); for the next *moderato mosso* (m.101, ♩ = 100) ♩ = 62 (Celibidache), ♩ = 86 (Coates, Ormandy, Giulini); for the end of the Scherzo movement ♩ = ca.186 instead of Tchaikovsky's ♩ = 152 (Martinon, Furtwängler); for the *andante* ♩ = 76 in the Finale (m.37) ♩ = ca.46 (Furtwängler, Monteux, Celibidache, Martinon, Ormandy), ♩ = 36 (Giulini); for the next *più mosso* (m.71) ♩ = ca.112 (Monteux, Furtwängler) instead of ♩ = 96 (Monteux, Furtwängler); for the *moderato assai* (m.116, ♩ = 88) ♩ = 144 (Furtwängler, Barenboim).

As I pointed out in connection with the widespread flouting of Beethoven's metronomizations, I would argue in the case of Tchaikovsky that, if a conductor has an antipathy for metronome markings and feels compelled to ignore or change them, that is one thing; it is quite another to read into the term *moderato assai* (a 'very moderate' tempo) a tempo that clocks in at ♩ = 144!

The ultimate irony of all this is that Tchaikovsky's tempo and metronome markings are perfect: the slow tempos are not too slow and the fast tempos not too fast. Indeed, if faithfully adhered to, they give his symphonies a more powerful expression, a tensile strength, and a formal clarity, unity, balance, and proportion, a sense of unfolding drama, that all the usual exaggerations, distortions, and distensions of form can never achieve. Analogously, the pervasive disregard of Tchaikovsky's dynamic markings is similarly destructive of the true quality of his work. The utter banality and crudeness of interpretation that result when the Sixth Symphony's dynamics are reduced to the usual one-dimensional three levels of 'as loud as possible,' 'relatively soft,' and a mediocre 'in-between,' are such that even this great work can barely withstand it.

Admittedly, Tchaikovsky's rarefied scale of dynamics is not so easy to achieve, certainly not without some real effort and concentration and change of playing habits. But what is so exasperating is that hardly any conductor seems even to recognize that there is a problem here, a very interesting one at that, technically and conceptually; exasperating also that these dynamics are either casually dismissed as unrealizable or as the meaningless aberrations of a composer given to excessive extremes. Nor is there any justification for the notion held by some conductors, who have at least noted Tchaikovsky's unconventional dynamic markings, that he was obviously intent on *expanding* the dynamic range of the orchestra, that he wanted to produce even louder climaxes and even more whisper-soft *pianissimos.*

The fact is that Tchaikovsky's use of five and six *p*'s (*ppppp* and *pppppp*) at one end of the scale and quadruple *f*'s (*ffff*) at the other end had nothing to do with playing louder or softer. Nor could it in reality. For all instruments have finite limitations of their dynamic range; and a composer could write fifteen *f*'s for all that any instrument could play any louder than its natural acoustic limitations permit—except, of course, through the bane of amplification.

Tchaikovsky was fascinated with something much more interesting than merely playing louder or softer: it was to achieve more refined, more discriminating gradations of dynamic levels, which he had noted the best players (in chamber music-playing, for example) could command, and in fact used quite naturally in giving subtler profile to their phrasing and musical expression. It is those in-between dynamics—between *f* and *mf*, between *p* and *mp*—which every good musician uses all the time, that he wanted to capture in notation.

Fig. 2

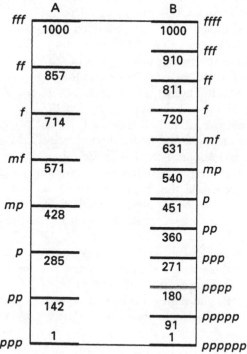

A = Usual dynamic levels (8)   B = Tchaikovsky Sixth Symphony's dynamic
level (12)   (on a scale of 1 to 1000)

To decode and correctly express this more refined notation requires some adjustment from the normal usage of dynamics. As Fig.2 shows, since the outer dynamic levels remain the same, all the dynamics in between take on new positions in the twelve-step scale.[2] This is only logical, for if Tchaikovsky's *ffff* is to correspond to the usual *fff*, and his *pppppp* is to equal the usual *ppp*, then a *f* in his notation, four levels away from the *ffff*, will have to be something more like a *mf*; conversely, a *p*, six levels above *ppppp*, will have to be more like an *mp*. I am not advocating that Tchaikovsky's twelve dynamic shadings should be—or in fact can be—produced with absolute scientific accuracy, but rather that the *f*'s in his score, for example, be not the usual all-out full-force power blasts (especially in the brass) that one usually hears, but a modified *f*, a *poco f*, allowing for three additional subtly delineated dynamic levels before reaching the maximum level. The same adjustment would apply to the lower dynamics where, in Tchaikovsky's score, the *p* would have to be more like a *mp*, to allow for another five or six dynamic gradations below that level to reach the softest possible sound.

When approached this way, Tchaikovsky's Sixth Symphony takes on classic proportions and a transcendent quality, a variety of expressive shadings and a glow of orchestral colors, that the usual 'slam-bang' 'get-the-audience-aroused-into-a frenzy-of-applause' performances can never achieve.

How detailed and careful Tchaikovsky could be in his notation—and therefore in his clear definition of how he intended a given passage to sound—is seen in our first excerpted example: the famous and justifiably popular *andante* theme from the Sixth Symphony's first movement (Ex.2). Even a cursory glance at the example reveals that *every* measure has the most exacting dynamic nuances. Analogously, in the twelve bars it comprises there are four tempo modifications, apart from the initial *andante* ($\downarrow$ = 69). That tempo marking is, alas, one of the most ignored. Indeed, most of Tchaikovsky's *andante*s are turned into *adagios*. If an orchestra and its conductor really faithfully observe Tchaikovsky's dynamics and tempos here, the music takes on a plasticity and subtle expressiveness that is far removed from the prevailing over-sentimentalized, maudlin interpretations one usually encounters. Note that the accompanying instruments (horns, bassoon, clarinets, second violins, violas, basses) are all marked one dynamic level lower than the main theme (in muted first violins and cellos).

If scrupulously followed, Tchaikovsky's tempo modifications also add tellingly to the originality of this passage: not merely some obvious sentimental popular tune. The *incalzando* (pressing forward) must be done gradually, stretched over two bars (mm.93–94), and then cancelled in perfect symmetry by two bars of *ritenuto* (mm.95–96). (The *come prima* in m.97 refers, of course, to the *incal-*

---

2. Actually, Tchaikovsky uses the full range of twelve dynamics in the Sixth Symphony, at the low level a six-fold *p* (*pppppp*), used only once for the four bass clarinet notes before the first movement's *allegro vivo*.

Ex. 2

zando of m.93). This passage returns two more times, once with the full orchestra (mm.130–41), and the last time as a soft clarinet solo accompanied only by quiet strings and timpani. It is intended to be a distant reminiscence of the music's first appearance, now marked *adagio mosso* (♩ = 60)—no longer *andante*—*and in the softest dynamics, gradually fading away* (*ppp* m.153, *pppp* m.157, *pppp* m.159, and finally six *p*'s in the bass clarinet in m.160).

A degree of caution and restraint is advisable in some of the more climactic brass-heavy moments of the work, as for example in mm.189–97 (Ex.3). It is

Ex. 3

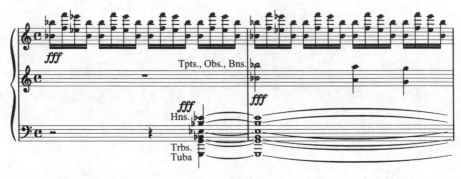

very tempting for the brass, seeing *fff* in their parts, to play their absolute loudest, not realizing at the moment that there are several even more climactic passages marked *ffff*, in the symphony. The same caution applies to mm.242–47 (Ex.4), where, despite the *ff*'s and *fff*'s, a balance must be achieved so that all three choirs of the orchestra (woodwinds, brass, and strings) are equalized. (In most performances brasses overwhelm the rest of the orchestra.)

A tempo question arises around m.220 to m.225, where hundreds of conductors have traditionally imposed a huge ritard, presumably under the influence of—and as a parallel to—the prolonged fourteen-bar diminuendo after the climax of m.214.[3] This automatic correlating of diminuendo and ritardando is, of

---

3. We should note, as an example of Tchaikovsky's meticulousness and conscientiousness in notation, that in m.216 he writes *dim. un poco*, not merely *dim.*, in the hopes of preventing too much and too quick a diminuendo, a very common fault among players and conductors. Tchaikovsky's

Ex. 4

course, one of the more universally practiced bad habits, and some conductors, like Furtwängler, for example, built an entire career on this mistaken notion.[4] When we consider that (a) here Tchaikovsky indicates no tempo change—there is no ritardando or ritenuto in the score—and (b) that in dozens of other places in this symphony there are numerous even more subtle tempo modifications clearly given, by all logic it docs raise the question whether we ought to allow ourselves to indulge in such unindicated re-interpretations. Moreover, shouldn't we pay some respect to the fact that the thematic material in this passage

(mm.230–37)  is the same as at the

intentions are confirmed by the cautionary *f* in m.217 and m.218. Note also that the composer, realizing that the second and fourth horns are resting in m.216, reiterates the *dim. un poco* especially for them in m.217. Not many composers of that era took care to be so explicitly clear.

4. Fürtwangler, in his many ways wonderful and remarkable 1938 Berlin Philharmonic recording of the Sixth Symphony, drops some forty metronome points between m.214 and m.229. The corresponding crescendo that follows is, of course, also accompanied by a tremendous accelerando!

beginning of the movement's first *allegro*

 , and for that reason alone ought to

retain some semblance of a lively *allegro* character, not the lugubrious *andante* one hears so often? I suppose that a slight relaxation of the tempo around m.229 is appropriate—although I would emphasize not absolutely necessary (the *pp* dynamics alone can give the real meaning and expression to the passage)[5]—but to drop from twenty to forty metronome points is a willful deconstruction of the music.

Another tempo distortion that has become traditional occurs at m.276. Again Tchaikovsky's score does not indicate any change (slowing) of tempo, and yet countless conductors indulge in a tremendous stretching of the tempo here. The fact is that Tchaikovsky has already composed a slower feeling into the music. From fast moving sixteenth-, eighth-, and quarter-notes—the basic pulse in quarter-notes—Tchaikovsky drops to half-notes and whole notes, a slowing of the momentum of the music by a factor of two or three at least (if one relates it to the preceding turbulent sixteenths). When mm.276–303 are kept in tempo, that is, *allegro*, but felt in ♩ or 𝅗𝅥, the music takes on an overwhelming power of expression that cannot be matched by the various *adagio* versions one usually hears. This is not some tearful maudlin exercise in bathos, but rather a music of extreme anguish, and of anger, a desperate outcry of pain, culminating in the quadruple *f* (*ffff*) of m.298.

What makes the usual over-distended tempo stretchings here so ludicrous—and I might add virtually impossible to perform from a technical, practical point of view, especially in the brass, who have to take untold numbers of breaths to get through the passage—is the fact that no player in the world can play mm.298–99 in that adopted slow tempo and still adhere to Tchaikovsky's dynamics | 𝅝 ___ | ♩ ___ ♪ ⅋ 𝄾 |. Dozens of recordings attest to this, in which one can hear either a sudden speeding up of the tempo in those two measures or a drop to *p* as early as the third beat of m.298. If kept in tempo, Tchaikovsky's extreme dynamic expression is difficult, but entirely possible.

The Scherzo of the Sixth Symphony is even today, a hundred years after it was first performed, technically and conceptually challenging enough that it is treated with considerable awe and respect and, as a result, usually played relatively well. I will therefore limit my comments to only three somewhat special interpretive/performance problems. The first of these—a misreading of the text—would hardly be worth mentioning if it weren't such an annoyingly careless mistake and so universally and blatantly misrendered. I have yet to hear a single recording on which the main theme heard near the beginning of the movement (mm.9–16) is played as written.

---

5. In fact, playing m.229–36 really *pp* but in full fast tempo has its own wonderful dramatic excitement, which the slowing-down interpretation can never achieve.

The usual rendition, successively in oboes, trombones, and horns, is

[musical notation].[6] Do not Slatkin and all the other famous
maestri who have the privilege of recording this great symphonic masterpiece
hear that their musicians are playing something other than what the composer
wrote? Evidently they don't, or else they don't care.

The second question—a much more interesting one—involves bowing in the
strings in m.97 (also mm.101, 105, 109–10). The score indicates successive
down-bows: [musical notation] and [musical notation]. If played
with that bowing, the actual acoustical result will not be [musical notation], but
[musical notation] or [musical notation] or [musical notation], depending on the
degree of laziness with which the performers approach this passage. (The 'retak-
ing' of the bow and the maintaining of the sound for the longest possible duration takes a lot of extra effort and concentration.) However, when the same
passage is recapitulated later on in the brass and woodwinds (mm.253, 257,
261), there obviously being no bowings involved, the half-notes on the face of
it [musical notation] are meant to be fully sustained. This is in my view
further confirmed by Tchaikovsky a few bars later when, developing and expanding upon this thematic fragment, he changes the rhythms (durations) unequivocally to [musical notation], maintaining that rhythmic version for an
entire six measures (mm.265–70). Clearly, Tchaikovsky meant to differentiate
between these two settings. But that being the case, how do we reconcile the
fully sustained half-notes in the brass with the impossible-to-sustain half-notes in
the strings earlier on? Did Tchaikovsky expect to hear a difference in the two
renditions? Or did he expect the brass to emulate the strings by shortening the
half-note durations? That makes little sense in view of his own varying of the
rhythm in m.265. It is hard to imagine that Tchaikovsky did not realize that it
is physically impossible fully to sustain those half-notes with successive down-
bows. Or did these down-bow markings originate not with Tchaikovsky, but with
the editor of the first publication?

There are no easy, obvious answers to these questions, and therefore no easy,
obvious ways to resolve the performance problem involved. My own suggestion
is to abandon the successive down-bow idea and bow the passage as it comes
(⊓∨⊓∨⊓∨); or, at the very least, conductors who wish to retain the special articulative feeling of the all-down-bow version, should exhort their string players to
not take full bows with each stroke. For if they do, and the more they do so, the
bigger the gaps between the individual notes will perforce become. Conversely,

6. Note that Tchaikovsky presents this theme in various durational alternatives, [musical notation]
[musical notation], for example, including [musical notation] (mm.214, 216), obviously wishing to
distinguish and discriminate between all these alternate ♩ versions.

the less length of bow one takes, the less time it will take to 'retake' the bow. And here one-tenth of a second makes a big difference. Nor should the strings play with all their force and might—the passage is marked (only) *ff*, with *fff* and *ffff* still in the offing.[7]

The just-mentioned brass and woodwinds passage (mm.265–70)—and its close relative, mm.301–08—is my third example of Scherzo passages that require special care and attention. As already noted, Tchaikovsky writes

| ♩ ♪ ᵧ ♩ ♪ ᵧ |. This is actually a relatively rarely used notation; most composers, for convenience's sake and to save time, write | ♩. ᵧ ♩. ᵧ |. However, the latter notation, although identical in meaning to the alternative version, is, for reasons I do not fully understand, usually played very sloppily, am-

biguously, imprecisely. The most common renditions are ♩. ᵧ and ♩.. ᵧ. Interestingly, if the pitch changes on the eighth-note, most players will, of course, make that note change audible, If, however, the pitch remains the same on the eighth-note (as in our Tchaikovsky example), it is usually dropped dynamically, losing whatever harmonic *and* rhythmic impact it is intended to have.[8] It is even worse when composers write ♩. That Tchaikovsky went to the trouble of writing | ♩ ♪ ᵧ twelve times in eighteen different parts means to me that he meant the tied-into eighth-note to be clearly heard harmonically and rhythmically. (The only notation I know to make that clearer is ♩ ♪ or ♩ ♩.) In any case, if so played, the passage in question takes on a tremendous firmness, a surging power, that the rather lame ♩. ᵧ version can never achieve.[9]

The Finale of the Sixth Symphony is full of performance challenges, especially in respect to tempo and dynamic questions. Taking the former first, it should be perhaps sufficient to reiterate that Tchaikovsky here marks his score with remarkable detail and precision, a subtlety of notation that one will find in very few (if any) 19th-century composers until the advent of Strauss and Mahler. In a mere nineteen score pages (comprising 171 measures) there are, apart from

---

7. Furtwängler and his Berlin Philharmonic strings manage this problem superbly, using as little bow as possible with a moderate *ff*.

Furtwängler's recording of the Tchaikovsky Sixth Symphony is an altogether astonishing achievement, in many ways one of the very finest recordings ever made. Although one cannot agree with some of his interpretive ideas, especially in regard to tempo, he presents his interpretation in such a convincing manner and with such a magical sense of line and continuity, that one is irresistibly drawn to his conception. But above all, it is the superb playing of the orchestra that makes this recording such a magnificent listening experience. The playing rivals that of the best American orchestras of the time (New York, Philadelphia, Boston), a fact difficult to contemplate and digest when one realizes that by 1938 the personnel of the Berlin Philharmonic consisted, in the majority, of Nazi Party members.

8. I have already referred to this problem in connection with similar rhythmic/articulative questions in the last movement of Beethoven's Seventh Symphony.

9. Similar performance/notational problems occur in mm.323–26.

the initial *Adagio lamentoso*, twenty-nine tempo markings (tempo changes, tempo modifications). To say that these, including the movement's basic tempo ♩ = 54, are generally ignored or rather arbitrarily treated is to understate the matter considerably.

It will be hard to convince the reader (and my conductor colleagues) that Tchaikovsky's tempo indications, inclusive of his metronome markings, are wonderful and perfect, because I cannot point to any recordings except (to some extent) by Gielen and Rozhdestvensky, in which the indicated tempos are respected and adhered to. I only know from my own performances of the work — and incidentally the positive (and amazed) reactions of the musicians involved — that Tchaikovsky's tempo markings work perfectly well — in fact, *really* perfectly.

As for the dynamics, ranging in this movement from one *ppppp* (m.146) to a number of *fff*'s: they also "work perfectly." Indeed, they give the music, with all its expression of anguish, sorrow, and resignation, a pliancy and nobility, a humaneness, that in more crudely undifferentiated performances simply cannot come to realization. Besides the normal control of dynamics that any great music requires of us performers, in the Sixth Symphony we have to be even more careful not to overplay the *f*'s and *ff*'s, for example — they should not incur the loudest possible playing — while the *p*'s and *pp*'s should have enough body and substance to allow for still softer and more refined lower dynamics.

Thus the very opening of the Finale with its remarkable, at the time very innovative, voice crossings in the strings,[10] need not be played 'as loud as possible,' but rather with a full-bodied, unforced sound and an intense *inner expression*. Other points for caution are mm.39, 47, 54, and 63 — *pp*, *p*, *mf*, and *f*, respectively — important expressive differentiations I have rarely heard made correctly.[11]

The entire opening passage (mm.1–18) is, of course, one of the most profoundly expressive moments in 19th-century music. It is also remarkably 'modern' in conception, not only in the twisting chromatic harmonies of m.1 and the plangent dissonances in mm.5, 7, 9–12, encircled in Ex.5. It is essential, in order to give these dissonances their full expressive meaning, to maintain an absolutely equal dynamic balance between strings and woodwinds. This can be achieved by the strings not being louder or heavier than the woodwinds and by the latter sustaining their suspension notes (the encircled notes in Ex.5) at the full required level. I also feel that, although the strings appear to have the dominant voice at first, by measure 5 the roles begin to change and the woodwinds gradually become primary, the strings gradually accompanimental. My reason for suggesting as much is that the woodwinds' rising line takes on increasing melodic character, until at m.12 it clearly dominates as the primary thematic material. Furthermore, in two of the three recapitulations of this subject, it is

10. Interestingly, in the otherwise exact recapitulation of these opening measures in m.90, Tchaikovsky relinquishes the voice crossings and uses instead the more conventional parallel voicing.
11. The annotation *con lenezza e devozione* is partly in error. There is no such word in Italian as *lenezza*; it was probably meant to be *lentezza* (slowness).

Ex. 5

again the winds (bassoons and horns respectively) that dominate melodically-thematically.[12]

Last, I should like to point to the remarkable dark, tortured harmonies in every alternate measure in the last sixteen bars of the movement, all the more anguished for being set in the lowest orchestral register (see Ex.6). The appropriately melancholy, gloomy mood of this most extraordinary ending (of a symphony) depends not only on rendering it in the right (Tchaikovsky's) dynamic levels, including its moderate, deeply expressive, singing *sf*'s, but also a dark sonic coloration in all the instruments as well as letting the harmonic dissonances (marked x in Ex.6) assert their most anguished expression.

While there is much, much more that could be said about the Tchaikovsky Sixth Symphony's interpretive problems and demands and how this great work has been cheapened and bowdlerized by generations of conductors in the indulgence of every possible *Effekthascherei*, the recounting of it would merely reiterate so much of what has already been set forth.

12. I recall when conducting the Sixth Symphony some years ago with the New York Philharmonic that some of the string players (especially in the first violins) were deeply offended at the suggestion that they give up their 'primary role' to the woodwinds around m.5.

Ex. 6

Even the few points I have raised in this brief discussion of the Sixth Symphony ought to awaken in us a deep concern for respecting Tchaikovsky's meticulous attention to detail, especially in regard to tempo and dynamics. To put it another way, how dare we question—and reject—what this great master wrote in the last anguished months of his life!

It can only be hoped that the overwhelming factual (not anecdotal) evidence presented herein will arouse conductors and performers of all stripes to rededicate themselves to serving, rather than using, the art of music—espousing the notion and principle that a great composer's creations ought to be inherently respected and cherished. Perhaps one could then supplant the motto "nobody gives a damn about the composer" with the more benign, gracious—and simple—"all for the composer."

Postscript

If the reader has faithfully followed me through the foregoing analyses of the recordings of eight major works of the symphonic repertory, he/she will have had to come to the sad conclusion (as I have had to) that in the cumulative we have hundreds (if not thousands) of recordings and performances whose main and common characteristic is that they ignore virtually all the basic information contained in the scores of the great composers. As orchestras have become during this century—since the advent of symphonic recordings in the teens and twenties—in increasing measure technically proficient and flexible, the over-all artistic result is the paradox of a plethora of incorrect, willful, *imperfect* interpretations, for the most part technically *perfectly* performed. The consummate skill with which orchestras, especially nowadays, follow and realize any interpretation, no matter how aberrant or unmusical or incorrect, is nothing short of amazing. It is also saddening. For it means in turn that orchestras have little or no interpretive say in the re-creation of a work, given the conditions of their employment, and that they for the most part don't care—or probably also don't know anymore—what a correct, appropriate, respectful-of-the-composer realization might be. Except for the occasional personalization of a solo passage by, say, a wind soloist or the concertmaster, orchestra musicians dutifully follow the conductor's every whim and wish, no matter how right or how wrong or illogical.

Given the hundreds of examples presented herein of incontrovertible, detailed evidence of how conductors mis-interpret, over-interpret, under-interpret, the great works of Beethoven, Brahms, Strauss, Tchaikovsky, Ravel, Schumann—and by extension the rest of the classical-Romantic repertory—the reader may now understand why I was originally tempted to entitle this book "Nobody Gives a Damn About the Composer," a title, I am sure, considered by most readers at first glance quite outrageous and off-putting. But if anything, the general knowledge I have had for a long time that conductors, by and large, tend not to know or choose to ignore what is actually stated in a composer's score, was painfully confirmed in the course of writing this book; and I now find my original title postulation to be rather mild. It hardly does justice to the situation and the actual state of affairs.

The major problem remains the same: the average audience, even many music professionals and 'experts,' has little or no idea what, say, a Brahms score *really* contains, what it *really* says. As a result, if the orchestra plays a performance technically well—meaning essentially no obvious wrong notes—an audience is likely to assume (how can it do otherwise?) that what they are hearing corresponds to what the composer wrote, especially if they have gotten some emotional charge from the performance. (As I have explained many times in the course of this book, the emotional charge an audience may get from a given performance may not be one intended by the composer, and may be quite extraneous to the music, may indeed be a substitution by the conductor.) The audience, the average listener, is left only with the ability to measure a performance by a sense of whether it was 'exciting' or 'not exciting.'

The writing of this book has been, of course, a tremendous, if somewhat depressing, learning experience for me. And how I wish I could have instead written a book filled with praise of conductors, of orchestras, of their music-making, and of their interpretations! On the one hand, even I became shocked at the depths of musical/interpretive debasement so many recordings revealed—including many recordings I had somehow not previously encountered. Even I was not prepared for the amazing range of misinterpretations that conductors are capable of.

On the other hand, I came to two interesting—and rather positive—realizations: 1) that by and large 20th-century music, especially newer contemporary music, is performed much better and much more respectfully than are the 19th-century classics; (2) my original sense that the conductors of the earlier part of this century—the conducting superstars of the 1930s, '40s, and '50s—were in general inherently and consistently superior to the reigning conductors of today is a notion I no longer find tenable. Indeed, many of my former heroes—Furtwängler, Toscanini, Klemperer, Walter—have fallen from their high pedestals on which, as a young musician, I had placed them. In light of the close, critical scrutiny to which I have subjected their recordings—as close and critical as I have scrutinized the work of today's conductors—only now substituting precise analysis for my earlier more generally impressionistic reactions, many of the 'great ones' of the past are revealed as being prey to the same foibles and weaknesses as the present generation(s) of maestri—or perhaps even more so.

Some will argue that evaluating the work of conductors solely on the basis of recordings gives a false picture of their actual quality as musicians. There is a kernel of truth in this, although only in respect to *some* conductors and, more important, not enough truth to negate the whole idea that official, commercial recordings are in the vast majority of cases accurate representations of a conductor's basic interpretation.[1] Also, some of the more famous conductors, past and

1. The recordings of Furtwängler are perhaps a case in point. It is surprising to realize how few studio recordings Furtwängler made in his lifetime, given his extraordinary fame. The majority of

present, have had a chance to record many of the major staples of the repertory several times, in the case of Karajan, Solti, Bernstein or Stokowski, for example, four or five times. To argue that none of these recordings are representative of their interpretive intentions is disingenuous.

Older readers with fond memories of Toscanini or Furtwängler or Walter performances and recordings may themselves be rather surprised to find in revisiting their recordings of the 1930s to '50s that these are not entirely unflawed, and that the nostalgia of the 'good old past' may indeed cloud their judgment. There is also the fact that in the concert hall most listeners are more intent on *watching* a conductor than just *hearing* the music. That, however, is what one is per force obliged to do in listening to a recording. I have often wondered why people are so drawn to watching a conductor, as if there were over a span of a forty-minute symphony something all that exciting to watch. The world would be a much better place—and conductors would not be seen as such god-like mythological creatures—if audiences didn't watch the conductor in such a mesmerized fashion, if they closed their eyes or bowed their heads and just *listened* to the music. Granted, Toscanini and Furtwängler were rather fascinating to watch, each in his totally different way, but I maintain that at the same time watching them could easily be distracting from fully, clearly hearing the music. The fact is that the average audience will hear mostly—or perhaps only—that to which the conductor is specifically, visually addressing his attention.[2]

In any case, I don't think the early decades of this century had some kind of monopoly on 'great conductors,' and that we are now in an era uniquely bereft of major conductorial talents. There is only more competition in the field today, because there are many more people conducting now than there were in the past; and many fine conductors are for one reason or another not in major positions with major orchestras and major recording contracts.

As for the point that the modern orchestral literature is played infinitely better than the big 19th- century repertory, I had a dim sense of this some time before writing this book. But listening to the hundreds of recordings that comprised the main research in this effort, and comparing them with a similar number of

---

Furtwängler's recordings available today are not studio recordings but tapings of 'live' performances, in many cases poorly or inadequately recorded, and/or not necessarily representative of his best work. His finest recordings are in fact some of his 'studio' recordings (Tchaikovsky's Sixth, Schubert's "Great" C major, Beethoven's Fifth), but it is known that he himself was dissatisfied with a number of others.

A similar situation exists with the recordings of Mitropoulos and DeSabata, with both of whom there are more 'live' concert performances than 'studio' recordings. And there is, of course, the special case of Celibidache, who has, since the early days of his career, refused to make any commercial recordings, which, however, hasn't prevented a flood of pirated recordings of his concert performances from being issued.

2. Television and the televising of concerts, that is, the pre-arranged, pre-determined visualization of performances, has made matters even worse. For now it is the television director who is directing and controlling the watching audience's attention to limited, specific, pre-selected aspects of a piece, not to others and not to *all* of the piece (except perhaps in certain camera long shots). And it is well known that the average listener hears mostly, or only, that which he sees.

recordings and performances of contemporary music in my regular life as a very active conductor, composer, and publisher (of new music), this point was unequivocally substantiated. The reason—or reasons—for this is actually quite simple. First of all, most of the famous maestri who travel all over the world with their big recording contracts and their memorized repertory of about thirty pieces don't touch any music of a certain modernity or complexity. They certainly avoid any contact with most music since 1945, unless it is an occasional easy-to-listen-to neo-classical or neo-Romantic piece, especially if it is scheduled to be a world premiere. By and large, they avoid like the plague the repertory of the Second Viennnese School and any of their stylistic offspring, as well as any truly complex and challenging modern music of any other school.[3] For them such music is indeed a plague, as it is also for most of their audiences, who are usually grateful to their music director if he doesn't do too much "of that modern stuff"—better if he does none.

The exceptions to this, like Abbado, Mehta, Dohnanyi, Levine, Slatkin (among a few others)—in earlier days the grand exceptions were Mitropoulos, Koussevitzky, and Stokowski—intelligently blend a fair amount of new and 'difficult' music into their programming. They do this partly out of conviction for the validity of the best of the contemporary repertory, and partly out of a sense that they owe it to their audiences and the young generations of musicians coming along, to inform them of the latest and best of the new music and its direct early 20th-century antecedents.

Except for these few just-mentioned valiant souls, the field of contemporary music—'contemptible music,' as some wags would have it—is left to the 'contemporary music specialists,' to those relatively numerous mostly younger conductors who labor in the orchestral minor leagues or who lead the hundreds of contemporary music ensembles, and who not only don't have the big recording

3. Bruno Walter's rationale for not conducting contemporary music is as lamentable as it is widespread among conductors. Walter, like so many of his colleagues, hides behind a series of specious arguments, which never even adjoin the real issue. Trying to justify his avoidance of new music, especially in his later post-1930 career, Walter invokes—rather self-righteously, it seems to me—ranking "my responsibility towards the art, to which my life is dedicated, higher than my duty towards the present; and it would have been wrong for me to make myself the advocate of tendencies which, I am convinced, are leading to its corruption." Ignoring the remarkable achievements in the realm of new music of conductors such as Koussevitzky, Reiner, Goossens, Stokowski and Mitropoulos, and blaming "atonality and dodecaphony" for causing "the decay of music," he indicts such music of "abstractness" and of being "divorced from life." He accuses composers of this persuasion of "lacking in fundamental musicality," substituting constructivism, intellectualism, "the artificial for the artistic," turning "from music to non-music." Zigzagging back and forth between a general condemnation of new music and disingenuous protestations of his "feelings of responsibility towards and profound interest in the present," he feels that he "could not bear to witness [music's] abuse, still less abet it." He paints the present as merely "transient," invoking his need to remain "mindful of the *unchanging* demands of yet higher powers" (Italics mine). Finally, in a holier-than-thou pose he claims that "in spite of my attempts to do justice to the present, I always endeavored to lift my eyes up"—one can see his halo—"above the distressing contemporary scene towards the region whence come to us those timeless sources of strength." (Walter, *Of Music and Music-making* (New York, 1957), pp.206–11.)

contracts but who will never get them from the Sony's, BMG's, EMI's, Telarc's, Philips's, etc., precisely *because* they do "too much modern music." They are then also presumed to have no abilities in the classical repertory. But these conductors have, by and large, a keen respect for the composer's score, realizing on the one hand that composers today for the most part are meticulously precise and conscientious in notating their works—Mr. Dohnanyi to the contrary—and on the other hand that, there being no tradition (bad or good) or previous knowledge of the work to fall back on, their only salvation is to do (as best their talents allow) what the composer has written.

I find that even the 'world-famous maestri,' when for some odd reason or other they are forced to venture forth into the really modern repertory, become all of a sudden very respectful of the composer's score—again to the limits of their abilities (which limits are often rather circumscribing). Their attitude suddenly changes; even their conducting, i.e.., their baton technique, becomes suddenly simpler and clearer, and less exhibitionistic, less directed towards audience pleasuring. As opposed to assuming that they know *everything* about the Beethoven Fifth or Brahms First, for the rare and occasional difficult new piece—I don't mean Barber's *Adagio*--they usually realize that they know *nothing*, and in a really new piece actually can, of course, have no prior knowledge or experience. Their only salvation, then, is, as best they can, to do what it says in the score. Suddenly Toscanini's *come è scritto* takes on real meaning for them, and becomes a kind of professional/philosophical life-raft.

In any case, I have noticed for years that the music of the early 20th century, say, the fine works of Bartók, Prokofiev, Milhaud, Honegger, Stravinsky (except for *The Firebird*), Walton, Nielsen, Szymanowski, Respighi, and the like—even late Debussy (like *Jeux* and *San Sebastian*)—are conducted with infinitely more deference and accuracy than any Beethoven or Brahms or Tchaikovsky symphony ever receives. And the more recent new music (of the last fifty years) is even more respectfully treated.[4] Again the reasons are quite simple. Most of the music of the 20th-century has not yet been betrayed by false traditions and false interpreters. What all this ironically comes down to is that the more a work is played, the more familiar it is, the more popular it becomes, the more likely it is to be bastardized and vulgarized, and its composer's intentions disregarded. The new challenging music of today has not yet had the misfortune of acquiring such an 'elevated' popular status.

There is another interesting, though disappointing, revelation that has emerged very clearly for me from the work and research on this book; it seems

---

4. It was not always thus, of course. I can easily recall that even fifty years ago, in my youth, performances of contemporary music—of almost any ilk—were quite dismal and inept. Not only were most conductors unable to grapple with the new music of the time, but many orchestral musicians also had great problems with it, from a technical point of view as well as dealing with the newer complex 'irrational' rhythms, the extravagant instrumental gestures and wide-ranging lines, with intonation problems, in short, with a whole spectrum of new musical concepts. Since then, of course, there has been a remarkable revolution in these respects, and most professionals today take all forms of contemporary demands pretty much in stride.

that the more explicit composers are with their notation, the more expressive marks they use, the more precisely they try to monitor the structure and continuity of a piece, the more room for error, for negligence, there seems to be in carrying out these intentions/instructions. I sometimes get the impression that, paradoxically, Bach's music, with its minimum of performance information— relatively few and rather imprecise dynamic indications, rarely even any tempo markings—is generally better served in performance than Beethoven or Brahms. Am I on slippery ground when I say that? Yes, of course, for we may never really *know* explicitly how Bach intended his music to be performed. We *believe* we have come a considerable way in recent times in understanding his music and grappling with its performance practice. But such information still remains rather conjectural, which—then—allows interpretive imaginations—healthy, strong ones—to flesh out the bare notation in myriad and often very compelling ways. And who is to say that one or the other Baroque interpretation/realization is wrong, is bad? Well, no one—really.

But when it comes to post-Mozart/Haydn literature with its much more precise notational paraphernalia, we can begin—rightly so—to invoke the composer's text as a specific guide to a certain kind of interpretation or realization— not one interpretation, not a definitive one (there is no such thing), but an interpretation which operates within certain prescribed/described limits. The problem here is that, while Bach forces you to use your best intelligence and musical imagination—failing that, performances of Bach are really deadly—Beethoven specifically challenges you as a re-creator in ways, as the record shows (pun intended), most interpreters are incapable of meeting, for all the reasons given in this text. Another way of stating this is to say the margin for error, neglect, abuse is much greater because it is more measurable—measurable against the more prescriptive text.

As I have stated, most 20th century music (say, Prokofiev, Bartók, Britten, Copland, Carter, et al.) is ultimately performed more accurately—at least respectfully—than the great literature of the 19th century. That is not to say that there isn't room for improvement on that front (especially the more modern, complex atonal literature); but as a general truism it is valid and stands up to close scrutiny. My argument on that point has been not so much that therefore there is more understanding of contemporary music—in specialized contemporary ensembles that is sometimes the case, but it cannot be broadened to an industry-wide generalization—just that newness and unfamiliarity forces a certain degree of respectfulness and blind reliance on the text, at least technically, while over-familiarity with the classics breeds a kind of willfulness, arbitrariness, interpretive self-indulgence, ego-driven interference, that is deeply disrespectful and damaging.

What is the answer to these problems? A group of answers come to mind: (1) a better, more thorough, more discriminating education—for performers as well as composers; (2) a better more painstaking, detailed education of composers in the meaning, the effect, the ramifications, and the possibilities of notation. Too many composers today know too little about the vast field of musical notation, or on the other hand are hell-bent on inventing new notational devices and/or systems, ignoring or erroneously duplicating what has already been given and used for decades, re-inventing the wheel as it were; (3) higher industry standards, more artistic integrity and honesty in the field; a more discriminating, (again) better educated, more culturally literate audience and critical fraternity—goals and dreams of which we have fallen severely short in recent times.

I should also add at this point, for the benefit particularly of those readers whose 'favorite conductor(s)' I have had to treat with something less than praise, that almost all the conductors referred to in this book have made both very good and very bad recordings—by the way, to reiterate my earlier point, only rarely poorly *played* by the orchestras. For it is a reality that almost all conductors are weak in some repertory areas and strong in others.[5] I also have not been able to discover among any of the finest conductors any who are (or were) entirely artistically consistent, either in their approach to certain works or across the range of their repertory.[6] There are many conductors who may be weak or feel uncomfortable in the classical German repertory, but who are quite excellent in the brilliant, splashy early 20th-century orchestral repertory (Ravel, Bartók, Prokofiev, etc.). And then there are their opposites. Thus, I would venture to say that if I criticized negatively a reader's favorite conductor hero in, say, a Beethoven or Brahms recording, the chances are that that conductor may have done a

5. A striking example of this kind of artistic dissociation is the case of Mengelberg, whose work does not come off very well in this book but whose recording of Strauss's *Ein Heldenleben*, made in 1928 with the New York Philharmonic, is unquestionably one of the greatest, most beautifully 'interpreted,' most perfect recordings ever made.

6. I am tempted to exclude from this list of fine but inconsistent conductors a few like Roger Désormière, Lovro von Matačić, Issay Dobrowen, Ataulfo Argento. But I may be on slippery ground here, for these conductors and others mentioned below either recorded relatively little—too small a sampling to make a fair judgment—or recorded none (or little) of the big major repertory, such as that dealt with in this book, by which ultimately any conductor must finally be judged.

This is a good moment to mention that I wish that I could have brought into the discussion a number of other fine conductors, who, like the four names mentioned directly above, recorded none of the eight works under consideration herein. Apart from Désormière, Matačić, Dobrowen (the latter two among other things superb accompanists), and Argento, my list would include Fritz Stiedry and Jonel Perlea (both of whom I worked with at the Metropolitan Opera), Bohdan Wodiczko, Walter Straram, Karl Muck, Sigmund Hausegger, Oswald Kabasta, Paavo Berglund, Vaclav Talich. It would also have been fascinating to include Henry Wood and Landon Ronald, two turn-of-the-century English conductors, who in the very earliest days of recordings contributed significantly to the establishment of criteria and higher standards in the performance and recording of classical music. Finally, in the category of more or less 'obscure' conductors, as far as popularity and eminence goes, there is Ernst von Schuch, a highly respected, even revered, conductor in the early years of this century who, however, never recorded at all.

Many readers will have noticed that there is no mention of women conductors in this book. The reason is simple: I know of no women conductors who have recorded (commercially) the eight works under discussion here, with the exception of Victoria Bond, who just recently has recorded Tchaikovsky's Sixth Symphony with the Shanghai Symphony Orchestra. Having had a number of talented women students in my conducting classes at Tanglewood and in my summer Festival in Idaho, I have no doubt that women conductors—many of them are as good or better than their male counterparts—will in the future occupy prominent conducting positions and be contracted by record companies to record the major works of the repertory.

superb recording of a Walton or Vaughn Williams symphony, or a Ravel *La Valse* or a Rimsky-Korsakov *Scheherezade,* or whatever—works that I happen not to have discussed in this book.

By the same token, the many musical interpretive misdemeanors committed by conductors described herein represent only a tiny sampling of what is perpetrated across the width and breadth of the entire repertory. What occurs time and time again in the performances of the eight works discussed in Part III can also be heard in the several hundred other works that round out the standard symphonic repertory. In other words, the examples cited herein must stand for all the others in the rest of the repertory; they are certainly *not* limited to the eight works discussed here. For the reader will have noted that the performance problems cited in this account are consistently the same and fall into a few broad categories: respecting tempos, dynamics, the full value of rhythms, details of articulation and phrasing, and respecting the meaning and feeling of meters and time signatures. But collectively and briefly stated, they all come down to a basic disrespect for and disloyalty to the score. And they are all a matter of attitude and of ethical conscience, not necessarily of capability. To put it another way—simple and dramatic: if these matters were attended to more respectfully and precisely by conductors and orchestras, this book would never have had to be written.

This is analogous to the thought I have often had that if musicians and conductors could be counted on to render accurately the first time—or at least the second time—what composers have written, there would be no need for extensive rehearsing. One could get almost immediately to the heart of the matter, namely, the essence, the true expression of the music at hand, and eliminate all those second, third, and fourth rehearsals most orchestras have per week, in which so much time is spent in dealing with merely technical matters and weeding out bad habits (and bad traditions). This way one could even make the daily life of orchestral musicians more enlivening, more inspiring, more resistant to the potential doldrums of day-in, day-out, year-in, year-out routine.

Am I assuming too much? And what about the technical difficulties contained in the music? What technical difficulties? I can hardly think of a single work in the standard repertory, up to and including the early 20th-century 'moderns,' which presents technical problems to the musicians sitting in our orchestras, especially the orchestras of today. There is nothing in any Beethoven or Brahms or Tchaikovsky or Dvořák symphony or Strauss tone poem, just to mention a few examples, that any musician in, say, the Chicago Symphony or Berlin Philharmonic or Concertgebouw Orchestra, or even the Cincinnati, BBC, Stuttgart, NHK, and Reykjavik symphonies cannot instantly render with technical ease. They have long ago—dozens if not hundreds of times—conquered those technical problems that perhaps still gave musicians trouble a hundred or even fifty years ago.

I personally even know—and have known in the past—quite a few musicians who play(ed) their individual parts perfectly, technically *and* musically, every time, especially in the standard repertory. And I know there are one or two such musicians in every orchestra worth mentioning in this context. And if they are not even remotely the majority of musicians, they are sufficient in number to prove that the totally accurate and respectful rendition of a given orchestral part is absolutely within the realm of possibility. (Such musicians occasionally even manage to play their parts absolutely correctly and perfectly *in defiance* of their conductors—a remarkable and laudable accomplishment, if one thinks about it and what that entails.) No, clearly, the problems with the standard repertory do not lie in the realm of instrumental technique and virtuosity, with digital and embouchuric skills. They *do* reside in all the other musical, interpretive, intellectual aspects of musical re-creation, including the fact that the reading of music, familiar or unfamiliar, by most orchestra musicians has been turned in recent decades into a *visual* rather than an *aural* skill. That is to say, musicians nowadays read music technically with a speed and note-accuracy unheard of, say, fifty years ago: definitely an enormous gain. However, this has been achieved at an enormous loss, for, at the same time, most musicians no longer hear harmonically, nor with any sensibility or accuracy regarding dynamics, let alone (all but the most obvious) ensemble aspects. There is now almost everywhere a singular (and impressive) ability to sight-read fast, to concentrate on 'getting the notes,' and getting them more or less technically 'perfect,' *but at the expense* of hearing any other aspects of music re-creation: harmony, dynamics, instrumental and sonoric balances, all manner of musical/structural interrelationships. Most players are little islands unto themselves—in the vast sea of the orchestra—concentrating solely on their own parts (linearly/technically), with virtually no interest in, curiosity about, or knowledge of what is going on in the rest of the orchestra around them, in voices above and below them, in what precedes or succeeds them. No wonder that in this challenge-less process and routine, many musicians get bored—*are* bored—and find life in an orchestra somewhat less than enlivening and inspiring. For all their technical prowess, most orchestra musicians see and hear only the merest surface of the music.

Add to that a plethora of bad or lazy habits, among which I list most prominently: (1) dropping last notes (rhythmically and/or dynamically) before a rest; (2) similarly, not sustaining long or held notes (especially string players), always making little holes (caesuras, *Luftpausen*) in phrases; (3) a general lack of attention to dynamics, above all to differentiations between, say, *p* and *pp* or *f* and *ff*; (4) rhythmic inaccuracies, and very little attention to the subtler details of rhythmic differentiations; (5) a remarkable unconcern for how one ends (leaves) a note, much attention being given to how one enters or attacks a note, but almost none to how one exits a note; and, finally, (6) making crescendos (as well as diminuendos) much too quickly and too early.

But, ultimately, the real problems lie not with the musicians but with the conductors. For it is they who have the obligation and responsibility to not let bad playing habits—of whatever kind—become the performance norm. And beyond that, it is they who finally shape and control a performance (and recording). It is they who either elicit from their musicians erroneous misinterpretations or tolerate the bad musical habits many musicians tend to acquire in their life in an orchestra, or, alas, tend to bring with them to an orchestra for lack of a proper and complete musical training.

The conductor's role, his mission, as it were, ought to be—and it so rarely is—to interpret accurately and respectfully, i.e. re-create the works they have chosen to conduct, and at the same time teach the musicians, to the extent that it becomes necessary, to treat their work with the same accuracy and respect.

And so we have come full circle to the sense of humility, love, and respect which conductor and musicians should bring to their respective tasks, the thought with which I began many, many pages ago. For I truly believe that the musical interpreter—conductor, instrumentalist, singer, whoever—is given a unique privilege: that of re-creating and re-producing the masterpieces greater geniuses than we interpreters can ever be, have given us. And with that privilege and honor come a profound obligation and commitment to humbly, *faithfully* 'realize' those works, to bring them accurately to acoustical life, and thereby to serve the art of music. For if we musicians serve the art of music and its great creative visionaries with humility, in the deepest sense we also best serve ourselves and those who put their faith in us—our audiences and fellow human beings.

Afterword

As a final denouement of the complex, almost encyclopedic plottings, analyses, critiquings presented in this book, I want to offer one illustration that represents, indeed epitomizes, in a summary fashion so much of what I have been, alas, constantly forced to assert in this book. It is a recording, a performance, of Strauss's *Till Eulenspiegel* by the world-famous Concertgebouw Orchestra of Amsterdam conducted by Bernard Haitink. Clearly, Haitink is one of the very best conductors around today—innately musical, well-trained, intelligent, not given to ego-driven extremes of over-personalized interpretations, a musician of taste and balance—obviously at the zenith of a most distinguished and celebrated career. The Concertgebouw is similarly one of the great, venerable orchestras of the world, with its own distinguished history reaching back well into the nineteenth century

Their recording of *Till* under Haitink, made in the glorious acoustics of the orchestra's home hall, the Concertgebouw, is one that every music lover, record collector, Strauss aficionado would want to own—and probably does. The playing of the orchestra is magnificent, the sound of the recording state-of-the-art: clean, rich, beautiful, thrilling—all-in-all a splendid sonic feast. I can well understand how anyone hearing this recording on the radio would feel immediately compelled to want to own this CD.

The problem for me—and for anyone who *really* knows the score and knows in detail what Strauss's score actually contains—is that this technically magnificently rendered performance simply ignores and/or rejects many, many important compositional/notational aspects of the work. Leaving aside the earlier-mentioned embarrassing digital/electronic distortion caused by the overly loud cymbal crash—a minor split-second technical blemish, of ultimately little consequence—the 'interpretation' by the orchestra and/or conductor represents exactly the dilemma that has been, in effect, the main thesis of this book: the relative interpretational willfulness and lack of respect for the work itself, the text, which embodies both the spirit and the letter of Strauss's remarkable cre-

547

ation. Many dynamic, rhythmic, balance, ensemble details are blithely, care-lessly, that is, incorrectly, rendered—and then *superbly played*. Unfortunately superb playing of something that is wrong or imperfectly interpreted does not make it right or good.

With it all, Haitink's recording of *Till* is still comparatively one of the best, if we allow, in this context, for a somewhat restricted meaning of that word 'best,' which is also to say that there are many much worse recordings. But it is em-blematic of what ails this conducting profession, even at the higher/highest lev-els, and the closely related activities of orchestra performance and recording. And I only wish it were otherwise.

Discography

This discography represents all the recordings studied and analyzed in the writing of this book. They are listed as CDs unless otherwise noted: + for 78s, * for LPs. For convenience sake, I have listed quite a few recordings as CDs, even though I may have listened to the works in question on 78s or LPs (because of their superior technical, sonoric quality).

Discographical listings have become quite complex in recent decades,* not only because the number of record companies has multiplied a hundredfold, but also because many record companies have merged in recent years, while others have split into various subsidiary labels; still others have changed names several times—the former Columbia Records first became CBS and then, more recently, Sony. Furthermore, in the vast reissue programs of the last three and a half decades, initially from 78s to LPs, and later from LPs to CDs, each new format acquired new numbering systems. To make matters more complicated, during the LP era many of the most popular recordings were reissued several times, each time with a new catalogue number. Thus, in the case of the five Beethoven Fifth recordings made by Karajan between 1949 and the 1980s, many were not only reissued several times on LP, always with a new number, but additionally reissued (in both LP boxed sets and CD formats) in various collections, comprising, for example, several (or sometimes all) Beethoven symphonies. I believe that if one were to list every reissue of all five Karajan Beethoven Fifth recordings, the listing would fill an entire page.

In addition, many previously available records (LPs as well as CDs) are presently no longer in stock. Also very few record companies list and sell recordings in perpetuity. Most labels take records out of their catalogue after a certain time, either dropping them altogether or, if reissued, giving them new numbers.

*In 1941 Irving Kolodin's *Guide to Recorded Music* listed only eight recordings of the Beethoven Fifth Symphony. Some recent Schwann Catalogues have listed as many as one hundred. By my estimate this work has been recorded commercially nearly 170 times.

Under all these circumstances, if one wants to keep a discography brief, it is hard to know by which of many possible numbers one should identify a certain much-reissued work. Even the latest numbering may be obsolete in a few months.

All of this is to say that the catalogue number by which a reader knows a certain recording may not be the one listed in this discography. Further, it would have been pointless, wastefully complex, and exhausting to list all past and present editions, versions, and reissues of any given performance.

It should also be noted that not every recording listed here is necessarily mentioned in the text, although most, of course, are.

It would have taken even more exhaustive research to accurately establish in what year a recording was made. I have contented myself with dating (I hope accurately) only certain recordings made before the 1960s and recordings of historical importance.

BEETHOVEN: FIFTH SYMPHONY

Conductor	Orchestra	Record Label/Number	Date
Abbado	BPO	DGG 427306–2	
Ancerl	Czech Phil	Supra 111937	
Ansermet	Suisse Romande	*Lon STS 15464	(1960)
Ashkenazy	Philharmonia	*Lon 71040	
Beecham	Royal Phil	*EMI 7644 6523	(1951)
Bernstein	NYP	CBS MYK 36719	
Bernstein	NYP	*Col M 31810	
Böhm	VPO	DGG 439681–2	
Böhm	BPO	*Decca 9942	(1960)
Boulez	New Philharmonia	*Col M 30085	
Boult	Philharmonia	*Vanguard 359	(1960)
Brüggen	Orch 18th Cent	Philips 434087–2	
Cluytens	BPO	*Seraphin 6071	
Davis, C.	BBC	*Philips 6500.462	
DeSabata	NYP	Melodram	(1960)
Dohnanyi	Cleve	*Telarc 80163	
Dorati	Minn	*Merc 50017	(1957)
Dorati	LSO	*Merc 14016	(1963)
Ferenczik	Hung State Orch	*Danube 11457	
Fricsay	BPO	*DGG 18813	(1962)
Furtwängler	BPO	+RCA 426	(1938)
Gardiner	Orch Revolutionaire	Archiv 439900–2	
Giulini	LA	*DGG 2532049	
Haitink	Concertgebouw	Philips 420540	
Haitink	LPO	*Philips 9500067	
Hanover Band	- - - - -	Nimbus 5007	
Harnoncourt	Ch Orch Europe	Teldec 2292–46452–2	

Conductor	Orchestra	Record Label/Number	Date
Herbig	BBC	IMP 9123	
Hickox	North Sinfonia (England)	ASV QS 6054	
Hogwood	Acad Anc't Music	Oiseau 425 644–2	
Jochum	Concertgebouw	Philips 6570166	
Jochum	Bavar (Munich)	DGG 427195–2	(1960)
Karajan	BPO	*DGG 419051	
Karajan	Philharmonia	*Angel 35231	(1957)
Kempe	Munich Phil	*Seraphim 6093–4	
Kleiber, C.	VPO	DGG 415861–2	
Kleiber, E.	Concertgebouw	Lon LL912	(1953)
Klemperer	Philharmonia	*EMI Classics CDM 63868	(1957)
Kletzki	Czech Phil	Supra 110619–2	(1967)
Knappertsbusch	BPO	*Arkadia 723	(1956)
Koussevitsky	BSO	+RCA LM 1021 [*Cam 103]	(1944)
Krauss	VPO	*Koch	
Krips	LSO	*Yorkshire 27000	
Kubelik	Bavar (Munich)	DGG 2535407	
Leibowitz	Royal Phil	Menuet 160 019–2	(1962)
Leinsdorf	BSO	*RCA 7745–2	(1960)
Maazel	Cleve	*CBS M5K 45532	
Maazel	VPO	*CBS IM 36711	
Markevitch	Lamoureux	Theorema 121219	(1959)
Masur	Leipzig	Philips 434156–2	
Mehta	NYP	*Col 35892	
Mengelberg	Concertgebouw	*Philips 6767003	(1940)
Mitropoulos	NYP	Musica Viva 90031 (cassette)	(1957)
Monteux	LSO	*Lon 443479–2	(1961?)
Munch	BSO	RCA 6803–2	(1957)
Muti	Philadelphia	Angel CDC 47447	
Nikisch	BPO	*DGG 2721 070	(1913)
Norrington	Lon Classical	EMI 749816–2; 74956–2	
Ormandy	Philadelphia	*Col M 31634	(1957)
Ozawa	BSO	*Telarc 10060	
Paita	Philharmonia	*Lodia 781	
Previn	LSO	*Angel 36927	
Prince's Orch	- - - - -	+Col A 5422 (Andante only)	(ca.1911)
Reiner	Chicago	*RCA LM 2343	(1960)
Rodzinski	PSO London	*West WST 14001	(1948?)
Sanderling	Berlin Symph Orch	Capriccio 10018	
Sawallisch	Concertgebouw	EMI Classics CDC 754504–2	
Sawallisch	NHK (Tokyo)	RCA 60534–2	
Schalk	VPO	+HMV 7105	(1928)
Scherchen	PSO London	*West XWN 18310	(1957)
Schuricht	Conservatoire (Paris)	*Trianon 33.335	(1959)
Schwarz	LSO	Delos 3027	
Solti	Chicago	Lon 21580–2	

BEETHOVEN: FIFTH SYMPHONY *(continued)*

Conductor	Orchestra	Record Label/Number	Date
Solti	VPO	*Lon 41016	
Steinberg	Pitts	*SQN 145/7	(1957)
Stokowski	All Amer Youth	Stokowski Soc LSS CD 4	(1940)
Stokowski	LPO	*Lon 430218–2	(1974)
Stransky	NYP	+Col A 5954 (Andante only)	(1917)
Strauss	Berlin State Opera	Koch 3–7115–241	(1928)
Suitner	Staatskapelle Berlin	Denon 8013	
Szell	Cleve	*Epic LC 3195	(1957)
Szell	Concertgebouw	*Philips 802769	(1967)
Thomas	Eng Ch Orch	*Col 37288	
Toscanini	NBC	*RCA LM 1757	
Toscanini	NBC	+RCA DM 640	(1951)
Van Otterloo	Sydney	*Chandos CBR 4001	
Victor Concert Orch	- - - - -	+RCA 18124 (Andante only)	(1920)
Walter	Col Symph	Col ML 5365	(1960)
Wand	NDR	RCA 09026–61930–2	
Weingartner	LPO	Music Memoria 30378	(1935?)

BEETHOVEN: SEVENTH SYMPHONY

Conductor	Orchestra	Record Label/Number	Date
Abbado	VPO	DGG 423364–2	
Ansermet	Suisse Romande	*Lon CM 9043	(1960)
Ashkenazy	Philharmonia	Lon 430701–2	
Barenboim	BPO	Sony SK45830	
Batiz	LSO	Varese 1000–160	
Beecham	Royal Phil	Music & Arts CD-281	(1959)
Bernstein	NYP	*Col ML 5438	
Bernstein	VPO	*DGG 2740216	
Böhm	VPO	*DGG 437928–2	
Boult	Promenade (Lon)	*Vanguard 2005	(1960)
Brüggen	Orch 18th Cent	Philips 426846–2	
Cantelli	Philharmonia	*EMI 2D2MB 88217	(1956)
Casals	Marlboro	Sony SMK 45893	
Celibidache	Stuttgart	Arkadia CDGI 737.1	(1964)
Cluytens	BPO	Angel 35526	(1960)

Conductor	Orchestra	Record Label/Number	Date
Collegium Aureum	- - - - -	*Pro Arte 123	
Davis, C.	LSO	*Philips 9500219	
DeBurgos	LSO	Collins 30192	
Dohnanyi	Cleve	Telarc 80163	
Dorati	Minn	*Mercury	
Ferencsik	Hungarian State Orch	*Qualiton 11791	
Fricsay	BPO	*DGG 138757	(1962)
Furtwängler	BPO	*DGG 427775–2660	
Furtwängler	BPO	*Fonit Cetra (Arkadia) FE 4	
Furtwängler	VPO	*Electrola 90016	(1950)
Gardiner	Orch Revolutionaire	Archiv 439900–2	
Giulini	Chicago	*Angel 36048	
Haitink	Concertgebouw	Philips 420540–2	
Haitink	LPO	*Philips 6747307	
Harnoncourt	Ch Orch of Europe	Teldec 2292 46452–2	
Hogwood	Acad Anc't Mus	Oiseau 425695–2	
Jochum	Bavar (Munich)	*Quint 7128	
Jochum	Concertgebouw	*Philips 7505010	
Karajan	BPO	DGG 415 121	
Karajan	Philharmonia	*Angel 35005	(1957)
Karajan	VPO	*RCA LM 2536	(1961)
Keilberth	BPO	*Tele 8040	(1961)
Kempe	Munich	*Seraphim 6093–6	
Kleiber, C.	VPO	DGG 415862–2	
Kleiber, E.	Concertgebouw	*Lon STS 15474	
Klemperer	Philharmonia	*Angel 35330	(1957)
Kletzki	Czech Phil	*Mus Her 169	
Krips	LSO	*Yorkshire 27000	
Kubelik	VPO	*DGG 2740155	
Leibowitz	Royal Phil	Menuet 160 020–2	(1962)
Leinsdorf	BSO	RCA 7997–2	(1963)
Maazel	Cleve	*Odyssey YT 42484	
Masur	Leipzig	*Philips 6570048	
Mehta	LA	*Lon CS 6870	
Mengelberg	Concertgebouw	*Philips 6767.003	(1940)
Monteux	LSO	Lon 443479–2	
Muti	Philadelphia	*Angel 37538	
Norrington	Lon Class	EMI CDC 749816–2	
Ormandy	Philadelphia	*Col 3ML 4011	(1957)
Paray	Detroit	*Merc 50022	(1957)
Previn	LSO	Angel 37116	
Previn	Royal Phil	RCA 7748–2	
Reiner	Chicago	RCA 6376–2	(1960)
Sanderling	Philharmonia	DuMaurier 5239	
Scherchen	V St Opera	*West 18319	(1957?)

BEETHOVEN: SEVENTH SYMPHONY (*continued*)

Conductor	Orchestra	Record Label/Number	Date
Schuricht	Conservatoire (Paris)	*Trianon 33337	(1953)
Solti	Chicago	Lon 425525–2	
Solti	VPO	*Lon 6093	(1960)
Steinberg	Pittsburgh	*SQN 145	
Stokowski	Philadelphia	+RCA 17	(1928)
Strauss	Berlin State Opera	Koch 3–7115–2	(1926)
Suitner	Staatskapelle Berlin	Denon 7032	
Szell	Cleve	Sony SBK 48158	
Tate	Dresden	Berlin Classics 1095	
Thomas	Eng Ch Orch	CBS MDK 44789	
Toscanini	NBC	*RCA LM 1991	
Toscanini	NYP	*Camden 352	(1936)
Walter	Col Symph	*CBS MK 42013	(1960)
Weingartner	VPO	+Col 260	(1938?)

BRAHMS: FIRST SYMPHONY*

Conductor	Orchestra	Record Label/Number	Date
Abbado	BPO	DGG 431790-2	
Abravanel	Utah	*Vanguard 10117	
Ancerl	Czech Phil	Supra 111941	(1962)
Anonymous	Anonymous	Critics Choice CCD 913	
Ansermet	Suisse Romande	*Lon STS 15144	(1957)
Barbirolli	VPO	*Angel SDC 3732	
Bernstein	VPO	DGG 431029-2	
Bernstein	NYP	*Col ML 5602	(1961)
Böhm	VPO	*DGG 2711017	
Boult	LPO	SQN 139	(1961)
Celibidache	Milano	Arkadia CDGI 764.3	(1959)
Chailly	Concertgebouw	Lon 421295-2	
Dohnanyi	Cleve	*Teldec 244972	
Dorati	LSO	*Merc 50268	
Furtwängler	BPO	DGG 415662-2	
Furtwängler	VPO	*Odeon 147 50336	(1947)

*A remarkably good performance of the Brahms First Symphony, issued on Critics Choice (CCD 913) in 1990, is played by an otherwise unidentified orchestra called "The English Philharmonic Orchestra." No conductor is listed (*sic*).

Conductor	Orchestra	Record Label/Number	Date
Giulini	LA	*DGG 2532056	
Haitink	Concertgebouw	*Philips 6514228	
Herbig	BBC	Collins 30492	
Horenstein	LSO	*Quint 7028	
Janowski	Liverpool	ASV CD 531	
Järvi	LSO	Chandos 8653	
Jochum	LPO	*Angel SDC 3845	
Karajan	BPO	DGG 423141-2	
Karajan	VPO	Lon STS 15194	
Karajan	Philharmonia	*Angel 35001	(1957)
Kempe	BPO	Cap G-7208	(1960)
Kertesz	VPO	Lon CS 6836	
Klemperer	Berlin State Opera	Koch Legacy 7053-2	(1928)
Klemperer	Philharmonia	*Angel 35481	(1960)
Kletzki	Royal Phil	*Angel 35619	(1960)
Kondrashin	Concertgebouw	*Philips 412071-1	
Krips	VPO	*Lon STS 15144	(1957)
Kubelik	Bavar (Munich)	*Orfeo 070834	
Lehel	Magyar Radio (Budapest)	*Hungar 12273	(1983)
Leinsdorf	BSO	*RCA LM 2711	(1964)
Levine	Chicago	*RCA ARL 1-1326	
Maazel	Cleve	*Lon 7007	
Masur	Leipzig	*Philips 6769009	
Mehta	VPO	*Lon CS 7017	
Mengelberg	Concertgebouw	Philips 416210-2	(1933)
Mravrinsky	Leningrad	*Memoria 991.006	(1950)
Munch	Orch de Paris	*Arabesque 8058	
Munch	BSO	RCA 7812-2	(1957)
Muti	Philadelphia	Philips 426299-2	
Norrington	Lon Class	EMI CDC 754286-2	
Ormandy	Philadelphia	Col 3 ML 4477	(1957)
Ozawa	BSO	*DGG 2530899	
Paita	Nat Phil	*Lodia 779	
Reiner	Chicago	*RCA	
Rowicki	Warsaw	*Mus Her ORB 251	
Sanderling	Dresden	Eurodisc 69220-2	
Sawallisch	LPO	EMI Classics 754359-2	
Scherchen	V St Opera	*West XWN 18448	(1957)
Skrowaczewski	Halle	*MCA 25188	
Solti	Chicago	*Lon CS 7198	
Steinberg	Pittsburgh	*SQN 7744	(1961)
Stokowski	Philadelphia	+RCA DM 301	(1927)
Suitner	Staatskapelle Berlin	Ars Vivendo	
Szell	Cleve	*Epic LC 3379	(1957)
Tennstedt	LPO	*Angel DS 38041	
Toscanini	NBC	RCA LM 1702	(1957)

BRAHMS: FIRST SYMPHONY *(continued)*

Conductor	Orchestra	Record Label/Number	Date
Van Beinum	Concertgebouw	*Epic LC 3603	(1959)
Walter	Col Symph	*Col MS 6389	(1960)
Wand	NDR	*Pro Arte SDS 626	
Weingartner	LSO	Centaur CRC 2124	(1939)

BRAHMS: FOURTH SYMPHONY*

Conductor	Orchestra	Record Label/Number	Date
Abbado	BPO	DGG 435349-2	
Abravanel	Utah	Vanguard SVC 1720	
Amsermet	Suisse Romande	*Lon STS 15383	
Anonymous	Anonymous	Critics Choice CCD913	
Barbirolli	VPO	Seraphim TOCE 7137	
Barenboim	Chicago	Erato 4509-95194-2	
Bernstein	VPO	DGG 410084-2	
Böhm	VPO	DGG 2711 017	
Boult	LPO	Angel 37034	
Celibidache	Milano	Arkadia CDGI 764.3	(1959)
Chailly	Concertgebouw	Lon 433151-2	
Davis, C.	Bavar (Munich)	RCA 60383-2	
DeSabata	BPO	*DGG 423715-2	(1939)
Dohnanyi	Cleve	Teldec 244972-2	
Dorati	LSO	*Merc SR 90503	
Fischer-Dieskau	Czech Phil	Quint PMC 7094	(1976)
Furtwängler	BPO	*Electrola 90995	
Giulini	VPO	*DGG 429403-2	
Haitink	BSO	Philips 434991-2	
Janowski	Liverpool	ASV DCA 533	
Järvi	LSO	Chandos CHAN 8595	
Jochum	LPO	*Angel SDC 3845	
Jochum	BPO	Memories HR 4246	(1951)
Karajan	BPO	DGG 431593-2	

*A remarkably good performance of the Brahms Fourth Symphony, issued on Critics Choice (CCD 913) in 1990, is played by an otherwise unidentified orchestra called "The English Philharmonic Orchestra." No conductor is listed *(sic)*.

Conductor	Orchestra	Record Label/Number	Date
Karajan	Philharmonia	*Angel 35298	
Kempe	Munich Phil	*BASF 2022394-9	
Kempe	Royal Phil	*Pantheon 18428	
Kertesz	VPO	*Lon CD 6838	
Kleiber, C.	VPO	DGG 400037-2	
Klemperer	Philharmonia	EMI Classics CDM 69649	
Krips	LSO	*Lon LL 208	
Kubelik	Bavar (Munich)	*Orfeo 070834	
Lehel	Magyar Radio (Budapest)	*Hungar 12276	(1982)
Leibowitz	Royal Phil	Chesky 6	(1963)
Leinsdorf	BSO	*RCA LSL 3010	
Levine	Chicago	*RCA ARL 1-2624	
Maazel	Cleve	*Lon CS 7096	
Masur	Leipzig	*Philips 6769009	
Mehta	NYP	Col 35832	
Mengelberg	Concertgebouw	Teldec 243 724-2	
Mitropoulos	NYP	Hunt 34020	
Mravinsky	Leningrad	Memoria 991006	
Munch	BSO	*RCA LM 2297	
Muti	Philadelphia	Philips 422377-2	
Ormandy	Philadelphia	*Col 31636	(1960?)
Paray	Detroit	*Merc 50057	
Previn	Royal Phil	Telarc 80155	
Reiner	Royal Phil	*Quint 7182	
Rowicki	Warsaw	Mus Her ORB 254	
Sanderling	Dresden	Eurodisc 69220-2	
Skrowaczewski	Halle	IMP Classics PCD 897	
Slatkin	St. Louis	*Telarc DG-10053	
Solti	Chicago	*Lon CS 7201	
Steinberg	Pittsburgh	SQN 139	
Stokowski	New Philharmonia	*RCA ARL 1-0719	
Suitner	Staatskapelle Berlin	Ars Vivendo 2100170	
Szell	Cleve	*Odyssey HB3X 45823	
Toscanini	NBC	*RCA LM 1713	
Van Beinum	Concertgebouw	Epic LC 3563	(1960)
Walter	Col Symph	Odyssey MBK 44776	(1959)
Wand	NDR	Har Mun 567 1695302	
Weingartner	LSO	Centaur CRC 2128	(1938)

STRAUSS: *TILL EULENSPIEGEL*

Conductor	Orchestra	Record Label/Number	Date
Abbado	LSO	*DGG 2532099	
Ashkenazy	Cleve	Lon 425112-2	
Barenboim	Chicago	Erato 2292-45621-2	
Bernstein	NYP	*Col MS 7165	
Blomstedt	Dresden	Denon 73801	
Böhm	BPO	*DGG 2535208	
Busch	BBC	+RCA 11724	(1936)
Celibidache	Stuttgart	Arkadia 487	(1962)
Coates	LPO	RCA 9272	
DeWaart	Minn	Virgin 59234	
Dohnanyi	Cleve	Lon 436 444-2	
Dorati	Detroit	*Lon 71025	
Dorati	Minn	Merc 434 348-2	(1955)
Fricsay	BPO	*Decca DL 9529	
Furtwängler	BPO	DGG 2740260 (rehearsal)	(1930)
Furtwängler	BPO	DGG 2740260	(1930)
Haitink	Concertgebouw	Philips 442 281-2	
Järvi	Scottish Nat	Chandos 8572	
Jochum	Concertgebouw	*Epic 3032	
Karajan	BPO	DGG 2530349	
Karajan	Philharmonia	*Angel 63316	
Kempe	Dresden	EMI 64342	
Koussevitsky	BSO	*Camden 101	(1945)
Krauss	VPO	*Lon 23208	
Lewis	Royal Phil	*Lon SPC 21054	
Maazel	Cleve	CBS MDK 44909	
Mackerras	LPO	*SON 2007	
Markevitch	Radio France	Angel D-35442	
Marriner	Stuttgart	Capriccio 10 369	
Masur	Leipzig	????	
Mata	Dallas	Pro Arte 403	
Munch	BSO	*RCA LSC 2565	
Ormandy	Philadelphia	*RCA AGL 1-1408	
Paternostro	Tokyo	Koch 311177	
Previn	VPO	Angel DS 37753	
Reiner	VPO	*Lon STS 15582	
Rodzinski	NYP	*Col ML 4884	(1946?)
Sawallisch	Philadelphia	EMI 55185	
Slatkin	LPO	*RCA 5959-1	
Solti	Chicago	*Lon CS 6978	
Solti	VPO	*Lon 26321	
Steinberg	BSO	*RCA LSC 3155	

Conductor	Orchestra	Record Label/Number	Date
Stock	Chicago	Chicago Symph (1st 100 years)	(1940)
Strauss	VPO	*Vanguard 325	(1944)
Szell	Cleve	*Col MY 36721	
Toscanini	NBC	RCA 09026-60296-2	(1952)

RAVEL: *DAPHNIS ET CHLOÉ*

Conductor	Orchestra	Record Label/Number	Date
Abbado	BSO	DGG 2530038	
Abbado	LSO	DGG 445519-2	
Ansermet	Suisse Romande	*Lon STS 15092	(1963)
Barbirolli	Halle	*Vanguard S-177	
Barenboim	Orch Paris	*DGG 2532041	
Bernstein	NYP	*Col MY 36714	
Boulez	Cleve	*CBS M 30651	
Cantelli	Philharmonia	Testament 1017	
Casadesus	Lille	Harm Mun 390064	
Chailly	Concertgebouw	Lon 443934-2	
Commissiona	Houston	*Vanguard 25022	
DeBurgos	New Philharmonia	*Angel S-36471	
Dervaux	Colonne (Paris)	*SQN 7774	
Dohnanyi	Cleve	Teldec 97439-2	
Dutoit	Montreal	*Lon 71028	
Furtwängler	BPO	DGG 427783-2	(1944)
Gaubert	Straram (Paris)	In Syne C-4134 (cassette)	(1928)
Haitink	BSO	Philips 426 260-2	
Inbal	Orch Nat France	Denon 1796	
Jansons	Oslo	EMI 749964-2	
Karajan	BPO	DGG 427250-2	
Kondrashin	Concertgebouw	*Philips 412071-1	
Koussevitsky	BSO	RCA DM	(1928)
Koussevitsky	BSO	RCA 09026-61392-2	(1942)
Levi	Atlanta	Telarc 80352	
Levine	VPO	DGG 415 360-2	
Maazel	Cleve	*Lon CS 6898	
Mackerras	LSO	Centaur 2090	
Martinon	Orch Paris	Quint PMC 7017	
Martinon	Chicago	RCA LM 2806	(1965)
Mata	Dallas	*RCA 3458	

RAVEL: *DAPHNIS ET CHLOÉ* (continued)

Conductor	Orchestra	Record Label/Number	Date
Mehta	LA	*Lon 6698	
Mengelberg	Concertgebouw	KICC 2061	
Monteux	LSO	*Lon STS 15090	
Munch	BSO	*RCA 11674	
Muti	Philadelphia	*Angel DS-37885	
Nagano	LSO	Erato 4509-91712-2	
Nowak	LSO	ASV DCA 536	
Ormandy	Philadelphia	*RCA ARD 10029	
Ozawa	BSO	*DGG 2530567	
Paray	Detroit	Merc 434306-2	(1961)
Previn	LSO	*Angel 37868	
Rattle	Birmingham	EMI 7 54303-2	
Rodzinski	Cleve	*Col 3ML 4884	(1957)
Rosenthal	Nat Opera Orch (Paris)	ADES 14074	
Schwarz	Seattle	Delos 3110	
Sinopoli	World Phil (Japan)	Valois 4601	
Skrowaczewski	Minn	*Vox Box CDX 5032	
Slatkin	St. Louis	Telarc 80052	
Solti	Chicago	Chicago Symph (1st 100 years)	
Stokowski	LSO	*Lon SPC 21059	
Szell	Cleve	*Odyssey Y 31928	
Tortelier	Ulster	Chandos 1504	
Toscanini	NBC	RCA 09026-60322-2	

SCHUMANN: SECOND SYMPHONY

Conductor	Orchestra	Record Label/Number	Date
Ansermet	Suisse Romande	*Lon 15285	
Barenboim	Chicago	*DGG 2530 939	
Bernardi	Calgary	CBC 5067	
Celibidache	Rome	Nova Era 013.6327	
Celibidache	Stockholm	Arkadia CDGI 373.1	
Commissiona	Houston	Pro Arte 394	
Haitink	Concertgebouw	*Philips 412852	
Janowski	Liverpool	ASV 6084	
Karajan	BPO	*2530170	
Klemperer	Philharmonia	*Angel S-36606	

Conductor	Orchestra	Record Label/Number	Date
Kubelik	BPO	*DGG 2535 117	
Mackerras	LSO	Centaur 1007	
Marriner	Stuttgart	Capriccio 10997	
Masur	LPO	Teldec 2292 46446-2	
Mehta	VPO	*Lon 7206	
Mitropoulos	Minn	Nickson NN 1008	(1940)
Muti	Philharmonia	*Angel 37602	
Paray	Detroit	*Merc 50102	(1957)
Patane	Hung St Orch	Qualiton 12278	
Pfitzner	Berlin State Opera	Koch 3-7039-2	(1928)
Sawallisch	Dresden	*Arabesque 8102	
Schuricht	Conservatoire (Paris)	*SQN 143	
Semkow	Warsaw	*Muza 1329	
Solti	VPO	*Lon 2310	
Szell	Cleve	*Epic 1159	
Weingartner	LPO	Musica Memoria	(1933)
Zinman	Baltimore	Telarc 80182	

TCHAIKOVSKY: SIXTH SYMPHONY

Conductor	Orchestra	Record Label/Number	Date
Bernstein	NYP	Sony SMK 47635	(1964)
Bond, Victoria	Shanghai	Protone 2205	
Bychkov	Concertgebouw	Philips 434150-2	
Celibidache	Milano	*Melodram MEL 217	(1959)
Coates	BPO	Beulah 1PD6	(1945)
Dorati	LSO	Merc 434353-2	(1960)
Furtwängler	BPO	+RCA 553 [Hist Perf HPS8]	(1938)
Gielen	SW German Radio	Gielen INT 860.923	
Giulini	Philharmonia	*Seraphim 72435 68531-2	
Jansons	Oslo	Chandos 8446	
Martinon	VPO	*Lon STS 15018	
Mengelberg	Concertgebouw	Teldec 4509-93673-2	(1948)
Monteux	BSO	*RCA Vics 1009	
Ormandy	Philadelphia	RCA 09026-60908-2	
Reiner	Chicago	RCA 09026-01246-2	(1955)
Rodzinski	PSO Lon	West XWN 18048	(1957)
Rozhdestvensky	LSO	IMP Classics PCD 878	
Slatkin	St. Louis	RCA 09026-60438-2	

Index

Note: **Boldface** indicates extensive discussion.

563